Global Sustainability in Energy, Building, Infrastructure, Transportation, and Water Technology

Md. Faruque Hossain

Global Sustainability in Energy, Building, Infrastructure, Transportation, and Water Technology

Md. Faruque Hossain
Kennesaw State University
Marietta, Atlanta, GA, USA

ISBN 978-3-030-62378-4 ISBN 978-3-030-62376-0 (eBook)
https://doi.org/10.1007/978-3-030-62376-0

This Springer imprint is published by the registered company Springer Nature Switzerland AG
The registered company address is: Gewerbestrasse 11, 6330 Cham, Switzerland

Contents

Part I Introduction

1 Design a Better World 3
Introduction 3
Materials and Methods 4
Results and Discussion 6
Conclusions 8
References 8

Part II Renewable Energy Technology

2 Solar Energy 13
Introduction 14
Material, Methods, and Simulation 15
Calculation of Net Solar Energy on Earth 15
Calculation of Net Electricity Energy Generation from Total Solar Irradiance on Earth 21
Results and Discussion 24
Calculation of Net Solar Energy on Earth 24
Calculation of Net Electricity Energy Generation from Total Solar Irradiance on Earth 33
Conclusion 36
References 37

3 Wind Energy 41
Introduction 41
Materials and Simulation 43
Results and Discussion 49
Conclusions 55
References 55

4 Geothermal Energy 59
Introduction 59
Materials and Methods 60
Results and Discussion 67
Conclusion 74
References 74

Part III Advanced Building Design Technology

5 Integrated Building Design Technology 81
Introduction 81
Methods and Simulation 82
Cooling Mechanism 82
Heating Mechanism 86
Results and Discussion 89
Cooling Mechanism 89
Heating Mechanism 93
Conclusions 101
References 102

6 Smart Building Technology 105
Introduction 105
Material, Methods, and Simulation 107
Design of PV Panel 111
Electrostatic Force Generation 113
Results and Discussion 116
Electricity Transformation 120
Electrostatic Force Analysis 123
In Situ Water Treatment 125
Conclusion 126
References 126

7 Advanced Green Building Technology 129
Introduction 129
Materials and Methods 130
Results and Discussion 137
Conclusion 142
References 143

Part IV Innovative Infrastructures and Transportation Engineering

8 Green Infrastructure 149
Introduction 149
Methods and Materials 150
Results and Discussion 152
Conclusion 154
References 154

9 Invisible Roads and Transportation Engineering 157
Introduction. 157
Methods and Materials 158
Results and Discussion 164
Construction Cost Estimate Comparison 167
Construction Cost Estimate Comparison 170
Cost of Maglev Infrastructure. 171
Cost of Traditional Road Infrastructure 171
Cost Saving. 172
Conclusions. 172
References. 173

10 Zero-Emission Vehicles. 177
Introduction. 177
Methodology and Materials 178
Turbine Modeling for Transportation Sector 178
Wind Energy Modeling. 181
Electrical Subsystem Modeling 182
Results and Discussions 186
Wind Turbine Modeling 186
Wind Energy Modeling. 188
Electrical Subsystem Modeling 189
Mathematic Experiment on Car 193
Battery Modeling 195
Conclusion 196
References. 196

11 Flying Transportation Technology. 201
Introduction. 201
Materials, Methods, and Simulation. 202
Solving the Concept Numerically. 203
The Flying Car's Wind Energy Modeling Sequence 204
Conversion of Wind Energy 207
Modeling the Generator 207
Battery Modeling 209
Results, Optimization, and Discussion. 209
Wind Energy Modeling for the Flying Vehicles. 212
Wind Energy Conversion 215
Electrical Subsystem. 216
Generator Modeling 217
Battery Modeling 219
Conclusion 220
References. 221

Part V Clean Water and Sanitation System

12 Potable Water 225
Introduction 225
Methods and Materials 226
Results and Discussion 228
Domestic Use 231
Agricultural Use 231
Industrial Use 232
Recreational Use 233
Environmental Use 233
Irrigation Use 233
Conclusion 235
References 235

13 Wastewater 237
Introduction 238
Current and Potential Applications for Water and Wastewater Treatment 239
Adsorption 239
Membranes and Membrane Processes 245
Photocatalysis 249
Disinfection and Microbial Control 252
Sensing and Monitoring 255
Conclusions 260
Recent Developments in Photocatalytic Water Treatment Technology 261
Introduction 261
Fundamentals and Mechanism of TiO_2 Photocatalysis 263
Advancements in Photocatalyst Immobilization and Supports 269
Challenges in the Development of Photocatalytic Water Treatment Process 269
Mesoporous Clays 270
Nanofibers, Nanowires, or Nanorods 271
Photocatalytic Membrane 272
Photocatalyst Modification and Doping 272
Photocatalytic Reactor Configuration 275
Operational Parameters of the Photocatalytic Reactor 279
Contaminants and Their Loading 285
Light Wavelength 286
Light Intensity 287
Response Surface Analysis 289
Kinetics and Modeling 290
Life Cycle Assessment of Photocatalytic Water Treatment Processes 303
Future Challenges and Prospects 305
References 306

14 Water Reduction Engineering 325
Introduction 325
Methods and Simulation 326
Static Electric Force Generation 326
In Situ Water Treatment 329
Results and Discussion 329
Conclusions 331
References 331

Part VI Sustainable Urban and Rural Development

15 Deurbanization and Rural Development 335
Introduction 335
Methods and Materials 336
Environment 336
Energy 337
Housing and Building 337
Infrastructure and Transportation 338
Water 338
Results and Discussion 339
Environment 339
Energy 340
Housing and Building 341
Infrastructure and Transportation 341
Water 342
Conclusion 343
References 343

16 Sustainable Cities and Communities 347
Introduction 347
Methods and Materials 349
Results and Discussion 355
Conversion of Electricity 357
Savings on Energy Cost 359
Conclusions 359
References 360

17 Rapid Connectivity Within the Urban and Rural Area 365
Introduction 365
Methods and Materials 366
Results and Discussion 371
Conclusion 374
References 375

Part VII Environment

18 Air 381
Introduction 381
Methods and Simulation 383
CO_2 Emissions from Fossil Fuel 383
CO_2 Emissions from the Land-Use Change (E_{LUC}) 384
Ocean CO_2 Sink 384
CO_2 Absorption by Terrestrial Vegetation and the Earth 385
Calculation of the Growth Rate of the Atmospheric CO_2 Concentration (G_{ATM}) 386
Results and Discussion 386
CO_2 Emissions from Fossil Fuels, Land-Use Change, and Other Factors 387
Ocean and Terrestrial Vegetation CO_2 Sinks 387
Cumulative CO_2 Emissions and Atmospheric Impact 389
Conclusion 392
References 392

19 Water 395
Introduction 395
Materials and Methods 396
Results and Discussion 397
Conclusions 401
References 401

20 Land 403
Introduction 403
Materials and Methods 405
Results and Discussion 406
Conclusion 411
References 412

Part VIII Sustainable Planet

21 Climate Control 417
Introduction 417
Methods and Simulation 418
Cooling Mechanism of Earth Surface 418
Heating Mechanism of Earth Surface 422
Results and Discussion 424
Cooling Mechanism of Earth Surface 424
Heating Mechanism of Earth Surface 428
Conclusions 435
References 436

22 Global Environmental Equilibrium 439
Introduction 439
Materials, Methodology, and Simulations 440
Generating Static Electric Forces 440
On-Site Water Treatment 442
Clean Energy Production 443
Results/Discussion 446
Electrostatic Force Analysis 446
Hydrogen Energy Production Through Electrolysis 449
Conclusion 451
References 452

Index 469

About the Author

Md. Faruque Hossain has more than 20 years of industry experience in the field of sustainability research, development, and project management under global top public agencies and fortune-listed companies. He worked and/or consulted in diverse global top tier companies to conduct research and development for million dollars to over billion dollars projects for ensuring global sustainability. Faruque also worked for the NYC Department of Citywide Administrative Services as Senior Management team and interacted with the heads of all public agencies, highest level government officials of local, state, federal, and international organization leaders for developing global sustainability policy. During his tenure in NYC Department of Environmental Protection as Acting Director, Faruque managed a world-class team of scientists, consultants, architects, engineers, and contractors from AECOM, Fluor, Skanska, and maintained highest level professional relationship to conduct global sustainability research and practice. Hossain received his Ph.D. from Hokkaido University, did post-graduate research in Chemical Engineering at the University of Sydney, and Executive Education in Architecture at Harvard University. He is an LEED-certified professional and editors of several International Journal of Global Sustainability-related field. Dr. Hossain is renowned as the industry leader and notable scientist to conduct innovative research and project development for energy, environment, building, infrastructure sustainability field for building a better Earth. He has hundreds of world-class publications in very high impact journals, and he wrote two books (Elsevier) and four book chapters (Francis and Taylor and Elsevier) in this field. Currently, he is working at the school of architecture and construction management at Kennesaw State University as an Assistant Professor and simultaneously running his own company "Green Globe Technology" with his motto to practice sustainability for building a better planet.

Part I
Introduction

Chapter 1
Design a Better World

Abstract COVID-19 is indeed a clear and present danger to the world which is due to conventional design in building technology to the world. Simply we need advanced building design technology that can be implemented everywhere in the world in order to kill all pathogens including COVID-19 inside the buildings and houses naturally before it invades into the human body. Simply photon energy is being proposed to be implemented using exterior glazing wall surface to transform *ultraviolet germicidal irradiation* (UVGI) from sunlight to kill all pathogens such as bacteria, protozoa, prion, viroid, fungus, molds, and virus including COVID-19 inside the buildings and houses around the globe in order to design a better world. Since killing pathogens with solar radiation will only require short-range wavelengths, 254–280 nm wavelength has been utilized to initiate UVGI light to destroy the pathogenic nucleic acids DNA and/or RNA, leaving all pathogens unable to perform their vital cellular functions, and letting them die in minutes. Simply, usage of exterior glazing wall surface in the building and house design as an acting photophysical reaction technology application will be an innovative field of science to kill all pathogens inside the buildings and houses naturally.

Keywords Advanced building design · Photonic wavelengths · Ultraviolet germicidal irradiation (UVGI) · Killing pathogens · Global sustainability

Introduction

The average microorganisms in indoors that occupy a building's room of 10′ × 10′ are typically around 8000 types of virus, bacteria, molds, and fungus and the concentrations are 10^5 pathogens per m^3 [1–3]. Pathogens, the infectious microorganism such as bacterium, protozoan, prion, viroid, fungus, and virus including COVID-19, are indeed the dangerous beings that threaten the entire human race to survive on earth in the near future. These tiny creatures can easily invade into the human body through absorbing, adsorbing, inhaling, ingesting, and producing

M. F. Hossain, *Global Sustainability in Energy, Building, Infrastructure, Transportation, and Water Technology*,
https://doi.org/10.1007/978-3-030-62376-0_1

mutants into the human body rapidly to form toxins which penetrate the tissues, hijack nutrients, immunosuppress the human body, and cause serious illness or death [4–6]. The recent COVID-19 is a deadly infectious disease that causes severe dreadful respiratory syndrome (SARS-CoV-2) which was first detected in November 2019. Since then it has spread to 210 countries and the World Health Organization (WHO) has declared "the 2019–2020 coronavirus outbreak" a pandemic [7, 8]. Despite this current COVID-19 pandemic crisis, we are not prepared to defeat any forthcoming deadly outbreak which could be much more deadly than COVID-19 that probably could completely wipe out the entire human race from earth in the future. The simple biochemistry is that some of the pathogens have the extremely brilliant ability to change their pathway frequently into the human body to deceive any vaccines to kill them into the host body which is indeed a clear and present danger for the entire human race [9–11]. Thus, prevention technology is an urgent demand to eliminate these deadly pathogens before it invades into the human body as a first level of defense. Interestingly, photonic ultraviolet germicidal irradiation (UVGI) has the excellent functionality to eliminate these deadly pathogens in a minute regardless of their ability of changing mutants and pathways. Therefore, in this research, photonic irradiance structure has been studied which has been formed by using exterior curtain wall of a building or house to release UVGI at short-range electromagnetic spectrum of wavelengths of 185–254 nm to destroy the nucleic acids and DNA and/or RNA structure of all pathogens in order to kill them eventually in order to secure a better world. Simply, in this research, photo-physical reaction application is being proposed to implement to design all buildings and homes globally to have the ability to create UVGI from sunlight in order to eradicate all pathogens inside the buildings and homes naturally before it invades into the human body.

Materials and Methods

The quantum field of the sunlight has been trapped for only 15 min in a day by using the outer glazing wall surface to emit UVGI light in between 254 and 280 nm for a limited time when nobody lives inside the building or house. Consequently, photonic electromagnetic field generation has been analyzed considering the penetration of solar irradiance rate into the outer glazing wall surface which is connected to the semiconductor to determine the required electromagnetic radiation (EM) from the sunlight to destroy pathogens naturally inside the building or house in order to design a better world. Subsequently, the electromagnetic radiation (EM radiation) and the waves of photonic quanta of the electromagnetic field have been clarified considering the electromagnetic radiant energy emission by the exterior **glazing wall surface** of the building or house. Since the electromagnetic waves are being transmitted by photon particles from the linearly polarized sinusoidal electromagnetic wave of the sunlight, EM waves will carry enough photon energy to accelerate its photonic momentum to release solar radiation continuously without any perturbation [12–14]. Interestingly, the effects of these radiations on biochemical functions of pathogenic cell have enough frequencies of photonic momentum energy to break down the biochemical bonds of the pathogens and cause severe malfunction of their cellular function.

Since the effects of UVGI on pathogens depend on the radiation's frequency and wavelength, the release of UVGI from sunlight is being measured and controlled by computerized PHOTO-ELECTRO-METER (Fig. 1.1). Then the effects of EM on pathogens, including coronavirus (COVID-19), are being determined using PCR analysis considering the radiation's frequency of wavelengths at the spectrum of 254–280 nm in order to break down the molecular bonds of microorganismal DNA and/or RNA producing thymine dimers ("T" of ATGC). Interestingly, the high frequency of EM with short range of light spectrum 254–280 nm utilization suggested that it is extremely effective to damage the pathogenic DNA and/or RNA thymine dimer, which results in elimination of their ability to reproduce and biochemical function of their ability to cause harm to its hosts.

Simply, the UVGI light-releasing mechanism using exterior glazing wall skin will certainly inactivate pathogens' cellular function by destroying their deoxyribonucleic acid (DNA) and ribonucleic acid (RNA). Just because when DNA and/or RNA of the pathogens absorb UVGI light from the sun, it breaks down the dimers (covalent bonds between the same nucleic acids) and their failure dimers cause the barrier in the transcription of information of DNA and/or RNA, which in turn results in disruption of pathogens' biochemical function and replication and lets them die immediately.

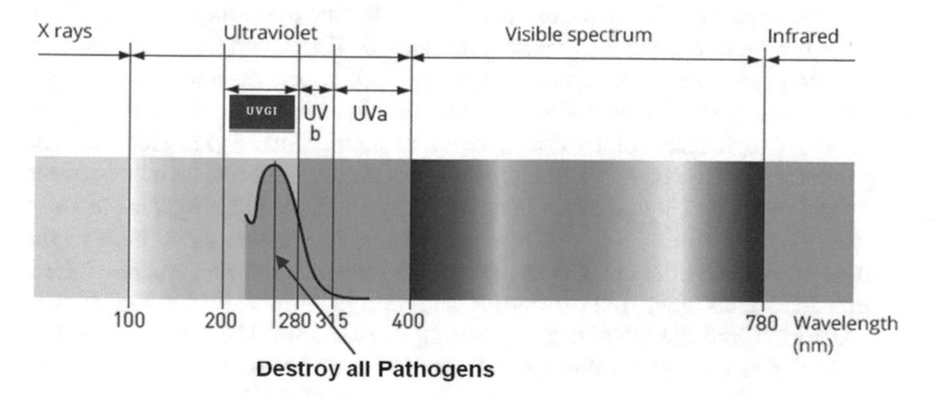

Fig. 1.1 The photo-physics radiation application for eliminating pathogens inside the building or house which suggests that once UVC which is UVGI radiation of 254–280 nm is applied inside the house or building, it stars to disinfect all pathogens immediately

Results and Discussion

The UVGI from sunlight can be trapped by using the outer glazing wall surface of a building or house considering the photon energy emission on the glazing wall surface area [15, 16, 6]. The glazing wall surface area has the ability to have consistent solar irradiance emission, and UVGI generation from sunlight has got a unique dynamic mode of solar energy penetration into the outer glazing wall surface. Simply, the UVGI is being trapped by the glazing wall surface from the electromagnetic radiation (EM) of the light spectrum in relation to short-wavelength photon-induced solar energy penetration on the glazing wall surface. Thus, this EM radiation is being determined by the wavelength of UVGI sinusoidal monochromatic waves, which in turn is being classified by the EM spectrum of the photon particles by using PHOTO-ELECTRO-METER.

Subsequently, the EM waveform is being determined using the spectral analysis of photonic frequency considering its energy content which is the spectral density of the solar energy. Thus, the random electromagnetic radiation of UVGI is being determined by the wideband forms of solar radiation from the unique sinusoidal wave field of the photon energy.

The EM field of the solar energy thus suggests that once the outer glazing wall surface emits light between 254 and 280 nm it allows to add a mass-term sunlight energy transformed into UVGI naturally. Since UVGI has the extraordinary ability to destroy the DNA and/or RNA of the fatal pathogens in seconds regardless of the pathogens' frequent changing mechanism of mutants and pathways, the UVGI light wavelength implantation suggests that EM field has been paved for the solar irradiance to penetrate uniformly into the pathogen bodies. Since DNA and/or RNA of the pathogens is composed of double- or single-strand bonds consisting of chemically coded nucleotide adenine (A), thymine (T), guanine (G), and cytosine (C) which are very sensitive to UBGI, UVGI light penetration into their bodies directly interferes these bonds between the nucleotides in the DNA and/or RNA and completely damages their cyclobutane pyrimidine dimers (CPD) and 6–4 pyrimidine pyrimidone photoproducts (6–4PPs), and its Dewar isomers [5, 9]. Simply it can be explained that once UVGI penetrates into the pathogen bodies, it forms the CPDs which are the two adjacent pyrimidine bases (thymine—TT or cytosine—CC) and it becomes covalently linked to produce a cyclic ring structure and thus 6–4PPs result from a single covalent bond formed between the 5′ end of C6 and 3′ end of C4 of adjacent pyrimidines [17, 18]. This leads to the formation of an unstable oxetane or azetidine intermediate end base of the thymine or cytosine dimers. Consequently, spontaneous rearrangement of these intermediate ends gave rise to 6–4PP and eventually the pyrimidine dimers caused a kink into the DNA strands, halting transcription and protein synthesis (Fig. 1.2). The 6–4 pyrimidine pyrimidone adducted isomerization and caused severe damage to cellular function (Fig. 1.2). Subsequently, UVGI radiation does additional function to cause damage to the DNA and/or RNA via absorption of photon energy by cellular chromophores to enforce to generate the reactive

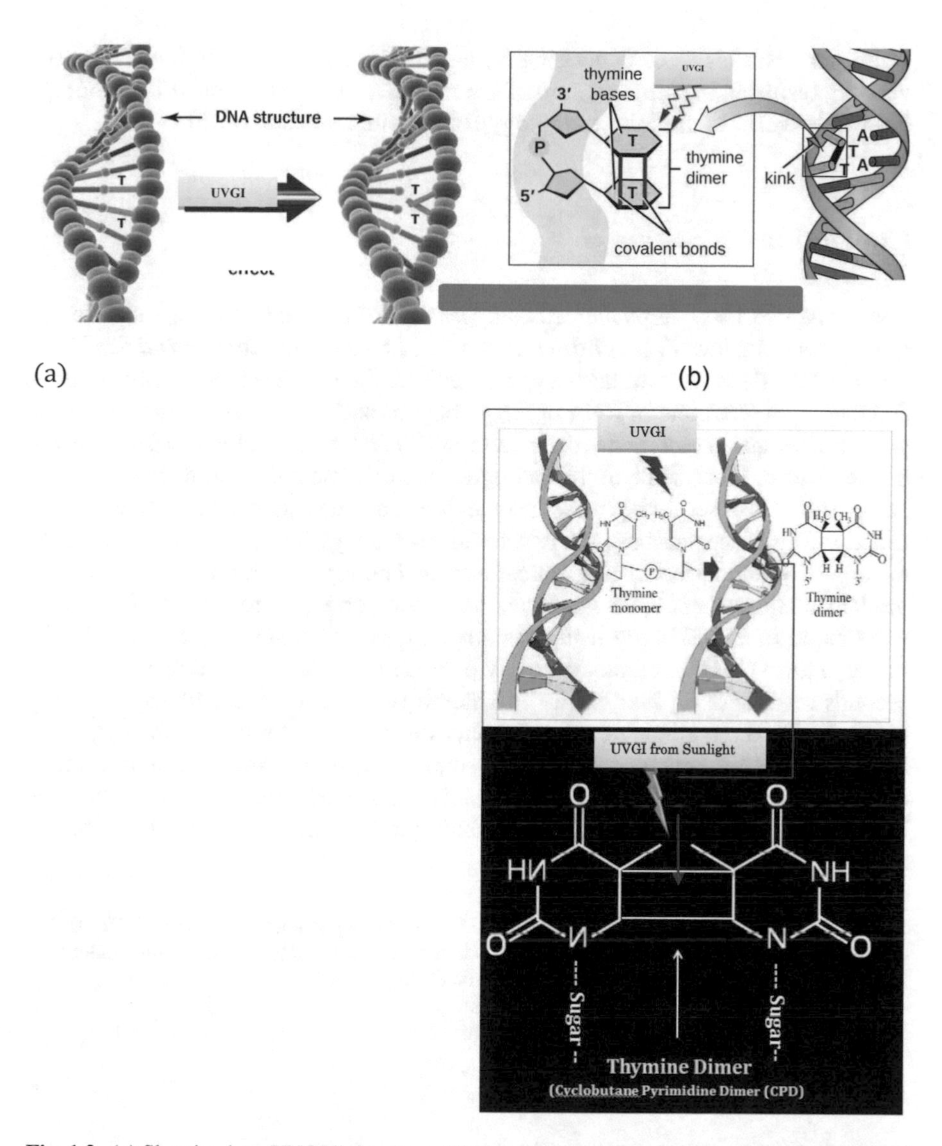

Fig. 1.2 (**a**) Showing how UVGI light affects DNA structure, (**b**) mechanism of the breakdown of DNA structure of pathogens by UVGI light, (**c**) detailed mechanism to destroy the thymine dimer of DNA of the pathogens due to the implementation of UVGI radiation which ultimately disables their biochemical function and reproduction

oxygen charge to oxidize the DNA bases causing them to prohibit the mutations just because most of the common mutations occurred at the G-T transversion process where guanine gets oxidized due to the reactive oxygen charge into the 8-oxo-7,8-dihydroguanine (8-oxoG), hindering its pairing with cytosine. Simply, UVGI penetration into the pathogen bodies completely interrupts this biochemical function by damaging the 8-oxoG to form pair with adenine during the strand synthesis, and thus 8-oxoG is being replaced with a thymine into the G-T transversion, resulting in complete stop of the mutation process which will encourage the whole world to use this technology for building a better world.

Conclusions

Solar energy has been implemented **using advanced design technology of exterior glazing wall surface of a building or house** to create ***ultraviolet germicidal irradiation*** (UVGI) to release short-range ultraviolet light to make extremely effective destruction of DNA and/or RNA of the pathogens such as bacteria, viroid, bacteria, protozoa, fungus, molds, and virus including COVID-19 naturally in order to design a better world. Since some of the pathogens are extremely dangerous to the human being, which also surprisingly can change their mutants and pathway to deceive the functionality of the vaccines, the prevention technology has been studied to defend these deadly pathogens rather than treatment technology. Therefore, in this research, implementation of photonic irradiance has been conducted to release short-range wavelength of 254–280 nm using the outer glazing wall surface of a building or house to form UVGI to eliminate deadly pathogens inside the building and house in seconds regardless of their ability to change the mutants and pathways. Simply using the UVGI to eliminate all pathogens inside the building by damaging their DNA and/or RNA structure and cellular function naturally would be an innovative design technology to eliminate these deadly pathogens before it invades into the human body and will mitigate future global outbreak challenges in order to secure a better world.

Acknowledgements Thanks to the GGT, Inc. for furnishing required fund (RD-02020-03) to conduct this research on time. Author confirms that this research does not have any conflicts of interest and welcomes any suitable journal to publish this research.

References

1. Chang, D. E., Sørensen, A. S., Demler, E. A. & Lukin, M. D. A single-photon transistor using nanoscale surface plasmons. Nat. Phys.3, 807–812 (2007).
2. Guo-an Yan, Hua Lu, Ai-xi Chen. “Single-photon router: Implementation of Information-Holding of Quantum States”, International Journal of Theoretical Physics, 2016.
3. Yuwen Wang, Yongyou Zhang, Qingyun Zhang, Bingsuo Zou, Udo Schwingenschlogl. “Dynamics of single photon transport in a one-dimensional waveguide two-point coupled with a Jaynes-Cummings system”, Scientific Reports, 2016.
4. Birnbaum, K. M. et al. Photon blockade in an optical cavity with one trapped atom. Nature 436, 87–90 (2005).
5. Dayan, B. et al. A photon turnstile dynamically regulated by one atom. Science 319, 1062–1065 (2008).

6. Tame, M. S., K. R. McEnery, Ş. K. Özdemir, J. Lee, S. A. Maier, and M. S. Kim. "Quantum plasmonics", Nature Physics, 2013.
7. Armani, D. K., Kippenberg, T. J., Spillane, S. M. & Vahala, K. J. Ultra-high-Q toroid microcavity on a chip. Nature 421, 925 (2003).
8. Hossain, Md. Faruque. "Solar energy integration into advanced building design for meeting energy demand and environment problem", International Journal of Energy Research, 2016.
9. Guerlin, C. et al. Progressive field-state collapse and quantum non-demolition photon counting. Nature 448, 889 (2007).
10. Joannopoulos, J. D., Villeneuve, P. R. & Fan, S. Photonic crystals: putting a new twist on light. Nature 386, 143 (1997).
11. Yao Xiao, Chao Meng, Pan Wang, Yu Ye, Huakang Yu, Shanshan Wang, Fuxing Gu, Lun Dai, Limin Tong. "Single-Nanowire Single-Mode Laser", Nano Letters, 2011.
12. Douglas, J. S., H. Habibian, C.-L. Hung, A. V. Gorshkov, H. J. Kimble, and D. E. Chang. "Quantum many-body models with cold atoms coupled to photonic crystals", Nature Photonics, 2015.
13. Hossain Md. Faruque. "Sustainable technology for energy and environmental benign building design", Journal of Building Engineering, 2019.
14. Hossain Md. Faruque. "Photon energy amplification for the design of a micro PV panel", International Journal of Energy Research, 2018.
15. Jianchao Hou, Haicheng Wang, Pingkuo Liu. "Applying the blockchain technology to promote the development of distributed photovoltaic in China", International Journal of Energy Research, 2018.
16. Sayrin, C. et al. Real-time quantum feedback prepares and stabilizes photon number states. Nature 477, 73 (2011).
17. Gleyzes, S. et al. Quantum jumps of light recording the birth and death of a photon in a cavity. Nature 446, 297 (2007).
18. Hossain Md. Faruque. "Green science: Advanced building design technology to mitigate energy and environment", Renewable and Sustainable Energy Reviews, 2018.

Part II
Renewable Energy Technology

Chapter 2
Solar Energy

Abstract The present level of consumption of the reverse fuel on earth for powering the world is taking fossil fuel to a limited level and all the while fossil fuel burning is causing deadly environmental vulnerability as well. Inevitably, alternative energy source must be needed in order to drive the modern world which is abundant and benign to the environment. The use of solar energy in every sectors of our modern life will be an interesting option to power the modern world which is renewable and abundant everywhere in the world. In this research, therefore, a calculative mechanism has been conducted to harvest the global solar energy in order to present a clean and renewable energy system for the world. Interestingly, the calculative result depicted that the average energy density of solar energy on the surface of earth is 1366 W/m^2 which has been determined by the calculated diameter of earth of 10,000,000 of meridian at the North Pole to the equator and the radius of earth is $2/\pi \times 10^7$ m. Thus, the net energy of solar radiation reaching earth is calculated as $1366 \times (4/\pi) \times 10^{14} \cong 1.73 \times 10^{17}$ W by computing that a day has 86,400 s, and a year has 365.2422 days in average. Therefore, the net solar energy reaching earth annually is $1.73 \times 10^{17} \times 86{,}400 \times 365.2422 \cong 5.46 \times 10^{24}$ J which is equal to 5,460,000 EJ energy which is 10,000 times higher than the net current energy demand on earth annually. Simply, utilization of solar energy could be an alternative source of energy to satisfy the net power demand of the whole world, which is clean and environmentally friendly.

Keywords Fossil fuel reserve limitation · Solar radiation · Alternative source of energy · Net solar energy on earth · Clean energy technology · Meeting global energy demand

M. F. Hossain, *Global Sustainability in Energy, Building, Infrastructure, Transportation, and Water Technology*,
https://doi.org/10.1007/978-3-030-62376-0_2

Introduction

The conventional energy consumption for powering the modern civilization throughout the world is indeed accelerating the finite level of current fossil fuel reserve of 36,630 EJ [1, 2]. The global fossil fuel energy consumption was 283 EJ/year in the year 1980, 347 EJ/year in 1990, 400 EJ/year in 2000, and 511 EJ/year in 2010 and it would be 607 EJ/year in 2020, 702 EJ/year in 2030, 855 EJ/year in 2040, and 988 EJ/year in 2050 [1, 3]. The utilization of fossil fuel globally in the year 2018 was 2.236×10^{20} EJ which was responsible for releasing 8.01×10^{11} tons of CO_2 into the atmosphere and accounted for acceleration of deadly climate change rapidly [4–6]. Consequently, adverse environmental impact such as acid precipitation, stratospheric ozone depletion, and massive diurnal temperature fluctuation is occurring unpredictably throughout the world [3, 7, 8].

Recent study shows that the concentration of atmospheric CO_2 amount is 400 ppm which needs to be lowered to a standard-level grade of 300 ppm CO_2 for clean breathing and healthy respiratory system for all mammals [9–11]. Another research revealed that climate change and rising global mean temperature (GMT) with associated consequences pose a serious threat to natural systems and mankind's well-being due to the burning of fossil fuel since it releases radioactive CO_2 into the atmosphere in a certain period of time [3, 12].

Unfortunately, the consumption of conventional energy currently is still accelerating rapidly throughout the world; the situation shall remain unchanged until a renewable source of energy is developed to utilize sustainable energy. Simply it is an urgent demand to develop sustainable energy technology to mitigate fossil fuel consumption where the "new source that fulfills the needs of the current without compromising the ability of future demand of energy to fulfill the complete needs for the future generations." Hence, solar energy utilization globally can be an interesting source to fulfill the net energy requirement throughout the world. It is a natural renewable energy source generated by the sun which is created by nuclear fusion that takes place in the sun [13–15]. Simply fusion occurs when protons of hydrogen atoms violently collide in the sun's core and fuse to create a helium atom and the process, known as a proton-proton (PP) chain reaction, emits an enormous amount of energy [3, 12]. Since nuclear fusion by the PP chain reaction releases tremendous amounts of energy in the form of waves and particles, solar energy is constantly flowing away from the sun to the solar system and part of it reaches earth which is a tremendous source of clean and renewable energy [5, 16–18]. If a mere 0.001% of the annual solar energy reaching earth is used, it will satisfy the net energy need for the entire world which is clean and abundant everywhere. In this study, therefore, a research has been performed to harvest the total global solar energy reaching earth in order to mitigate global net energy need which is clean and environmentally friendly.

Material, Methods, and Simulation

Calculation of Net Solar Energy on Earth

The total earth surface is being clarified by characterizing various directional angles considering the Cartesian coordinate system, where x denotes skyline convention, y denotes east-west, and z denotes zenith in order to measure the total solar irradiance during the day the entire year (Fig. 2.1). The position of the celestial body in this

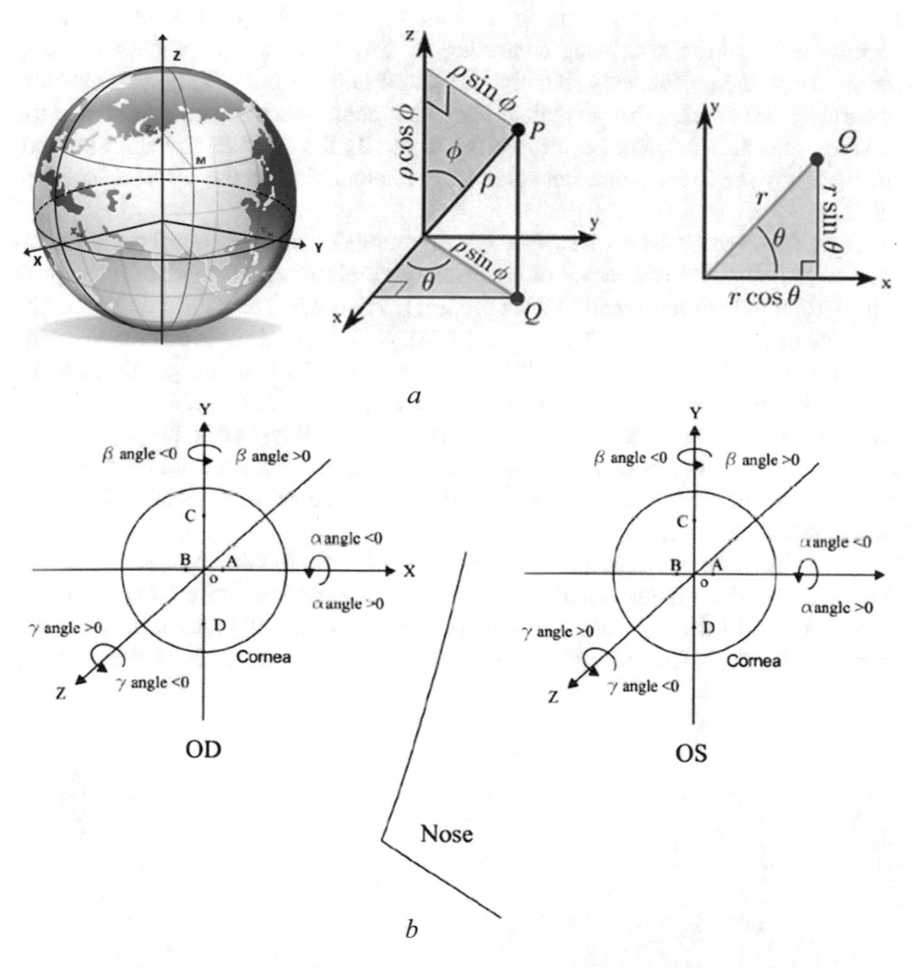

Fig. 2.1 (**a**) Cartesian coordinate clarification of south is x, west is y, and zenith is z that has been clarified in order to calculate the total solar energy reaching earth considering that the average energy density of solar energy on the surface of earth is 1366 W/m^2 by implementing the diameter of the earth as 10,000,000 of meridian at the North Pole to the equator and the radius of earth as $2/\pi \times 10^7$ m. The location of this celestial body is analyzed by determining two angles of sinθ and cosθ. (**b**) The longitudinal and latitudinal equatorial angles have been clarified where the convention z-axis point denotes the North Pole and the east-west axis and y-axis denote the identical angles of the horizon

framework is thus chosen by h which denotes height and A which denotes the azimuth angle while the central framework is utilized as the convention factor which is z hub. It focuses toward the North Pole, the y hub indistinguishably focuses on the horizon of the skylight, and x pivot is opposite to both the North Pole and horizon. Therefore, the angles and coordinate frequencies are being encountered mathematically by calculating the latitude and longitude in order to implement correct angles to trap the solar irradiance most efficiently. Here, the zero point of latitude is considered the primary meridian which controls the function of meridian of Eastern Hemisphere and Western Hemisphere angle of the earth surface and so the north of the equator is the Northern Hemisphere and south of the equator is the Southern Hemisphere that are also being controlled by this earth surface modeling to trap solar energy more efficiently. Finally, the δ and ω point hours are being clarified accurately considering this analytical Cartesian coordinates in order to decide the position of solar irradiance vector in order to clarify the solar energy emission into the earth surface to determine net solar energy calculation on the earth surface [16, 19, 20].

Once the angle of the earth surface is being modeled, the earth surface is considered as the net areal dimension of solar energy emission by considering the peak hours' solar radiation generation from the sun [17, 21, 22]. Thus, the radiation of the solar energy flux emitted by the sun and intercepted by the earth is computed by the solar constant which is defined by the measurement of the solar energy flux density perpendicular to the ray direction per unit area per unit time [23–25]. Thus, the calculation of this amount of net solar energy includes all types of radiations of scattered and reflected ones that are being sent to the earth surface from all directions modeled by using MATLAB software in order to calculate total global solar radiation emission on earth (Fig. 2.2).

Then, the sunlight is clarified as the motion of the photon flux by considering the first function of the fundamental solar thermal energy and antireflective coatings of solar cells and then it is modified into the second order of function of the solar energy [26–28]. The integration of these two functions is computed by

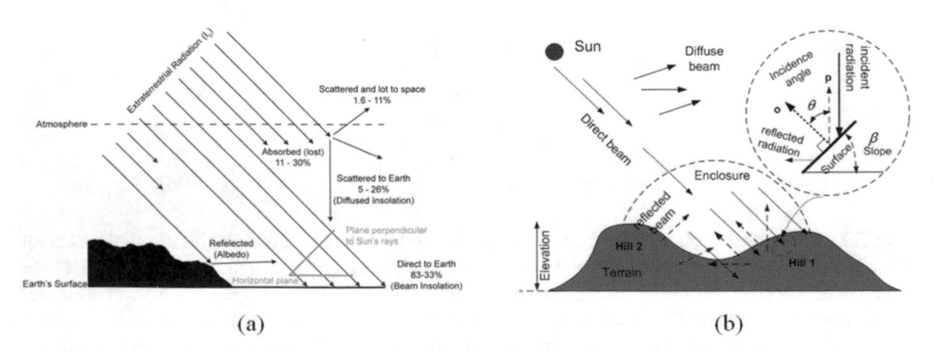

Fig. 2.2 (**a**) Shows the emission of solar energy on the earth surface, (**b**) different types of radiation on the earth surface; direct beam, reflected beam, diffuse beam at various angles

implementing the solar quantum dynamics which is clarified as the most acceptable quantum technology to calculate the net solar energy emission on earth [10, 20, 30]. This is because the earth surface can emit solar irradiance accurately at a given temperature of approximately 700 °C where the energy density of the solar radiation is derived from the maximum solar energy generation from a single solar photon excitation [3, 12, 31].

The amount of global solar radiation calculation on the earth surface is further clarified considering the three background solar data calculation by using **pyrheliometer** to measure direct beam radiation coming from the sun and radius of the earth surface [24, 32, 33]. Then, the **pyranometer** is also used to measure total hemispherical radiation beam plus diffusion on a horizontal surface and the net global total irradiance (W/m^2) is measured on a horizontal surface by a pyranometer and then expressed as follows:

$$I_{\text{tot}} = I_{\text{beam}} \cos\theta + I_{\text{diffuse}}$$

where θ is the zenith angle (i.e., angle between the incident ray and the normal to the horizontal instrument plane) which has been implemented to calculate the net solar energy reaching earth by the clarification of electron energy level of hydrogen (Fig. 2.3).

This measurement is then calibrated against standard pyrheliometers with the thermocouple detectors and with photovoltaic detectors considering the wavelength of the solar spectrum and angle of incidence [4, 23]. Eventually **photoelectric sunshine recorder** has been used for the natural solar radiation which is notoriously intermittent and varying in intensity by clarifying the most potent radiation that creates the highest potential for concentration and conversion in the bright sunshine [2, 34, 35]. Since solar radiation is related to the photon charge, the attributes of photon energy on earth surface are computed considering the quantum flow of photon radiation in global scale by using MATLAB 9.0 Classical Multidimensional Scaling [36–38]. Consequently, a computational model of photon radiation is quantified to demonstrate the solar energy generation from sunlight considering radiation emission. Thereafter, the mode of the solar quantum absorbance by earth surface is determined by the peak solar radiation output tracking into the earth surface [24, 39, 40]. Naturally, the induced solar irradiance is, thereafter, computed by the earth surface area by implementing the parameters of solar energy proliferation on it, and transformation rate of solar energy into electricity energy generation. Thus, the accurate calculation of the current–voltage (*I*–*V*) characteristic is subsequently conducted by the conceptual model of net solar radiation intake into the earth surface by computing the net active solar volt (I_{v+}) generation into the earth surface [15, 41, 42].

Then, the mathematical determination of the net current formation via I_{pv} on earth surface has been modeled out, by calculating *I*–*V*–*R* relationship within the earth surface in order to use this energy commercially throughout the world (Fig. 2.4).

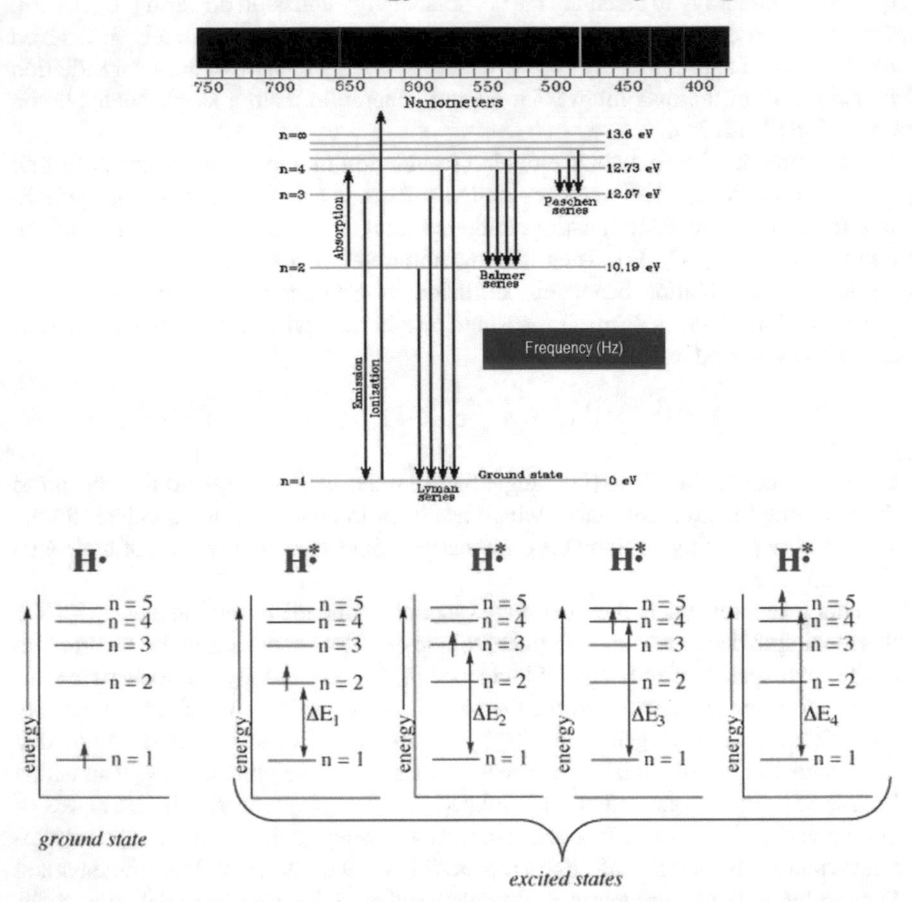

Fig. 2.3 The solar energy-state hydrogen depicting the absorption and emission modes, and energy deliberation rate revealing the energy density by clarifying the optimum solar irradiance deliberation from a photon particle at various wavelengths and frequencies of 10^{17}–10^{14} Hz. Then the electron-state hydrogen energy clarification is expressed by considering a downward transition involving emission of a solar energy with respect to ground state and excited state of solar energy

Hence, the following equation is computed as the energy deliberation from the earth surface, which origin is the photon irradiance and ambient temperature of the solar energy:

$$P_{\mathrm{PV}} = \eta_{\mathrm{pvg}} A_{\mathrm{pvg}} G_{\mathrm{t}} \tag{2.1}$$

Here, η_{pvg} denotes the earth surface performance rate, A_{pvg} denotes the earth surface array (m^2), and G_{t} denotes the photon irradiance intake rate on the plane (W/m^2) of earth surface and thus η_{pvg} could be rewritten as follows:

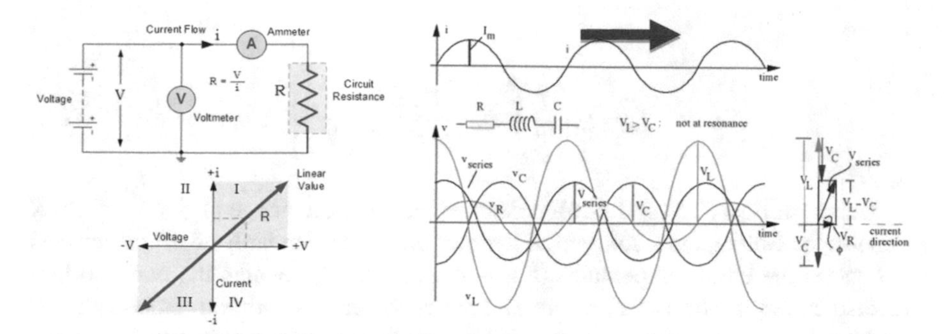

Fig. 2.4 The conceptual circuit diagram of the whole earth surface depicting the net photo-physical current generation into the earth surface by detailing the model of *I–V–R* relationship in order to use electricity throughout the world, respectively

$$\eta_{\text{pvg}} = \eta_{\text{r}}\eta_{\text{pc}}\left[1-\beta\left(T_{\text{c}}-T_{\text{cref}}\right)\right] \tag{2.2}$$

η_{pc} denotes the energy formation efficiency, when maximum power point tracking (MPPT) is implemented which is close to 1; here β denotes the temperature cofactor (0.004–0.006 per °C); η_{r} denotes the mode of energy efficiency; and T_{cref} denotes the condition of temperature at °C. The reference earth temperature (T_{cref}) can be rewritten by calculating from the equation below:

$$T_{\text{c}} = T_{\text{a}} + \left(\frac{\text{NOCT}-20}{800}\right)G_{\text{t}} \tag{2.3}$$

T_{a} denotes the ambient temperature in °C, G_{t} denotes the solar radiation on earth surface (W/m^2), and thus it denotes the modest optimum earth temperature in °C. Considering this temperature condition, the net solar radiation on earth surface can be calculated by the equation below:

$$I_{\text{t}} = I_{\text{b}}R_{\text{b}} + I_{\text{d}}R_{\text{d}} + \left(I_{\text{b}} + I_{\text{d}}\right)R_{\text{r}} \tag{2.4}$$

The solar energy here is necessarily working as a conceptual P-N junction superconductor in order to form electricity through the earth surface, which is interlinked in a parallel series connection [25, 43, 44]. Thus, a unique conceptual circuit model, as shown in Fig. 2.4, with respect to the N_{s} series of earth surface and N_{p} parallel arrays has been computed by the following earth surface solar energy equation based on current and volt relationship:

$$I = N_{\text{p}}\left[I_{\text{ph}} - I_{\text{rs}}\left[\exp\left(\frac{q\left(V+IR_{s}\right)}{AKTN_{\text{s}}}-1\right)\right]\right] \tag{2.5}$$

where

$$I_{\mathrm{rs}} = I_{\mathrm{rr}} \left(\frac{T}{T_{\mathrm{r}}} \right)^{3} \exp \left[\frac{E_{\mathrm{G}}}{AK} \left(\frac{1}{T_{\mathrm{r}}} - \frac{1}{T} \right) \right] \tag{2.6}$$

Hence, in Eqs. (2.5) and (2.6), q denotes the electron charge (1.6×10^{-19} C), K denotes the Boltzmann's constant, A denotes the diode standardized efficiency, and T denotes the earth temperature (K). Accordingly, IR_s denotes the earth surface reverse current motion at T, where T_r denotes the earth condition temperature, I_{rr} denotes the reverse current at T_r, and E_G denotes the photonic bandgap energy of the superconductor utilized for the earth surface. Thus, the photonic current I_{ph} will be generated in accordance with the earth surface temperature and radiation condition which can be expressed by

$$I_{\mathrm{ph}} = \left[I_{\mathrm{SCR}} + k_i \left(T - T_{\mathrm{r}} \right) \frac{S}{100} \right] \tag{2.7}$$

Here, I_{SCR} denotes the current motion considering the optimum temperature of the earth and solar radiation dynamic on the earth surface, k_i denotes the short-circuited current motion, and S denotes the solar radiation calculation in a unit area (mW/cm^2). Subsequently, the I–V features of the earth surface shall be deformed from the conceptual model of the circuit which can be expressed by the following equation:

$$I = I_{\mathrm{ph}} - I_{\mathrm{D}} \tag{2.8}$$

$$I = I_{\mathrm{ph}} - I_{0} \left[\exp \left(\frac{q \left(V + R_{\mathrm{s}} I \right)}{AKT} - 1 \right) \right] \tag{2.9}$$

I_{ph} denotes the photonic current dynamic (A), I_D denotes the diode-originated current dynamic (A), I_0 denotes the inversed current dynamic (A), A denotes the diode-induced constant, q denotes the charge of the electron (1.6×10^{-19} C), K denotes Boltzmann's constant, T denotes the earth temperature (°C), R_s denotes the series resistance (ohm), R_{sh} denotes the shunt resistance (Ohm), I denotes the cell current motion (A), and V denotes the earth voltage motion (V). Therefore, the net current flow into the earth surface can be determined by conducting the following equation:

$$I = I_{pv} - I_{\mathrm{D1}} - \left(\frac{V + IR_{\mathrm{s}}}{R_{\mathrm{sh}}} \right) \tag{2.10}$$

where

$$I_{D1} = I_{01}\left[\exp\left(\frac{V + IR_s}{a_1 V_{T1}}\right) - 1\right] \tag{2.11}$$

Here, I and I_{01} denote the reverse current flow into the conceptual circuit, respectively, and V_{T1} and V_{T2} denote the optimum thermal voltages into the circuit. Thus, the circuit standard factor is presented by a_1 and a_2 and then it is normalized by the mode of earth surface by expressing the following equation:

$$v_{oc} = \frac{V_{oc}}{cKT / q} \tag{2.12}$$

$$P_{max} = \frac{\frac{V_{oc}}{cKT/q} - \ln\left(\frac{V_{oc}}{cKT/q} + 0.72\right)}{\left(1 + \frac{V_{oc}}{KT/q}\right)}\left(1 - \frac{V_{oc}}{\frac{V_{oc}}{I_{SC}}}\right)\left(\frac{V_{oc0}}{1 + \beta \ln \frac{G_0}{G}}\right)\left(\frac{T_0}{T}\right)^{\gamma} I_{sc0}\left(\frac{G}{G_o}\right)^{\alpha} \tag{2.13}$$

where v_{oc} denotes the standard point of the open-circuit voltage, V_{oc} denotes the thermal voltage $V_t = nkT/q$, c denotes the constant current motion, K denotes Boltzmann's constant, T denotes the temperature into the earth surface PV cell in Kelvin, α denotes the function which represents the nonlinear motion of photocurrents, q denotes the electron charge, γ denotes the function acting for all nonlinear temperature-voltage currents, while β denotes the earth surface mode for specific dimensionless function for enhancing current flowing rate. Subsequently, Eq. (2.13) represents the peak energy generation from the earth surface module which is interlined in both series and parallel connection. Thus, the equation for the net energy formation in the array of N_s has been interlinked in series and N_p has been interlinked in parallel considering the power P_M of each mode of connection and which is finally expressed by using the following equation:

$$P_{array} = N_s N_p P_M \tag{2.14}$$

Calculation of Net Electricity Energy Generation from Total Solar Irradiance on Earth

To convert global solar energy into electricity energy, a model is also being prepared by integrating global Albanian symmetries of scalar gauge field [40, 45, 46]. Naturally, the net solar energy particle will functionally be acted as the dynamic photons of particle T^{α} at the global symmetrical array of earth surface by initiating the gauge field of $A_{\mu}^{\alpha}(x)$ and then the local Albanian will subsequently be started to activate at the global U(1) phase symmetry to deliver net electricity energy [30, 47]. Thus, the model is being considered as a complex vector field of $\Phi(x)$ of earth surface where electric charge q will couple with the EM field of $A^{\mu}(x)$ and thus the equation can be expressed by $\mathfrak{H}$:

$$\mathfrak{H} = -\frac{1}{4}F_{\mu\nu}F^{\mu\nu} + D_{\mu}\Phi^{*}D^{\mu}\Phi - V(\Phi^{*}\Phi) \tag{2.15}$$

where

$$D_{\mu}\Phi(x) = \partial_{\mu}\Phi(x) + iqA_{\mu}(x)\Phi(x)$$

$$D_{\mu}\Phi^{*}(x) = \partial_{\mu}\Phi^{*}(x) - iqA_{\mu}(x)\Phi^{*}(x) \tag{2.16}$$

And

$$V(\Phi^{*}\Phi) = \frac{\lambda}{2}(\Phi^{*}\Phi)^{2} + m^{2}(\Phi^{*}\Phi) \tag{2.17}$$

Here >0 $m^2 < 0$; therefore $\Phi = 0$ is a local optimum vector quantity, while the minimum form of degenerated scalar circle is clarified as $\Phi = \frac{v}{\sqrt{2}} * \mathrm{e}^{i\theta}$,

$$v = \sqrt{\frac{-2m^{2}}{\lambda}}, \text{any real}\,\theta \tag{2.18}$$

Subsequently, the vector field Φ of the global earth surface will form a nonzero functional value $\langle\Phi\rangle \neq 0$, which will simultaneously determine the U(1) symmetrical net solar energy generation. Therefore, the global U(1) net symmetrical electrical energy of $\Phi(x)$ will be delivered as expected value of $\langle\Phi\rangle$ by confirming the x-dependent state of the symmetrical $\Phi(x)$ array of earth surface and can be expressed by the following equation:

$$\Phi(x) = \frac{1}{\sqrt{2}}\Phi_{\mathrm{r}}(x) * \mathrm{e}^{i\Theta(x)}, \text{real}\,\Phi_{\mathrm{r}}(x) > 0, \text{real}\,\Phi(x) \tag{2.19}$$

Thus, the net calculation of the electricity energy generation from the net earth surface solar energy is being determined considering the vector $\Phi(x) = 0$, and it is first-order function of $\langle\Phi\rangle \neq 0$, considering the peak level of solar energy emission on the earth surface of $\Phi\langle x\rangle \neq 0$ [1, 12, 39]. Thus, the net electricity energy generation is done from the global solar energy calculation $\phi_{\mathrm{r}}(x)$ and $\Theta(x)$, and its vector on the earth surface field ϕ_{r} has been confirmed by conducting the following equation:

$$V(\phi) = \frac{\lambda}{8}\left(\phi_{\mathrm{r}}^2 - v^2\right)^2 + \text{const}, \tag{2.20}$$

or the resultant electricity energy generation is shifted by its VEV, $\Phi_{\mathrm{r}}(x) = v + \sigma(x)$,

$$\phi_{\mathrm{r}}^2 - v^2 = (v + \sigma)^2 - v^2 = 2v\sigma + \sigma^2 \tag{2.21}$$

$$V = \frac{\lambda}{8}\left(2v\sigma - \sigma^2\right)^2 = \frac{\lambda v^2}{2} * \sigma^2 + \frac{\lambda v}{2} * \sigma^3 + \frac{\lambda}{8} * \sigma^4 \tag{2.22}$$

Simultaneously, the functional derivative $D_\mu\phi$ will become

$$D_\mu\phi = \frac{1}{\sqrt{2}}\left(\partial_\mu\left(\phi_{\mathrm{r}}\mathrm{e}^{i\Theta}\right) + iqA_\mu * \phi_{\mathrm{r}}\mathrm{e}^{i\Theta}\right) = \frac{\mathrm{e}^{i\Theta}}{\sqrt{2}}\left(\partial_\mu\phi_{\mathrm{r}} + \phi_{\mathrm{r}} * i\partial_\mu\Theta + \phi_{\mathrm{r}} * iqA_\mu\right) \tag{2.23}$$

$$\begin{aligned} \left|D_\mu\phi\right|^2 &= \frac{1}{2}\left|\partial_\mu\phi_{\mathrm{r}} + \phi_{\mathrm{r}} * i\partial_\mu\Theta + \phi_{\mathrm{r}} * iqA_\mu\right|^2 \\ &= \frac{1}{2}\left(\partial_\mu\phi_{\mathrm{r}}\right) + \frac{\phi_{\mathrm{r}}^2}{2} * \left(\partial_\mu\Theta qA_\mu\right)^2 \\ &= \frac{1}{2}\left(\partial_\mu\sigma\right)^2 + \frac{(v+\sigma)^2}{2} * \left(\partial_\mu\Theta + qA_\mu\right)^2 \end{aligned} \tag{2.24}$$

Altogether,

$$\mathfrak{h} = \frac{1}{2}\left(\partial_\mu\sigma\right)^2 - v(\sigma) - \frac{1}{4}F_{\mu\nu}F^{\mu\nu} + \frac{(v+\sigma)^2}{2} * \left(\partial_\mu\Theta + qA_\mu\right)^2 \tag{2.25}$$

To determine the formation of this net electricity generation referred to as ($\mathfrak{h}_{\mathrm{sef}}$) on the earth surface, the function of the electrostatic fields has been quantified by conducting the quadratic calculation and described by the following equation:

$$\mathfrak{h}_{\mathrm{sef}} = \frac{1}{2}\left(\partial_\mu\,\sigma\right)^2 - \frac{\lambda v^2}{2} * \sigma^2 - \frac{1}{4}F_{\mu\nu}F^{\mu\nu} + \frac{v^2}{2} * \left(qA_\mu + \partial_\mu\Theta\right)^2 \tag{2.26}$$

Here this net electricity generation ($\mathfrak{h}_{\mathrm{free}}$) function certainly will admit a realistic vector particle of positive mass^2 = λv^2 integrating the areal $A_\mu(x)$ function and the electricity energy generation fields $\Theta(x)$ to determine the net electricity energy from the global solar energy calculation on the earth surface [2, 48].

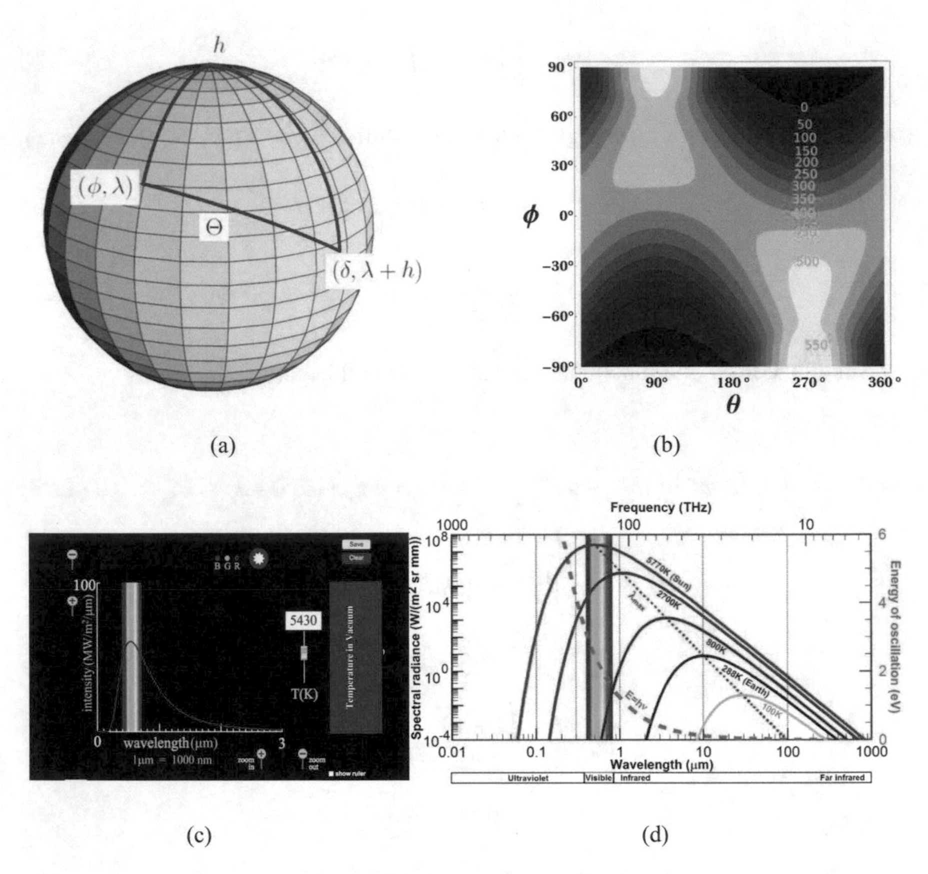

Fig. 2.5 (**a**) Spherical triangle for application of the spherical law of cosines for the calculation of solar zenith angle Θ for observer at latitude φ and longitude λ from knowledge of the hour angle h and solar declination δ. (δ is the latitude of subsolar point, and h is the relative longitude of subsolar point.) (**b**) The theoretical daily-average irradiation at the top of the atmosphere, where θ is the polar angle of the earth's orbit, and $\theta = 0$ at the vernal equinox, and $\theta = 90°$ at the summer solstice; φ is the latitude of the earth. The calculation assumed conditions appropriate for 2000 A.D.: a solar constant of $S_0 = 1367$ W/m^2, obliquity of $\varepsilon = 23.4398°$, longitude of perihelion of $\varpi = 282.895°$, and eccentricity of $e = 0.016704$. Contour labels (green) are in units of W/m^2; (**c**) shows the solar irradiance at various frequencies and (**d**) shows the peak temperature which suggest the calculative power to determine the net solar energy

Results and Discussion

Calculation of Net Solar Energy on Earth

To calculate the net solar energy on the earth surface, the net irradiance of photon emission has been calculated by integrating Eqs. (2.22) and (2.23). Necessarily, the functional earth surface area $J(\omega)$, the photonic quantum field, and the unit area $J(\omega)$ are being calculated considering the constant irradiance coupling point, and the Weisskopf-Wigner approximation mechanism in order to confirm the accurate solar energy emission on the earth surface (Fig. 2.5).

The computed results show that the distribution of solar radiation at the top of the earth's sphericity and orbital parameters is the application of the unidirectional beam incident to a rotating sphere of Milankovitch cycles from spherical earth law of cosines:

$$\cos(c) = \cos(a)\cos(b) + \sin(a)\sin(b)\cos(C) \tag{2.27}$$

where a, b, and c are arc lengths, in radians, of the sides of a spherical triangle. C is the angle in the vertex opposite the side which has arc length c. Applied to the calculation of solar zenith angle Θ, the following applies to the spherical law of cosines:

$$C = h$$

$$c = \Theta$$

$$a = \frac{1}{2}\pi - \phi$$

$$b = \frac{1}{2}\pi - \delta$$

$$\cos(\Theta) = \sin(\phi)\sin(\delta) + \cos(\phi)\cos(\delta)\cos(h) \tag{2.28}$$

In order to simplify this equation, it has been further clarified as a general one derived as follows:

$$\begin{aligned}\cos(\theta) = {} & \sin(\phi)\sin(\delta)\cos(\beta) + \sin(\delta)\cos(\phi)\sin(\beta)\cos(\gamma) + \\ & \cos(\phi)\cos(\delta)\cos(\beta)\cos(h) - \\ & \cos(\delta)\sin(\phi)\sin(\beta)\cos(\gamma)\cos(h) - \\ & \cos(\delta)\sin(\beta)\sin(\gamma)\sin(h)\end{aligned}$$

where β is an angle from the horizontal and γ is an azimuth angle.

The sphere of earth from the sun here is denoted as R_E and the mean distance is denoted as R_0, with approximation of one astronomical unit (AU). The solar constant is denoted as S_0. The solar flux density (insolation) onto a plane tangent to the sphere of the earth, but above the bulk of the atmosphere (elevation of 100 km or greater), is calculated as

$$Q = \begin{cases} S_0 \dfrac{R_0^2}{R_E^2}\cos(\theta) & \cos(\theta) > 0 \\ 0 & \cos(\theta) \le 0 \end{cases}$$

The average of Q over a day is the average of Q over one rotation, or the hour angle progressing from $h = \pi$ to $h = -\pi$: Thus, the equation has been rewritten as

$$Q^{-\text{day}} = -\frac{1}{2\pi}\int_{\pi}^{-\pi} Q dh$$

Since h_0 is the hour angle when Q becomes positive, it could occur at sunrise when $\Theta = 1/2\pi$, or for h_0 as a solution of

$$\sin(\phi)\sin(\delta) + \cos(\phi)\cos(\delta)\cos(h_0) = 0$$

or

$$\cos(h_0) = -\tan(\phi)\tan(\delta)$$

Once $\tan(\varphi)\tan(\delta) > 1$, then the sun does not set and the sun is already risen at $h = \pi$, so $h_0 = \pi$. Then the $\tan(\varphi)\tan(\delta) < -1$, the sun does not rise, and

$$Q^{-\text{day}} = 0$$

$\frac{R_0^2}{R_E^2}$ is nearly constant over the course of a day, and can be taken outside the integral

$$\int_{\pi}^{-\pi} Q dh = \int_{h_0}^{-h_0} Q dh = S_0 \frac{R_0^2}{R_E^2} \int_{h_0}^{-h_0} \cos(\theta) dh$$

$$= S_0 \frac{R_0^2}{R_E^2}\left[h \sin(\phi)\sin(\delta) + \cos(\phi)\cos(\delta)\sin(h)\right]_{h=h_0}^{h=-h_0}$$

$$= -2S_0 \frac{R_0^2}{R_E^2}\left[h_0 \sin(\phi)\sin(\delta) + \cos(\phi)\cos(\delta)\sin(h_{0)}\right]$$

Therefore

$$Q^{-\text{day}} = \frac{S_0}{\pi}\frac{R_0^2}{R_E^2}\left[h_0 \sin(\phi)\sin(\delta) + \cos(\phi)\cos(\delta)\sin(h_{0)}\right]$$

Since θ is being considered as the conventional polar angle describing a planetary orbit, $\theta = 0$ at the vernal equinox and the declination δ as a function of orbital position would be

$$\delta = \varepsilon \sin(\theta)$$

where ε is the obliquity and the conventional longitude of perihelion ϖ shall be related to the vernal equinox, so for the elliptical orbit it can be rewritten as

$$R_E = \frac{R_0}{1 + e\cos(\theta - \omega)}$$

or

$$\frac{R_0}{R_E} = 1 + e\cos(\theta - \omega)$$

With knowledge of ϖ, ε, and e from astrodynamical calculations [34] and S_0 from a consensus of observations or theory, $Q^{-\text{day}}$ can be calculated for any latitude φ and θ. Because of the elliptical orbit, and as a consequence of Kepler's second law. Nevertheless, $\theta = 0°$ is considered as exactly the time of the vernal equinox, $\theta = 90°$ is exactly the time of the summer solstice, $\theta = 180°$ is exactly the time of the autumnal equinox, and $\theta = 270°$ is exactly the time of the winter solstice. Therefore, the equation can be simplified for irradiance on a given day as follows:

$$Q = S_0\left(1 + 0.034\cos\left(2\pi\frac{n}{365.25}\right)\right)$$

where n is the number of a day of the year and thus the solar characteristics for both theoretical function of optimum and modular to generate electricity can be shown per unit area (Fig. 2.6).

Eventually, a peak high-frequency cutoff Ω_C of solar irradiance is calculated to keep away the bifurcation of DOS from the earth surface. Necessarily, a tipped high-frequency cutoff of earth surface Ω_d is determined by controlling the positive DOS in 2D and 1D of the photon irradiance (Table 2.1). Hence $\text{pi}_2(x)$ acts as an algorithm function and $e_{\text{rfc}}(x)$ acts as an additional function [28, 49]. Thus, the DOS of earth surface, here represented as $\varrho_{PC}(\omega)$, is determined by calculating photonic energy frequencies of Maxwell's rules on the earth surface [2, 12, 32]. For a 1D on earth surface, the represented DOS is thus being expressed as $\varrho_{PC}(\omega) \propto \frac{1}{\sqrt{\omega - \omega_e}}\Theta(\omega - \omega_e)$, where $\Theta(\omega - \omega_e)$ represents the Heaviside step function and ω_e expresses the frequency of the net solar energy generation [14, 23].

This DOS is thus determined to confirm a 3D isentropic function on the earth surface to acquire an accurate net qualitative state of solar energy by inducing the non-Weisskopf-Wigner mode of photons on the earth surface [14, 17, 50]. Naturally, this 3D state will be the functional DOS into the PBE area of DOS: $\varrho_{PC}(\omega) \propto \frac{1}{\sqrt{\omega - \omega_e}}\Theta(\omega - \omega_e)$, and thus, it has been integrated to the net electricity (EF) vector of earth surface to determine the net electricity energy generation accurately on the earth surface [15, 51]. Considering the 2D and 1D, the photonic energy DOS is clarified by the pure algorithm of divergence which is close to the PBE, and is thus expressed as $\varrho_{PC}(\omega) \propto -[\ln|(\omega - \omega_0)/\omega_0| - 1]\Theta(\omega - \omega_e)$, where ω_e denotes the midpoint of tip algorithm (Table 2.1). The functional area $J(\omega)$ is thus clarified

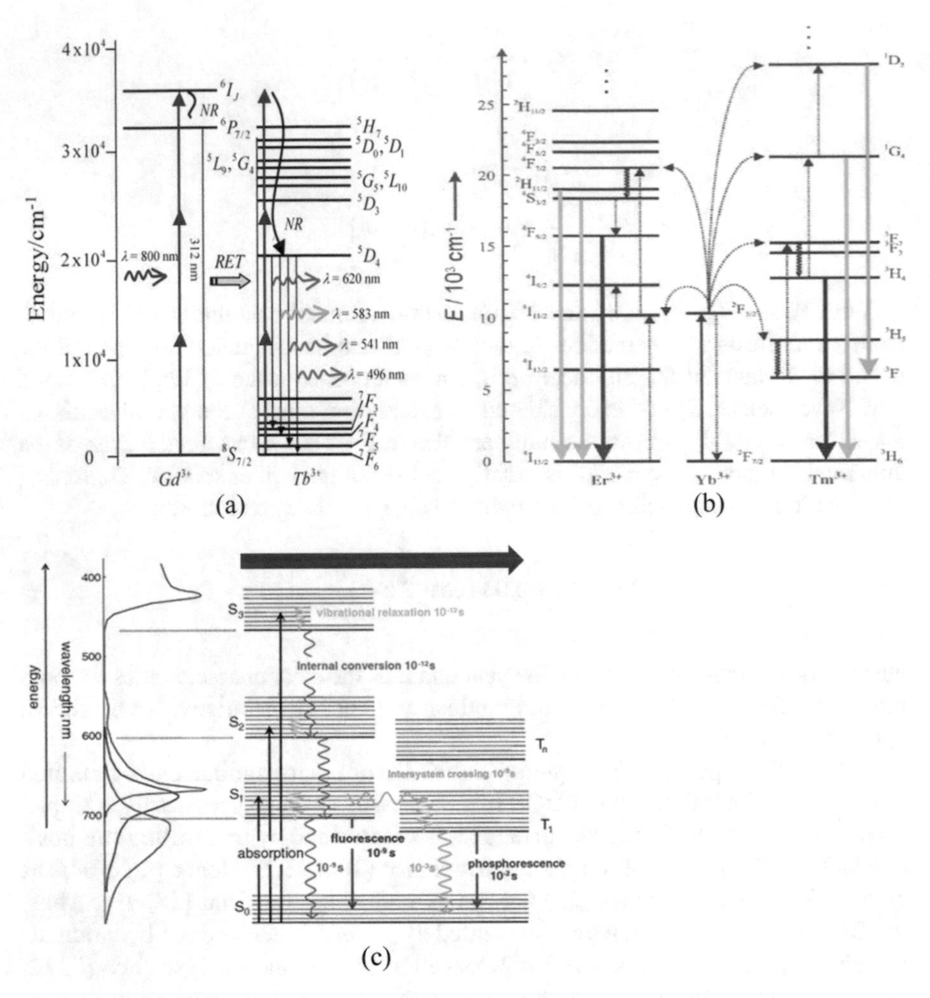

Fig. 2.6 (**a**) Shows the above stands for the solar characteristics for theoretical function of optimum working point for energy generation per unit area, (**b**) depicts the theoretical function of modular point of energy generation per unit area, and (**c**) shows energy generation at various wavelengths of photon spectrum

as the photon energy generation on the earth surface where the solar energy generation of $V(\omega)$ depends on the total solar irradiance on the earth surface [27, 42],

$$J(\omega) = \varrho(\omega)\left|V(\omega)\right|^2 \tag{2.29}$$

Hence, the PB frequency ω_c and proliferative solar energy are considered as the function $u(t, t_0)$ for photon energy generation in the relation $\langle a(t)\rangle = u(t, t_0)\langle a(t_0)\rangle$. It is therefore determined using the functional integral equation and is expressed as

Table 2.1 The photonic structures in different DOS dimensional modes in the earth surface (ES). They correspond with different unit area $J(\omega)$ and self-energy induction at reservoir $\Sigma(\omega)$, which is determined by the photon dynamics into the extreme relativistic earth surface. The variables C, η, and χ function like coupled forces between the solar energy on the earth surface of 1D, 2D, and 3D areal surface

Solar energy (ES)	Unit area $J(\omega)$ for different DOS	Solar energy correction on earth surface $\Sigma(\omega)$
1D	$\frac{\mathrm{ES}}{2\pi r}\frac{1}{\sqrt{\omega-\omega_e}}\Theta(\omega-\omega_e)$	$-\frac{\mathrm{ES}}{\sqrt{2\omega_e-\omega}}$
2D	$-\mathrm{ES}\left[\ln\left\lvert\frac{\omega-\omega_0}{2\omega_0}\right\rvert-1\right]\Theta(\omega-\omega_e)\Theta(\Omega_d-\omega)$	$\mathrm{ES}\left[\mathrm{Li}_2\left(\frac{\Omega_d-\omega_0}{\omega-\omega_0}\right)-\mathrm{Li}_2\left(\frac{\omega_0-\omega_e}{\omega_0-\omega}\right)-\ln\frac{\omega_0-\omega_e}{\Omega_d-\omega_0}\ln\frac{\omega_e-\omega}{\omega_0-\omega}\right]$
3D	$\mathrm{ES}\sqrt{\frac{2\omega-\omega_e}{\Omega_C}}\exp\left(-\frac{\omega-\omega_e}{\Omega_C}\right)\Theta(\omega-\omega_e)$	$\mathrm{ES}\left[\pi\sqrt{\frac{\omega_e-\omega}{\Omega_C}}\exp\left(-\frac{2\omega-\omega_e}{\Omega_C}\right)e_{\mathrm{rfc}}\sqrt{\frac{\omega_e-\omega}{\Omega_C}}-\sqrt{2\pi r}\right]$

$$u(t,t_0)=\frac{1}{1-\Sigma'(\omega_b)}e^{-i\omega(t-t_0)}+\int_{\omega_e}^{\infty}d\omega\frac{J(\omega)e^{-i\omega(t-t_0)}}{\left[\omega-\omega_c-\Delta(\omega)\right]^2+\pi^2J^2(\omega)} \tag{2.30}$$

where $\Sigma'(\omega_b)=\left[\partial\Sigma(\omega)/\partial\omega\right]_{\omega=\omega_b}$ and $\Sigma(\omega)$ denote the storage-induced PB photonic energy proliferations,

$$\Sigma(\omega)=\int_{\omega_e}^{\infty}d\omega'\frac{J(\omega')}{\omega-\omega'} \tag{2.31}$$

Here, the frequency ω_b in Eq. (2.17) denotes the photon energy frequency module in the PBG ($0 < \omega_b < \omega_e$) and thus it is calculated using the areal condition $\omega_b - \omega_c - \Delta(\omega_b) = 0$, where $\lesssim\Delta(\omega)=\mathcal{P}\left[\int d\omega'\frac{J(\omega')}{\omega-\omega'}\right]$ is a primary-value integral.

Therefore, the net photon energy, considering the proliferation magnitude $|u(t, t_0)|$, has been calculated and is shown in Table 2.1 for 1D, 2D, and 3D of earth surface with respect to PBG function [26, 42, 52]. The solar energy dynamic rate $\kappa(t)$ is depicted in Fig. 2.4b, neglecting the function $\delta = 0.1\omega_e$. The result revealed that emitted photons are generated at a high rate once ω_c crosses from the PBG to PB area. Because the range in $u(t, t_0)$ is $1 \geq |u(t, t_0)| \geq 0$, the crossover area as related to the condition is denoted as $0.9 \gtrsim |u(t \to \infty, t_0)| \geq 0$ where this corresponds to $-0.025\omega_e \lesssim \delta \lesssim 0.025\omega_e$, with a production rate $\kappa(t)$ within the PBG ($\delta < -0.025\omega_e$) and in the area of the PBE($-0.025\omega_e \lesssim \delta \lesssim 0.025\omega_e$) of the earth surface.

The generation of solar energy emission is almost exponential for $\delta \gg 0.025\omega_e$, which is a Markov factor. It is shown in Fig. 2.6 as the dash-dotted black curves with $\delta = 0.1\omega_e$. In the crossover area ($-0.025\omega_e \lesssim \delta \lesssim 0.025\omega_e$), the PB frequency of the PBE of earth surface sharply increases the mode of emission of photon energy generation [53, 54]. Thus, this proliferation of emitted solar photon confirms the net energy-state photon on the earth surface of the PBG where the photons are in a nonequilibrium photonic energy state [4, 11].

Then, the solar irradiance on the entire earth surface is clarified considering thermal variation with respect to the solar energy concentration function $v(t, t)$ by determining the nonequilibrium solar energy scattering and reflecting calculation globally [5, 55]:

$$v(t,t)=\int_{t_0}^{t}dt_1\int_{t_0}^{t}dt_2u^*(t_1,t_0)\tilde{g}(t_1,t_2)u(t_2,t_0) \tag{2.32}$$

Here, the two-time correlation function of earth surface $\tilde{g}(t_1,t_2)=\int d\omega J(\omega)\bar{n}(\omega,T)e^{-i\omega(t-t')}$ reveals the solar energy generation variations induced by the thermal relativistic condition of earth surface, where $\bar{n}(\omega,T)=1/\left[e^{\hbar\omega/k_BT}-1\right]$ is the proliferation of the photon energy emission on the earth surface at the optimum temperature T and is expressed as

$$v(t,t\to\infty)=\int_{\omega_e}^{\infty}d\omega\mathcal{V}(\omega)$$

with

$$\mathcal{V}(\omega)=\bar{n}(\omega,T)\left[\mathcal{D}_1(\omega)+\mathcal{D}_d(\omega)\right] \tag{2.33}$$

Here, Eq. (2.18) is simplified to determine the nonequilibrium condition $\mathcal{V}(\omega)=\bar{n}(\omega,T)\mathcal{D}_d(\omega)$. Under low-temperature conditions on earth surface, Einstein's photon energy fluctuation dissipation not only is dynamically viable at the PB on earth surface, but also connects the photonic energy state which has been measured as the field intensity of solar energy induction $n(t)=\langle a^{\dagger}(t)a(t)\rangle=|u(t,t_0)|^2n(t_0)v(t,t)$, where $n(t_0)$ represents the primary PB of earth surface. Therefore, in Fig. 2.6, the plotted net amount of photon energy s versus temperature on earth surface has been clarified as the nonequilibrium proliferated photon energy generation, as shown by the solid-blue curve (Fig. 2.6). To be more specific, the first PB of earth surface has been considered as the Fock state photon number n_0,i.e., $\rho(t_0)=|n_0\rangle\langle n_0|$, which is obtained mathematically through the quantum dynamics of the photon energy and then by solving Eq. (2.36), with respect to the state of net photon energy production at time t:

$$\rho(t)=\sum_{n=0}^{\infty}\mathcal{P}_n^{(n_0)}(t)|n_0\, n_0| \tag{2.34}$$

$$\mathcal{P}_n^{(n_0)}(t)=\frac{\left[v(t,t)\right]^n}{\left[1+v(t,t)\right]^{n+1}}\left[1-\Omega(t)\right]^{n_0}\times\sum_{k=0}^{\min\{n_0,n\}}\binom{n_0}{k}\binom{n}{k}\left[\frac{1}{v(t,t)}\frac{\Omega(t)}{1-\Omega(t)}\right]^k \tag{2.35}$$

where $\Omega(t)=\dfrac{|u(t,t_0)|^2}{1+v(t,t)}$. Therefore, the result reveales that an electron-state photon energy will evolve into different Fock states of $|n_0\rangle$ is $\mathcal{P}_n^{(n_0)}(t)$ on the earth surface. The proliferation of net photon energy dissipation $\mathcal{P}_n^{(n_0)}(t)$ in the primary state $|n_0=5\rangle$ and steady-state limit, $\mathcal{P}_n^{(n_0)}(t\to\infty)$, is thus shown in Fig. 2.7. Therefore, the generation of net photon energy on earth surface will ultimately reach the thermal nonequilibrium state which is expressed as

$$\mathcal{P}_n^{(n_0)}(t\to\infty)=\frac{\left[\bar{n}(\omega_c,T)\right]^n}{\left[1+\bar{n}(\omega_c,T)\right]^{n+1}} \tag{2.36}$$

To probe this huge photon energy generation on earth surface, a further calculation of the photon energy distribution within the quantum field of earth surface has

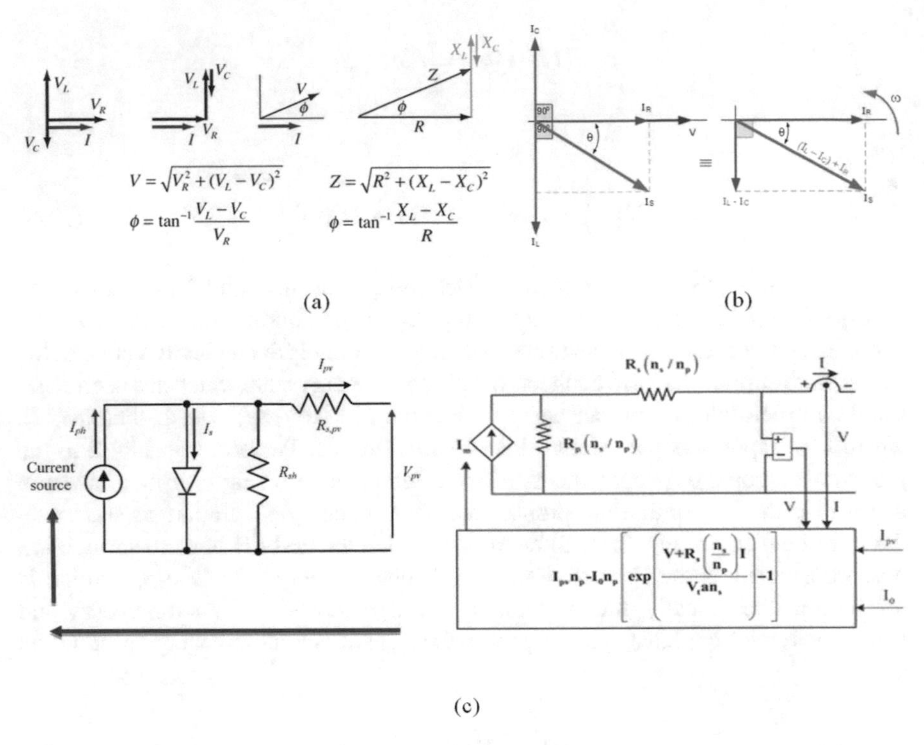

Fig. 2.7 (**a**) The scalar field of the earth surface, (**b**) solar energy scalar field on earth surface, (**c**) net electricity current energy generation on earth from the total solar energy on earth [1, 17]

been conducted through the high-temperature coherent states and solving Eq. (2.17) considering the energy state of photons, and it is expressed by

$$\rho(t) = \mathcal{D}\left[\alpha(t)\right]\rho_{\mathrm{T}}\left[v(t,t)\right]\mathcal{D}^{-1}\left[\alpha(t)\right] \tag{2.37}$$

where $\mathcal{D}\left[\alpha(t)\right] = \exp\{\alpha(t)a^{\dagger} - \alpha^{*}(t)a\}$ denotes the displacement functions with $\alpha(t) = u(t, t_0)\alpha_0$ and

$$\rho_{\mathrm{T}}\left[v(t,t)\right] = \sum_{n=0}^{\infty}\frac{\left[v(t,t)^{n}\right]}{\left[1+v(t,t)\right]^{n+1}}\ nn| \tag{2.38}$$

Here, ρ_{T} denotes a thermal state with an average particle quantum $v(t, t)$, where Eq. (2.11) suggests that the peak point photon energy generation state will be evolved into a thermal state [11, 47], which is considered as the functional state of the photon $\mathcal{D}\left[\alpha(t)\right]|n$ on the earth surface. Thus, the net photon energy generation calculation is represented by the following equation:

$$m\|\rho(t)\|n = J(\omega) = \mathrm{e}^{-\Omega(t)|\alpha_0|^2} \frac{[\alpha(t)]^m [\alpha^*(t)]^n}{[1+v(t,t)]^{m+n+1}}$$
$$= \sum_{k=0}^{\min\{m,n\}} \frac{\sqrt{m!n!}}{(m-k)!(n-k)!k!} \left[\frac{v(t,t)}{\Omega(t)|\alpha_0|^2}\right]^k \tag{2.39}$$

where the emission of the net photon energy $(\langle m|\rho(t)|n\rangle)$ into the earth surface, and its conversion of photon energy into electricity $[1 + v(t, t)]^{m+n+1}$ and nonequilibrium condition $[\alpha(t)]^m[\alpha^*(t)]^n$ of the earth surface, has been calculated.

Calculation of Net Electricity Energy Generation from Total Solar Irradiance on Earth

To transform this tremendous amount of photon energy into electricity energy, the net solar energy is being computed on a conceptual model of series and parallel circuit of earth surface. The conceptual earth surface is then hypothetically implemented into the *I–V* single-diode circuit of earth surface in order to get the precise *I–V* relationship of the net solar energy reaching on earth surface by calculating from the following equation:

$$I = I_\mathrm{L} - I_\mathrm{o}\left\{\exp^{\left[\frac{q(V+I_\mathrm{Rs})}{AkT_\mathrm{c}}\right]} - 1\right\} - \frac{(V+I_\mathrm{Rs})}{R_\mathrm{Sh}} \tag{2.40}$$

Here, I_L denotes the photon formation current, I_o denotes the ideal current flow into the diode, R_s denotes the resistance in a series, A denotes the diode function, k (= 1.38×10^{-23} W/m^{2}K) denotes the Boltzmann's constant, q (= 1.6×10^{-19} C) denotes the charge amplitude of the electron, and T_C denotes the earth temperature. Consequently, the *I–q* linked in the earth surface is varied in the diode cell which is expressed as the dynamic current as follows [5, 56]:

$$I_\mathrm{o} = \mathrm{I}_\mathrm{Rs}\left(\frac{T_\mathrm{c}}{T_\mathrm{ref}}\right)^3 \exp\left[\frac{qE_\mathrm{G}\left(\frac{1}{T_\mathrm{ref}} - \frac{1}{T_\mathrm{c}}\right)}{KA}\right] \tag{2.41}$$

where IR_s denotes the dynamic current representing the functional transformation of solar radiation and qE_G denotes the bandgap solar radiation into the conceptual earth surface at different DOS dimensional modes of 1D, 2D, and 3D (Table 2.1).

Here, considering this conceptual earth surface, the *I–V* relationship with the exception of *I–V* curve, a calculative result of linked *I–V* curves among all of the conceptual solar cells has been determined [49, 51]. Thus, the equation is being rewritten as follows in order to determine the *V–R* relationship much more accurately:

$$V = -IR_s + K\log\left[\frac{I_L - I + I_o}{I_o}\right] \tag{2.42}$$

where K denotes the constant $\left(= \frac{AkT}{q}\right)$ and I_{mo} and V_{mo} denote the net current and voltage in the conceptual earth surface. Subsequently, the relationship among I_{mo} and V_{mo} shall remain motional in the *I–V* earth surface which can be written as

$$V_{mo} = -I_{mo}R_{Smo} + K_{mo}\log\left(\frac{I_{Lmo} - I_{mo} + I_{omo}}{I_{omo}}\right) \tag{2.43}$$

where I_{Lmo} denotes the photon-induced current, I_{omo} denotes the dynamic current into the diode, R_{smo} denotes the resistance in series, and K_{mo} denotes the factorial constant.

Once all non-series (Ns) cells are being interlinked in the series, then the series resistance is calculated as the sum of each solar cell series resistance $R_{smo} = N_s \times R_s$ current considering the functional coefficient of the constant factor $K_{mo} = N_s \times K$. Since the flow of current dynamics into the circuit is lined to the cells in a series connection, the current dynamics in Eq. (2.40) remains the same in each part of $I_{omo} = I_o$ and $I_{Lmo} = I_L$. Thus, the mode of I_{mo}–V_{mo} relationship for the N_s series of connected cells can be expressed by

$$V_{mo} = -I_{mo}N_sR_s + N_sK\log\left(\frac{I_L - I_{mo} + I_o}{I_o}\right) \tag{2.44}$$

Naturally, the current–voltage relationship can be further modified considering all parallel links in N_p cell connection in all parallel modes and can be described as follows [10, 39]:

$$V_{mo} = -I_{mo}\frac{R_s}{N_p} + K\log\left(\frac{N_{sh}I_L - I_{mo} + N_pI_o}{N_pI_o}\right) \tag{2.45}$$

Since the photon-induced current primarily depends on the solar radiation and optimum temperature configuration, the net current dynamic is being calculated as

$$I_L = G\left[I_{SC} + K_I\left(T_c - T_{ref}\right)\right] \times V_{mo} \tag{2.46}$$

where I_{sc} denotes the current at 25 °C and KW/m^2, K_I denotes earth surface coefficient factor, T_{ref} denotes the optimum temperature, and G denotes the solar energy in mW/m^2 [30, 54].

Finally, the electricity energy generation around the earth surface has been computed in order to confirm the net emitted photon utilization by integrating local Albanian electric fields; thus, the global U(1) gauge field will allow to add a mass term of the functional particle of $\varnothing' \rightarrow e^{i\alpha(x)}\varnothing$. It is then further clarified by explaining the variable derivative of transformation law of scalar field using the following equation [12, 51]:

$$\begin{aligned} &\partial_\mu \rightarrow D_\mu = \partial_\mu = ieA_\mu \ [\text{covariant derivatives}] \\ &A'_\mu = A_\mu + \frac{1}{e}\partial_\mu \alpha \left[A_\mu \text{ derivatives} \right] \end{aligned} \tag{2.47}$$

Here, the global U(1) gauge denotes the invariant local Albanian for a complex scalar field which is further expressed as

$$\mathfrak{h} = (D^\mu)^\dagger (D_\mu \varnothing) - \tfrac{1}{4} F_{\mu\nu} F^{\mu\nu} - V(\varnothing) \tag{2.48}$$

The term $\frac{1}{4}F_{\mu\nu}F^{\mu\nu}$ is the dynamic term for the gauge field of the earth surface and $V(\varnothing)$ denotes the extra term in the local Albanian which is $V(\varnothing^*\varnothing) = \mu^2(\varnothing^*\varnothing) + \lambda\,(\varnothing^*\varnothing)^2$.

Therefore, the generation of local Albanian ($\mathfrak{h}$) under the perturbational function of the quantum field of the earth surface has been confirmed by the calculation of mass scalar particles $\phi_1$and ϕ_2 along with a mass variable of μ. In this condition $\mu^2 < 0$ had an infinite number of quantum which is clarified by $\phi_1^2 + \phi_2^2 = -\mu^2 / \lambda = v^2$ and the $\mathfrak{h}$ through the variable derivatives using further shifted fields η *and* ξ defined the quantum field as $\phi_0 = \frac{1}{\sqrt{2}}\left[(\upsilon + \eta) + i\xi\right]$.

$$\begin{aligned} \text{Kinetic term: } \mathfrak{h}(\eta,\xi) &= (D^\mu\phi)^\dagger (D^\mu\phi) \\ &= (\partial^\mu + ieA^\mu)\phi^*(\partial_\mu - ieA_\mu)\phi \end{aligned} \tag{2.49}$$

Thus, this expanding term in the $\mathfrak{h}$ associated to the scalar field of the earth surface suggests that the net earth surface field is prepared to initiate the net electricity energy generation into its quantum field of induced photon energy, respectively, at the normal, normalized, and normal modes [1, 17].

To determine this electricity energy, hereby, a non-variable function of ready dynamics has been implemented for the calculation of $\overline{\varphi}\left[S_0\right]$ to confirm the expected value of S_0 considering the earth surface [32, 51]. Thus, the corrective functional asymptotic formulas are used as follows:

$$\begin{aligned} \overline{\varphi}\left[S_0\right] = 2S_0\left(\ln 4S_0 - 2\right) + \ln 4S_0\left(\ln 4S_0 - 2\right) - \\ \frac{\left(\pi^2 - 9\right)}{3} + s_0^{-1}\left(\ln 4S_0 + \frac{9}{8}\right) + \cdots\left(S_0 \gg 1\right) \end{aligned} \tag{2.50}$$

$$\bar{\varphi}[S_0] = \left(\frac{2}{3}\right)(S_0 - 1)^{\frac{3}{2}} + \left(\frac{5}{3}\right)(S_0 - 1)^{\frac{5}{2}} - \left(\frac{1507}{420}\right)(S_0 - 1)^{\frac{7}{2}}(1/2\,\text{instead of}\,1). \quad (2.51)$$

Then the final equation can be rewritten where S_0 is the areal value of electricity energy generation on the earth surface (1 m^2):

$$\bar{\varphi}[S_0] = \left(\frac{2}{3}\right)(S_0 - 1)^{\frac{3}{2}} + \left(\frac{5}{3}\right)(S_0 - 1)^{\frac{5}{2}} - \left(\frac{1507}{420}\right)(S_0 - 1)^{\frac{7}{2}} \quad (2.52)$$

The function $\bar{\varphi}[S_0]$ is thus determined by the net electricity energy generation from the total solar energy into the atmosphere by calculation of earth's cross-sectional area of 127,400,000 km^2, and the total Sun's power intercepted by the earth of 1.740×10^{17} W. Since no energy is received during the night and the Sun's energy is distributed across the earth's entire surface area, the average insolation is only one-quarter of the solar constant or about 342 W/m^2. Taking into account the seasonal and climatic conditions the actual power reaching the ground generally averages less than 200 W/m^2; thus, the average power intercepted at any time by the earth's surface is around $127.4 \times 10^6 \times 10^6 \times 200 = 25.4 \times 10^{15}$ W or 25,400 TW. Integrating this power over the whole year the total solar energy received by the earth is 25,400 TW × 24 × 365 = 222,504,000 TWh. To put this into perspective, the total annual electrical energy (not the total energy) consumed in the world from all sources in 2018 was 22,126 TWh and thus the available solar energy is over 10,056 times the world's consumption.

Conclusion

Since the fossil fuel energy utilization throughout the world is getting to a finite level and is a major contributor to climate change, usage of solar energy as the renewable clean energy source will indeed be an interesting source to mitigate the global energy demand and environmental perplexity. Simply, energy from the Sun could play a key role in decarbonizing the global economy alongside improvements in energy efficiency and imposing costs on greenhouse gas emitters. Inevitably, solar energy, the radiant energy from the Sun, is thus proposed to use as the natural source of renewable energy to capture and convert it into electricity energy to meet the global energy demand. Simply the development of affordable, inexhaustible, and clean solar energy technologies will have huge longer term benefits since it will increase world's energy security through reliance on an indigenous, inexhaustible, and mostly import-independent resource, and indeed will enhance sustainability, reduce pollution, and mitigate the climate change.

Acknowledgements This research was supported by Green Globe Technology under the grant RD-02017-07 for building a better environment. Any findings, predictions, and conclusions described in this chapter are solely performed by the authors and it is confirmed that there is no conflict of interest for publishing in a suitable journal.

References

1. Guerlin, C. et al. Progressive field-state collapse and quantum non-demolition photon counting. Nature 448, 889 (2007).
2. Md. Faruque Hossain. "Natural mechanism to console global water, energy, and climate change crisis", Sustainable Energy Technologies and Assessments, 2019.
3. Md. Faruque Hossain. "Green building complexes", Elsevier BV, 2019.
4. Dayan, B. et al. A photon turnstile dynamically regulated by one atom. Science 319, 1062–1065 (2008).
5. Md. Faruque Hossain. "Sustainable technology for energy and environmental benign building design", Journal of Building Engineering, 2019.
6. Md. Faruque Hossain. "Green science: Independent building technology to mitigate energy, environment, and climate change", Renewable and Sustainable Energy Reviews, 2017.
7. Joannopoulos, J. D., Villeneuve, P. R. & Fan, S. Photonic crystals: putting a new twist on light. Nature 386, 143 (1997).
8. Ramzi Ben Messaoud. "Extraction of Uncertain Parameters of Double-Diode Model of a Photovoltaic Panel Using Simulated Annealing Optimization", The Journal of Physical Chemistry C, 2019.
9. Leijing Yang, Sheng Wang, Qingsheng Zeng, Zhiyong Zhang, Tian Pei, Yan Li & Lian-Mao Peng (2011). Efficient photovoltage multiplication in carbon nanotubes – Nature Photonics pp 672 – 676.
10. Pregnolato, T., Lee, E., Song, J., Stobbe, D., Lodahl, P. Single-photon non-linear optics with a quantum dot in a waveguide. *Nat. Commun.* 6, 8655 (2015).
11. Yan, Wei-Bin, Jin-Feng Huang, and Heng Fan. "Tunable single-photon frequency conversion in a Sagnac interferometer", Scientific Reports, 2013.
12. Md. Faruque Hossain. "Photon energy amplification for the design of a micro PV panel", International Journal of Energy Research, 2018.
13. Bresar, B. "Quasi-median graphs, their generalizations, and tree-like equalities", European Journal of Combinatorics, 2003.
14. Douglas, J. S., H. Habibian, C.-L. Hung, A. V. Gorshkov, H. J. Kimble, and D. E. Chang. "Quantum many-body models with cold atoms coupled to photonic crystals", Nature Photonics, 2015.
15. Md. Faruque Hossain. "Advanced Building Design", Elsevier BV, 2019 Crossref.
16. Md. Faruque Hossain. "Photonic thermal control to naturally cool and heat the building", Applied Thermal Engineering, 2018.
17. Hossain, Md. Faruque. "Solar energy integration into advanced building design for meeting energy demand and environment problem: Climate change, photoenergy, solar panel, and clean energy", International Journal of Energy Research, 2016.
18. Julie R. Newell. "A story of things yet-to-be: the status of geology in the United States in 1807", Geological Society, London, Special Publications, 2009.
19. Md. Faruque Hossain. "Theoretical mechanism to breakdown of photonic structure to design a micro PV panel", Energy Reports, 2019.
20. Peter Fratzl. "Biomaterial systems for mechanosensing and actuation", Nature, 11/26/2009.
21. Md. Faruque Hossain. "Water", Elsevier BV, 2019.

22. Zhu, Y., Xiaoyong, H., Hong, Y., Qihuang, G. On-chip plasmon-induced transparency based on plasmonic coupled nanocavities. *Sci. Rep*. 4, 3752 (2014).
23. Besharat, Fariba, Ali A. Dehghan, and Ahmad R. Faghih. "Empirical models for estimating global solar radiation: A review and case study", Renewable and Sustainable Energy Reviews, 2013.
24. Birnbaum, K. M. et al. Photon blockade in an optical cavity with one trapped atom. Nature 436, 87–90 (2005).
25. Chang, D. E., Sørensen, A. S., Demler, E. A. & Lukin, M. D. A single-photon transistor using nanoscale surface plasmons. Nature Physics. 3, 807–812 (2007).
26. Md. Faruque Hossain. "Transforming dark photons into sustainable energy", International Journal of Energy and Environmental Engineering, 2018.
27. Md. Faruque Hossain. "Applied energy technology", Elsevier BV, 2019.
28. Md. Faruque Hossain. "Green science: Advanced building design technology to mitigate energy and environment", Renewable and Sustainable Energy Reviews, 2018.
29. Reed, M., Maxwell, L. Connections between groundwater flow and transpiration partitioning. *Sci*. 353, 377-380 (2015).
30. Yan, W., Heng, F. Single-photon quantum router with multiple output ports. *Sci. Rep*. 4, 4820 (2014).
31. H. Z. Shen, Shuang Xu, H. T. Cui, X. X. Yi. "Non-Markovian dynamics of a system of two-level atoms coupled to a structured environment", Physical Review A, 2019.
32. Andreas Reinhard. "Strongly correlated photons on a chip", Nature Photonics, 2011.
33. Mir Sayed Shah Danish, Tomonobu Shah Senjyu. "Chapter 6 Green Building Efficiency and Sustainability Indicators", IGI Global, 2020.
34. Gleyzes, S. et al. Quantum jumps of light recording the birth and death of a photon in a cavity. Nature 446, 297 (2007).
35. Ronald Vargas, David Carvajal, Lorean Madriz, Benjamín R. Scharifker. "Chemical kinetics in solar to chemical energy conversion: The photoelectrochemical oxygen transfer reaction", Energy Reports, 2019.
36. Armani, D. K., Kippenberg, T. J., Spillane, S. M. & Vahala, K. J. Ultra-high-Q toroid microcavity on a chip. Nature 421, 925 (2003).
37. Md. Faruque Hossain. "Power systems", Elsevier BV, 2019.
38. Yi Guo, Ali Al-Jubainawi, Zhenjun Ma. "Performance investigation and optimisation of electrodialysis regeneration for LiCl liquid desiccant cooling systems", Applied Thermal Engineering, 2018.
39. Jaivime, E., Scott, J. McDonnell. Global separation of plant transpiration from groundwater and streamflow. *Nat*. 525, 91–94 (2015).
40. Kimin Park, Petr Marek, Radim Filip. "Qubit-mediated deterministic nonlinear gates for quantum oscillators", Scientific Reports, 2017.
41. Faruque Hossain. "Photon application in the design of sustainable buildings to console global energy and environment", Applied Thermal Engineering, 2018.
42. Md. Faruque Hossain. "Best Management Practices", Elsevier BV, 2019.
43. Md. Faruque Hossain. "Green Technology: Transformation of Transpiration Vapor to Mitigate Global Water Crisis", Polytechnica, 2019.
44. Tobias, D., Wheeler & Abraham, Stroock, D. The transpiration of water at negative pressures in a synthetic tree. *Nat*. 455, 208–212 (2008).
45. Lang, C. et al. Observation of resonant photon blockade at microwave frequencies using correlation function measurements. Phys. Rev. Lett. 106, 243601 (2011).
46. Yan, Wei-Bin, and Heng Fan. "Single-photon quantum router with multiple output ports", Scientific Reports, 2014.
47. Tu, M. W. Y. & Zhang, W. M. Non-Markovian decoherence theory for a double-dot charge qubit. Phys. Rev. B 78, 235311 (2008).
48. Langer, L., Poltavtsev, S., Bayer, M. Access to long-term optical memories using photon echoes retrieved from semiconductor spins. *Nat. Phot*. 8, 851–857 (2014).

49. Scott, J., Zachary, D. Terrestrial water fluxes dominated by transpiration. *Nat.* 496, 347–350 (2013).
50. Md. Faruque Hossain. “Energy”, Elsevier BV, 2019.
51. Sayrin, C. et al. Real-time quantum feedback prepares and stabilizes photon number states. Nature 477, 73 (2011).
52. Tame, M. S., K. R. McEnery, Ş. K. Özdemir, J. Lee, S. A. Maier, and M. S. Kim. “Quantum plasmonics”, Nature Physics, 2013.
53. G. Baur, K. Hencken, D. Trautmann. Revisiting unitarity corrections for electromagnetic processes in collisions of relativistic nuclei. Phys. Rep. 453, 1 (2007).
54. Xiao, Y. F. et al. Asymmetric Fano resonance analysis in indirectly coupled microresonators. Phys. Rev. A 82, 065804 (2010).
55. Zhang, W. M., Lo, P. Y., Xiong, H. N., Tu, M. W. Y. & Nori, F. General Non-Markovian Dynamics of Open Quantum Systems. Phys. Rev. Lett. 109, 170402 (2012).
56. Md. Faruque Hossain. “Water delivery systems”, Elsevier BV, 2019.

Chapter 3
Wind Energy

Abstract Global environmental vulnerability has become the crucial issue due to greenhouse gas concentration in the atmosphere by burning fossil fuel. It is causing the sea-level rising, ocean acidification, and ozone layer depletion and threating the public and animal health. Hence, the evolution of renewable energy over the past decade has shown great promising since it is abundantly available anywhere in the world and can be utilized as an alternative clean energy source in every sector of our daily lives. This chapter, therefore, provides a model for simulating wind energy abundance and its application as a source of clean energy to mitigate global energy demand and environmental vulnerability and perplexity. Simply, in this research, the wind energy abundance has been analyzed mathematically in order to determine the use of wind energy to confirm a realistic assessment of this energy for application commercially in every sector by appropriate operation and control strategy and technology. This integrated model is thus statistically clarified by mathematical simulation where results suggest that wind energy indeed would be the alternative source of renewable energy to mitigate global energy crisis which is benign to the environment.

Keywords Wind energy · Theoretical modeling · Optimization · Alternative energy technology

Introduction

On the outside of the earth, wind comprises the mass development of air. Wind is ordinarily ordered by their spatial scale, their speed, the sorts of powers that cause them, the areas where they occur, and the impact on its surrounding area [1, 2]. Winds have different dynamics: speed (wind speed); the thickness of the gas in question; vitality substance; or wind vitality. In meteorology, winds are regularly alluded to as indicated by their quality, and its heading from which the breeze is blowing. Its short explosions of rapid breeze are named blasts while solid breezes of

M. F. Hossain, *Global Sustainability in Energy, Building, Infrastructure, Transportation, and Water Technology*,
https://doi.org/10.1007/978-3-030-62376-0_3

halfway span (around one moment) are named gusts [3, 4]. Long-term winds have different names related with their normal quality, for example, breeze, hurricane, tempest, and tropical storm. Wind happens on a scope of scales, from tempest streams enduring several minutes to neighborhood breezes created by warming of land surfaces and enduring a couple of hours and to worldwide breezes coming about because of the distinction in ingestion of Sun-oriented vitality between the atmosphere zones on earth [5, 6]. The two primary causes of large-scale atmospheric wind circulation are the differential heating between the equator and the poles, and the rotation of the planet (Coriolis effect). Within the tropics, thermal low circulations over terrain and high plateaus can drive monsoon circulations. In beach front zones the ocean breeze/land breeze cycle can characterize neighborhood twists; in zones that have variable landscape, mountain and valley breezes can overwhelm nearby breezes [7, 8]. Notwithstanding, the breeze is brought about by contrasts in the air pressure. At the point when a distinction in environmental weight exists, air moves from the higher to the lower pressure region, bringing about breezes of different rates [9–11]. On a pivoting planet, air will likewise be diverted by the Coriolis impact, aside from precisely on the equator. All inclusive, the two significant driving elements of enormous scope wind energy are the most promising renewable energy source which is everywhere in differential warming between the equator and the posts (contrast in retention of sunlight-based vitality prompting lightness powers) and the pivot of the planet. Outside the tropics and overtop from frictional impacts of the surface, the enormous scope twists will in general methodology create geostrophic balance. Close to the earth's surface, contact makes the breeze slower and thus surface contact likewise makes winds blow all the more internal into low-pressure territories. Since the breeze is characterized by a harmony of physical powers in this way, it tends to be helpful by rearranging the air conditions of movement by making subjective investigation about the level and vertical appropriation of winds. The geostrophic wind part in this way happens from the consequence of the harmony between Coriolis power and weight inclination power of the environment. It streams corresponding to isobars and approximates the stream over the barometrical limit layer in the midlatitudes [12–14]. The ageostrophic wind part is the contrast among real and geostrophic wind, which is liable for air "topping off" violent winds after some time and the slope wind is like the geostrophic twist yet in addition incorporates radial power (or centripetal speeding up which could be utilized as the gigantic wellspring of clean vitality). In any case, this breeze is a colossal wellspring of sustainable power source which can be utilized in each division as the elective wellspring of clean vitality which is copious all over the place. However, this wind is a tremendous source of renewable energy which can be used in every sector as the alternative source of clean energy which is abundant everywhere. Thus, the aim of this chapter is to present the analysis of wind energy modeling and its commercial application as an alternative clean energy source in every sector in our daily lives to mitigate the global energy demand and environmental crisis.

Materials and Simulation

Since wind energy is the kinetic energy of air in motion, the kinetic energy of the air of mass m is calculated here with its velocity v which is governed by $\frac{1}{2}mv^2$. To determine the mass of the air passing through any area A perpendicular to its velocity, it is multiplied by its volume considering the time t within the air density ρ, which will be denoted as $m = Avt\rho$, and then the wind energy is calculated as

$$E = \frac{1}{2}\rho A \upsilon^3 t \tag{3.1}$$

Consequently, differentiating with respect to time, the rate of increase of energy has been calculated as wind power:

$$P = dE / dt = \frac{1}{2}\rho A \upsilon^3 \tag{3.2}$$

Here, P, the wind power is thus denoted as proportional to the third power of the wind velocity.

Since the wind energy is achievable from the speed of wind considering its air mass flow into the biosphere, in this chapter a model is proposed using turbine implementation to convert wind into electricity energy [8, 15, 16]. Since this wind energy is eventually delivered from the kinetic force, the mechanism of this energy conversion modeling has been described by the chain reaction of interaction of air dynamics. Since the air dynamic mechanism is to govern the wind velocity, it has been analyzed by using the real determinations of the velocity considering wind speed in the atmosphere where both deterministic effects and stochastic variations of turbulence are calculated. Consequently, the characteristics of wind speed in the biosphere have been modeled considering the air dynamic deterministic approaches in the wind turbine in order to confirm the net wind energy output as

$$P_{\mathrm{w}} = \eta_{\mathrm{pvg}} A_{\mathrm{pvg}} G_{\mathrm{t}} \tag{3.3}$$

where η_{pvg} is the wind energy generation efficiency, A_{pvg} is the wind energy generation area (m^2), and G_{t} is the wind energy in tilted module plane (W/m^2). η_{pvg} is further defined as

$$\eta_{\mathrm{pvg}} = \eta_{\mathrm{r}} \eta_{\mathrm{pc}} \left[1 - \beta\left(T_{\mathrm{c}} - T_{\mathrm{cref}}\right)\right] \tag{3.4}$$

where η_{pc} is the power conditioning efficiency which is equal to one when MPPT is used, β is the temperature coefficient ((0.004–0.006) per °C), η_{r} is the reference module efficiency, and T_{cref} is the reference cell temperature in °C [17, 18]. Reference temperature (T_{cref}) can be obtained by the relation

$$T_c = T_a + \left(\frac{\text{NOCT} - 20}{800}\right) G_t \tag{3.5}$$

where T_a is the ambient temperature in °C, NOCT is the nominal temperature in °C, and G_t is the tilted module plane (W/m^2). Thus, the potential total energy output can be estimated as below in order to implement this energy commercially by using wind turbine:

$$I_t = I_b R_b + I_d R_d + (I_b + I_d) R_r \tag{3.6}$$

Since the wind turbine is used in this modeling, variable wind speed driven into the multipole permanent magnet synchronous generator (PMSG) effect has been extensively calculated because of its higher efficiency, low weight, less maintenance, and easier controllability to get much electricity energy from the wind [7, 19]. Using direct driven PMSG will not only increase reliability but also decrease the weight in nacelle to confirm much wind energy deliberation which is calculated as voltage:

$$V_d = -R_s i_d - L_d \frac{di_q}{dt} + \omega L_q i_q \tag{3.7}$$

$$V_q = -R_s i_q - L_q \frac{di_q}{dt} + \omega L_d i_d + \omega \lambda_m \tag{3.8}$$

Here, the energy torque is governed by the speed of the wind turbine which ultimately acts as the driving force to produce energy:

$$T_e = 1.5\rho \left[\lambda i_q + (L_d - L_q) i_d i_q \right] \tag{3.9}$$

where L_q is the q axis inductance, L_d is the d axis inductance, i_q is the q axis current, i_d is the d axis current, V_q is the q axis voltage, V_d is the d axis voltage, ω_r is the angular velocity of rotor, λ is the amplitude of flux induced, and p is the number of pairs of poles. In case of squirrel cage induction generator (SCIG) the following equation in stationary d–q frame of reference can be used for dynamic modeling:

$$\begin{bmatrix} V_{qs} \\ V_{ds} \\ V_{qr} \\ V_{dr} \end{bmatrix} = \begin{bmatrix} R_s + pL_s & 0 & pL_m & 0 \\ 0 & R_s + pL_s & 0 & pL_m \\ pL_m & -\omega_r L_m & R_r + pL_r & -\omega_r L_r \\ \omega_r L_m & pL_m & \omega_r L_r & R_r + pL_r \end{bmatrix} \begin{bmatrix} i_{qs} \\ i_{ds} \\ i_{qr} \\ i_{dr} \end{bmatrix} \tag{3.10}$$

From stator side the equation is

$$\begin{aligned}
\lambda_{ds} &= L_s i_{ds} + L_m i_{dr} \\
\lambda_{qs} &= L_s i_{qs} + L_m i_{dr} \\
L_s &= L_{ls} + L_m \\
L_r &= L_{lr} + L_m \\
V_{ds} &= R_s i_{ds} + \frac{d}{dt}\lambda_{ds} \\
V_{qs} &= R_s i_{qs} + \frac{d}{dt}\lambda_{qs}
\end{aligned} \tag{3.11}$$

From rotor side the equation is

$$\begin{aligned}
\lambda_{dr} &= L_r i_{dr} + L_m i_{ds} \\
\lambda_{qr} &= L_r i_{qr} + L_m i_{qs} \\
V_{dr} &= R_r i_{dr} + \frac{d}{dt}\lambda_{dr} + \omega_r \lambda_{qr} \\
V_{qr} &= R_r i_{qr} + \frac{d}{dt}\lambda_{qr} - \omega_r \lambda_{dr}
\end{aligned} \tag{3.12}$$

For the air gap flux linkage, the equations are

$$\begin{aligned}
\lambda_{dm} &= L_m \left(i_{ds} + i_{dr}\right) \\
\lambda_{qr} &= L_m \left(i_{qr} + i_{qs}\right)
\end{aligned} \tag{3.13}$$

where R_s, R_r, L_m, L_{ls}, L_{lr}, ω_r, i_d, i_q, V_d, V_q, λ_d, and λ_q are the stator winding resistance, motor winding resistance, magnetizing inductance, stator leakage inductance, rotor leakage inductance, electrical rotor angular speed, current, voltage, and fluxes, respectively, of the *d–q* model, respectively [20, 21]. Simply, the net wind energy power considering the torque of turbine (T_t) in terms of rotational speed is calculated as

$$P_w = \frac{1}{2}\rho A C_p(\lambda,\beta)\left(\frac{R\omega_{opt}}{\lambda_{opt}}\right)^3 \tag{3.14}$$

$$T_t = \frac{1}{2}\rho A C_p(\lambda,\beta)\left(\frac{R}{\lambda_{opt}}\right)^3 \omega_{opt} \tag{3.15}$$

The power coefficient (C_p) is a nonlinear function expressed by the fitting equation in the form

$$C_p(\lambda,\beta) = c_1\left(c_2\frac{1}{\lambda_i} - c_3\beta - c_4\right)e^{-c_5\frac{1}{\lambda_i}} + c_6\lambda \tag{3.16}$$

with

$$\frac{1}{\lambda_i} = \frac{1}{\lambda + 0.08\beta} - \frac{0.035}{\beta^3 + 1} \tag{3.17}$$

The value of constants c_1–c_6 has been explained in later section.

Since wind energy is a semiconductor capable of producing energy, in this model its high efficiency has been calculated to produce electricity power from the interconnected series-parallel configuration of diode model to convert wind energy into electricity energy that can be supplied in the national grid (Fig. 3.1).

Consequently, using the ideal diode for an array with N_s series-connected cells and N_p parallel-connected cells, the array current is calculated as voltage as follows:

$$I = N_p\left[I_{ph} - I_{rs}\left[\exp\left(\frac{q(V + IR_s)}{AKTN_s} - 1\right)\right]\right] \tag{3.18}$$

where

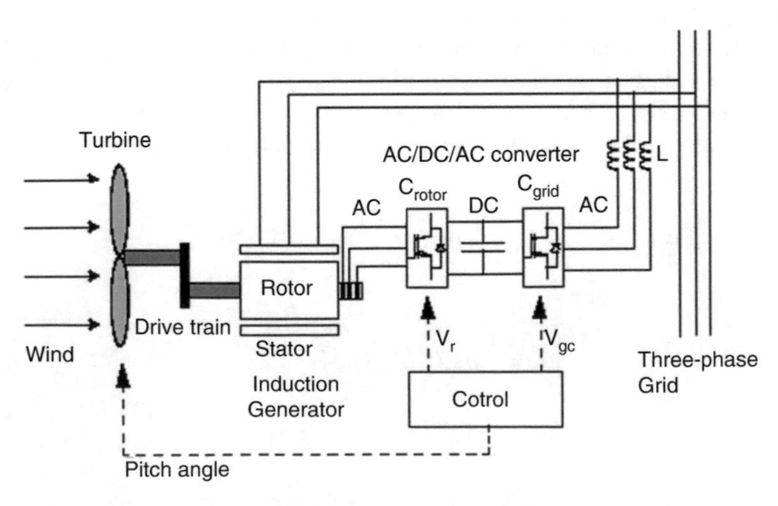

Fig. 3.1 Modeling and simulation of wind energy that is converted into electricity energy from the process of drivetrain to control system which finally can supply the three-phase grid

$$I_{rs} = I_{rr}\left(\frac{T}{T_r}\right)^3 \exp\left[\frac{E_G}{AK}\left(\frac{1}{T_r} - \frac{1}{T}\right)\right] \quad (3.19)$$

and q is the electron charge (1.6×10^{-9} C), K is Boltzmann's constant, A is the diode ideality factor, and T is the cell temperature (K). IR_s is the cell reverse saturation current at T, T_r is the cell referred temperature, I_{rr} is the reverse saturation current at T_r, and E_G is the bandgap energy of the semiconductor used in the cell. The photocurrent I_{ph} varies with the cell's temperature and radiation as follows:

$$I_{ph} = \left[I_{SCR} + k_i\left(T - T_r\right)\frac{S}{100}\right] \quad (3.20)$$

where I_{SCR} is the cell short-circuit current at reference temperature and radiation, k_i is the short-circuit current temperature coefficient, and S is the solar radiation in (mW/cm^2).

Here, the diode model is used as an additional shunt resistance in parallel to ideal shunt diode model. I–V characteristics of energy are as follows:

$$I = I_{ph} - I_D \quad (3.21)$$

$$I = I_{ph} - I_0\left[\exp\left(\frac{q\left(V + R_s I\right)}{AKT} - 1\right) - \frac{V + R_s I}{R_{sh}}\right] \quad (3.22)$$

where I_{ph} is the current (A), I_D is the diode current (A), I_0 is the inverse saturation current (A), A is the diode constant, q is the charge of the electron (1.6×10^{-9} C), K is Boltzmann's constant, T is the areal temperature (°C), R_s is the serics resistance (ohm), R_{sh} is the shunt resistance (Ohm), I is the areal current (A), and V is the real voltage (V). Thus, the output of current of the energy using diode model can be written as follows:

$$I = I_{pv} - I_{D1} - I_{D2} - \left(\frac{V + IR_S}{R_{sh}}\right) \quad (3.23)$$

where

$$I_{D1} = I_{01}\left[\exp\left(\frac{V + IR_s}{a_1 V_{T1}}\right) - 1\right] \quad (3.24)$$

$$I_{D2} = I_{02}\left[\exp\left(\frac{V + IR_s}{a_2 V_{T2}}\right) - 1\right] \quad (3.25)$$

I_{01} and I_{02} are reverse saturation current of diode 1 and diode 2, and V_{T1} and V_{T2} are thermal voltage of respective diode. a_1 and a_2 represent the diode ideality constants.

Simplified model for energy system modeling is thus presented here as

$$v_{oc} = \frac{V_{oc}}{cKT/q} \tag{3.26}$$

$$P_{max} = \frac{\frac{V_{oc}}{cKT/q} - \ln\left(\frac{V_{oc}}{cKT/q} + 0.72\right)}{\left(1 + \frac{V_{oc}}{nKT/q}\right)} \left(1 - \frac{V_{oc}}{\frac{V_{oc}}{I_{sc}}}\right) \left(\frac{V_{oc0}}{1 + \beta \ln \frac{G_0}{G}}\right) \left(\frac{T_0}{T}\right)^{\gamma} I_{sc0} \left(\frac{G}{G_0}\right)^{\alpha} \tag{3.27}$$

where v_{oc} is the normalized value of the open-circuit voltage V_{oc} with respect to the thermal voltage $V_t = nkT/q$, n is the ideality factor ($1 < n < 2$), K is Boltzmann constant, T is the energy module temperature in kelvin, q is the electron charge, α is the factor responsible for all the nonlinear effects that the photocurrent depends on, β is an energy module technology-related dimensionless coefficient, and γ is the factor considering all the nonlinear temperature-voltage effects (Fig. 3.2). Equation (3.27) represents the maximum power output of a single energy module. A real system consists of the number of energy modules connected in series and parallel. The total wind energy power output for an array with N_s series and N_p parallel with P_M power is finally calculated as

$$P_{array} = N_s N_p P_M \tag{3.28}$$

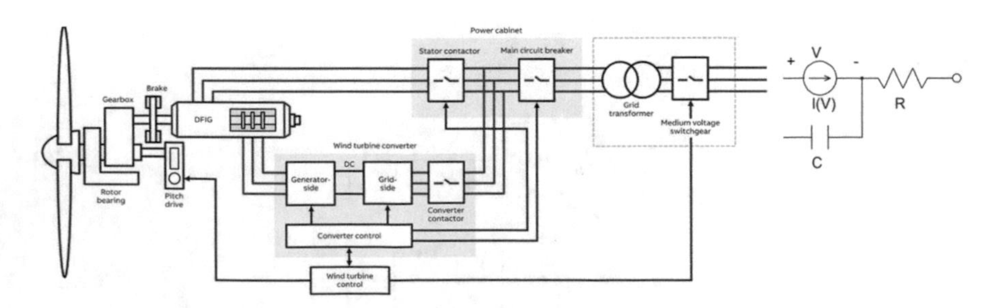

Fig. 3.2 The diagram of wind energy output from the module array of continuous function of operating conditions, which is derived from physical principles of wind turbine generation

Results and Discussion

The use of variable wind speed in the turbine has confirmed the rotational speed of the wind turbine relative to the wind speed in a way the turbine operates which ultimately confirms the production of maximum power point in varying wind speed [22, 23]. Thus, maximum power point can be captured in a varying wind speed by keeping the tip ratio in its optimal value in a variable speed generation system of the wind where the fundamental equation governing the mechanical power of this wind turbine is confirmed by

$$P_{w} = \frac{1}{2} C_{p}(\lambda, \beta) \rho A V^{3} \tag{3.29}$$

where ρ is the air density (kg/m^3), C_p is the power coefficient, A is the intercepting area of the rotor blades (m^2), V is the average wind speed (m/s), and λ is the tip speed ratio. Here, the maximum value of the power coefficient C_p is 0.593, also known as Betz's coefficient. The tip speed ratio (TSR) for wind turbine is defined as the ratio of rotational speed of the tip of a blade to the wind velocity and thus it has been described mathematically as

$$\lambda = \frac{R\omega}{V} \tag{3.30}$$

where R is the radius of the turbine (m), ω is the angular speed (rad/s), and V is the average wind speed (m/s).

The energy generated by wind can be obtained by

$$Q_{w} = P \times (\text{Time}) \ [\text{kWh}] \tag{3.31}$$

Simply, the various factors like the velocity of wind considering height are also calculated by direct measurement of wind speed considering the error proneness due to vegetation, shading, and obstacles in the vicinity [24]:

$$v(Z) \ln\left(\frac{Z_{r}}{Z_{0}}\right) = v(Z_{r}) \ln\left(\frac{Z}{Z_{0}}\right) \tag{3.32}$$

where Z_r is the reference height (m), Z is the height where wind speed is to be determined, Z_0 is the measure of surface roughness (0.1–0.25 for crop land), $v(Z)$ is the wind speed at height Z (m/s), and $v(Z_r)$ is the wind speed at reference height z (m/s).

Consequently, the change of the wind velocity considering the optimum turbine generator speed corresponding to max power is determined to extract the maximum power point tracking (MPPT) and is expressed by the optimum rotor speed of the turbine drivetrain:

$$\omega_{opt} = \frac{\lambda_{opt}}{R} V_{wn} \tag{3.33}$$

which gives

$$V_{wn} = \frac{R\omega_{opt}}{\lambda_{opt}} \tag{3.34}$$

where ω_{opt} is the optimum rotor angular speed in rad/s, λ_{opt} is the optimum tip speed ratio, R is the radius of turbine in meters, and V_{wn} is the wind speed in m/s.

Simply the drivetrain of the turbine, here, transfers high air dynamics torque at rotor to low-speed shaft of generator through gearbox which is directly coupled with the rotor to maximize the energy production [25, 26]. Subsequently, the drivetrain result is calculated using mass model based on the torsional multibody dynamic model of the turbine as follows:

$$\begin{bmatrix} \dot{\omega}_I \\ \dot{\omega}_g \\ \dot{T}_{Ix} \end{bmatrix} = \begin{bmatrix} -\frac{K_I}{J_I} & 0 & -\frac{1}{J_I} \\ 0 & -\frac{K_g}{J_g} & \frac{1}{n_g J_g} \\ \left(B_{Ix} - \frac{K_{Ix}K_r}{J_r}\right) & \frac{1}{n_g}\left(\frac{K_{Ix}K_r}{J_g} - B_{1x}\right) & -K_{Ix}\left(\frac{J_r + n_g^2 J_g}{n_g^2 J_g J_r}\right) \end{bmatrix} \begin{bmatrix} \omega_I \\ \omega_g \\ T_{Ix} \end{bmatrix} + \begin{bmatrix} \frac{1}{J_r} \\ 0 \\ \frac{K_{Ix}}{J_r} \end{bmatrix} T_m + \begin{bmatrix} 0 \\ -\frac{1}{J_g} \\ \frac{K_{Ix}}{n_g J_g} \end{bmatrix} T_g \tag{3.35}$$

Here, the *mass model* is a perfectly rigid low-speed shaft where a turbine is analyzed to calculate its energy production rate as its rotational speed as follows:

$$J_t \dot{\omega}_t = T_a - K_t \omega_t - T_g \tag{3.36}$$

and

$$\begin{aligned} J_t &= J_r + n_g^2 J_g \\ K_t &= K_r + n_g^2 K_g \\ T_g &= n_g T_{em} \end{aligned} \tag{3.37}$$

where J_t is the turbine rotor moment of inertia in [kg m^2], ω_t is the low shaft angular speed in [rad/s^2], K_t is the turbine damping coefficient in [Nm/rad/s] representing aerodynamic resistance, and K_g is the generator damping coefficient in [Nm/rad/s] representing mechanical friction and windage.

Subsequently, the mass model has also been further calculated considering the precise function of wind turbine system to produce energy relating to the rotational speed of turbine of which the rotor-side inertia J_r is

$$J_t \frac{d\omega_t}{dt} = T_m - T_{ls} - K_t \omega_t \tag{3.38}$$

The low-speed shaft torque is calculated as

$$T = B_{ls}(\theta_t - \theta_{ls}) + K_{ls}(\omega_t - \omega_{ls}) \tag{3.39}$$

The generator inertia J_g is driven by the high-speed shaft and braked by the electromagnetic torque T_g of the generator:

$$J_g \frac{d\omega_g}{dt} = T_{hs} - K_g \omega_g - T_g \tag{3.40}$$

If we assume the ideal gearbox with ratio n, then

$$n = \frac{T_{ls}}{T_{hs}} = \frac{\omega_g}{\omega_t} = \frac{\theta_g}{\theta_{ls}} \tag{3.41}$$

where the notations are the same as those of one mass model. K_{ls} is the low-speed shaft damping coefficient in [Nm/rad/s], ω_g is the high-speed shaft angular speed in [rad/s^2], T_m is the turbine torque in [Nm], T_{ls} is the low-speed shaft torque in [Nm], J_g is the generator rotor moment of inertia in [kg m^2], and T_{hs} is the high-speed shaft torque in [Nm]. After eliminating T_{ls} time derivative from (3.39) and using (3.40) and (3.41), the following dynamical system is derived:

$$\frac{dT_{ls}}{dt} = \left(B_{ls} - \frac{K_{ls}K_l}{J_l}\right)\omega_l + \frac{1}{n}\left(\frac{K_{ls}K_l}{J_g} - B_{ls}\right)\omega_g - K_{ls}\left(\frac{J_l + n^2 J_g}{n^2 J_l J_g}\right)T_{ls} + \frac{K_{ls}}{J_l}T_\alpha + \frac{K_{ls}}{nJ_g}T_g \tag{3.42}$$

where
$K_{ls} = IG/L_{ls}$
$D_{ls} = \xi D_s$

$$\xi = \sqrt{1 - \left(\frac{\omega}{\omega_n}\right)^2} \tag{3.43}$$

$$D_s = 2\sqrt{K_{ls} m} \tag{3.44}$$

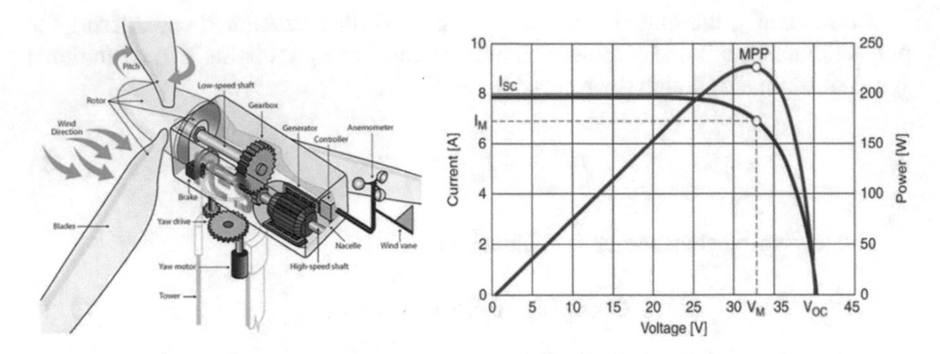

Fig. 3.3 Maximum power point tracking when isolation changes in wind turbine keep forcing to attain the peak point using generator control system to produce power in relation to the current and voltage generation into the maximum point of the wind turbine

and ω/ω_n is the ratio of shaft frequency of oscillation to the undamped natural frequency of shaft, m is the mass of shaft, I is the second momentum of area about the axis of rotation, L_{ls} is the shaft length, G is the modulus of rigidity, D_s is the critical damping of shaft, and ξ is the damping ratio of shaft [27–29].

Eventually, in order to extract maximum energy from wind in a varying speed condition the rotational speed has been calculated considering the optimal value of TSR by calculating the voltage-current relation and voltage-power relation as non-linear as shown in Fig. 3.3. Thus, the maximum power point (MPP) is tracked for extraction of maximum power from maximum power point tracking (MPPT) from the turbine's optimum torque control (Fig. 3.3).

Consequently, the optimum TSR control is where optimum TSR (λ_{opt}) is confirmed by the maximum exploitation of available wind energy which is related to the torque of wind turbine at a given wind speed as

$$V = \frac{\omega R}{\lambda} \tag{3.45}$$

Using Eq. (3.45) in Eq. (3.16) yields

$$P = \frac{1}{2}\rho\pi R^5 \frac{\omega^3}{\lambda^3} C_p \tag{3.46}$$

When rotor is rotating at λ_{opt}, $C_p = C_{p\,max}$. So, Eq. (3.46) becomes

$$P_{opt} = \frac{1}{2}\rho\pi R^4 \frac{C_{pmax}}{\lambda_{opt}^3}\omega^3 = K_{opt}\omega^3 \tag{3.47}$$

Power is also defined as

$$P = \omega T \tag{3.48}$$

Rearranging (3.48) we get

$$T = \frac{P}{\omega} \tag{3.49}$$

Using Eqs. (3.49) and (3.47) we can obtain

$$T_{\text{opt}} = \frac{1}{2} \rho \pi R^5 \frac{C_{\text{pmax}}}{\lambda_{\text{opt}}^3} \omega^2 = K_{\text{opt}} \omega^2 \tag{3.50}$$

Thus, this analytical value is obtained by the above equation considering the torque of turbine which is measured directly from the calculation of wind speed to confirm maximum power point tracking (MPPT).

Simply, the MPPT is being done by reading the current power output in order to determine the control mechanism of the wind turbine to clarify the maximum power obtained from wind turbine (Fig. 3.4). Thus, a functional approach is being used where the power coefficient is expressed as the function of TSR considering the wind turbine rotor coefficients c_1–c_6. The coefficients (c_1–c_6) and parameter (λ_i) here are used for the measurement of the power coefficient as $c_1 = 0.53$, $c_2 = 151$, $c_3 = 0.58$, $c_4 = 0.002$, and $c_5 = 13.2$, $c_6 = 18.4$ and x is 2.14. Thus, the parameter (λ_i) is dcfined as

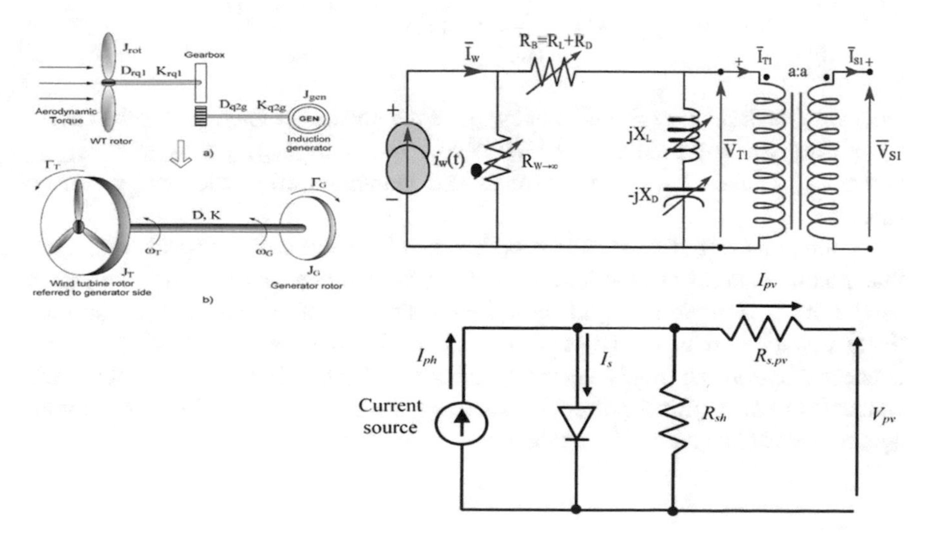

Fig. 3.4 The diagram of wind turbine synchronous mechanism which shows the dynamic mode of wind energy production in the nonlinear open-circuit voltage

$$\frac{1}{\lambda_i} = \frac{1}{\lambda - 0.02\beta} - \frac{0.03}{1+\beta^3} \tag{3.51}$$

Consequently, the dynamic mode of wind energy production is shown in a circuit diagram considering the nonlinear nature of the open-circuit voltage (V_{oc}) where R_p represents the self-discharge resistance, R_{ic} and R_{id} are the internal resistance which compensates resistance of the electrolyte, R_{co} and R_{do} are the voltage drop for the charge and discharge process, and C_o represents the double-layer capacitance behavior of the battery during charge and discharge [30–32].

Simply, this open-circuit voltage is a nonlinear function of the temperature (T) and discharge current (I_B) and thus this energy can also be stored as (E_{cd}). Thus, this model used here initially calculates current (I_B) to confirm the maximum available energy stored which is determined by V_{oc} and SOC, calculated as the power generation by wind energy as $P_{wg}(t) \geq P_{load}(t)$, and the storage capacity at any given time (t) is expressed as

$$C_{bat}(t) = C_{bat}(t-1) + \left(P_{PV}(t) + P_{wg}(t) - P_{load}(t)\eta_{cad}\right)\Delta t \eta_{cha} \tag{3.52}$$

where $C_{bat}(t)$ and $C_{bat}(t-1)$ are the available battery capacity at time (t) and ($t-1$).

P_{PV} is the power generation rate, P_{wg} is the power generated by wind turbine generator, $P_{load}(t)$ is the power consumed at load t, t is the simulation time step ($\Delta t = 1$ h), η_{cad} is the efficiency of AC/DC converter, and η_{cha} is the energy efficiency.

Here, wind energy has been finally estimated as, i.e., $(\eta_{inv}P_{wg}(t) + P_{PV}(t)) \geq P_{load}$ (t), the extra storage of energy as

$$C_{bat}(t) = C_{bat}(t-1) + \frac{1}{\eta_{dech}}\left(P_{PV}(t) - \left(\frac{P_{load}(t) - P_{wg}(t)}{\eta_{inv}}\right)\right)\Delta t \tag{3.53}$$

where η_{dech} is the energy rate efficiency, η_{inv} is the inverter efficiency, C is the capacity of energy at any time t, and $C_{bat}(t)$ is constrained by $C_{bat\ min} \leq C_{bat}(t) \leq C_{bat\ max}$ where $C_{bat\ min}$ and $C_{bat\ max}$ are minimum and maximum allowable energy storage capacity.

Simply, this stored wind energy system can be stand-alone, or grid connected. Stand-alone systems need to have generation and storage capacity large enough to handle the load while in a grid-connected system the storage device can be relatively smaller as deficient power can be obtained from the grid [24, 33]. A grid-connected hybrid can supply electricity to both load and utility grid. However, when connected to grid, proper power electronic controllers are required to control voltage, frequency and harmonic regulations, and load sharing.

Conclusions

Energy crisis, ever-increasing oil prices, and climate changes due to the greenhouse gas emission have increased people's attention toward effective, efficient, sustainable, and pollution-free renewable energy systems where wind energy could be the promising source of alternative energy. Since wind power is widely available and not confined to the banks of fast-flowing streams, the development of wind power will play a vital role in meeting the global energy demand while safeguarding the environment. Simply optimizing the renewable wind energy is clarified in this chapter by a mathematical model which summarizes that the nonlinear power characteristics of wind energy require simple techniques to extract maximum power from the air. Here, the simplicity of the system increases with maximum power point tracking (MPPT) techniques; thus it is clarified by mathematical modeling of MPPT techniques which indeed would be an innovative technology to meet the global energy demand. Simply the utilization of wind energy power system will be a great relief for the current finite level of fossil fuel and a means for the reduction of greenhouse gas emission. Thus, the presented simulation study of global wind energy clearly depicts that the proposed abundant wind energy can be an interesting source of renewable energy to meet the global energy demand which is also very much environmentally friendly.

Acknowledgements This research was supported by Green Globe Technology under the grant RD-02017-07 for building a better environment. Any findings, predictions, and conclusions described in this chapter are solely performed by the author. The author confirms that there is no conflict of interest for publishing this chapter in a suitable journal or book.

References

1. Binayak Bhandari, Shiva Raj Poudel, Kyung-Tae Lee, Sung-Hoon Ahn. "Mathematical modeling of hybrid renewable energy system: A review on small hydro-solar-wind power generation", International Journal of Precision Engineering and Manufacturing-Green Technology, 2014.
2. Youcef Saidi, Abdelkader Mezouar, Yahia Miloud, Mohammed Amine Benmahdjoub. "A robust control strategy for three phase voltage t source PWM rectifier connected to a PMSG wind energy conversion system", 2018 International Conference on Electrical Sciences and Technologies in Maghreb (CISTEM), 2018.
3. Abdullah Asuhaimi B. Mohd Zin, Mahmoud Pesaran H. A, Azhar B. Khairuddin, Leila Jahanshaloo, Omid Shariatu, An overview on doubly fed induction generators' controls and contributions to wind-based electricity generation, *Renewable and Sustainable Energy Rreviews, vol. 27*, 2013, pp. 692-708.
4. Farhad Ilahi Bakhsh, Dheeraj Kumar Khatod. "A new synchronous generator-based wind energy conversion system feeding an isolated load through variable frequency transformer", Renewable Energy.
5. Huang Ligang, Wang Xiangdong, Yan Kang. "Optimal speed tracking for double fed wind generator via switching control", The 27th Chinese Control and Decision Conference (2015 CCDC), 2015.

6. K. Kerrouche, A. Mezouar, Kh. Belgacem. "Decoupled control of doubly fed induction generator by vector control for wind energy conversion system", Energy Procedia, 2013.
7. Majid A. Abdullah, A.H.M. Yatim, Chee Wei Tan. "A study of maximum power point tracking algorithms for wind energy system", 2011 IEEE Conference on Clean Energy and Technology (CET), 2011.
8. Md. Faruque Hossain. "Green science: Independent building technology to mitigate energy, environment, and climate change", Renewable and Sustainable Energy Reviews, 2017.
9. Abdelhamid Loukriz, Mourad Haddadi, Sabir Messalti. "Simulation and experimental design of a new advanced variable step size Incremental Conductance MPPT algorithm for PV systems", ISA Transactions, 2016.
10. Junyent-Ferre, A. "Modeling and control of the doubly fed induction generator wind turbine", Simulation Modelling Practice and Theory, 201010.
11. Md. Faruque Hossain. "Theoretical modeling for hybrid renewable energy: An Initiative to Meet the Global Energy", Journal of Sustainable Energy Engineering, 2016.
12. Ahmed M. Othman, Mahdi M.M. El-arini, Ahmed Ghitas, Ahmed Fathy. "Real-world maximum power point tracking simulation of PV system based on Fuzzy Logic control", NRIAG Journal of Astronomy and Geophysics, 2019.
13. Mario Di Nardo, Mosè Gallo, Teresa Murino, Liberatina Carmela Santillo. "System dynamics simulation for fire and explosion risk analysis in home environment", International Review on Modelling and Simulations (IREMOS), 2017.
14. Sara Mohamed Ismail, Ahmed Ali Daoud, Kamel Ahmed El-Serafi, Sobhy Serry Dessouky. "A novel control strategy for improving the power quality of an isolated microgrid under different load conditions", 2018 Twentieth International Middle East Power Systems Conference (MEPCON), 2018.
15. Loucif, Mourad, and Abdelmadjid Boumediene. "Modeling and direct power control for a DFIG under wind speed variation", 2015 3rd International Conference on Control Engineering & Information Technology (CEIT), 2015.
16. Md. Faruque Hossain. "Photon application in the design of sustainable buildings to console global energy and environment", Applied Thermal Engineering, 2018.
17. Getachew Bekele, Getnet Tadesse. "Feasibility study of small hydro/PV/wind hybrid system for off-grid rural electrification in Ethiopia", Applied Energy, 2012.
18. Sakthivel, B. Kanaga, and D. Devaraj. "Modelling, simulation and performance evaluation of solar PV/wind hybrid energy system", 2015 International Conference on Electrical Electronics Signals Communication and Optimization (EESCO), 2015.
19. Phan, Dinh-Chung, and Shigeru Yamamoto. "Rotor speed control of doubly fed induction generator wind turbines using adaptive maximum power point tracking", Energy, 2016.
20. K. Kerrouche, A. Mezouar, L. Boumedien. "A simple and efficient maximized power control of DFIG variable speed wind turbine", 3rd International Conference.
21. Md. Faruque Hossain. "Advanced building design", Elsevier BV, 2019.
22. H. Amimeur, D. Aouzellag, R. Abdessemed, K. Ghedamsi. "Sliding mode control of a dual-stator induction generator for wind energy conversion systems", International Journal of Electrical Power & Energy Systems, 2012.
23. Zohoori, Alireza, Abolfazl Vahedi, Mohammad Ali Noroozi, and Santolo Meo. "A new outer-rotor flux switching permanent magnet generator for wind farm applications: Flux switching permanent magnet generator for wind farm applications", Wind Energy, 2016.
24. Manfred Stieber, *Wind Energy System for Electric Power Generation* (Springer, Verlag Berlin Heidelberg, 2008).
25. Gunasekaran Nallappan, Young-Hoon Joo. "Robust sampled-data fuzzy control for nonlinear systems and its applications: Free-weight matrix method", IEEE Transactions on Fuzzy Systems, 2019.
26. Ouled Amor, Walid, A. Ltifi, and M. Ghariani. "Study of a wind energy conversion systems based on doubly-fed induction generator", International Review on Modelling and Simulations (IREMOS), 2014.

27. Jogendra Singh, Mohand Ouhrouche. "Chapter 15 MPPT Control Methods in Wind Energy Conversion Systems", InTech, 2011.
28. K. Ghedamsi, D. Aouzellag. "Improvement of the performances for wind energy conversions systems", International Journal of Electrical Power & Energy Systems, 2010.
29. Nabil Taib, Brahim Metidji, Toufik Rekioua. "Performance and efficiency control enhancement of wind power generation system based on DFIG using three-level sparse matrix converter", International Journal of Electrical Power & Energy Systems, 2013.
30. Kamal Anoune, Mohsine Bouya, Mokhtar Ghazouani, Abdelali Astito, Abdellatif Ben Abdellah. "Hybrid renewable energy system to maximize the electrical power production", 2016 International Renewable and Sustainable Energy Conference (IRSEC), 2016.
31. Kiflom Gebrehiwot, Md. Alam Hossain Mondal, Claudia Ringler, Abiti Getaneh Gebremeskel. "Optimization and cost-benefit assessment of hybrid power systems for off-grid rural electrification in Ethiopia", Energy, 2019.
32. Kheira Belgacem, Abelkader Mezouar, Najib Essounbouli. "Design and analysis of adaptive sliding mode with exponential reaching law control for double-fed induction generator based wind turbine", International Journal of Power Electronics and Drive Systems (IJPEDS), 2018.
33. Mohammed Ouassaid, Kamal Elyaalaoui, Mohammed Cherkaoui. "Reactive power capability of squirrel cage asynchronous generator connected to the grid", 2015 3rd International Renewable and Sustainable Energy Conference (IRSEC), 2015.

Chapter 4
Geothermal Energy

Abstract The geothermal energy is abundant everywhere in the world which can be the alternative source of energy once it is implemented by houses and buildings in situ in order to console the global energy and environmental vulnerability. Therefore, in this research, the application of geothermal energy production and consumption technology has been proposed *in situ* by the process of geothermal energy pumping from *Earth Magna* and conversion of it into electricity energy by heat exchanger by implementing advanced design technology for houses and buildings. Simply, production and application of geothermal energy in situ by the houses and buildings itself would be an interesting technology since the high-temperature resources (>150 °C) of Earth Magna provide baseload generating excellent capacity of geothermal energy while lower temperature resources of Earth Magna also provide energy significantly. Besides, the application of *in situ* geothermal energy will not only be the self-generating energy source for houses and buildings but also have a number of positive characteristics of simplicity, safety, and capability of providing continuous baseload energy flow with negligible emissions of CO_2, SO_2, and NO_2, and thus this innovative technology will play a vital role in mitigating global environmental vulnerability.

Keywords Housing and building sector energy · Greenhouse gases · Climate change · Geothermal energy · Environmental sustainability

Introduction

Global environmental crisis became severe since 1900s due to the perceived risk by releasing billion tons of greenhouse gas emission of CO_2 into the atmosphere every year from houses and building sectors due to the consumption of traditional energy supplies [1, 2]. It has been calculated that nearly 40% of CO_2 gas emissions in the year of 2019 are triggered by the house's and building's utilization of fossil fuel energy [3–5]. The estimate shows that global CO_2 releases by burning fossil energy

M. F. Hossain, *Global Sustainability in Energy, Building, Infrastructure, Transportation, and Water Technology*,
https://doi.org/10.1007/978-3-030-62376-0_4

certainly have increased tremendously since year zero (0) and the emissions of CO_2 jumped significantly within the year zero (255 ppm) and the year 1900 (300 ppm) and then it has increased again at 400 ppm in the year 2019 [6–8]. Simply, this heat-trapping gas CO_2 which is the derivative formation of fossil energy got fully depleted into radioactive form of ^{14}C into the atmosphere, causing the changes of diurnal temperature, melting polar ices, increasing water level, and eventually changing the climate [9–11]. The simple solar radiation principle is that once sunlight reaches earth, 70% of it is absorbed by earth surface, vegetation, and ocean surface [12–14]. The rest of the sunlight is reflected back to the atmosphere due to the unsaturated condition of the earth and biosphere which causes diurnal temperature changing globally [6, 15, 16]. Naturally, it can be explained that the sunlight-colorful objects and surfaces, like snow and clouds, tend to reflect most sunlight, while darker objects and surfaces, like the ocean, forests, or soil, tend to absorb more sunlight and yet nearly 30% of radiation is reflected back and creates climate change [3, 17, 18]. Subsequently, this climate changing impacts the global ecosystem dangerously and causes flood, rainfall, and draught; influences agricultural crop yields; affects human health; and influences changes in forests and other ecosystems occurring throughout the world [19–21].

Since the unique aspects of geothermal heat are available everywhere in the world which is also noted as a “clean natural” power supply which anyone can take advantage of, this research discusses about drilling the earth to a certain depth to reach the expectant-level temperature to get geothermal energy in situ [22–24]. Since the Earth Magna (which controls its geothermal gradient) varies, earth’s crust has to be determined to have an average temperature of 3745 °C degrees with the maximum depth of drilling to produce geothermal heat at a pressure of at least 220 bar to mitigate the net energy demand for a standard house or building daily.

Materials and Methods

In this research, an *in situ* technology of capturing the geothermal heat from Earth Magna has been proposed to trap by the installation of “hydrothermal convection pipe” systems by houses and buildings, where its water force is made to seep into Earth Magna in order to force to rise up the heated water in the surface elevation of houses and buildings [25–27]. Once this heated water reaches the surface elevation of houses and buildings, it is then captured as steam and used to drive electric generators to deliver energy which is represented by *P-h* diagram (Fig. 4.1). Simply the following diagram has been generated in order to effectively use geothermal energy *in situ* for houses and buildings where a tube runs under the ground of the houses and buildings in order to circulate into a house’s or a building’s heat pumping system.

Thereafter, the detailed energy generation has been calculated by further *P-h* diagram plot analysis in order to clarify the intuitive and qualitative analysis of the steam for evaluating the enthalpy of both kinetic and potential energy flow by

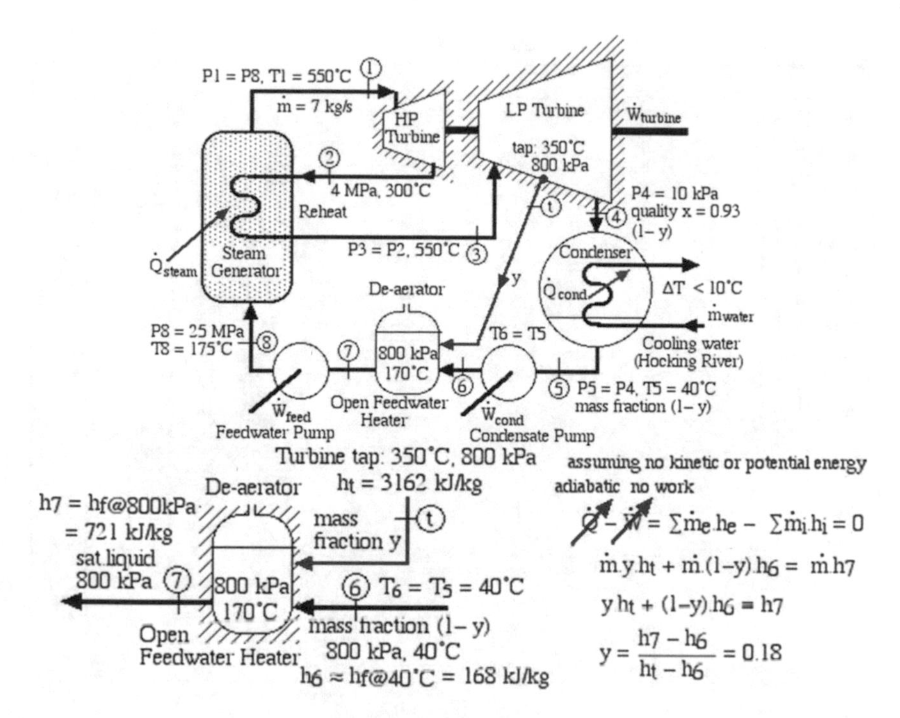

Fig. 4.1 The mechanism technology is referred to as a regenerative heating cycle, and describes the excellent geothermal energy generation showed in a *P-h* diagram and its conversion process through open feedwater heater

(1) clarifying the open feedwater heater adiabatic condition to confirm the mass fraction of steam y requirement to bleed off the LP turbine of the generator, (2) determining the saturated liquid state in the deaerator, and (3) confirming both the condensate pump and the feedwater pump adiabatic condition in order to determine the energy obtained from geothermal source to drive the pumps to general energy into the house or building (Fig. 4.2).

Subsequently, the mass fraction of $(1 - y)$ is obtained at less 0 power output due to a reduced mass flow rate in part of the turbine from the tap (t) to station (4), and then implemented into the generator installed in the home building in order to determine the net electricity energy generation by conducting mathematical model analysis using MATLAB software [28–30]. Here, the model is clarified in order to confirm that the transfer mechanism of the fluid dynamic torque of the rotor is quantified in order to determine the speed of the gearbox of the generator to confirm the net energy generation rate. Since the rotor is directly connected to the ground-source heat pump, the model is calculated based on the torsional force of the rotor and is expressed as

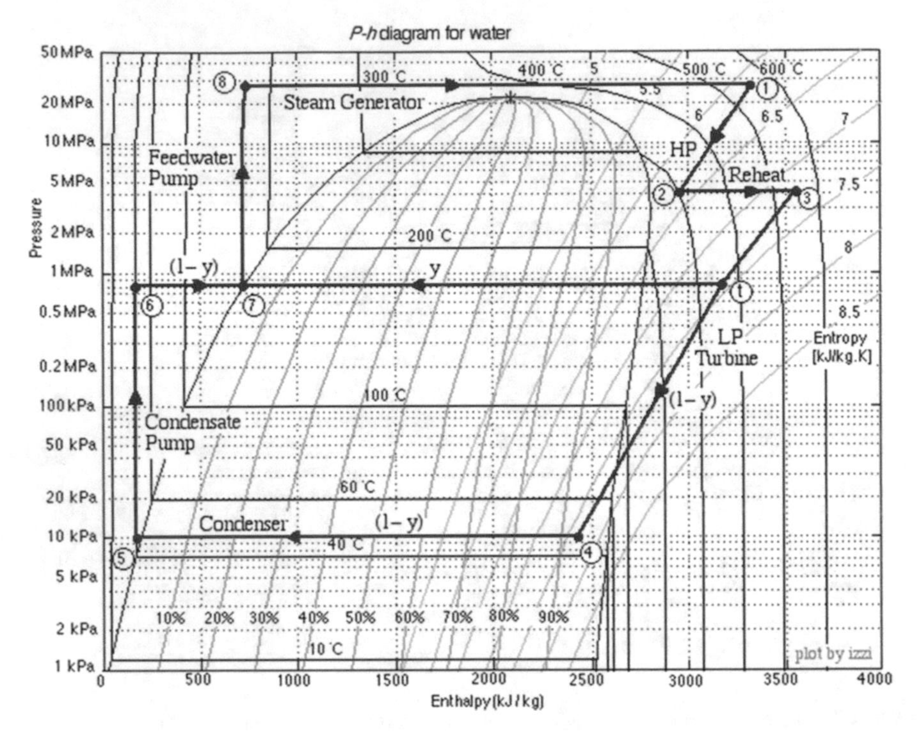

Fig. 4.2 The detailed *P-h* diagram plot where the mass fractional steam *y* is trapped from the turbine tap (*t*) which mixes with (1 − *y*) of the steam H_2O at station (6) where the water reaches the saturated form at point (7), and the feed H_2O pump is used at point (8) in order to regenerate the tremendous amount of heat energy from the H_2O steam of point (8) which is connected to the turbine inlet point (1)

$$\begin{bmatrix}\dot{\omega}_I\\ \dot{\omega}_g\\ \dot{T}_{Ix}\end{bmatrix} = \begin{bmatrix} -\frac{K_1}{J_1} & 0 & -\frac{1}{J_I}\\ 0 & -\frac{K_g}{J_g} & \frac{1}{n_g J_g}\\ \left(B_{Ix}-\frac{K_{Ix}K_r}{J_r}\right) & \frac{1}{n_g}\left(\frac{K_{Ix}K_r}{J_g}-B_{Ix}\right) & -K_{Ix}\left(\frac{J_r+n_g^2 J_g}{n_g^2 J_g J_r}\right)\end{bmatrix}\begin{bmatrix}\omega_I\\ \omega_g\\ T_{Ix}\end{bmatrix} + \begin{bmatrix}\frac{1}{J_r}\\ 0\\ \frac{K_{Ix}}{J_r}\end{bmatrix} T_m + \begin{bmatrix}0\\ -\frac{1}{J_g}\\ \frac{K_{Ix}}{n_g J_g}\end{bmatrix} T \tag{4.1}$$

Since ground-source heat pumps are the primary functional equipment to confirm the draw-up of the heat from the ground constantly year-round, the air or antifreeze liquid is used in the pump throughout all piping system which runs underground, and redirected into the house or building by the application of mass force of the generator turbine which is expressed by

$$J_t\dot{\omega}_t = T_a - K_t\omega_t - T_g \tag{4.2}$$

and

$$\begin{aligned} J_t &= J_r + n_g^2 J_g \\ K_t &= K_r + n_g^2 K_g \\ T_g &= n_g T_{em} \end{aligned} \tag{4.3}$$

where J_t is the turbine rotor moment of inertia in kg m^2; ω_t denotes the minimum shaft angular velocity considering rad/s^2; K_t denotes the rotor damping cofactor in Nm/rad/s; and K_g denotes the generator damping cofactor in Nm/rad/s, considering the mechanical function.

Since the generator system is shown mathematically, the rotor-side inertia J_r is presented by the following calculation:

$$J_t\frac{d\omega_t}{dt} = T_m - T_{ls} - K_t\omega_t \tag{4.4}$$

where the minimum speed of shaft force is represented by

$$T = B_{ls}\left(\theta_t - \theta_{ls}\right) + K_{ls}\left(\omega_t - \omega_{ls}\right) \tag{4.5}$$

and the moment of inertia of rotor J_g is represented by the high-speed shaft and braked by the electromagnetic force of T_g of the generator:

$$J_g\frac{d\omega_g}{dt} - T_{hs} - K_g\omega_g \quad T_g \tag{4.6}$$

Necessarily, the ideal gearbox ratio n of the generator is then determined as

$$n = \frac{T_{ls}}{T_{hs}} = \frac{\omega_g}{\omega_t} = \frac{\theta_g}{\theta_{ls}} \tag{4.7}$$

Here, the rotation is calculated as a one-mass model calculation where K_{ls} represents the minimum velocity of shaft damping cofactor in Nm/rad/s, ω_g represents the maximum velocity of angular shaft in rad/s^2, T_m represents the turbine force in Nm, T_{ls} represents the minimum velocity force in Nm, J_g represents the rotational moment of inertia in kg m^2, and T_{hs} represents the maximum velocity of shaft force in Nm. After cancelling the T_{ls} time function, the equation can be derived as

$$\frac{dT_{ls}}{dt} = \left(B_{ls} - \frac{K_{ls}K_l}{J_l}\right)\omega_l + \frac{1}{n}\left(\frac{K_{ls}K_l}{J_g} - B_{ls}\right)\omega_g - K_{ls}\left(\frac{J_l + n^2 J_g}{n^2 J_l J_g}\right)T_{ls} + \frac{K_{ls}}{J_l}T_\alpha + \frac{K_{ls}}{nJ_g}T_g \tag{4.8}$$

where
$K_{ls} = IG/L_{ls}$
$D_{ls} = \xi D_s$

$$\xi = \sqrt{1 - \left(\frac{\omega}{\omega_n}\right)^2} \tag{4.9}$$

$$D_s = 2\sqrt{K_{ls} m} \tag{4.10}$$

where ω/ω_n is calculated as the mean of value of shaft frequency of oscillation, m represents the net mass of the shaft, I represents the secondary momentum of the rotation of the shaft, L_{ls} denotes the length of the shaft, G represents the rigid module, D_s represents the critical factor of damping, and ξ is the damping rate of the shaft (Fig. 4.3).

Since the generator acts as either an induction or a synchronous of the rotor, the variable speed direct-driven synchronous force is also clarified [31–33]. Thus, the presence of a rotor in the generator is calculated in order to transform the thermal energy into electricity which is shown as

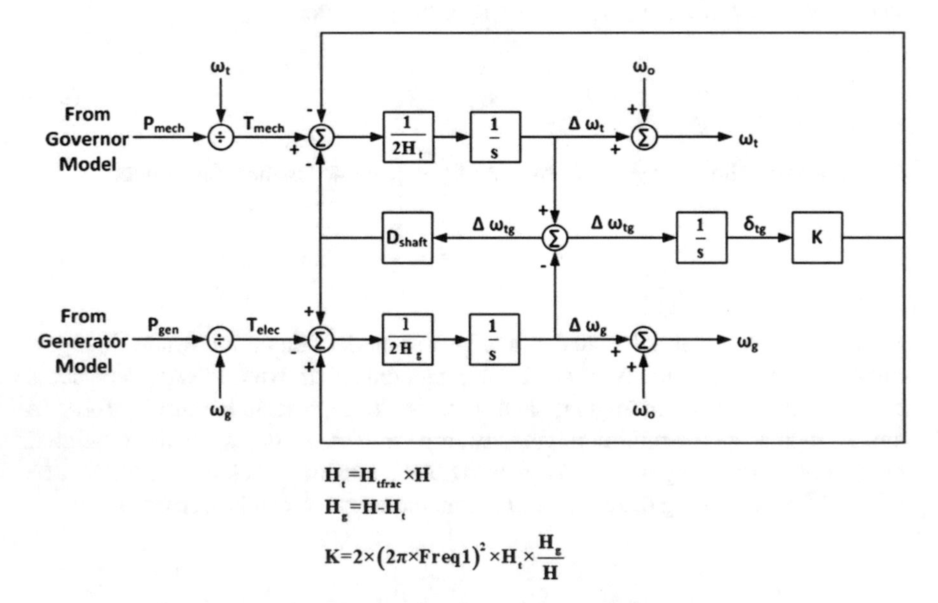

Fig. 4.3 Conceptual sketches of the full-scale conversion of geothermal energy generation from the synchronous mechanism of energy via rotor of the generator

$$V_d = -R_s i_d - L_d \frac{di_q}{dt} + \omega L_q i_q \tag{4.11}$$

$$V_q = -R_s i_q - L_q \frac{di_q}{dt} + \omega L_d i_d + \omega \lambda_m \tag{4.12}$$

Here, the torque is represented by [5]

$$T_e = 1.5\rho\left[\lambda i_q + \left(L_d - L_q\right) i_d i_q\right] \tag{4.13}$$

where L_q represents the q axis induction, L_d represents the d axis induction, i_q represents the q axis charge, i_d represents the d axis charge, V_q represents the q axis volts, V_d represents the d axis volts, ω_r represents the angular speed of the rotor, λ represents flux induction, and p represents the number of pairs of poles for electricity energy generation and thus the below equation is calculated in order to integrate it with a stationed d–q frame of dynamic modeling to confirm the production of net electricity energy:

$$\begin{bmatrix} V_{qs} \\ V_{ds} \\ V_{qr} \\ V_{dr} \end{bmatrix} = \begin{bmatrix} R_s + pL_s & 0 & pL_m & 0 \\ 0 & R_s + pL_s & 0 & pL_m \\ pL_m & -\omega_r L_m & R_r + pL_r & -\omega_r L_r \\ \omega_r L_m & pL_m & \omega_r L_r & R_r + pL_r \end{bmatrix} \begin{bmatrix} i_{qs} \\ i_{ds} \\ i_{qr} \\ i_{dr} \end{bmatrix} \tag{4.14}$$

From the rotor force, the equation is calculated as

$$\begin{aligned} \lambda_{ds} &= L_s i_{ds} + L_m i_{dr} \\ \lambda_{qs} &= L_s i_{qs} + L_m i_{dr} \\ L_s &= L_{ls} + L_m \\ L_r &= L_{lr} + L_m \\ V_{ds} &= R_s i_{ds} + \frac{d}{dt}\lambda_{ds} \\ V_{qs} &= R_s i_{qs} + \frac{d}{dt}\lambda_{qs} \end{aligned} \tag{4.15}$$

and from the rotor side, the equation is calculated as

$$\begin{aligned} \lambda_{dr} &= L_r i_{dr} + L_m i_{ds} \\ \lambda_{qr} &= L_r i_{qr} + L_m i_{qs} \\ V_{dr} &= R_r i_{dr} + \frac{d}{dt}\lambda_{dr} + \omega_r \lambda_{qr} \\ V_{qr} &= R_r i_{qr} + \frac{d}{dt}\lambda_{qr} - \omega_r \lambda_{dr} \end{aligned} \tag{4.16}$$

Simultaneously, from the air gap flux link, the equation is calculated as

$$\begin{aligned}\lambda_{dm} &= L_{m}\left(i_{ds}+i_{dr}\right)\\ \lambda_{qr} &= L_{m}\left(i_{qr}+i_{qs}\right)\end{aligned} \tag{4.17}$$

where R_s, R_r, L_m, L_{ls}, L_{lr}, ω_r, i_d, i_q, V_d, V_q, λ_d, and λ_q represent the rotor force, motor rotation rate, functional induction of the rotor, rotor leak induction, motor leak induction, rotor angular velocity, charge, volts, and fluxes, respectively. The output electricity energy due to the torque force of the turbine (T_t) in terms of rotational velocity is thus calculated by the flowing equation:

$$P_{w} = \frac{1}{2}\rho A C_{p}\left(\lambda,\beta\right)\left(\frac{R\omega_{opt}}{\lambda_{opt}}\right)^{3} \tag{4.18}$$

$$T_{t} = \frac{1}{2}\rho A C_{p}\left(\lambda,\beta\right)\left(\frac{R}{\lambda_{opt}}\right)^{3}\omega_{opt} \tag{4.19}$$

Here, the energy cofactor (C_p) is represented as a nonlinear function which is denoted by a fitting equation in the form

$$C_{p}\left(\lambda,\beta\right) = c_{1}\left(c_{2}\frac{1}{\lambda_{i}} - c_{3}\beta - c_{4}\right)e^{-c_{5}\frac{1}{\lambda_{i}}} + c_{6}\lambda \tag{4.20}$$

with

$$\frac{1}{\lambda_{i}} = \frac{1}{\lambda + 0.08\beta} - \frac{0.035}{\beta^{3}+1} \tag{4.21}$$

Then the next step is the modeling of energy flow calculation, which is conducted by the application of house's or building's catchment area and is expressed by

$$Q_{site} = K\left[\frac{A_{site}}{A_{gauge}}\right]Q_{gauge} \tag{4.22}$$

where A_{site} represents the net area of the house or building (m^2), A_{gauge} represents the functional net area of the gauge (m^2), Q_{site} represents the energy charge rate (m^3/s), Q_{gauge} represents the energy flow rate at the gauge (m^3/s), and K represents the scale consistence and thus the net electricity power generation is calculated as

$$P = \eta_{total}\rho g Q H \tag{4.23}$$

where P is the electricity power output, η_{total} represents the efficiency of the energy power rate, ρ represents the density of energy (kg/m^3), g represents the acceleration caused by gravity (9.81 m/s^2), and H represents the effective pressure head (m) in order to determine the net electricity energy generation.

Results and Discussion

The results of the *P-s* diagram suggested that the feedwater pumping pressure is no longer incompressible, and thus it is easy to capture the geothermal heat from Earth Magna by installation of water force through "hydrothermal convection pipe" systems in houses and buildings to force the heated water to rise up to the surface elevation of houses and buildings (Fig. 4.4).

Here, the *P-s* diagram confirms that both turbines are adiabatic which is determined by the integrated power output results from both of the turbines as suggested to be 10.6 MW if no steam is bled from the turbine (9.65 MW) to supply the required energy demand for a house or building (Fig. 4.5).

Thereafter, the net heat energy is clarified, and the result suggests that the power output is subsequently transferred by function of thc fccdwater pump velocity, and its associated deaerator function of the water mass flow rate, respectively [34, 35].

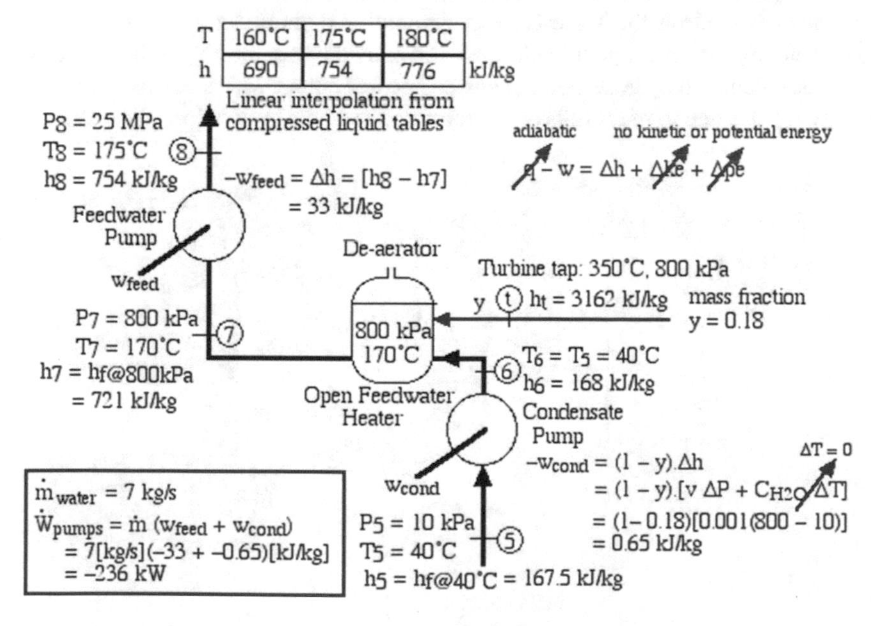

Fig. 4.4 The *P-s* diagram shows the feedwater pumping pressure mechanism to capture the geothermal heat from hydrothermal convection pipe and its conversion process into energy through the turbine

Then, the net heat transformation into the steam generator is determined in order to confirm the achievement of the maximum heat transfer rate which is found to be approximately 22.2 MW (Fig. 4.6).

Subsequently, the determination of the net thermal efficiency (η_{th}) of 132 is confirmed by the calculation of the net work done by the heat exchanger which is 133 to the steam generator which has been calculated as the maximum thermal efficiency of 42% [32, 36, 37]:

$$\eta_{th} = \left[\frac{\dot{W}_{turbines} + \dot{W}_{pumps}}{\dot{Q}_{steam}} \right] = \left[\frac{(9.65 - 0.24)\ \text{MW}}{22.2\ \text{MW}} \right] = 42\% \tag{4.24}$$

Since the generation controls the fuel inflow rate which is extracted from the geothermal heat, the maximum power peak (MPP) is traced in accordance with the extraction of geothermal energy throughout the feedwater pump piping (Fig. 4.7). Here, the maximum geothermal heat transformation rate is calculated considering the speed of the *rotor* in two different modes of fixed speed and variable speed which shows that fixed-speed generators are interconnected to an electric circuit in order to measure precisely the conversion of energy generation rate [17, 38, 39].

Therefore, the maximum geothermal energy generation considering various control strategies has been calculated to track the maximum energy supply for the house or building by conducting the algorithm analysis of mechanical sensor speed that forces to produce energy steadily [26, 40, 41]. The results show that this algorithm continuously controls the highest energy generation point under the conditions of the rotor velocity which frequently changes and thus the maximum geothermal energy traced is calculated by generation of current-power output which delivers the highest energy release due to the results of the rotor speed of the generation (Fig. 4.8).

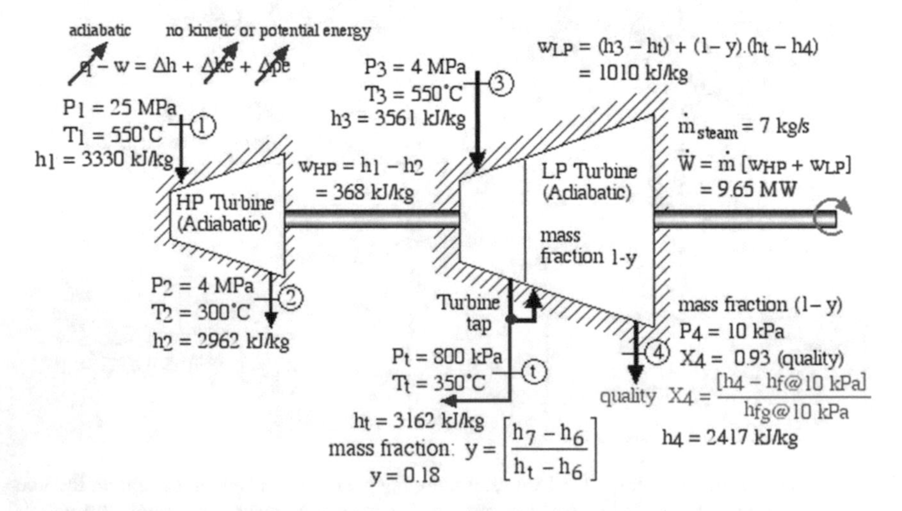

Fig. 4.5 The detail energy calculation of captured geothermal energy into the turbine by the process of adiabatic

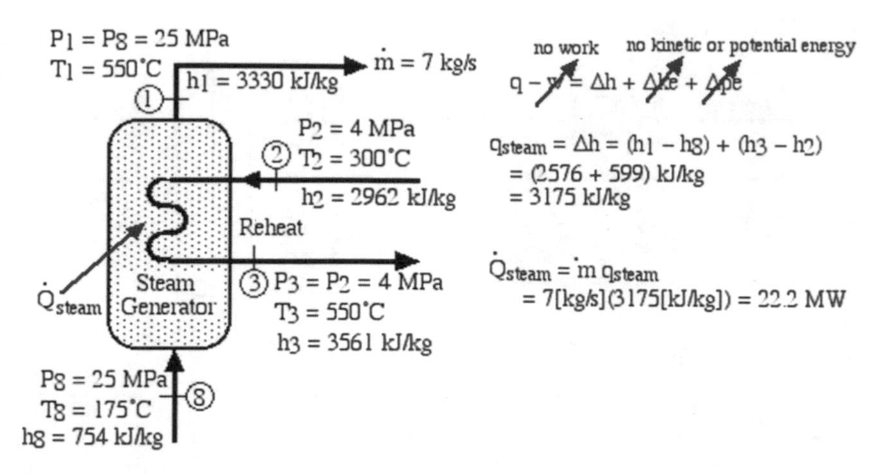

Fig. 4.6 The rate of energy generation calculated from the geothermal energy in the generator by the reheat process

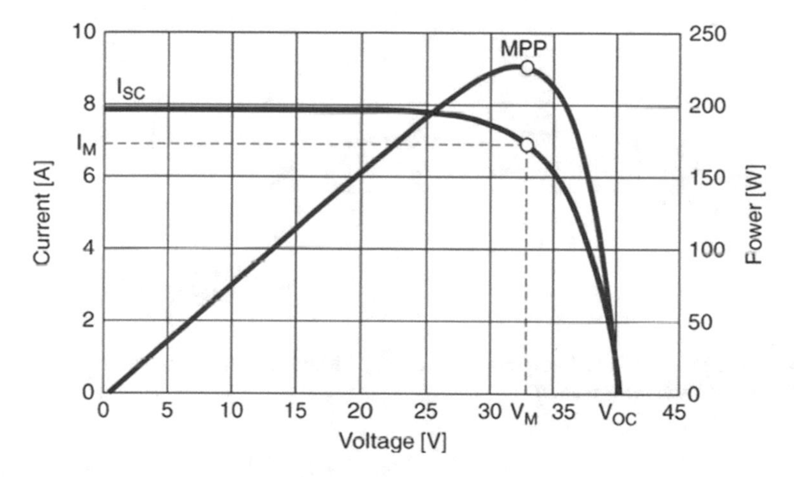

Fig. 4.7 The current-charge and power-voltage generation in the generator cell in relation to the MPP and the cell's current-charge generation rate

Subsequently, the control mechanism of the rotor which is artificial neural network (ANN) has revealed the maximum geothermal energy point tracking that is connected into the electric circuit [42–44]. Interestingly, the implemented electric circuit suggests that the produced volts near the MPP within changed thermal conditions deliver the maximum energy (I = current) which is characterized by the V–I relationship as follows neglecting the internal shunt resistance in order to use it as the electricity energy [45–47]:

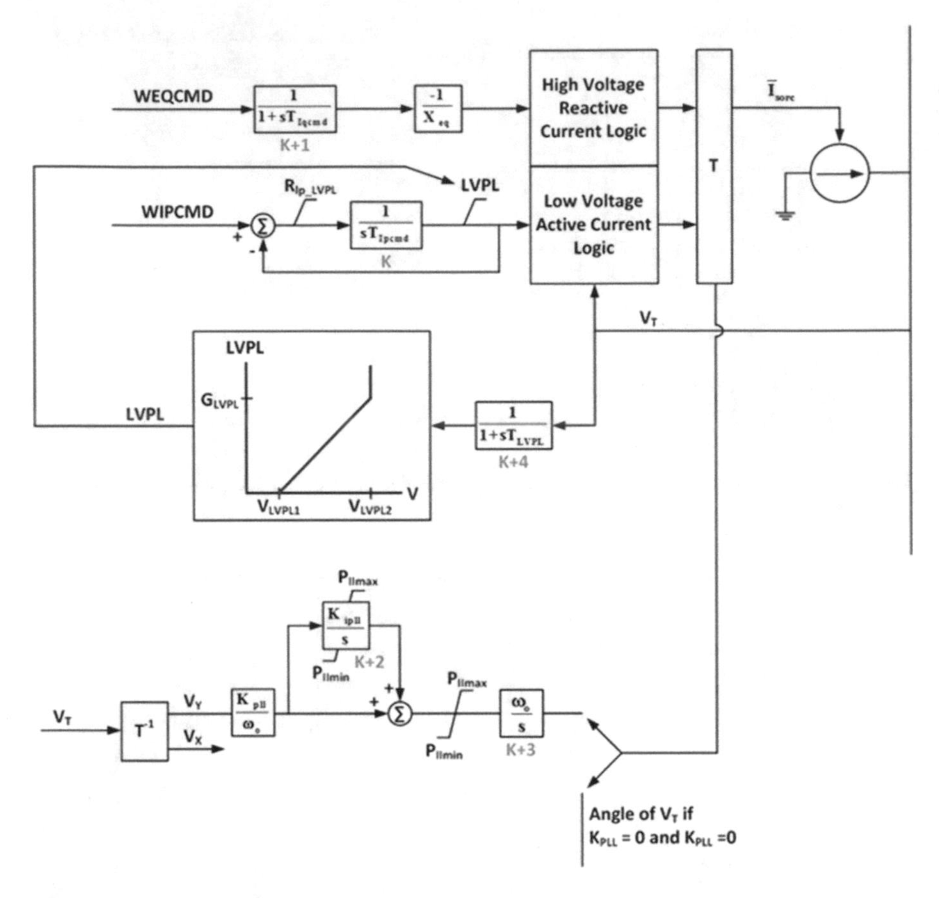

Fig. 4.8 The functional block diagram of net energy functional current and min-volt reaction current formation from the generator

$$I_0 = I_g - I_{sat}\left\{\exp\left[\frac{q(V_o + I_0 R_s)}{AKT}\right] - 1\right\} \quad (4.25)$$

where A represents the dimensionless factor, K represents the kinetic energy, T is the standard temperature at kelvin, q represents the charges, V represents the volts, and R represents the resistance under varying weather conditions.

Then the *incremental conductance* (*IC*) is calculated to determine the maximum volts which is configured in accordance to the MPP volts by determining the conductance values of *V–I* circuit (*dI*/*dV* and *I*/*V*, calculatively) (Fig. 4.9). Thus, this result suggests overcoming the limitations of IC to calculate the sign of *dP*/*dV* to deliver net current flow from the geothermal energy of $dP/dV = 0$, and is expressed as follows:

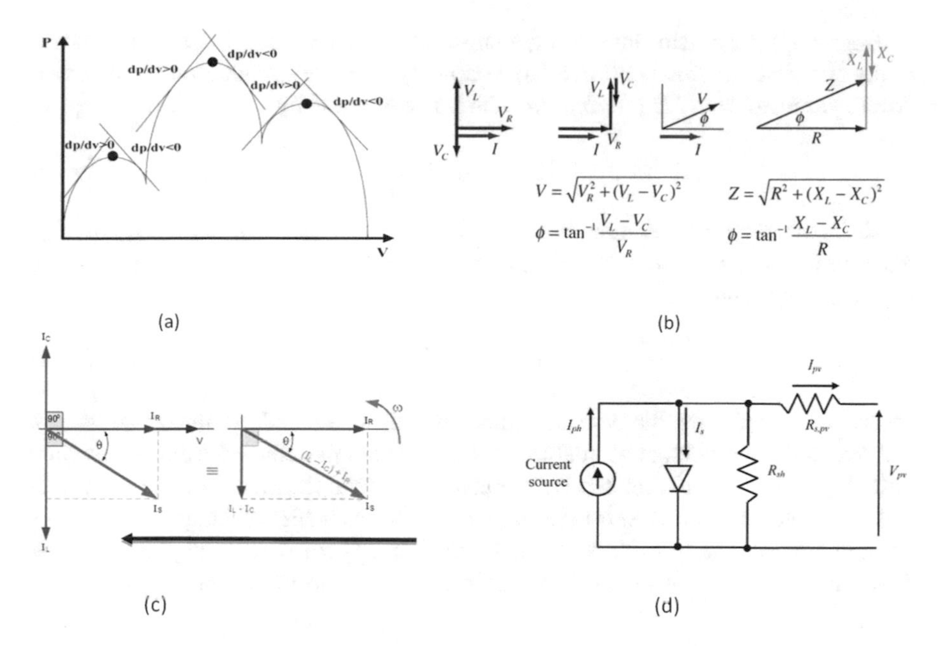

Fig. 4.9 (**a**) Shows the deviation of *dP*/*dV* in both functional MPP of P&O, (**b**) shows the proliferation and flexibility of electricity generation, (**c**) shows the scalar field of the electricity generation, (**d**) shows the net electricity current energy generation

$$\frac{dP}{dV} = I\frac{dV}{dV} + V\frac{dI}{dV} = 0 \tag{4.26}$$

$$\frac{dI}{dV} = -\frac{1}{V} \tag{4.27}$$

Here, the mathematical results suggest that *dP*/*dV* and net electricity generation are the functional current operating points that are up to the right of the MPP which is quite applicable as a source electricity energy (Fig. 4.9). Simply, the results confirm that the capability of determining MPP has been reached at the maximum current level constantly in order to deliver the sufficient energy to satisfy the net energy need for a house or building as a self-generated energy source which is defined by the power-voltage (*P*-*V*) function as follows:

$$P_{PV} = aV_{PV}^3 + bV_{PV}^2 + cV_{PV} + d \tag{4.28}$$

Here, *a*, *b*, *c*, and *d* represent the cofactors calculated by the functional results of the volts which are calculated by conducting the following equation:

$$V_{\mathrm{MPP}} = \frac{-b \pm \sqrt{b^2 - 3ac}}{3a} \tag{4.29}$$

Here, the mechanism simplifies the physical characteristics of energy in relation to the circuit in order to confirm a large capacity of energy generation by the *open-circuit voltage-based MPP technique* which is expressed by

$$P_{\text{MPP}} = I_{\text{MPP}} V_{\text{MPP}} \tag{4.30}$$

Here, the energy generation rate is clarified as the highest of either I_{MPP} or V_{MPP} and thus the net power generation P_{MPP} can be calculated easily by implementing the following equation:

$$V_{\text{MPP}} \approx K_{\text{oc}} V_{\text{oc}} \tag{4.31}$$

where K_{oc} represents the volt function, which is related to the rotor speed. Subsequently, net voltage estimation of V_{oc} has been determined from the module volts V_{oc}, and the open-circuited V_{oc} is transformed with the measurement of V_{MPP} in order to interconnect to *short-circuit current-based MPP* technique. Thus, this open-circuit voltage transformation will confirm a linear dependency between the M_{PP} current and the short-circuit current in order to confirm the generation of electricity as follows:

$$I_{\text{MPP}} \approx K_{\text{sc}} I_{\text{sc}} \tag{4.32}$$

where K_{sc} represents the current flow, which is calculated by analyzing the rotor speed of the generator.

Naturally, the dynamic electricity energy delivery constantly suggests the non-linear function of current proliferation via the open circuit which is also functional and is expressed as

$$C_{\text{bat}}(t) = C_{\text{bat}}(t-1) + \left(P_{\text{PV}}(t) + P_{\text{wg}}(t) - P_{\text{load}}(t)\eta_{\text{cad}}\right)\Delta t \eta_{\text{cha}} \tag{4.33}$$

where $C_{\text{bat}}(t)$ and $C_{\text{bat}}(t-1)$ confirm the released energy densities in times (t) and $(t-1)$. P_{PV} denotes the net energy generation, P_{wg} is the net energy generation load, $P_{\text{load}}(t)$ confirms the energy released at load t where t represents the simulated time sequence ($\Delta t = 1$ h), η_{cad} represents the efficiency of the AC/DC conversion ratio, and η_{cha} represents the energy releasing efficiency [5, 46].

Here, the total energy generated shall meet the net energy load demand of a house or building by the process of discharging (i.e., $\eta_{\text{inv}}\, P_{\text{wg}}(t) + P_{\text{PV}}(t) < P_{\text{load}}(t)$) which is expressed as

$$C_{\text{bat}}(t) = C_{\text{bat}}(t-1) + \frac{1}{\eta_{\text{dech}}}\left(P_{\text{PV}}(t) - \left(\frac{P_{\text{load}}(t) - P_{\text{wg}}(t)}{\eta_{\text{inv}}}\right)\right)\Delta t \tag{4.34}$$

where η_{dech} represents the energy release efficiency, η_{inv} represents the conversion rate, C represents the capability of the energy delivery in time t, and $C_{\text{bat}}(t)$ is constrained by $C_{bat\,min} \leq C_{bat}(t) \leq C_{bat\,max}$, where $C_{bat\,min}$ and $C_{bat\,max}$ are the nominal and peak energy delivery capabilities, respectively, and thus the maximum capability (C_{batn}) is determined by

$$C_{batmin} = \text{DOD} \cdot C_{batn} \tag{4.35}$$

where DOD represents the depth of discharge energy and thus the net energy bank considering the maximum DOD and temperature that affect the energy deliberation has been calculated and expressed by

$$C_{bat} = \frac{P_{load} \times D_s}{\text{DOD}_{max} \times \eta_t} \tag{4.36}$$

where C_{bat} is the energy delivery capability in amp h, and P_{load} is the load in amp h and it has been confirmed by the following mode to calculate net energy charge determination in time "t":

$$E_B(t) = E_B(t-1)(t-\sigma) + \left(E_{GA}(t) - \frac{E_L(t)}{\eta_{inv}} \right) \cdot \eta_{batt} \tag{4.37}$$

and while charging in relation to energy discharge efficiency it is calculated as

$$E_B(t) = E_B(t-1)(t-\sigma) - \left(\frac{E_L(t)}{\eta_{inv}} - E_{GA}(t) \right) \cdot \eta_{batt} \tag{4.38}$$

where $E_B(t)$ and $E_B(t-1)$ represent the ratio of energy delivery in times t and $(t-1)$, σ represents the hourly self-generated energy release rate, $E_{GA}(t)$ represents the net energy generation, $E(L)$ represents the total energy loading in time t, and η_{inv} and η_{batt} denote the efficiency rate of the energy generation in minimum level which is expressed by

$$E_{Bmin} \leq E_{B(t)} \leq E_{Bmax} \tag{4.39}$$

Since the maximum quantity of energy in the energy bank $E_{B\,max}$ is related to the minimum efficiency of energy generation $E_{B\,min}$, the minimum efficiency of energy release has been confirmed in relation to the DOD considering maximum area of the house or building:

$$E_{Bmin} = (1 - \text{DOD}) \cdot C_{batt} \tag{4.40}$$

Simply, this minimum electricity energy generation shall meet the net energy demand of a house or building which is clean and environmentally friendly.

Conclusion

We are the first generation with the devices to perceive how global environmental framework is running toward risk because of the traditional energy utilization by the houses and buildings; simultaneously, we are the last generation with chances to forestall this peril. Simply, we do *can* independently by applying cleaner energy for example *in situ* geothermal energy for all houses and buildings and aggregately assisting by making worldwide view for the advancement of this innovative technology in order to secure cleaner and greener earth. Simply, geothermal energy implementation *in situ* in the house or building sectors can fully replace a natural gas and oil heating energy system which indeed will be helpful for mitigating climate change. Since greenhouse gases are depleting exponentially due to the utilization of traditional energy supply by houses and buildings, renewable energy technology such as geothermal energy capture by houses and buildings in situ is an urgent demand to console the global energy demand and environmental perplexity in order to assist in building a cleaner and greener earth.

Acknowledgements This research was supported by Green Globe Technology under the grant RD-02017-07 for building a better environment. Any findings, predictions, and conclusions described in this chapter are solely performed by the authors and it is confirmed that there is no conflict of interest for publishing in a suitable journal.

References

1. Chatterjee, Arunava, Krishna Roy, and Debashis Chatterjee. "A gravitational search algorithm (GSA)-based photo-voltaic (PV) excitation control strategy for single phase operation of three phase wind-turbine coupled induction generator", Energy, 2014.
2. Chihming Shen. "Comparative study of peak power tracking techniques for solar storage system", APEC 98 Thirteenth Annual Applied Power Electronics Conference and Exposition APEC-98, 1998.
3. De Carli, Michele, Antonio Galgaro, Michele Pasqualetto, and Angelo Zarrella. "Energetic and economic aspects of a heating and cooling district in a mild climate based on closed loop ground source heat pump", Applied Thermal Engineering, 2014.
4. Md. Faruque Hossain. "Green science: Advanced building design technology to mitigate energy and environment", Renewable and Sustainable Energy Reviews, 2018.
5. Mani P, Lee J-H, Kang K-W, Joo YH. Digital controller design via LMIs for direct-driven surface mounted PMSG-based wind energy conversion system. IEEE Trans Cybern 2019. https://doi.org/10.1109/TCYB.2019.2923775. [Epub ahead of print].
6. Elmansouri A, El-mhamdi J, Boualouch A. Wind energy conversion system using DFIG controlled by back-stepping and RST controller. Proceedings of the International Conference on Electrical and Information Technologies (ICEIT); 2016 May 4-7; Tangiers, Morocco; https://doi.org/10.1109/EITech.2016.7519612.
7. Ligang H, Xiangdong W, Kang Y. Optimal speed tracking for double fed wind generator via switching control. Proceeding of the 27th Chinese Control and Decision Conference (2015 CCDC); 2015 May 23-25; Qingdao, China; https://doi.org/10.1109/CCDC.2015.7162368.

8. M. Kalantar, S.M. Mousavi G.. "Dynamic behavior of a stand-alone hybrid power generation system of wind turbine, microturbine, solar array and battery storage", Applied Energy, 2010.
9. Abderrahmane El Kachani, El Mahjoub Chakir, Tarik Jarou, Anass Ait Laachir, Jamal Zerouaoui, Abdelkader Hadjoudja. "Robust model predictive control applied to a WRIG-based wind turbine", International Review of Automatic Control (IREACO), 2016.
10. Abulizi M, Peng L, Francois B, Li Y. Performance analysis of a controller for doubly-fed induction generators based wind turbines against parameter variations. Int Rev Electr Eng. 2014;9:262–9. https://doi.org/10.15866/iree.v9i2.1797.
11. Haiying Li, Hao Liu, Aimin Ji, Feng Li, Yongli Jia. "Design of a hybrid solar-wind powered charging station for electric vehicles", 2013 International Conference on Materials for Renewable Energy and Environment, 2013.
12. Bento F, Cardoso AJM. A comprehensive survey on fault diagnosis and fault tolerance of DC-DC converters. Chin J Electr Eng. 2018;4:1–12 https://doi.org/10.23919/CJEE.2018.8471284.
13. Byrne, John, Job Taminiau, Lado Kurdgelashvili, and Kyung Nam Kim. "A review of the solar city concept and methods to assess rooftop solar electric potential, with an illustrative application to the city of Seoul", Renewable and Sustainable Energy Reviews, 2015.
14. Ouassaid M, Elyaalaoui K, Cherkaoui M. Reactive power capability of squirrel cage asynchronous generator connected to the grid. Proceedings of the 3rd International Renewable and Sustainable Energy Conference (IRSEC); 2015 Dec. 10-13; Marrakech, Morocco. https://doi.org/10.1109/IRSEC.2015.7455003.
15. Loucif M, Boumediene A. Modeling and direct power control for a DFIG under wind speed variation. Proceedings of the 2015 3rd International Conference on Control Engineering & Information Technology (CEIT); 2015 May 15-27; Tlemcen, Algeria. https://doi.org/10.1109/CEIT.2015.7233042.
16. Ramasamy Subramaniam, Young Hoon Joo. "Passivity-based fuzzy ISMC for wind energy conversion systems with PMSG", IEEE Transactions on Systems, Man, and Cybernetics: Systems, 2019.
17. Bekele, Getachew, and Getnet Tadesse. "Feasibility study of small hydro/PV/wind hybrid system for off-grid rural electrification in Ethiopia", Applied Energy, 2012.
18. Rad MAV, Ghasempour R, Rahdan P, Mousavi S, Arastounia M. Techno-economic analysis of a hybrid power system based on the cost-effective hydrogen production method for rural electrification, a case study in Iran. Energy. 2019:116421 https://doi.org/10.1016/j.energy.2019.116421.
19. Chao Zhang, Guangzheng Jiang, Xiaofeng Jia, Shengtao Li, Shengsheng Zhang, Di Hu, Shengbiao Hu, Yibo Wang. "Parametric study of the production performance of an enhanced geothermal system: A case study at the Qiabuqia geothermal area, northeast Tibetan plateau", Renewable Energy, 2018.
20. Heydari M, Smedley K. Comparison of maximum power point tracking methods for medium to high power wind energy systems. Proceedings of the 20th Conference on Electrical Power Distribution Networks Conference (EPDC); 2015 April 28-29; Zahedan, Iran. https://doi.org/10.1109/EPDC.2015.7330493.
21. Youcef Saidi, Abdelkader Mezouar, Yahia Miloud, Kamel Djamel Eddine Kerrouche, Brahim Brahmi, Mohammed Amine Benmahdjoub. "Advanced non-linear backstepping control design for variable speed wind turbine power maximization based on tip-speed-ratio approach during partial load operation", International Journal of Dynamics and Control, 2019.
22. Ander Ordono, Eneko Unamuno, Jon Andoni Barrena, Julen Paniagua. "Interlinking converters and their contribution to primary regulation: a review", International Journal of Electrical Power & Energy Systems, 2019.
23. Boumassata A, Kerdoun D, Madaci M. Grid power control based on a wind energy conversion system and a flywheel energy storage system. IEEE EUROCON 2015: Proceedings of the International Conference on Computer as a Tool (EUROCON); 2015 Sept. 8–11; Salamanca, Spain. https://doi.org/10.1109/EUROCON.2015.7313699.

24. Rani MD, Kumar MS. Development of doubly fed induction generator equivalent circuit and stability analysis applicable for wind energy conversion system. Proceedings of the International Conference on Recent Advances in Electronics and Communication Technology (ICRAECT); 2017 March 16-17; Bangalore, India. https://doi.org/10.1109/ICRAECT.2017.34.
25. Saeed MSR, Mohamed EEM, Sayed MA. Design and analysis of dual rotor multi-tooth flux switching machine for wind power generation. Proceedings of the 18th International Middle East Power Systems Conference (MEPCON); 2016 Dec. 27-29; Cairo, Egypt. https://doi.org/10.1109/MEPCON.2016.7836937.
26. Van Tan Nguyen, Duong Hung Hoang, Huu Hieu Nguyen, Kim Hung Le, The Khanh Truong, Quoc Cuong Le. "Analysis of Uncertainties for the Operation and Stability of an Islanded Microgrid", 2019 International Conference on System Science and Engineering (ICSSE), 2019.
27. Yu Shi, Xianzhi Song, Zhonghou Shen, Gaosheng Wang, Xiaojiang Li, Rui Zheng, Lidong Geng, Jiacheng Li, Shikun Zhang. "Numerical investigation on heat extraction performance of a CO2 enhanced geothermal system with multilateral wells", Energy, 2018.
28. Xing Luo, Xu Zhu, Eng Gee Lim. "A hybrid model for short term real-time electricity price forecasting in smart grid", Big Data Analytics, 2018.
29. Lakhdar Saihi, Brahim Berbaoui, Hachemi Glaoui, Larbi Djilali, Slimani Abdeldjalil. "Robust sliding mode H∞ controller of DFIG Based on variable speed wind energy conversion system", Periodica Polytechnica Electrical Engineering and Computer Science, 2019.
30. Li Peng, Zhou Jun, Yu Xiaozhou. "Design and on-orbit verification of EPS for the world's first 12U polarized light detection CubeSat", International Journal of Aeronautical and Space Sciences, 2018.
31. Feng Zhou, Shibo Sun, Mingyu Xu, Wenbo Hao, Bing Wang, Luxin Wang, Peng Jiang. "Research on the control strategy of maximum energy utilization and disordered disturbance suppression in distributed generation system", Journal of Physics: Conference Series, 2020.
32. Mahato SN, Singh SP, Sharma MP. Dynamic behavior of a single-phase self-excited induction generator using a three-phase machine feeding single-phase dynamic load, Int J Electr Power Energy Syst. 2013;47:1–12. https://doi.org/10.1016/j.ijepes.2012.10.067.
33. Saidi Y, Mezouar A, Miloud Y, Benmahdjoub MA. A robust control strategy for three phase voltage t source PWM rectifier connected to a PMSG wind energy conversion system. Proceedings of the International Conference on Electrical Sciences and Technologies in Maghreb (CISTEM); 2018 Oct. 28-31; Algiers, Algeria. https://doi.org/10.1109/CISTEM.2018.8613359.
34. Upadhyay VC, Sandhu KS. Reactive power management of wind farm using STATCOM. Proceedings of the International Conference on Emerging Trends and Innovations In Engineering And Technological Research (ICETIETR); 2018 July 11-13; Ernakulam, India. https://doi.org/10.1109/ICETIETR.2018.8529090.
35. Kasra Mohammadi, Jon G. McGowan. "Thermodynamic analysis of hybrid cycles based on a regenerative steam Rankine cycle for cogeneration and trigeneration", Energy Conversion and Management, 2018.
36. Geddam Kiran Kumar, Devaraj Elangovan. "Review on fault-diagnosis and fault-tolerance for DC–DC converters", IET Power Electronics, 2020.
37. Venkatesan C, Sundararaman K, Gopalakrishnan M. Grid integration of PMSG based wind energy conversion system using variable frequency transformer. Proceedings of the International Conference on Intelligent Computing, Instrumentation and Control Technologies (ICICICT); 2017 July 6-7; Kannur, India. https://doi.org/10.1109/ICICICT1.2017.8342783.
38. Md. Faruque Hossain. "Theoretical modeling for hybrid renewable energy: An initiative to meet the global energy", Journal of Sustainable Energy Engineering, 2016.
39. Md. Faruque Hossain. "In situ geothermal energy technology: an approach for building cleaner and greener environment", Journal of Ecological Engineering, 2016.
40. Timothy Beatley. "Chapter 1033-1 Biophilic Cities", Springer Science and Business Media LLC, 2018.

41. Lalia Merabet, Abdelkader Chaker, Abdellah Kouzou, Houari Merabet Boulouiha, Mohamed Elaguab. “Investigation on the control of DFIG used in power generation based on sliding mode control and SV-PWM”, International Journal on Energy Conversion (IRECON), 2019.
42. Hoda A. El-Sattar, Salah Kamel, Francisco Jurado. “Fixed bed gasification of corn stover biomass fuel: Egypt as a case study”, Biofuels, Bioproducts and Biorefining, 2019.
43. Huang, Xiaoxue, Jialing Zhu, Jun Li, Chengyu Lan, and Xianpeng Jin. “Parametric study of an enhanced geothermal system based on thermo-hydro-mechanical modeling of a prospective site in Songliao Basin”, Applied Thermal Engineering, 2016.
44. Md. Faruque Hossain. “Design and construction of ultra-relativistic collision PV panel and its application into building sector to mitigate total energy demand”, Journal of Building Engineering, 2017.
45. Junyent-Ferre A, Gomis-Bellmunt O. Wind turbine generation systems modeling for integration in power systems. In: Zobaa AF, Bansal R, editors. Handbook of Renewable Energy Technology, Singapore, Singapore: World Scientific; 2011, p. 53-68.
46. Kerrouche K, Mezouar A, Boumedien L. A simple and efficient maximized power control of DFIG variable speed wind turbine. Proceedings of the 3rd International Conference on Systems and Control; 2013 Oct. 29-31; Algiers, Algeria. https://doi.org/10.1109/ICoSC.2013.6750963.
47. Md. Faruque Hossain. “Bose-Einstein (B-E) photon energy reformation for cooling and heating the premises naturally”, Applied Thermal Engineering, 2018.

Part III
Advanced Building Design Technology

Chapter 5
Integrated Building Design Technology

Abstract The thermal state of photon is controlled by inducing Bose-Einstein (*B-E*) discrete photon mechanism and Higgs boson ($H \rightarrow \gamma\gamma^-$) electro-quantum charge application to cool and heat the building naturally. To cool the building naturally, helium (*He*)-aided curtain wall is utilized to capture the solar energy to cool the photons by employing Bose-Einstein (*B-E*) photonic bandgap to form a cooling-state photon. This cooling-state photon is named as the *Hossain cooling photon* (HcP^-). When needed this HcP^- can be reformed into a heating-state photon here signified as *Hossain thermal photon* (HtP^-) which is created by Higgs boson ($H \rightarrow \gamma\gamma^-$) electromagnetic quantum empowered by a single-diode semiconductor. It is because the Higgs boson ($H \rightarrow \gamma\gamma^-$) quantum is proposed to be instigated into the curtain wall through the extreme low-range weak force which regulates the HcP^- quantum to get agitated in order to convert it into an HtP^-. The creation of HcP^- and the reformation of HtP^- have been clarified with the use of a set of computational mathematics, which revealed the feasibility of reformed photons (HcP^- and HtP^-) that are positively doable in the exterior curtain wall to cool and heat the building naturally.

Keywords Higgs boson BR ($H \rightarrow \gamma\gamma^-$) quantum mechanics · Bose-Einstein discrete photonic structure · Hossain cooling photon (HcP^-) · Hossain thermal photon (HtP^-) · Natural cooling-heating mechanism

Introduction

The application of conventional heating and cooling mechanism in buildings is the primary cause for environmental vulnerability. These old-fashioned technologies to cool the buildings discharge chlorofluorocarbons (CFCs) which cause holes in the ozone layer. The ozone layer, lying between 18.6 and 9.3 miles over the earth's surface, acts as the shielding blanket layer to block majority of the UV rays of the Sun to protect the humans and all mammals from skin cancer and reproductive disorder [1–3]. Simultaneously, conventional heating technology concurrently uses the

M. F. Hossain, *Global Sustainability in Energy, Building, Infrastructure, Transportation, and Water Technology*,
https://doi.org/10.1007/978-3-030-62376-0_5

conventional energies and discharges CO_2, which is the main subsidizer for climate change. Emissions of such excessive CO_2 are a real threat to the environment which cause disastrous natural disasters for the Mother Earth.

Though conventional heating and cooling technologies have been studied comprehensively, the natural mechanism for cooling and heating technologies is not yet researched to resolve the global energy demand and environmental vulnerability [4–6]. In this study therefore, an innovative mechanism of natural heating and cooling technology has been proposed by the utilization of Higgs boson ($H \rightarrow \gamma\gamma^-$) electro-quantum and Bose-Einstein discrete photonic distribution in the building's exterior curtain wall. The proposed mechanism transforms the solar irradiance into the cooling photons and thermal photons. Hence, photon will be formed into cooling-state photon by the nano-point breaks and waveguides of photonic bandgaps, prompted by helium of the exterior curtain wall by the induction of quantum electrodynamics (QED) [7, 8]. Then, through the implementation of single-diode semiconductor into the curtain wall skin, the cooling-state photon is proposed to reform into a thermal state photon through the agitation of Higgs boson ($H \rightarrow \gamma\gamma^-$) electro-quantum by forming electromagnetic field to create the thermal photon in order to heat the building naturally [9–11]. This natural mechanism for cooling and hearing the building by controlling solar energy shall indeed be a novel approach to console global energy demand and climate change perplexity.

Methods and Simulation

Cooling Mechanism

Solar energy is transformed into the cooling-state photons by inducing nano-point breaks of waveguides of the helium (*He*)-aided curtain walls, by the formation of point defects in the B-E photon discretion field [12–14]. Simply, the structure of photonic bandgap (PBG) waveguide gets defected once it is integrated into the walls of the curtain, and then the defective PBG waveguide deforms the quantum mechanics of solar photons under helium (*He*)-aided curtain wall to generate a cooling-state photon [15–17]. To confirm this formation of cooling-state photon, a mathematical test using MATLAB v. 9.0 has been conducted where helium (*He*)-embedded waveguides of the curtain walls are used as photon storage and expressed by the following Hossainian [18–20]:

$$H = \Sigma \omega_{ci} a_i^\dagger a_i + \sum_K \omega_k b_k^\dagger b_k + \sum_{ik} \left(V_{ik} a^\dagger b_k + V_{ik} b_k^\dagger a_i \right), \tag{5.1}$$

where $b_k \left(b_k^\dagger \right)$ and $a_i \left(a_i^\dagger \right)$ stand for the drivers of the photodynamic modules and the nano-point-break modules of the photon nanostructure, correspondingly, and the coefficient, V_{ik}, stands for the order of photonic modules among the photon nanostructures and nano-breakpoints to show the transmissivity contours of photon (Fig. 5.1).

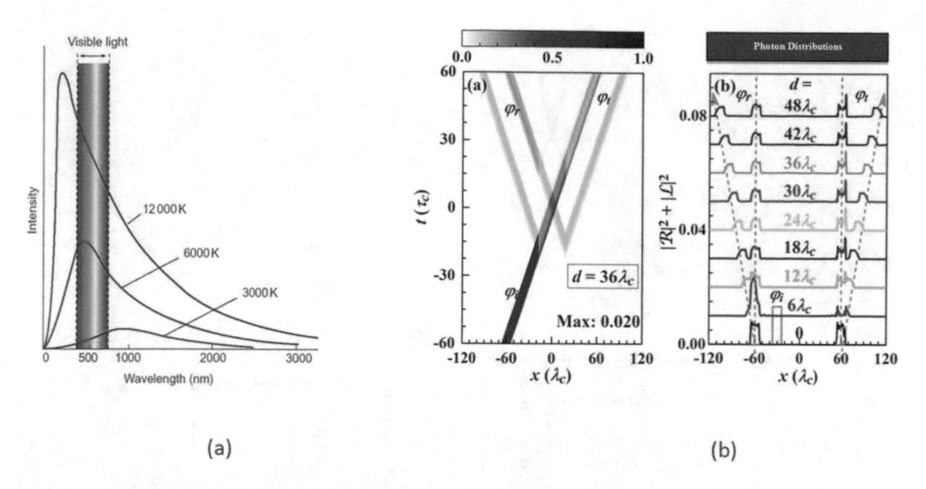

Fig. 5.1 (**a**) The distributions of the emission and absorption of transmitted solar irradiance. (**b**) The photon distribution density's contour map (regulates the supreme rate of 0.020) as the action of x and t [gray solid line]

Therefore, the proposed photon model that generates HcP^{-} comprises helium-assisted point breaks addressed by the converted current module to be used to cool a building (Fig. 5.2).

The current–voltage (I–V) features of the photonic structures into the single-diode module are illustrated by

$$I = I_{\mathrm{L}} - I_{\mathrm{o}}\left\{\exp^{\left[\frac{q(V+I_{\mathrm{Rs}})}{AkT_{\mathrm{c}}}\right]} - 1\right\} - \frac{(V + I_{\mathrm{Rs}})}{R_{\mathrm{S}}}, \tag{5.2}$$

whereby I_{L} stands for the photon generation current, I_{o} stands for the induced current, and R_{s} is the series resistance. A denotes the diode's inactive mechanism, k (= 1.38 × 10^{-23} W/m^{2} K) represents Boltzmann's constant, q (= 1.6 × 10^{-19} C), and T_{c} stands for the activated curtain wall temperature.

Consequently, the I–q affiliation in the photonic structure differs within the diode or induced current, which is given by [21–23]

$$I_{\mathrm{o}} = I_{\mathrm{Rs}}\left(\frac{T_{\mathrm{c}}}{T_{\mathrm{ref}}}\right)^{3} \exp\left[\frac{qE_{\mathrm{G}}\left(\frac{1}{T_{\mathrm{ref}}} - \frac{1}{T_{\mathrm{c}}}\right)}{KA}\right]. \tag{5.3}$$

where IR_{s} is the saturation current, which is determined by solar irradiance speed and working temperature, and qE_{G} stands for the bandgap energy of the electrons for each photon cell's unit area. I–V characteristics of two-diode dissimilar models are presented in Fig. 5.3.

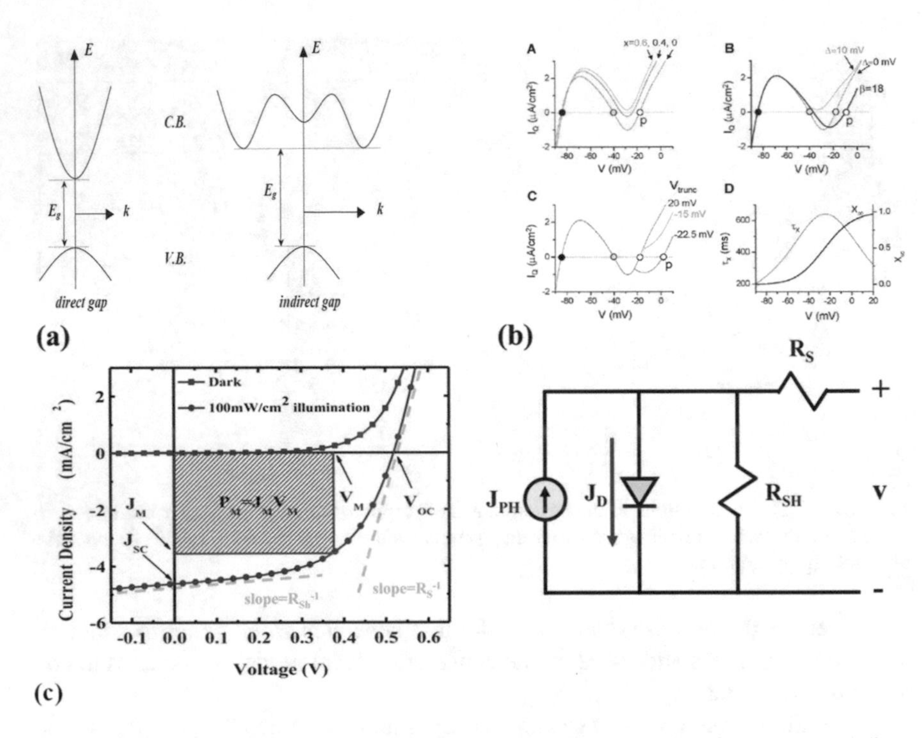

Fig. 5.2 *I–V* feathers of a single-diode cooling state: (**a**) Curtain wall at the indirect and direct photon gap state. (**b**) Curtain wall under energy mode. (**c**) Current density formation in the curtain wall by helium (He)-aided point breaks

In the photon element, the *I–V* equation joins the *I–V* curves of every cell in the photonic radiation panel. The *V–R* correlation at the section is governed by

$$V = -IR_S + K\log\left[\frac{I_L - I + I_o}{I_o}\right], \tag{5.4}$$

where V_{mo} and I_{mo} are the voltage and current in the PV panel, respectively, and K is a constant (AkT/q). Consequently, the link between I_{mo} and V_{mo} is the same as the *I–V* correlation in the PV cell:

$$V_{mo} = -I_{mo}R_{Smo} + K_{mo}\log\left(\frac{I_{Lmo} - I_{mo} + I_{omo}}{I_{omo}}\right), \tag{5.5}$$

whereby R_{Smo} is the series resistance, I_{omo} symbolizes the induced current into the diode, I_{Lmo} symbolizes the photon-produced current, and K_{mo} is a constant. Once the resistances of all non-series (Ns) cells are arranged in series, the constant is

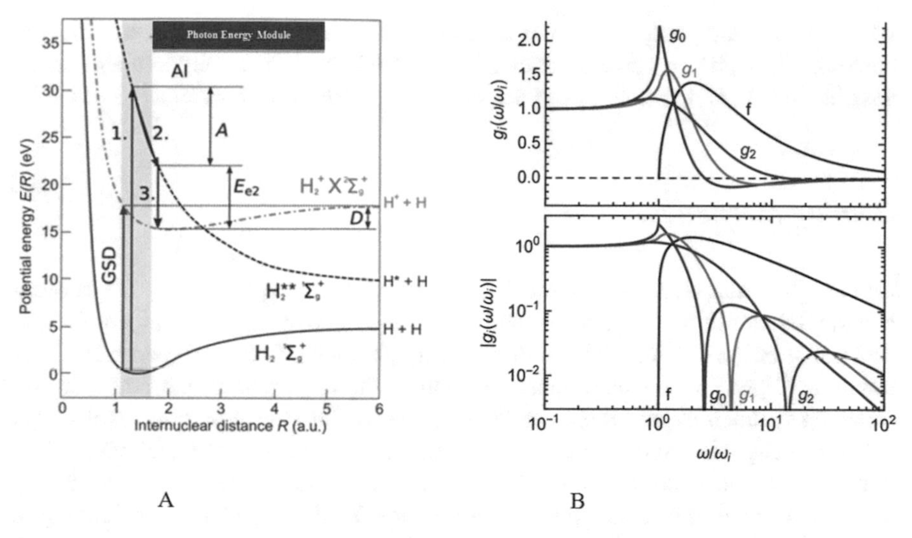

A B

Fig. 5.3 (**a**) *I–V* clarification of the single diode where curtain wall is in the standard condition to deform energy considering photonic internuclear distance. (**b**) Photon-photon spreading as a factor energy. Black line is an absorptive region $f(\omega/\omega_0)$ as illustrated in Eq. (5.10). Blue line is the dispersive region for related photons with static energy and static direction, $g_0(\omega/\omega_0)$, as illustrated in Eq. (5.12). Green line is the angle average for isotropic dispersal with static energy ωB, $g_1(\omega/\omega_1)$ as illustrated in Eq. (5.13). Red line represents thermal, isotropic average, $g_2(\omega/\omega_2)$, as illustrated in Eq. (5.14) [10, 24, 25]

$K_{mo} = N_s \times K$ and the total resistance is $R_{smo} = N_s \times R_s$. The current flows into the series cell arrangement are equal in all components, that is, $I_{Lmo} = I_L$ and $I_{omo} = I_o$. Therefore, the I_{mo}–V_{mo} correlation in the N_s arranged cells is denoted by

$$V_{mo} = -I_{mo} N_s R_s + N_s K \log\left(\frac{I_L - I_{mo} + I_o}{I_o}\right). \tag{5.6}$$

In the same way, when all N_p cells are arranged in parallel, the I_{mo}–V_{mo} connection is expressed by [24, 26]

$$V_{mo} = -I_{mo} \frac{R_S}{N_p} + K \log\left(\frac{N_{sh} I_L - I_{mo} + N_p I_o}{N_p I_o}\right). \tag{5.7}$$

Since the photon-generated current is primarily reliant on the photon emission panel's relativistic temperature conditions and the solar irradiance, the current could be computed as follows:

$$I_L = G\left[I_{SC} + K_I \left(T_{cool}\right)\right] V_{mo} \tag{5.8}$$

whereby I_{sc} is the photonic current for every unit area at 25 °C, K_I represents the corresponding photon panel coefficient, T_{cool} stands for the cooling temperature of the photon cell, and G exemplifies the solar energy for every unit area [23, 26–28].

Heating Mechanism

The Higgs boson electro-quantum dynamics have been exploited and the photon-heating relationship's parameters and have been accurately determined for the deforming of the cooling photons into heating photons [29, 30]. To generate a surrounding Higgs boson electro-quantum field in the curtain wall, the albanian surrounding symmetries have therefore been replicated. Since the emission of solar light splits the gauge field symmetry, the Goldstone scalar particles shall become the longitudinal module vector boson [24, 26]. In the corresponding gauge field of $A_{\mu}^{\alpha}(x)$ in the albanian, the surrounding symmetry particle T^{α} of every photon shall therefore be shaken up spontaneously. The Higgs quantum field shall start operating in local U(1)-stage symmetries [28, 31]. The quantum mode will therefore consist of a composite scalar field $\Phi(x)$ of electrically charged q combined to the EM field $A^{\mu}(x)$ to generate heat, which could be represented by the Lagrangian function as below:

$$L = \frac{-1}{4} F_{\mu\nu} F^{\mu\nu} + D_{\mu}\Phi D^{\mu}\Phi - V(\Phi\Phi), \tag{5.9}$$

whereby

$$D_{\mu}\Phi(x) = \partial_{\mu}\Phi(x) + iqA_{\mu}(x)\Phi(x)$$

$$D_{\mu}\Phi(x) = \partial_{\mu}\Phi(x) - iqA_{\mu}(x)\Phi(x), \tag{5.10}$$

However

$$V(\Phi\Phi) = \frac{\lambda}{2}(\Phi\Phi)^2 + m^2(\Phi\Phi). \tag{5.11}$$

Assuming that $\lambda > 0$ but $m^2 < 0$ such that $\Phi = 0$ is the scalar potential's local maximum, the minimum formation of a degenerate circle is $\Phi = \frac{v}{\sqrt{2}} * e^{i\theta}$ with

$$v = \sqrt{\frac{-2m^2}{\lambda}} \text{for any real } \theta. \tag{5.12}$$

Subsequently, the scalar field Φ shall be developed as a nonzero vacuum value $\langle\Phi\rangle \neq 0$ that simultaneously forms the U(1) symmetries of the electric field. From the phase of the complex field $\Phi(x)$, this symmetry's breakdown creates a massless

Goldstone scalar. In local U(1) symmetries however, the phase of $\Phi(x)$ is the x-corresponding phase of the dynamic $\Phi(x)$ field instead of the phase of the expected value $\langle\Phi\rangle$.

For this mechanism to be confirmed, it has been expressed by the scalar field space in polar coordinates:

$$\Phi(x)=\frac{1}{\sqrt{2}}\Phi_{\mathrm{r}}(x)*\mathrm{e}^{i\Theta(x)},\mathrm{real}\,\Phi_{\mathrm{r}}(x)>0,\mathrm{real}\,\Phi(x). \tag{5.13}$$

Since this creation of the field is singular at $\Phi(x) = 0$, it is not applicable to theories with $\langle\Phi\rangle \neq 0$ though it is sufficient for theories that are impulsively broken, whereby $\Phi\langle x\rangle \neq 0$ is expected nearly everywhere. When it comes to the real fields $\phi_{\mathrm{r}}(x)$ and $\Theta(x)$, the scalar potential is just reliant on the radial field ϕ_{r},

$$V(\phi)=\frac{\lambda}{8}\left(\phi_{\mathrm{r}}^2-v^2\right)^2+\mathrm{const.} \tag{5.14}$$

In a situation where the radial field is shifted by a variable scalar, $\Phi_{\mathrm{r}}(x) = v + \sigma(x)$, then

$$\phi_{\mathrm{r}}^2-v^2=(v+\sigma)^2-v^2=2v\sigma+\sigma^2 \tag{5.15}$$

$$V=\frac{\lambda}{8}\left(2v\sigma-\sigma^2\right)^2=\frac{\lambda v^2}{2}*\sigma^2+\frac{\lambda v}{2}*\sigma^3+\frac{\lambda}{8}*\sigma^4. \tag{5.16}$$

In the interim, the covariant derivative $D_\mu\phi$ will be

$$D_\mu\phi=\frac{1}{\sqrt{2}}\left(\partial_\mu\left(\phi_r\mathrm{e}^{i\Theta}\right)+iqA_\mu*\phi_r\mathrm{e}^{i\Theta}\right)=\frac{\mathrm{e}^{i\Theta}}{\sqrt{2}}\left(\partial_\mu\phi_{\mathrm{r}}+\phi_{\mathrm{r}}*i\partial_\mu\Theta+\phi_{\mathrm{r}}*iqA_\mu\right) \tag{5.17}$$

$$\begin{aligned}\left|D_\mu\phi\right|^2&=\frac{1}{2}\left|\partial_\mu\phi_{\mathrm{r}}+\phi_{\mathrm{r}}*i\partial_\mu\Theta+\phi_{\mathrm{r}}*iqA_\mu\right|^2=\frac{1}{2}\left(\partial_\mu\phi_{\mathrm{r}}\right)+\frac{\phi_r^2}{2}*\left(\partial_\mu\Theta qA_\mu\right)^2\\&=\frac{1}{2}\left(\partial_\mu\sigma\right)^2+\frac{(v+\sigma)^2}{2}*\left(\partial_\mu\Theta+qA_\mu\right)^2\end{aligned} \tag{5.18}$$

The Lagrangian is then given by

$$L=\frac{1}{2}\left(\partial_\mu\sigma\right)^2-v(\sigma)-\frac{1}{4}F_{\mu\nu}F^{\mu\nu}+\frac{(v+\sigma)^2}{2}*\left(\partial_\mu\Theta+qA_\mu\right)^2. \tag{5.19}$$

For the heating (Lllheat) to be incorporated into the electric field properties of this Lagrangian therefore, I enlarged L_{heat} as a power series (and their derivatives) and extracted the quadratic part that describes the free particles:

$$L_{\text{heat}} = \frac{1}{2}\left(\partial_{\mu}\sigma\right)^{2} - \frac{\lambda v^{2}}{2} * \sigma^{2} - \frac{1}{4}F_{\mu\nu}F^{\mu\nu} + \frac{v^{2}}{2} * \left(qA_{\mu} + \partial_{\mu}\Theta\right)^{2} \tag{5.20}$$

Evidently, to start high heating in the quantum field of the curtain wall, the free particles (with Lagrangian L_{free}) should be real scalar particles with affirmative $m^2 = \lambda v^2$ (in which m represents the particle mass; check Fig. 5.4).

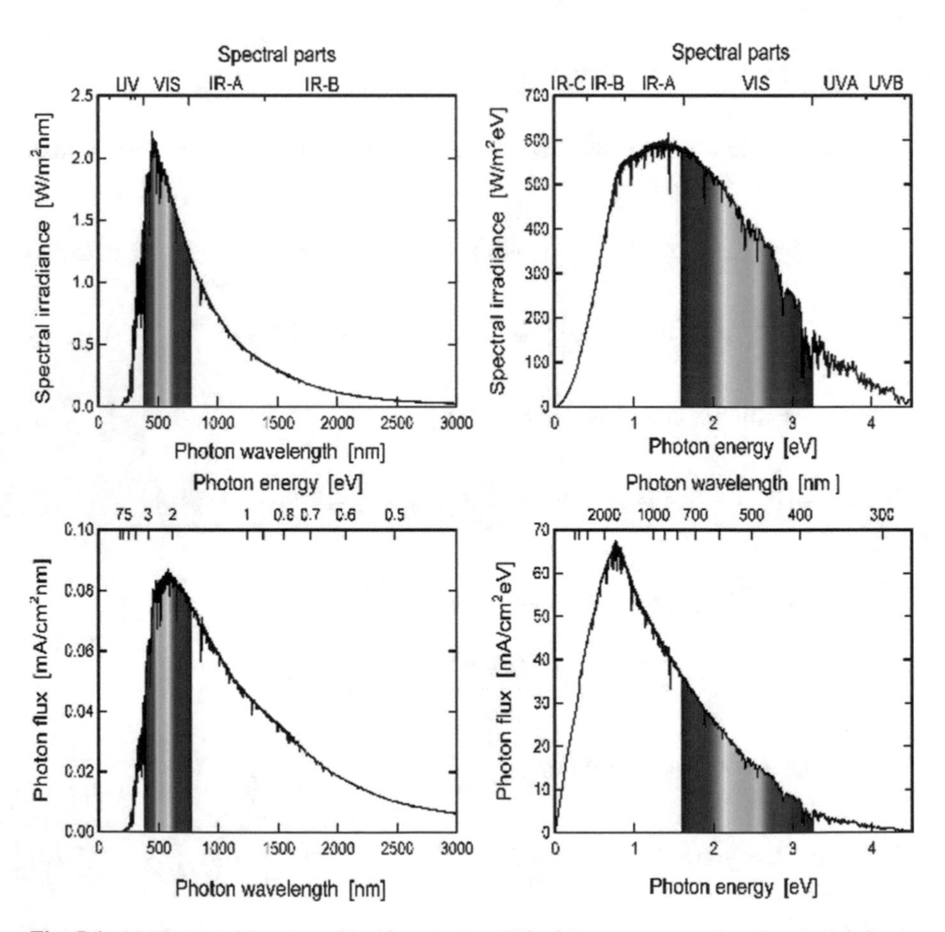

Fig. 5.4 (**a**) The consideration of heating photon (GHz-fs2) energy regarding wavelength (ym). (**b**) A photon's transformation mechanism at the level of energy from the electrons' quantum intensity spectra. The photon absorbance is demonstrated taking wavelength into consideration

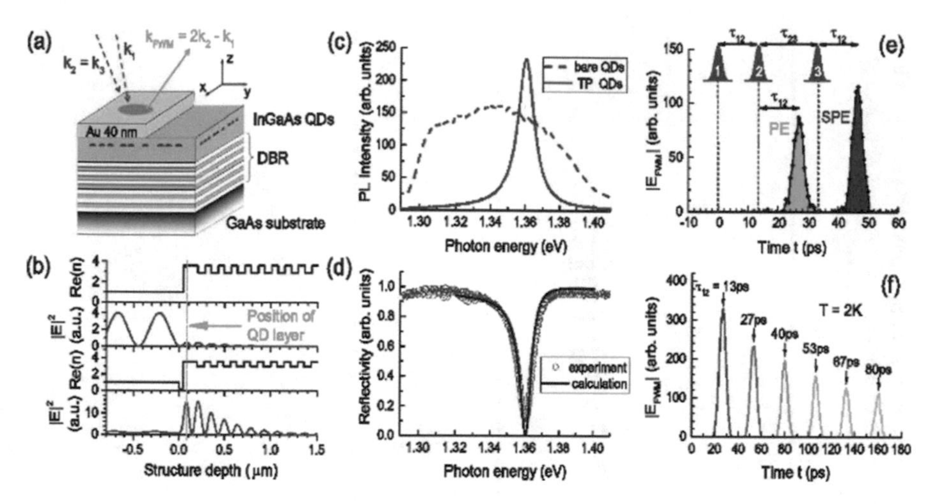

Fig. 14.4 (continued)

Results and Discussion

Cooling Mechanism

To computationally demonstrate the creation of cooling photons by the helium-aided curtain wall, the dynamic photon proliferation has been determined by integrating Eqs. (5.15) and (5.16). Owing to the cool-state region $J(\omega)$ and the consistent weak coupling, the curtain wall surface is expected to proliferate photons [28, 32]. Hence, $J(\omega)$ represents the quantum area defining the density of state (DOS) area generated in the photovoltaic cell through the standard cooling photon mode $V(\omega)$ within the photonic band (PB) and the PV cell [33, 34]. Moreover, photon production should follow the Weisskopf-Wigner assumption. Subsequently, the proliferated *HcP*s will conduct a dynamic mode (A, B, and C) in the curtain wall, as described in Table 5.1 [25, 35].

Within the C curtain wall, Ω_C represents a fine frequency cutoff, which evades the bifunctional DOS. The A and B curtain wall in the same way necessitates a fine hertz cutoff at Ω_d to evade adverse DOS (Fig. 5.5). $e_{rfc}(x)$ and $Li_2(x)$ are therefore additive and di-logarithmic variables in that order. The DOS here, noted as $\varrho_{PC}(\omega)$, is therefore calculated through the photon eigenfunctions and eigenfrequencies of Maxwell's rules [38, 39]. In the A curtain wall, the DOS is provided by $\varrho_{PC}(\omega) \propto \frac{1}{\sqrt{\omega - \omega_e}} \Theta(\omega - \omega_e)$, where $\Theta(\omega - \omega_e)$ is the Heaviside step function, and ω_e stands for the PBE's frequency at the provided DOS.

The DOS is needed for precisely foreseeing the non-Weisskopf-Wigner mode's qualitative state as well as the photon cell's photon cooling state in a C calculation in the curtain wall. Projected DOS (PDOS) and the DOS are revealed in Fig. 5.6. In a 3D curtain wall, the DOS next to the PBE is illustrated as $\varrho_{PC}(\omega) \propto \frac{1}{\sqrt{\omega - \omega_e}} \Theta(\omega - \omega_e)$.

Table 5.1 The DOS (density of states) of photonic structures in diverse dimensional approaches of the curtain wall skin. Self-generated energy-induced reservoir $\Sigma(a)$ and the unit area $J(a)$. The C, η, $\wedge$ χ function as combined forces amid PV and the point break in the curtain wall in dimensions of A, B, and C

Photonic structure	Area $J(a)$ for photonic density of states	Energy-induced self-energy $\Sigma(a)$
A	$\frac{X-a}{2e}\frac{1-a}{\sqrt{(2a-a_e)\Omega}}*\Theta-C(2a-4a_e)$	$X-a-C\Omega/\sqrt{(a_e-a)}$
B	$X-\eta\left[\ln\left\lvert\frac{a-a_0}{a_0}\right\rvert-1\right]\Theta*\Omega(a-a_e)\Theta(\Omega_d-a)$	$\eta\Omega\left[\mathrm{Li}_2\left(\frac{\Omega_d-a_0}{a-a_0}\right)*\Omega-\mathrm{Li}_2\left(\frac{a_0-a_e}{a_0-a}\right)-\ln\frac{a_0-a_e}{\Omega_d-a_0}\ln\frac{a_e-a}{a_0-a}\right]$
C	$\chi\Omega\sqrt{((a-a_e)/\Omega_C)}\exp(-(a-a_e)/\Omega_C)\Theta(\omega-a_e)$	$\chi\left[\pi\Omega\sqrt{\frac{a_e-a}{\Omega_C}}\exp\left(\frac{-a-a_e}{\Omega_C}\right)e_{rfc}\sqrt{\frac{a_e-a}{\Omega_C}}-\sqrt{\pi}\right]$

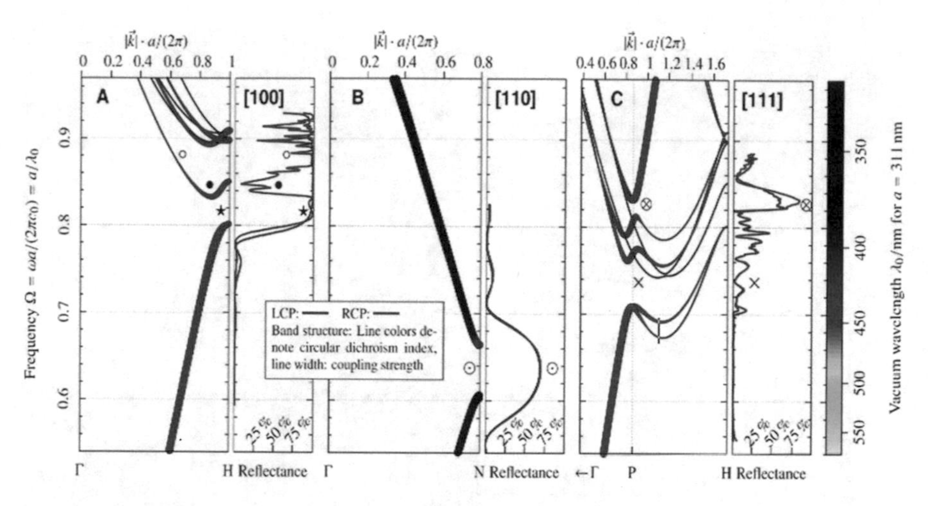

Fig. 5.5 Energy conversion modes and photonic band structure into the units against frequencies at different DOS in A, B, and C curtain wall. The frequencies of photon module magnitudes to discharge energy computed by Eq. (5.2) taking into account the state illustrate the moment frequency *vc* transmuted through a PBG field to a PB area [36, 37]

Subsequently, this DOS is illuminated in relation to the electric field vector [1, 40, 41]. In A curtain wall skins and B, the cooling photon DOS shows a pure anti-logarithmic anti-divergence next to the PBE, which is estimated as $\varrho_{PC}(\omega) \propto -[\ln|(\omega - \omega_0)/\omega_0| - 1]\Theta(\omega - \omega_e)$, whereby ω_e exemplifies the peak's central point in the distribution of DOS.

As aforementioned, $J(\omega)$ describes the DOS field manufactured in the PV cell with the fine cooling photonic magnitude $V(\omega)$ in the PV cell and the PB [7, 22]:

$$J(\omega) = \varrho(\omega)\left|V(\omega)\right|^2. \tag{5.21}$$

Henceforth, I deliberate the PB frequency ω_c and the proliferative photon dynamics $\langle a(t)\rangle = u(t, t_0)\langle a(t_0)\rangle$, whereby the function $u(t, t_0)$ describes the photon structure. $u(t, t_0)$ is computed with the use of the dissipative integral–differential equation that Eq. (5.18) describes:

$$u(t,t_0) = \frac{1}{1-\Sigma'(\omega_b)} e^{-i\omega(t-t_0)} + \int_{\omega_e}^{\infty} d\omega \frac{J(\omega) e^{-i\omega(t-t_0)}}{\left[\omega - \omega_c - \Delta(\omega)\right]^2 + \pi^2 J^2(\omega)}, \tag{5.22}$$

whereby $\Sigma'(\omega_b) = \left[\partial\Sigma(\omega)/\partial\omega\right]_{\omega=\omega_b}$ and $\Sigma(\omega)$ stand for the PB photon self-energy correction that is tempted in the reservoir:

$$\Sigma(\omega) = \int_{\omega_e}^{\infty} d\omega' \frac{J(\omega')}{\omega - \omega'}. \tag{5.23}$$

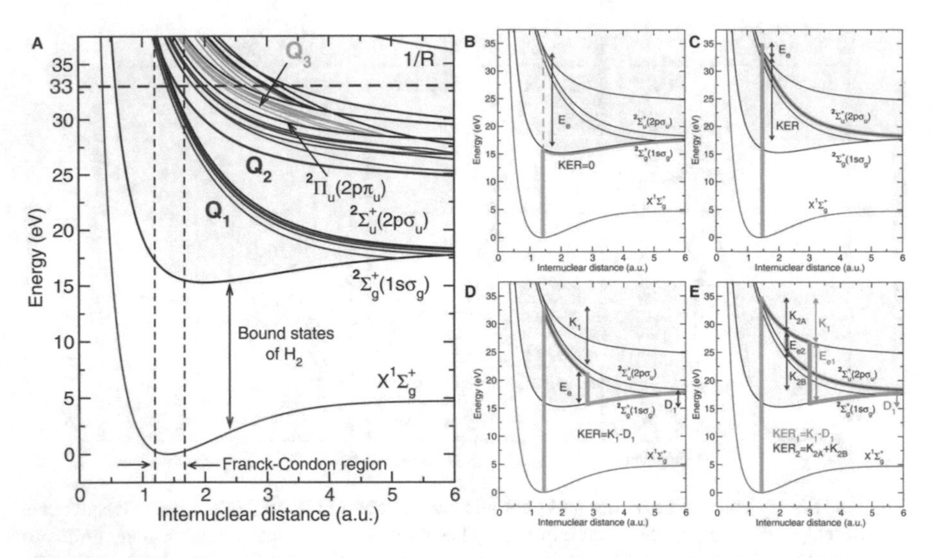

Fig. 5.6 (**a**) Total DOS and the PDOS (projected density of states) of deformed photons for conversion into the cool condition. Panel (**a**): (1) all DOS (T) and DOS extrapolated onto the *s*, *p*, and *d* orbitals; (2) *d* orbitals' PDOS on the fourth level of Mo particles; and (3) PDOS of *d* orbitals on the Mo atoms. Panel (**b**): like in panel (**a**) though for targeted DOS of Mo particles. Panel (**c**): (1–3) like in panel (**a**) and (4) *p* orbitals' PDOS of O particles. Panel (**d**): (1–3) like in panel (**b**) and (4) *p* orbitals' PDOS of exterior S particles. (**b**) Diagram showing energy levels and paths to dissociating ionization. (**a**) Whole H_2 and $H^+{}_2$ structure energy as a function of internuclear distance (a.u., atomic units). The two lowest series of doubly agitated states of H_2 with $^1\Pi_u$symmetry are red and blue. At bigger internuclear distances, the Q_1 terms break into $H(n = 1) + H(n = 2, \ldots, \infty)$ while the Q_2 terms into $H(n = 2, l = 1) + H(n = 2, \ldots, \infty)$, whereby n and l are the principal and angular momentum quantum numbers of the state, respectively. (**b–e**) are semiclassical paths for dissociative ionization by absorbing one 33-eV photon. (**b**) Direct ionization resulting in $H^+{}_2(1s\sigma_g)$ (Eq. 5.2). (**c**) Direct ionization resulting in $H^+{}_2(2p\sigma_u)$ (Eq. 5.3). (**d**) The resonant ionization by means of the lowest $Q_{1,}$ doubly excited states resulting in $H^+{}_2(1s\sigma_g)$ (Eq. 5.4). (**e**) Resonant ionization through the lowest Q_2 doubly excited states resulting in $H^+{}_2(1s\sigma_g)$ (Eq. 5.5) or $H^+{}_2(2p\sigma_u)$ (Eq. 5.6) [10, 19]

The frequency ω_b in Eq. (5.2) here stands for the cool photon frequency module in the PBG ($0 < \omega_b < \omega_e$), computed under the pole condition $\omega_b - \omega_c - \Delta(\omega_b) = 0$, in which $\lesssim \Delta(\omega) = P\left[\int d\omega' \frac{J(\omega')}{\omega - \omega'}\right]$ is a very important principal value.

In Fig. 5.7a, the cool photon motion of the formation magnitude $|u(t, t_0)|$ is computed in A, B, and C photonic structures for several structure δ and incorporated into the PB area from the PBG area [38, 39]. The rates of cool photon motion $\kappa(t)$ are generated in Fig. 5.7b. The outcomes point out that the moment ω_c is transferred into the PB area from the PBG area; the rate of generating dynamic photons becomes very high. Since the range of $u(t, t_0)$ is $1 \geq |u(t, t_0)| \geq 0$, I have defined the crossover area to fulfill $0.9 \gtrsim |u(t \rightarrow \infty, t_0)| \geq 0$. This represents $-0.025\omega_e \lesssim \delta \lesssim 0.025\omega_e$, at a cool photonic motion rate $\kappa(t)$ at the PBG ($\delta \leftarrow 0.025\omega_e$) and close to the PBE ($-0.025\omega_e \lesssim \delta \lesssim 0.025\omega_e$).

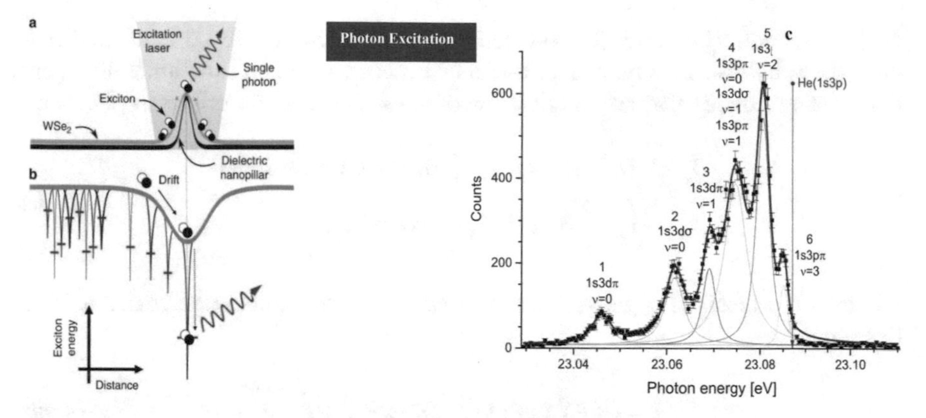

Fig. 5.7 Formation of photonic motion into the acting curtain wall PV cells. (**a**) Photon excitation field of wSe_2 <*a*(*t*)> = 5*u*(*t*, t_0) <*a*(t_0)>; (**b**) photonic motion rate *k*(*t*), formed for (1) A, (2) B, and (3) C acting as curtain wall PV cells (Table 5.1); (**c**) photon energy (eV) production taking into consideration the agitated photons and area wSe_2 <*a*(*t*)> = 5*u*(*t*, t_0) <*a*(t_0)> in that order [24]

To be specific, I first of all took into consideration the PB as the Fock cooling determination n_0, i. e. , $\rho(t_0) = |n_0\rangle\langle n_0|$, which is theoretically acquired via quantum field [2, 24, 36], and at that point, through solving Eq. (5.1), taking into account the cool photonic induction state at time *t*:

$$\rho(t) = \sum_{n=0}^{\infty} P_n^{(n_0)}(t) |n_0\, n_0| \tag{5.24}$$

$$P_n^{(n_0)}(t) = \frac{\left[v(t,t)\right]^n}{\left[1+v(t,t)\right]^{n+1}}\left[1-\Omega(t)\right]^{n_0} x \sum_{k=0}^{\min\{n_0,n\}} \binom{n_0}{k}\binom{n}{k}\left[\frac{1}{v(t,t)}\frac{\Omega(t)}{1-\Omega(t)}\right]^k, \tag{5.25}$$

whereby $\Omega(t) = \frac{|u(t,t_0)|^2}{1+v(t,t)}$. This outcome recommends a prompt into dynamic states $P_n^{(n_0)}(t)$ of n_0. Figure 5.7 in fact plots the photon dissipation proliferation $P_n^{(n_0)}(t)$ in the primary state $n_0 = 5\rangle$ as well as in the steady-state limit $P_n^{(n_0)}(t \to \infty)$. The deliberation of the formed cool photons will therefore eventually reach a state that is a nonequilibrium cooling state to cool the building.

Heating Mechanism

In this suggested heating mechanism, the electric field is formed into a semiconductor by the activation of Higgs boson electrodynamics. Hence, the local U(1) acts as gauged invariant QED as a result of quantitative weight of gauged particles formed at $\varnothing' \rightarrow e^{i\alpha(x)}\varnothing$. Therefore, the induced cool-state photons will be converted into heat-state photons. This procedure could be described using a functional divergent considering a distinctive reformation law of the scalar field, described in [14, 17]:

$$\begin{aligned} &\partial_{\mu} \rightarrow D_{\mu} = \partial_{\mu} = ieA_{\mu} \ \left[\text{covariant derivatives}\right] \\ &A'_{\mu} = A_{\mu} + \frac{1}{e}\partial_{\mu}\alpha \ \left[A_{\mu} \text{ derivatives}\right], \end{aligned} \tag{5.26}$$

whereby the local U(1) gauged variant Lagrangian for a complex scalar field is described by

$$L = \left(D^{\mu}\right)^{\dagger}\left(D_{\mu}\varnothing\right) - \frac{1}{4}F_{\mu\nu}F^{\mu\nu} - V\left(\varnothing\right). \tag{5.27}$$

The term $\frac{1}{4}F_{\alpha\nu}F^{\alpha\nu}$ is the motion field at the gauged area and $V(\varnothing)$ is an extra form that is defined as $V(\varnothing\varnothing) = \mu^2(\varnothing\varnothing) + \lambda(\varnothing\varnothing)^2$.

In accordance with the Lagrangian L, activation of quantum into the scalar fields ϕ_1 and ϕ_2 will form a heating mass as μ where $\mu^2 < 0$ induced in this situation confesses with an infinite number of quanta, whereby each fulfills $\phi_1^2 + \phi_2^2 = -\mu^2 / \lambda = v^2$. When it comes to the shifted fields $\eta \wedge \xi$, the quantum field is described as $\phi_0 = \frac{1}{\sqrt{2}}\left[\left(\upsilon + \eta\right) + i\xi\right]$, while the Lagrangian's covariant derivatives will be as follows:

$$\begin{aligned} \text{Kinetic term: } L_{\text{kin}}(\eta,\xi) &= (D^{\mu}\phi)^{\dagger}(D^{\mu}\phi) \\ &= (\partial^{\mu} + ieA^{\mu})\phi(\partial_{\mu} - ieA_{\mu})\phi \end{aligned} \tag{5.28}$$

Possible term (to second order): $V(\eta,\xi) = \lambda\upsilon^2\eta^2$. Thus the full Lagrangian could hence be written as

$$\begin{aligned} &L_{\text{kin}}\left(\eta,\xi\right) = \frac{1}{2}\left(\partial_{\mu}\eta\right)^2 - \lambda\upsilon^2\eta^2 + \frac{1}{2}\left(\partial_{\mu}\xi\right)^2 - \\ &\frac{1}{4}F_{\mu\nu}F^{\mu\nu} + \frac{1}{2}e^2\upsilon^2A_{\mu}^2 - e\upsilon A_{\mu}\left(\partial^{\mu}\xi\right) + \int.\text{terms}. \end{aligned} \tag{5.29}$$

η here is mass, while ξ is massless (as in the past), μ refers to the mass form into the quantum field, while A_{μ} is fixed up to a term $\partial_{\mu}\alpha$, as is illustrated in Eq. (5.27). Generally, A_{μ} and ϕ concurrently change; therefore Eq. (5.28) could be reformed to put up the heat-photon particles into the quantum field:

$$\begin{aligned} L_{\text{scalar}} &= \left(D^{\mu}\phi\right)^{\dagger}\left(D^{\mu}\phi\right) - V\left(\phi^{\dagger}\phi\right) \\ &= \left(\partial^{\mu} + ieA^{\mu}\right)\frac{1}{\sqrt{2}}\left(v+h\right)\left(\partial_{\mu} - ieA_{\mu}\right)\frac{1}{\sqrt{2}}\left(v+h\right) - V\left(\phi^{\dagger}\phi\right) \end{aligned} \tag{5.30}$$

$$= \frac{1}{2}\left(\partial_{\mu}h\right)^2 + \frac{1}{2}e^2A_{\mu}^2\left(v+h\right)^2 - \lambda v^2h^2 - \lambda vh^3 - \frac{1}{4}\lambda h^4 + \frac{1}{4}\lambda h^4. \tag{5.31}$$

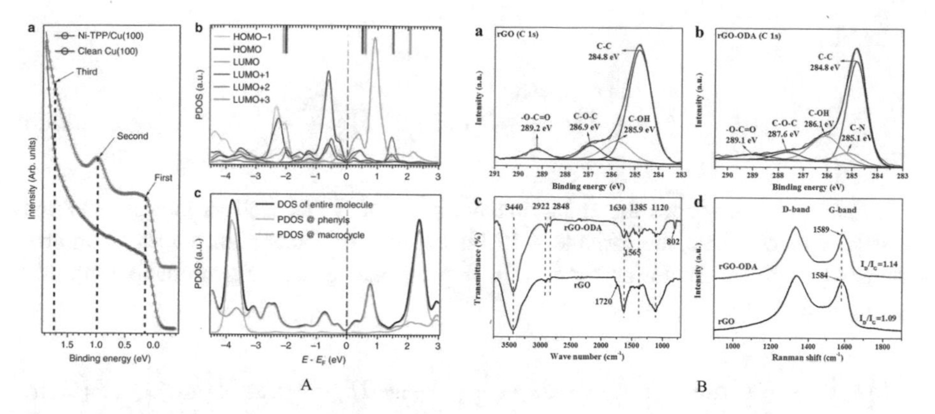

Fig. 5.8 The mechanisms for the production of electrically formed photon. (**a**) Heat photon is concurrently combined into the functional module into the quantum dynamics taking into consideration the rate of coincidence of the fundamental mode of production. (**b**) Functional photonic proliferation rates of heat photon in the electromagnetic field of photonic band structure [7, 26]

Hence, the Lagrangian scalar quantity is used as the extended term which recommends that the Higgs boson quantum field could be induced to form heating photons (Fig. 5.8).

To confirm this heat-photon deformation mechanism, a mathematical computing has been performed by computing the isotropic distribution of movement on the distinction cone pertaining to the angle θ from the vertical axis. The difference between θ and $\theta + d\theta$ is $\frac{1}{2}\sin\theta d\theta$. The differential photon density at energy and angle θ is therefore given by

$$dn = \frac{1}{2} n(\ ) \sin\theta d d\theta. \tag{5.32}$$

Therefore, the active velocity photonic energy transformation is computed as $c(1 - \cos\theta)$, while the emission considered at every unit path of length is expressed by

$$\frac{d\tau_{\text{abs}}}{dx} = \int\int \frac{1}{2} \sigma n(\in)(1 - \cos\theta)\sin\theta d d\theta. \tag{5.33}$$

Re-expression of these active variables as integrals over s rather than of θ, by Eqs. (5.31) and (5.33), is given as

$$\frac{d\tau_{\text{abs}}}{dx} = \pi r_0^2 \left(\frac{m^2c^4}{E}\right)^2 \int_{\frac{m^2c^4}{E}}^{\infty} {}^{-2}\, n(\in)\bar{\phi}\left[s_0(\ )\right] de, \tag{5.34}$$

whereby

$$\bar{\phi}\left[s_0\left(\ \right)\right]=\int_1^{s_0(\in)} s\bar{\sigma}(s)ds,\ \bar{\sigma}(s)=\frac{2\sigma(s)}{\pi r_0^2}. \tag{5.35}$$

This result describes the functional variable of $\bar{\phi}$ as well as nondimensional particles of $\bar{\sigma}$. Here, the variable $\bar{\phi}$ $[S_0]$ is computed in accordance with a comprehensive graphical frame for $1 < S_0 < 10$. I computed $\bar{\phi}$ by a functional asymptotic computation:

$$\bar{\phi}\left[S_0\right]=\frac{1+\beta_0^2}{1-\beta_0^2}\ln\omega_0-\beta_0^2\ln\omega_0-\ln^2\omega_0-\frac{4\beta_0}{1-\beta_0^2}+2\beta_0+4\ln\omega_0\ln\left(\omega_0+1\right)-L\left(\omega_0\right),$$

whereby $S_0 - 1 \ll 1$ or $S_0 \gg 1$,

$$\beta_0^2=\frac{1-1}{S_0},\omega_0=\frac{\left(1+\beta_0\right)}{\left(1-\beta_0\right)},\text{and}\,L\left(\omega_0\right)=\int_1^{\omega_0}\omega^{-1}\ln\left(\omega+1\right)d\omega. \tag{5.36}$$

The final integral could be expressed as

$$\left(\omega+1\right)=\omega\left(\frac{1+1}{\omega}\right),L\left(\omega_0\right)=\frac{1}{2}\ln^2\omega_0+L'\left(\omega_0\right),$$

whereby

$$L'\left(\omega_0\right)=\int_1^{\omega_0}\omega^{-1}\ln\left(1+\frac{1}{\omega}\right)d\omega, \tag{5.37}$$

$$=\frac{\pi^2}{12}-\sum_{n=1}^{\infty}\left(-1\right)n^{-1}n^{-2}\omega_0^{-n}.$$

This corrected form of heat photon proliferation instantly gives room for a precise computation of $\bar{\phi}$ $[S_0]$ to the accuracy that is anticipated for the estimated value of S_0. The below equation describes how the counteractive function of variable is quantified:

$$\bar{\phi}\left[S_0\right]=2S_0\left(\ln4S_0-2\right)+\ln4S_0\left(\ln4S_0-2\right)-$$
$$\frac{\left(\pi^2-9\right)}{3}+S_0^{-1}\left(\ln4S_0+\frac{9}{8}\right)+\ldots\left(S_0\gg1\right); \tag{5.38}$$

$$\bar{\phi}[S_0]=\left(\frac{2}{3}\right)(S_0-1)^{\frac{3}{2}}+\left(\frac{5}{3}\right)(S_0-1)^{\frac{5}{2}}-\left(\frac{1507}{420}\right)(S_0-1)^{\frac{7}{2}}+\ldots(S_0-1\ll 1). \quad (5.39)$$

The function $\frac{\bar{\phi}[S_0]}{(S_0-1)}$ is illustrated in Fig. 5.5 for $1 < S_0 < 10$; at larger S_0, it turns out to be a normal logarithmic function of s_0. The heating photons' power-law spectrum is conveyed into newly decoded photonic structure of $n()^m$ at two parameters in the curtain wall.

Therefore, the tendency of the high-energy cutoff photonic energy $m > 0$ can be developed as the heat-photon structure to release the high heat energy by taking into consideration a spectrum of heat-photon counts as expressed below:

$$n(\ )=D^{\beta},<m, \qquad \beta 0 \quad (5.40)$$

$$=0,>m. \quad (5.41)$$

For this spectrum, it has been further computed as

$$\frac{d\tau_{abs}}{dx}=\pi r_0^2 D\left(\frac{m^2c^4}{E}\right)^{1+\beta}\times\{0,|\ E<E_m,| \quad (5.42)$$

in which

$$\sigma_m=\frac{E}{E_m}=\frac{{}_mE}{m^2c^4}, \quad (5.43)$$

$$F_{\beta}(\sigma_m)=\int_1^{\sigma m} s_0^{\beta-2}\bar{\phi}[s_0]ds_0. \quad (5.44)$$

Again, by Eqs. (5.40) and (5.41), we could attain the asymptotic forms:

$$\begin{aligned}&\beta=0: F_{\beta}(\sigma_m)\to A_{\beta}+\ln^2\sigma_m-4\ln\sigma_m+\cdots,\\&\beta\quad 0: F_{\beta}(\sigma_m)\to A_{\beta}+2\beta^{-1}\sigma_m^{\beta}\left(\ln 4\sigma_m-\beta^{-1}-2\right)+\cdots,\sigma_m>10\end{aligned} \quad (5.45)$$

$$\text{all }\beta: F_{\beta}(\sigma_m)\to\left(\frac{4}{15}\right)(\sigma_m-1)^{\frac{5}{2}}+\left[\frac{2(2\beta+1)}{21}\right](\sigma_m-1)^{\frac{7}{2}}+\cdots,\sigma_m-1\ll 1. \quad (5.46)$$

Figure 5.6 plots $\sigma_m^{-\beta}\ F_{\beta}(\sigma_m)$ for β = 0–3.0 A_{β} in 0.5–A_{β} variables that acts as the most important part in the area [12, 23]. The function is computed as A_{β} = 8.111 (β = 0), 13.53 (β = 0.5), 9.489 (β = 1.0), 15.675 (β = 1.5), 34.54 (β = 2.0), 85.29

(β = 2.5), and 222.9 (β = 3.0). Successively, the terms of heat photon have been computed considering the spectra for both positive and negative variables:

$$n(\ ) = 0, \quad < \epsilon_0 \tag{5.47}$$

$$= C^{-\alpha} \text{ or } D^{\beta}, \quad 0 \ll m \tag{5.48}$$

$$= 0, > m. \tag{5.49}$$

Therefore the following was acquired:

$$\left(\frac{d\tau_{\text{abs}}}{dx}\right)_{\alpha} = \pi r_0^2 C \left(\frac{m^2 c^4}{E}\right)^{1-\alpha}$$

$$\times \begin{cases} 0 & , \quad E < E_m \\ \left[F_\alpha(1) - F_\alpha(\sigma_m)\right], & E_m < E < E_0 \\ \left[F_\alpha(\sigma_0) - F_\alpha(\sigma_m)\right], & E > E_0; \end{cases} \tag{5.50}$$

$$\left(\frac{d\tau_{\text{abs}}}{dx}\right)_{\beta} = \pi r_0^2 D \left(\frac{m^2 c^4}{E}\right)^{1+\beta}$$

$$\times \begin{cases} 0 & , \quad E < E_m \\ [F_\beta(\sigma_m) & , \quad E_m < E < E_0 \\ \left[F_\beta(\sigma_m) - F_\beta(\sigma_0)\right], & E > E_0. \end{cases} \tag{5.51}$$

The heating photon spectrum in this case could be sufficiently described using asymptotic formulas. The term $\Gamma_\gamma^{\text{LPM}}$ represents the photonic emission to the discharged irradiance for every unit volume as a result of the processes of bremsstrahlung [33, 40]:

$$\Gamma_\gamma \equiv \frac{dn_\gamma}{dVdt}. \tag{5.52}$$

Following the summing of the contributions $\Gamma_\gamma^{\text{LPM}}$, the rate was long established as $O(\alpha_{\text{EM}}\, \alpha_s)$. Therefore, it has been expressed below with the form of light emission $\Gamma_\gamma^{\text{LPM}}$ considering the temperature T that reacts under the photo-physical condition μ:

$$\frac{d\Gamma_\gamma^{\text{LPM}}}{d^3k} = \frac{d_{\text{F}} q_{\text{s}}^2 \alpha_{\text{EM}}}{4\pi^2 k} \int_{-\infty}^{\infty} \frac{dp_\parallel}{2\pi} \int \frac{d^2 p_\perp}{(2\pi)^2} A(p_\parallel, k) \Re\{2p_\perp \cdot f(p_\perp; , p_\parallel; , k)\}, \tag{5.53}$$

whereby d_F denotes the active photonic particle strategy [N_c in SU(N_c)], q_s stands for the albanian photon charge, and $k \equiv |k|$ represents the dynamic function $A(p_{\|}, k)$ of the discharged photonic particle, which is expressed as

$$A\left(p_{\|},k\right) \equiv \begin{cases} \dfrac{n_b\left(k+p_{\|}\right)\left[1+n_b\left(p_{\|}\right)\right]}{2p_{\|}\left(p_{\|}+k\right)}, \text{scalars} \\ \dfrac{n_f\left(k+p_{\|}\right)\left[1-n_f\left(p_{\|}\right)\right]}{2\left[p_{\|}\left(p_{\|}+k\right)\right]^2}\left[p_{\|}^2+\left(p_{\|}+k\right)^2\right], \text{fermions} \end{cases} \tag{5.54}$$

and

$$n_b\left(p\right) \equiv \frac{1}{\exp\left[\beta\left(p-\mu\right)\right]-1}, n_f\left(p\right) \equiv \frac{1}{\exp\left[\beta\left(p-\mu\right)\right]+1}. \tag{5.55}$$

The variable $f(p_{\perp}; p_{\|}, k)$ in Eq. (5.51) is presented here to resolve the functional integration below that approves numerous photon deliberation [33, 34, 46]:

$$2p_{\perp} = i\delta Ef\left(p_{\perp};,p_{\|};,k\right)$$
$$+\frac{\pi}{2}C_F g_s^2 m_D^2 \int \frac{d^2q_{\perp}}{\left(2\pi\right)^2}\frac{dq_{\|}}{2\pi}\frac{dq^0}{2\pi}2\pi\delta\left(q^0-q_{\|}\right)x\frac{T}{|q|}\left[\frac{2}{\left|q^2-\Pi_L\left(Q\right)\right|^2}+\frac{\left[1-\left(q^0/|q_{\|}|\right)^2\right]^2}{\left|\left(q^0\right)^2-q^2-\Pi_T\left(Q\right)\right|^2}\right] \tag{5.56}$$
$$[f\left(\left(p_{\perp};,p_{\|};,k\right)-f\left(q+p_{\perp};,p_{\|};,k\right)\right].$$

In Eq. (5.54), C_F represents a photonic particle [$C_F = (N_c^2 - 1)/2N_c = 4/3$ in QCD], m_D represents the Debye mass, and δE denotes the energy variation among photon particles that takes into consideration the photon emission:

$$\delta E \equiv k^0 + E_p sign\left(p_{\|}\right) - E_{p+k} sign\left(p_{\|}+k\right). \tag{5.57}$$

For N_f Dirac fermions and an SU(N) gauge theory with N_s complex scalars, the Debye mass in the ordinal demonstration is expressed as [38]

$$m_D^2 = \frac{1}{6}\left(2N+N_s+N_f\right)g^2T^2 + \frac{N_f}{2\pi^2}g^2\mu^2. \tag{5.58}$$

To precisely define the rate of photon irradiance emission in the electromagnetic area $p_{\|} > 0$, the permutation of $n(k + p_{\|})$ [$1 \pm n(p_{\|})$] has been computed as a fundamental function that contains $A(p_{\|}, k)$ in Eq. (5.51), which controls photon emission, with the use of the equation below:

$$n_b(-p) = -\left[1 + n_b(p)\right], n_f(-p) = \left[1 - n_f(p)\right], \tag{5.59}$$

whereby $n(p) \equiv 1/[e\beta(p + \mu) \mp 1]$ is the suitable antiparticle variable mechanism. As a result, the function $A(p_{||}, k)$ in this function could be expressed as

$$A(p_{||}, k) \equiv \begin{cases} \dfrac{n_b(k - |p_{||}|)\, n_b(|p_{||}|)}{2|p_{||}|(k - |p_{||}|)}, \text{scalars} \\ \dfrac{n_f(k - |p_{||}|)\, n_f(|p_{||}|)}{2\left[|p_{||}|(k - |p_{||}|)\right]^2}\left[p_{||}^2 + (k - |p_{||}|)^2\right], \text{fermions} \end{cases}. \tag{5.60}$$

The energy E_p of a dynamic photon quantum $|p|$ is therefore expressed as

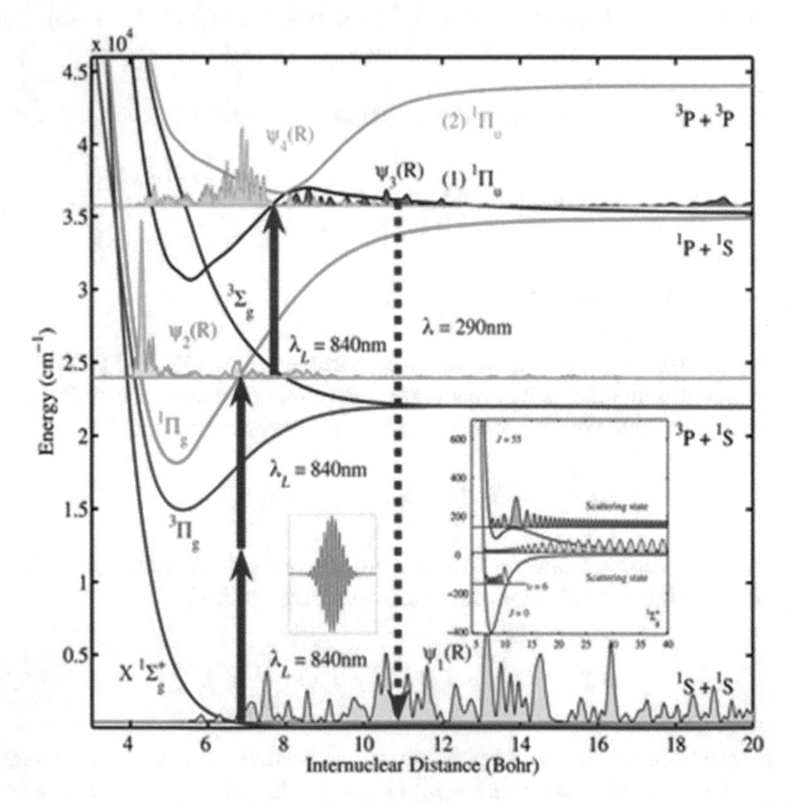

Fig. 5.9 Heating photon energy formation against the internuclear distance (Bohr) at different variable frequencies where the photon particle carries out the energy measurements (DQD) by integrating Higgs boson quantum field

$$E_p = \sqrt{p^2 + m_\infty^2} \simeq |p| + \frac{m_\infty^2}{2|p|} \simeq |p_\parallel| + \frac{p_\perp^2 + m_\infty^2}{2|p_\parallel|}, \tag{5.61}$$

whereby the asymptotic thermo "mass" is

$$m_\infty^2 = \frac{C_f g^2 T^2}{4}. \tag{5.62}$$

Reforming the explicit form of E_p in the description (Eq. 5.60), the following has been obtained:

$$\delta E = \left[\frac{p_\perp^2 + m_\infty^2}{2}\right]\left[\frac{k}{p_\parallel\left(k + p_\parallel\right)}\right]. \tag{5.63}$$

Therefore, overt terms have been derived from Eqs. (5.52) and (5.55) above. Thus, the heating rate of photon emission has confirmed the active energy deliberation through the photonic electromagnetic field of the curtain wall plane (Fig. 5.9).

Conclusions

Conventional cooling and heating technologies implemented in buildings are constituting a primary cause for climate change and distortion of the ozone layer. In order to alleviate these disastrous impacts, this research has proposed an innovative technology to deform the solar energy into cool-state photons through the implementation of Bose-Einstein (*B-E*) discrete photonic mechanics into the helium (*He*)-aided curtain walls to cool the premises naturally. Contemporarily, the cooling-state photon has been proposed to reform into heat-state photons by inducing Higgs boson [BR ($H \rightarrow \gamma\gamma^-$)] electro-quantum using single-diode semiconductor in the curtain wall to conduct a natural mechanism to heat the building. It can be simply summarized that cool-state photons (HcP^-) could be produced by the deformation of the solar energy using functional helium (*He*) application and then it could be reformed into heat-state photons (HtP^-) by the activation of Higgs boson quantum's electrical charge to cool and heat the building naturally, which indeed would be a new field of science to mitigate global energy, environment, and ozone layer vulnerability dramatically.

Acknowledgements This research has been conducted under the support of Green Globe Technology through the grant of RD-02019-06. The author declares that the chapter has no conflict of interest to publish this research in an appropriate scientific journal.

References

1. A.K. Agger, A.H. Sørensen, Atomic and molecular structure and dynamics, Phys. Rev. A 55 (1997) 402–413.
2. Alexandra Dobrynina, Alexander Kartavtsev, Georg Raffelt. "Photon-photon dispersion of TeV gamma rays and its role for photon-ALP conversion", Phys. Rev. D, 2015.
3. T. Pregnolato, E.H. Lee, J.D. Song, S. Stobbe, P. Lodahl, Single-photon non-linear optics with a quantum dot in a waveguide, Nat. Commun. 6 (2015) 8655.
4. A.N. Celik, N. Acikgoz, Modelling and experimental verification of the operating current of mono-crystalline photovoltaic modules using four- and five-parameter models, Appl. Energy 84 (2007) 1–15.
5. A. Reinhard, T. Volz, M. Winger, A. Badolato, K.J. Hennessy, E.L. Hu, A. Imamoğlu, Strongly correlated photons on a chip, Nat. Photonics 6 (2012) 93–96.
6. B. Najjari, A.B. Voitkiv, A. Artemyev, A. Surzhykov, Simultaneous electron capture and bound-free pair production in relativistic collisions of heavy nuclei with atoms, Phys. Rev. A 80 (2009) 012701.
7. G. Baur, K. Hencken, D. Trautmann, S. Sadovsky, Y. Kharlov, Dense laser-driven electron sheets as relativistic mirrors for coherent production of brilliant X-ray and γ-ray beams, Phys. Rep. 364 (2002) 359–450.
8. J. Eichler, T. Stöhlker, Radiative electron capture in relativistic ion-atom collisions and the photoelectric effect in hydrogen-like high-Z systems, Phys. Rep. 439 (2007) 1–99.
9. J. Fernández. "Electron and ion angular distributions in resonant dissociative photoionization using linearly polarized light", N J. Phys., 04/15/2009.
10. J.S. Douglas, H. Habibian, C.-L. Hung, A.V. Gorshkov, H.J. Kimble, D.E. Chang, Quantum many-body models with cold atoms coupled to photonic crystals, Nat. Photonics 9 (2015) 326–331.
11. K. Hencken, G. Baur, D. Trautmann, Transverse momentum distribution of vector mesons produced in ultraperipheral relativistic heavy ion collisions, Phy. Rev. Lett. 96 (2006) 012303.
12. L. Hong Idris Lim, Zhen Ye, Jiaying Ye, Dazhi Yang, Hui Du. "A Linear Identification of Diode Models from single I–V characteristics of PV panels", IEEE Transactions on Industrial Electronics, 2015.
13. L. Langer, S.V. Poltavtsev, I.A. Yugova, M. Salewski, D.R. Yakovlev, G. Karczewski, T. Wojtowicz, I.A. Akimov, M. Bayer, Access to long-term optical memories using photon echoes retrieved from semiconductor spins. Nat. Photonics 8 (2014) 851–857.
14. W.B. Yan, H. Fan, Single-photon quantum router with multiple output ports, Sci. Rep. 4 (2014) 4820.
15. L. Yang, S. Wang, Q. Zeng, Z. Zhang, T. Pei, Y. Li, L.M. Peng, Efficient photovoltage multiplication in carbon nanotubes, Nat. Photonics 5 (2011) 672–676.
16. M. F. Hossain, Solar energy integration into advanced building design for meeting energy demand and environment problem, Int. J. Energy Res. 17 (2016) 49–55.
17. W. De Soto, S.A. Klein, W.A. Beckman, Improvement and validation of a model for photovoltaic array performance, Sol. Energy 80 (2006) 78–88.
18. B. Igor, Jan M. Pawlowski, and Sebastian Diehl. "Ultracold atoms and the Functional Renormalization Group", Nuclear Physics B – Proceedings Supplements, 2012.
19. F. Martin. "Single photon-induced symmetry breaking of H2 dissociation", Science, 02/02/2007.
20. G. Baur, K. Hencken, D. Trautmann, Revisiting unitarity corrections for electromagnetic processes in collisions of relativistic nuclei, Phys. Rep. 453 (2007) 1–27.
21. M. F. Hossain (2017). Design and construction of ultra-relativistic collision PV panel and its application into building sector to mitigate total energy demand. Journal of Building Engineering. 9, 147-154.
22. M. F. Hossain (2018). Photonic thermal control to naturally cool and heat the building. Applied Thermal Engineering. 131, 576–586.

23. M.S. Tame, K.R. McEnery, Ş.K. Özdemir, J. Lee, S.A. Maier, M. S. Kim, Quantum plasmonics, Nat. Phys. (2013).
24. M. F. Hossain, "Bose-Einstein (B-E) Photon Energy Reformation for Cooling and Heating the Premises Naturally", Applied Thermal Engineering, 2018.
25. S.R. Valluri, U. Becker, N. Grün, W. Scheid. Relativistic Collisions of Highly-Charged Ions, J. Phys. B: At. Mol. Phys. 17 (1984) 4359–4370.
26. M. F. Hossain. "Transforming dark photons into sustainable energy", International Journal of Energy and Environmental Engineering, 2018.
27. M. F. Hossain. "Photon application in the design of sustainable buildings to console global energy and environment", Applied Thermal Engineering, 2018.
28. M.W.Y. Tu, W.M. Zhang, Non-Markovian decoherence theory for a double-dot charge qubit, Phys. Rev. B 78 (2008) 235311.
29. M. F. Hossain (2017). Green science: independent building technology to mitigate energy, environment, and climate change. Renewable and Sustainable Energy Reviews. 73; 695-705.
30. M. F. Hossain (2017). Green science: advanced building design technology to mitigate energy and environment. Renewable and Sustainable Energy Reviews. 81; 3051-3060.
31. P. Arnold, G.D. Moore, L.G. Yaffe, Photon Emission from Ultrarelativistic Plasmas, J. High Energy Phys. 11 (2001) 057.
32. R.J. Gould, GP Schréder, Pair Production in Photon-Photon Collisions, Phys. Rev. 155 (1967) 1404–1407.
33. Y.T. Tan, D.S. Kirschen, N. Jenkins, A model of PV generation suitable for stability analysis, IEEE Trans. Energy Convers. 19 (2004) 748–755.
34. Y. Zhu, X. Hu, H. Yang, Q. Gong, On-chip plasmon-induced transparency based on plasmonic coupled nanocavities, Sci. Rep. 4 (2014).
35. S.A. Klein, Calculation of flat-plate collector loss coefficients, Sol. Energy 17 (1975) 79–80.
36. V. Cardoso. "Quasinormal modes of Schwarzschild black holes in four and higher dimensions", Physical Review D, 02/2004.
37. W.M. Zhang, P.Y. Lo, H.N. Xiong, M.W.Y. Tu, F. Nori, General Non-Markovian Dynamics of Open Quantum Systems. Phys. Rev. Lett. 109 (2012) 170402.
38. S. Robert, and Andrzej Czarnecki (2016). "High-energy electrons from the muon decay in orbit: Radiative corrections", Physics Letters B.
39. U. Becker, N. Grün, W. Scheid, K-shell ionisation in relativistic heavy-ion collisions, J. Phys. B: At. Mol. Phys. 20 (1987) 2075.
40. Y.F. Xiao, M. Li, Y.C. Liu, Y. Li, X. Sun, Q. Gong, Asymmetric Fano resonance analysis in indirectly coupled microresonators, Phys. Rev. A 82 (2010) 065804.
41. Yi Guo, Ali Al-Jubainawi, Zhenjun Ma. "Performance investigation and optimisation of electrodialysis regeneration for LiCl liquid desiccant cooling systems", Applied Thermal Engineering, 2018.

Chapter 6
Smart Building Technology

Abstract The burning of fossil fuels for powering the building sector around the world is putting our Mother Earth in a vulnerable condition; fossil fuels are running out due to the present level of consumption of this reverse fuel by the building sector. Therefore, the use of solar energy to generate electricity energy by building's exterior curtain wall skin shall be an interesting technology to power the building naturally. On the other hand, traditional water supply into the building sector is lowering the groundwater strata which is dangerous for the survival of future generation due to shortage of groundwater. Hence, the cloud water molecule has been proposed to catch using static electricity force forming insulator water tank of a building and treat it in situ by implementing UV application to fulfill the day-to-day water need for a building. Simply, the combination of these two technologies to capture sunlight and cloud water by designing a smart building technology to satisfy the net energy and water need for a building will certainly be an innovative technology to mitigate the global energy demand and environmental and climate vulnerability.

Keywords Photon energy · Building skin solar panel · Clean energy · Urban cloud · Electrostatic force · UV technology · Potable water · Smart building technology

Introduction

The mass development of building construction throughout the world is seriously impacting the environment since it consumes conventional energy to meet its daily energy demand. Naturally, the energy requirement in the building sector in the whole world will be increased accordingly and it is expected to be doubled up by the year 2050 [1, 2]. The global fossil fuel energy consumption was 283 EJ/year in the year 1980, 347 EJ/year in 1990, 400 EJ/year in 2000, and 511 EJ/year in 2010 and it would be 607 EJ/year in 2020, 702 EJ/year in 2030, 855 EJ/year in 2040, and 988 EJ/year in 2050 where the building sector alone consumes 40% of this total

M. F. Hossain, *Global Sustainability in Energy, Building, Infrastructure, Transportation, and Water Technology*,
https://doi.org/10.1007/978-3-030-62376-0_6

conventional energy. Simply the utilization of fossil fuel by the building sector in the year 2018 was 2.236×10^{20} EJ which was responsible for creating 8.01×10^{11} tons of CO_2 into the atmosphere and accordingly accounted for 40% of climate change [3–5]. Consequently, adverse environmental impacts such as acid precipitation, stratospheric ozone depletion, and massive diurnal temperature fluctuation are occurring unpredictably [6, 7]. Recent study shows that the concentration of atmospheric CO_2 amount is 400 ppm which needs to be lowered to a standard-level grade of 300 ppm CO_2 for clean breathing and healthy respiratory system for all mammals [8, 9]. Unfortunately, the consumption of conventional energy by the building sector is still accelerating rapidly throughout the world; the situation shall remain unchanged until an advanced technology is developed to utilize sustainable energy. Simply it is an urgent demand to develop sustainable building technology to mitigate fossil fuel consumption where the "building sector that fulfills the needs of the current without compromising the ability of future demand of energy to fulfill the complete needs." Hence, the solar radiation capture by a smart building design can be an interesting idea to fulfill the net energy requirement for a building. Recent study revealed that the average energy density of solar energy on the surface of the earth is 1366 W/m^2 where the diameter of the earth is one over 10,000,000 of meridian, respectively, at the North Pole to the equator and the radius of the earth is $(2/\pi) \times 10^7$ m, respectively [10, 11]. Therefore, the net energy of solar radiation reaching earth is $1366 \times (4/\pi) \times 10^{14} \cong 1.73 \times 10^{17}$ W [10, 12]. In detail it can be explained that a day has 86,400 s, and a year has 365.2422 days in average; thus, the net solar energy reaching the earth annually is $1.73 \times 10^{17} \times 86{,}400 \times 365.2422 \cong 5.46 \times 10^{24}$ J which is equal to 5,460,000 EJ energy annually [10, 13]. If a mere 0.004% of the annual solar energy reaching the earth is used by designing a smart building technology, it will satisfy the net energy need for the building sector globally which is clean. In this research, therefore, a zero emission of greenhouse gas (GHG) formation technology by a building has been proposed to level its net energy need by using building's exterior curtain wall skin to capture solar energy and convert it into electricity energy. Simultaneously, water supply into the buildings is a vital part of our daily lives in both urban and rural areas. The tradition water supply from groundwater and reservoir and filtration systems are indeed technologically correct but not an interesting technology since groundwater level has been getting lower during the past several decades which threatens the survival of all living beings on earth in the near future [14–16]. In this concept, therefore, a natural water capture mechanism has been proposed to catch urban cloud by applying static electricity force technology to satisfy the daily water demand by designing a smart building technology. Simply electrostatic force-generating plastic tank is proposed to construct at all building roofs to catch the cloud by its attraction force. In more detail it can be explained that H_2O molecule consists of positive and negative charges and its electrons are ended up with electrostatic force which is a positive charge; while the H_2O electrostatic force bears the positive charge (+ve) and H_2O molecules contain the negative charge (−ve) on one side, the +ve charge and −ve negative charge pull each other resulting in the +ve charge tugging the water to come down to accumulate into the tank. This cloud water can be treated in situ to satisfy the daily water needs for buildings and houses. Simply application of these two technologies to design a smart building technology would indeed be an interesting idea to mitigate net energy and water crisis for the building sector globally.

Material, Methods, and Simulation

Since the photoreaction is correlated to the photon charge, the attributes of photon electromagnetic wave and the dynamic of the photon particle are induced into the building curtain wall skin [12, 17]. Then the photon energy density frequency into the curtain wall surface is computed considering the quantum flow of photon radiation [18, 19]. Consequently, a computational model of photon radiation is quantified to demonstrate the photon energy generation from sunlight considering radiation intake dynamics by the active PV panel of building exterior curtain wall (Fig. 6.1). Thereafter, the mode of the solar quanta absorbance by curtain wall skin is determined by the peak solar radiation output tracking in the acting PV panel [20–22]. The induced solar irradiance is, thereafter, collected by the acting photovoltaic array of the curtain wall skin considering the parameters of solar energy proliferation, transformation rate, and PVVI curves. The accurate calculation of the current–voltage (*I*–*V*) characteristic is subsequently done by the mode of radiator intake into a single-diode circuit connected to the curtain wall PV cell in order to confirm the active solar volt (I_{v+}) generation from the curtain wall PV panel (Fig. 6.1) [23–25].

Then, the determination of the PV current formation via I_{pv} considering single diode has been carried out (Fig. 6.2a), by calculating *I*–*V*–*R* relationship within the PV panel vicinity in order to transform DC to AC for the usage of this energy domestically (Fig. 6.2b).

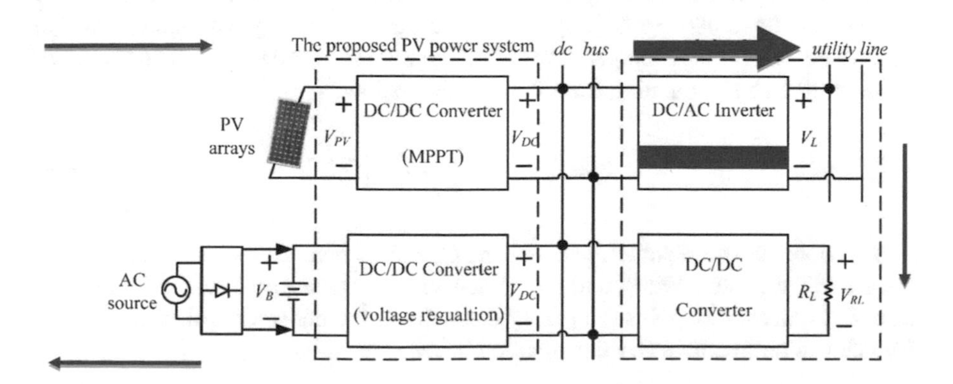

Fig. 6.1 The model diagram of the acting PV panel energy deliberation mechanism depicted that the proposed PV power system module flow mechanism is conducted via maximum power point track (MPPT) and DC/DC converter to get finally usable AC power from the PV panel

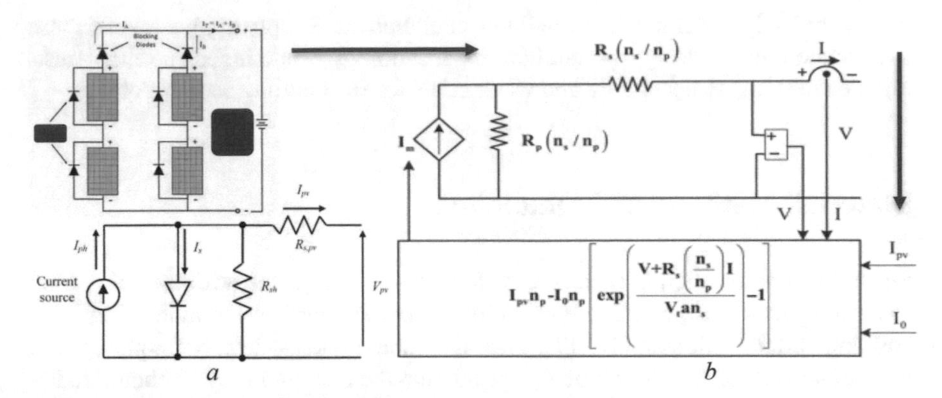

Fig. 6.2 The one-diode circuit diagram of a photovoltaic (PV) cell depicted the (**a**) photo-physical current generation by the active PV panel, and (**b**) shows the mode of *I–V–R* relationship in order to conduct transforming process of DC into AC for the use of electricity locally for the building

Hence, the following equation is calculated as the energy deliberation from a photovoltaic (PV) panel, which origin is the photon irradiance and ambient temperature of the acting PV panel:

$$P_{\mathrm{pv}} = \eta_{\mathrm{pvg}} A_{\mathrm{pvg}} G_{\mathrm{t}} \tag{6.1}$$

Here, η_{pvg} denotes the PV performance rate, A_{pvg} denotes the PV generator array (m^2), and G_{t} denotes the photon irradiance intake rate on the plane (W/m^2) of acting PV panel and thus η_{pvg} could be rewritten as follows:

$$\eta_{\mathrm{pvg}} = \eta_{\mathrm{r}} \eta_{\mathrm{pc}} \left[1 - \beta \left(T_{\mathrm{c}} - T_{\mathrm{cref}}\right)\right] \tag{6.2}$$

η_{pc} denotes the energy formation efficiency, when maximum power point tracking (MPPT) is implemented which is close to 1. Here β denotes the temperature cofactor (0.004–0.006 per °C); η_{r} denotes the mode of energy efficiency; and T_{cref} denotes the condition of temperature at °C. The reference cell temperature (T_{cref}) can be determined by calculating from the equation below:

$$T_{\mathrm{c}} = T_{\mathrm{a}} + \left(\frac{\mathrm{NOCT} - 20}{800}\right) G_{\mathrm{t}} \tag{6.3}$$

T_{a} denotes the ambient temperature in °C, and G_{t} denotes the solar radiation in the active PV panel (W/m^2), and thus it denotes the modest optimum cell temperature in °C. Considering this temperature condition, the net solar radiation in acting PV cell can be calculated by the equation below:

$$I_{\mathrm{t}} = I_{\mathrm{b}} R_{\mathrm{b}} + I_{\mathrm{d}} R_{\mathrm{d}} + \left(I_{\mathrm{b}} + I_{\mathrm{d}}\right) R_{\mathrm{r}} \tag{6.4}$$

The solar panel cells, here, necessarily work as a P-N junction superconductor in order to form electricity through the acting PV panels, which are interlinked in a parallel series connection [26–28]. Thus, a unique one-diode circuit, as shown in Fig. 6.2, with respect to the N_s series-connected arrays and N_p parallel-connected arrays has been determined by the following cell current and volt relationship:

$$I = N_p \left[I_{ph} - I_{rs} \left[\exp\left(\frac{q(V + IR_s)}{AKTN_s} - 1 \right) \right] \right] \tag{6.5}$$

where

$$I_{rs} = I_{rr} \left(\frac{T}{T_r} \right)^3 \exp\left[\frac{E_G}{AK} \left(\frac{1}{T_r} - \frac{1}{T} \right) \right] \tag{6.6}$$

Hence, in Eqs. (6.5 and 6.6), q denotes the electron charge (1.6×10^{-19} C), K denotes the Boltzmann's constant, A denotes the diode standardized efficiency, and T denotes the cell temperature (K). Accordingly, IR_s denotes the PV cell reverse current motion at T, where T_r denotes the cell condition temperature, I_{rr} denotes the reverse current at T_r, and E_G denotes the photonic bandgap energy of the superconductor utilized for the PV cell. Thus, the photonic current I_{ph} will vary in accordance with the PV cell's temperature and radiation condition which can be expressed by

$$I_{ph} = \left[I_{SCR} + k_i (T - T_r) \frac{S}{100} \right] \tag{6.7}$$

Here, I_{SCR} denotes the cell's short-circuit current motion considering the optimum temperature of the cell and solar radiation dynamic on the plane of the active PV panel, k_i denotes the short-circuited current motion, and S denotes the solar radiation calculation in a specific area (mW/cm^2). Subsequently, the I–V features of the active PV cell shall be deformed from the single-diode mode of the cell which can be expressed by the following equation:

$$I = I_{ph} - I_D \tag{6.8}$$

$$I = I_{ph} - I_0 \left[\exp\left(\frac{q(V + R_s I)}{AKT} - 1 \right) \right] \tag{6.9}$$

I_{ph} denotes the photonic current dynamic (A), I_D denotes the diode-originated current dynamic (A), I_0 denotes the inverse current dynamic (A), A denotes the diode-induced constant, q denotes the charge of the electron (1.6×10^{-19} C), K denotes the Boltzmann's constant, T denotes the cell temperature (°C), R_s denotes the series resistance (ohm), R_{sh} denotes the shunt resistance (Ohm), I denotes the cell current motion (A), and V denotes the cell voltage motion (V). Therefore, the net current flow into the acting PV cell can be determined by conducting the following equation:

$$I = I_{pv} - I_{\mathrm{D1}} - \left(\frac{V + IR_{\mathrm{s}}}{R_{\mathrm{sh}}}\right) \tag{6.10}$$

where

$$I_{\mathrm{D1}} = I_{01}\left[\exp\left(\frac{V + IR_{\mathrm{s}}}{a_1 V_{T1}}\right) - 1\right] \tag{6.11}$$

Here, I and I_{01} denote the reverse current flow into the diode, respectively, and V_{T1} and V_{T2} denote the optimum thermal voltages into the diode. Thus, the diode standard factor is presented by a_1 and a_2 and then the mode of photovoltaic (PV) panel has been normalized as expressed in the following equation:

$$v_{\mathrm{oc}} = \frac{V_{\mathrm{oc}}}{cKT / q} \tag{6.12}$$

$$P_{\max} = \frac{\dfrac{V_{\mathrm{oc}}}{cKT/q} - \ln\left(\dfrac{V_{\mathrm{oc}}}{cKT/q} + 0.72\right)}{\left(1 + \dfrac{V_{\mathrm{oc}}}{KT/q}\right)}\left(1 - \frac{V_{\mathrm{oc}}}{\dfrac{V_{\mathrm{oc}}}{I_{\mathrm{SC}}}}\right)\left(\frac{V_{\mathrm{oc0}}}{1 + \beta \ln \dfrac{G_0}{G}}\right)\left(\frac{T_o}{T}\right)^{\gamma} I_{\mathrm{sc0}}\left(\frac{G}{G_{\mathrm{o}}}\right)^{\alpha} \tag{6.13}$$

where ν_{oc} denotes the standard point of the open-circuit voltage, V_{oc} denotes the thermal voltage $V_{\mathrm{t}} = nkT/q$, c denotes the constant current motion, K denotes the Boltzmann's constant, T denotes the temperature in the PV cell in Kelvin, α denotes the function which represents the nonlinear motion of photocurrents, q denotes the electron charge, γ denotes the function acting for all nonlinear temperature-voltage currents, while β denotes the photovoltaic (PV) mode for specific dimensionless function for enhancing current flowing rate. Subsequently, Eq. (6.13) represents the peak energy generation from a single photovoltaic (PV) module which is interlinked in both series and parallel connection. Thus, the equation for the net energy formation in the array of N_{s} cells has been interlinked in series and N_{p} cell has been interlinked in parallel considering the power P_{M} of each mode of connection and which is finally expressed by using the following equation:

$$P_{\mathrm{array}} = N_{\mathrm{s}} N_{\mathrm{p}} P_{\mathrm{M}} \tag{6.14}$$

Design of PV Panel

To determine the highest transformation rate of solar power, structurally sound and durable photovoltaic (PV) panel consisting of nanocrystalline materials and films is utilized which is in fact the exterior curtain wall skin of the building [29–31]. Necessarily, both uniform and concentrated wind load-bearing capacity resistances are acquainted to install a technically sound photovoltaic (PV) panel to confirm that it is sufficiently able to ensure itself and work normally under tempest condition. In order to confirm that the tornado-resistant PV panel is installed, the structural calculation has been conducted considering the accurate the wind load protection capacity (F6 tornado level; 379 mile/h) as expressed in the following equation:

$$p_{\mathrm{w}} = 0.5 \rho C_{\mathrm{p}} v_{\mathrm{r}}^{2} \quad (6.15)$$

Here p_{w} denotes the wind load in pascal (Pa), ρ denotes the air density (kg/m^3), C_{p} denotes the wind force cofactor, and v_{r}^{2} denotes the wind speed (m/s) at building height.

Since the optimum air density is 1.2 kg/m^3, the wind force cofactor is 1.00, and stagnation force of the wind is half (½) of the density of the air, the final equation of $P_{\mathrm{w}} = 0.5 \times 1.2$ kg/m^3 $\times$ 379^2 m/s is 86,185 Pa and the total wind force can be expressed as F = area × drag coefficient × stagnation pressure by the following equation:

$$F = 1\,\mathrm{m}^{2} \times 1.0 \times 86{,}185 = 86{,}185\,\mathrm{N}\left(8788\,\mathrm{kgf}\right) = 19{,}375\,\mathrm{ibf} \quad (6.16)$$

Once the tornado type 6 level wind load resistance capability of acting PV panel limit has been finalized, then the PV is acquainted with sophistication by characterizing various directional angles considering the Cartesian coordinate system, where x denotes the skyline convention, y denotes the east-west, and z denotes the zenith in order to trap the maximum solar irradiance during the day the entire year (Fig. 6.3). The position of the celestial body in this framework is thus chosen by h which denotes height and A denotes the azimuth angle while the central framework is utilized as the convention factor which is z hub. It focuses toward the North Pole, and the y hub indistinguishably focuses on the horizon of the skylight, and x pivot opposite to both of North Pole and horizon. Therefore, the angles and coordinate frequencies are encountered mathematically by calculating the latitude and longitude in order to implement correct angles to trap the solar irradiance most efficiently. Here, the zero point of latitude is considered the primary meridian which controls the function of meridian of Eastern Hemisphere and Western Hemisphere angle of the active PV panel and so the north of the equator is the Northern Hemisphere and south of the equator is the Southern Hemisphere that are also controlled by this PV panel to trap solar energy more efficiently. Finally, the δ and ω point hours are clarified accurately considering this analytical Cartesian coordinates in order to decide the position of solar irradiance vector in order to clarify the light

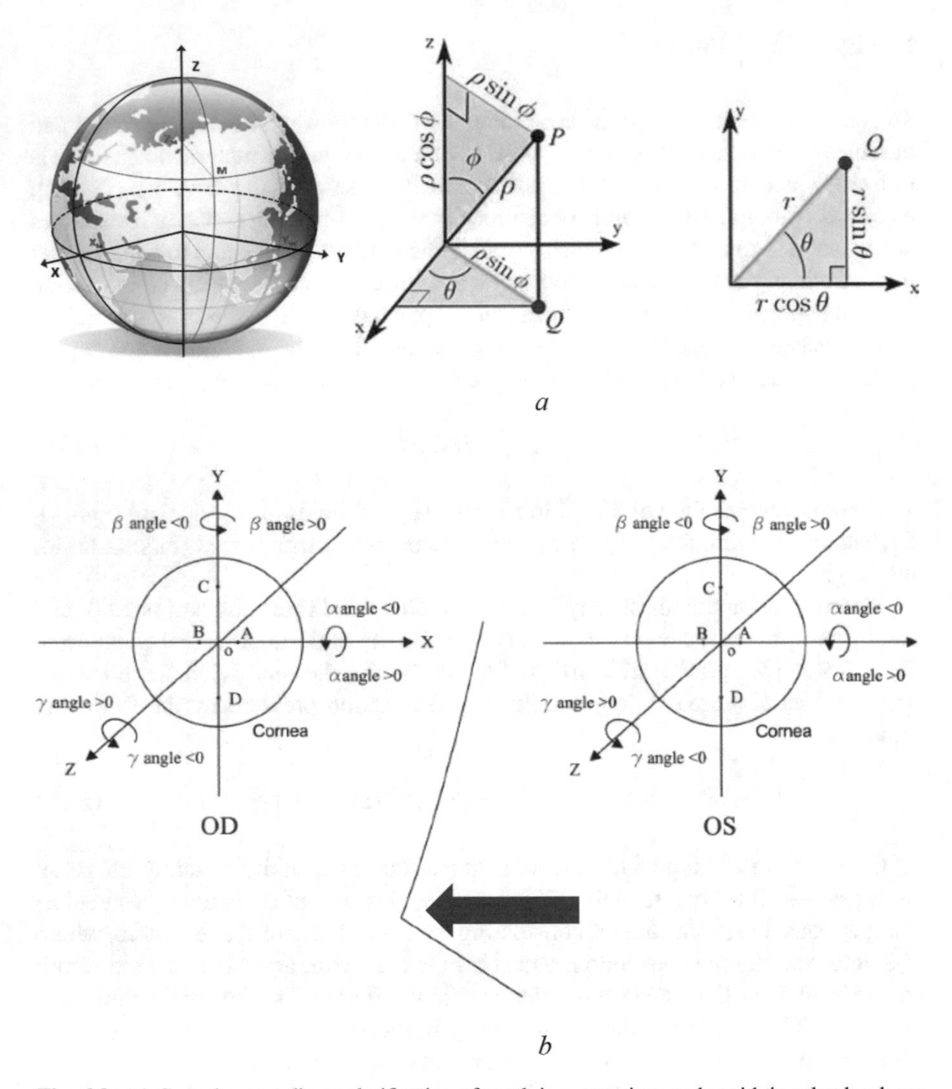

Fig. 6.3 (**a**) Cartesian coordinate clarification of south is *x*, west is *y*, and zenith is *z* that has been performed in order to implement this system for designing the active PV panel using curtain wall skin. The location of this celestial body is analyzed by determining two angles of sinθ and cosθ. (**b**) The longitudinal and latitudinal equatorial angles have been clarified where the convention *z*-axis point denotes the North Pole and the east-west axis *y*-axis denotes the identical angle of the horizon

flow dynamic into the active PV panel to meet 100% energy need for a building [32–34]. Simply, the building exterior wall is then designed smartly, where 25% of outside curtain wall skin is developed considering Cartesian coordinates in order to trap much solar irradiance by this active PV panel [13, 35, 36].

Once the angle of the solar panel is determined as per the above description, the panel wiring is then implemented to design an electric circuit in order to confirm that the peak solar power can be transported through the photoelectric system

[37–39]. Simply, the sunlight is clarified as the motion of the photon flux considering the first function of the fundamental solar thermal energy and antireflective coating of solar cells is then considered as the second function of the solar cells' photoelectro-physics [40–42]. The combination of the two functions is conducted by solar quantum dynamics which is considered the most acceptable quantum technology in modern physics to trap the solar electricity [9, 43, 44]. This is because this active PV panel can emit solar irradiance accurately at a given temperature of approximately 700 °C where the energy density of the solar radiation can be calculated by using the maximum solar energy formation from a single photon excitation at a rate of 1.4 eV with an energy value of 27.77 MW/m^2 eV (Fig. 6.4).

Electrostatic Force Generation

To trap the cloud from air, a model is implemented to form *Hossain static electric force (HSEF* = $\mathfrak{h}$) by installing friction-oriented insulator tank to tug down the cloud water into the tank on a building roof [45, 46]. To create *HSEF* into the insulator tank, albanian local symmetries have been calculation by using MATLAB software with respect to the gauge field symmetrical scalar quantity at the mode functional vector [47, 48]. Therefore, the photon particle will functionally act as the dynamic particle T^a at the symmetrical array by initiating the gauge field of $A_\mu^a(x)$ and then the *HSEF* will subsequently be started to activate a local U(1)-phase symmetry to deliver static electricity force [48, 50]. Thus, the model will be considered as a complex vector field of $\Phi(x)$ where static electric charge q will couple with the EM field of $A^\mu(x)$ and the equation can be expressed by $\mathfrak{h}$:

$$\mathfrak{h} = -\frac{1}{4} F_{\mu\nu}F^{\mu\nu} + D_\mu\Phi^* D^\mu\Phi - V(\Phi^*\Phi) \tag{6.17}$$

where

$$D_\mu\Phi(x) = \partial_\mu\Phi(x) + iqA_\mu(x)\Phi(x)$$
$$D_\mu\Phi^*(x) = \partial_\mu\Phi^*(x) - iqA_\mu(x)\Phi^*(x) \tag{6.18}$$

and

$$V(\Phi^*\Phi) = \frac{\lambda}{2}(\Phi^*\Phi)^2 + m^2(\Phi^*\Phi) \tag{6.19}$$

Here >0 $m^2 < 0$; therefore $\Phi = 0$ is a local optimum vector quantity, while the minimum form of degenerated scalar circle is clarified as $\Phi = \frac{v}{\sqrt{2}} * e^{i\theta}$:

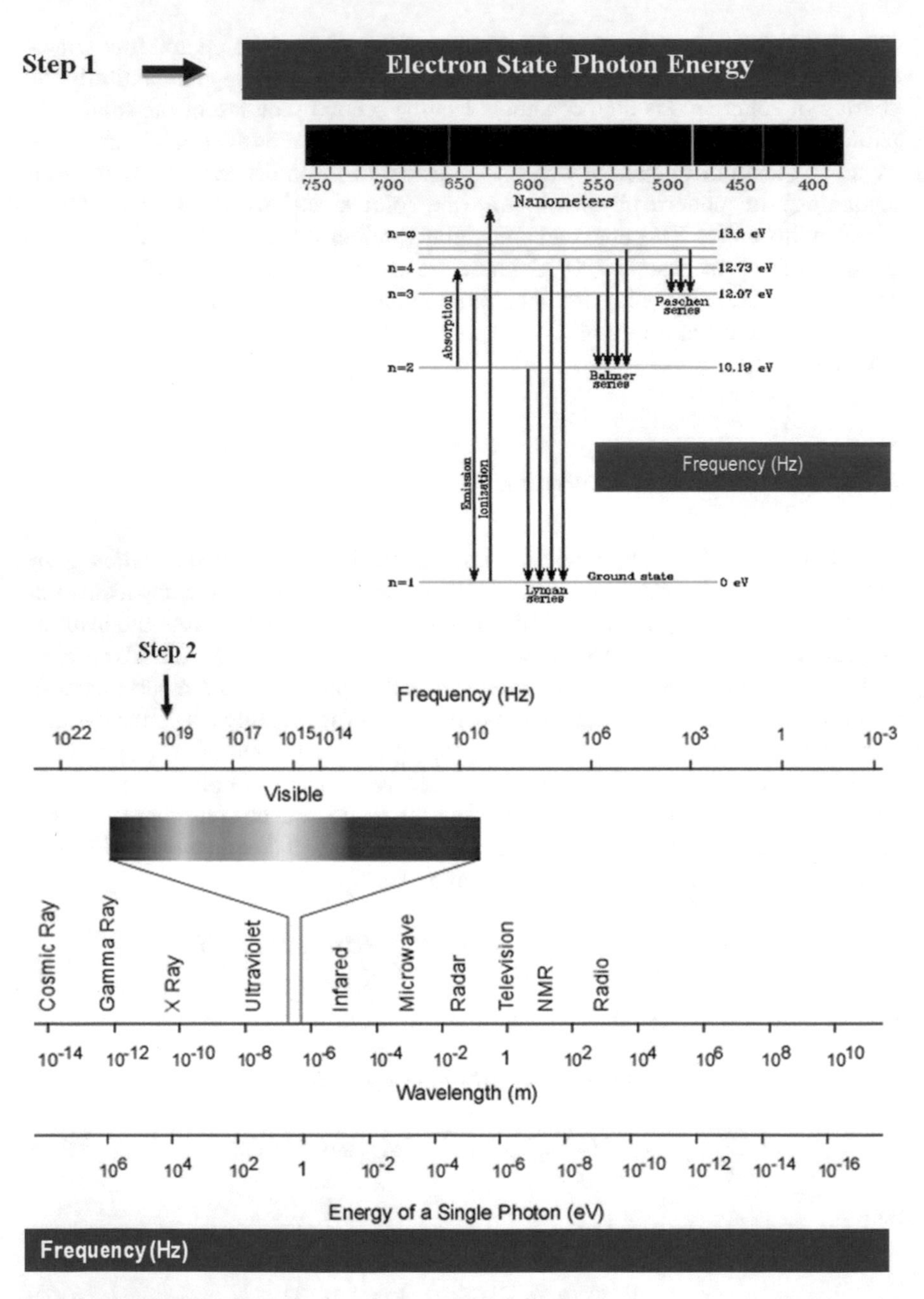

Fig. 6.4 The photonic electron-state energy (step 1) depicting the absorption and emission modes, and energy deliberation rate (step 2) revealing the energy density (1.4 eV at an energy of 27.77 MW/ m^2 eV) by clarifying the optimum solar irradiance deliberation from a photon particle at various wavelengths and frequencies of 10^{17} to 10^{14} Hz

$$v = \sqrt{\frac{-2m^2}{\lambda}}, \text{any real } \theta \tag{6.20}$$

Subsequently, the vector field Φ will form a nonzero functional value $\langle\Phi\rangle \neq 0$, which will simultaneously create the U(1) symmetrical electrostatic field. Therefore, the local U(1) symmetrical phase of $\Phi(x)$ will not only just deliver the expected value $\langle\Phi\rangle$ but also confirm the x-dependent state of the symmetrical $\Phi(x)$ array to confirm an electrostatic field. To determine this electrostatic force formation, the polar coordinates of insulator tank have been clarified by implementing the vector quantity and are expressed by the following equation:

$$\Phi(x) = \frac{1}{\sqrt{2}} \Phi_r(x) * e^{i\Theta(x)}, \text{real } \Phi_r(x) > 0, \text{real } \Phi(x) \tag{6.21}$$

This field of static electricity force mechanism is induced when the vector $\Phi(x) = 0$, and thus the mechanism is used for the identification of $\langle\Phi\rangle \neq 0$ as the peak level of static force generation where it can be expected that the calculative maximum value of this force is $\Phi\langle x\rangle \neq 0$ which is everywhere in the insulator tank. In terms of the realistic static electricity fields $\phi_r(x)$ and $\Theta(x)$, its vector quantitative force on the radial field ϕ_r on insulator tank has been calculated and is expressed by the following equation:

$$V(\phi) = \frac{\lambda}{8}\left(\phi_r^2 - v^2\right)^2 + \text{const}, \tag{6.22}$$

or the expectant radial static electricity force is shifted by its VEV, $\Phi_r(x) = v + \sigma(x)$:

$$\phi_r^2 - v^2 = (v + \sigma)^2 - v^2 = 2v\sigma + \sigma^2 \tag{6.23}$$

$$V = \frac{\lambda}{8}\left(2v\sigma - \sigma^2\right)^2 = \frac{\lambda v^2}{2} * \sigma^2 + \frac{\lambda v}{2} * \sigma^3 + \frac{\lambda}{8} * \sigma^4 \tag{6.24}$$

Simultaneously, the functional derivative $D_\mu\phi$ will become

$$D_\mu\phi = \frac{1}{\sqrt{2}}\left(\partial_\mu\left(\phi_r e^{i\Theta}\right) + iqA_\mu * \phi_r e^{i\Theta}\right) = \frac{e^{i\Theta}}{\sqrt{2}}\left(\partial_\mu\phi_r + \phi_r * i\partial_\mu\Theta + \phi_r * iqA_\mu\right) \tag{6.25}$$

$$\begin{aligned}|D_\mu \phi|^2 &= \frac{1}{2}|\partial_\mu \phi_r + \phi_r * i\partial_\mu \Theta + \phi_r * iqA_\mu|^2 \\ &= \frac{1}{2}(\partial_\mu \phi_r) + \frac{\phi_r^2}{2} * (\partial_\mu \Theta q A_\mu)^2 \\ &= \frac{1}{2}(\partial_\mu \sigma)^2 + \frac{(v+\sigma)^2}{2} * (\partial_\mu \Theta + qA_\mu)^2\end{aligned} \tag{6.26}$$

Altogether,

$$\mathfrak{h} = \frac{1}{2}(\partial_\mu \sigma)^2 - v(\sigma) - \frac{1}{4} F_{\mu\nu}F^{\mu\nu} + \frac{(v+\sigma)^2}{2} * (\partial_\mu \Theta + qA_\mu)^2 \tag{6.27}$$

To determine the formation of this *HSEF* electrostatic force referred to as ($\mathfrak{h}_{\mathrm{sef}}$) in the insulator tank, the function of the electrostatic fields has been quantified by conducting the quadratic calculation and is described by the following equation:

$$\mathfrak{h}_{sef} = \frac{1}{2}(\partial_\mu \sigma)^2 - \frac{\lambda v^2}{2} * \sigma^2 - \frac{1}{4} F_{\mu\nu}F^{\mu\nu} + \frac{v^2}{2} * (qA_\mu + \partial_\mu \Theta)^2 \tag{6.28}$$

Here this *HSEF* ($\mathfrak{h}_{\mathrm{free}}$) function certainly will admit a realistic vector particle of positive mass$^2 = \lambda v^2$ integrating the areal $A_\mu(x)$ function and the electric fields $\Theta(x)$ to confirm to form tremendous amount of electrostatic force within the electric field of the insulator tank (Fig. 6.5).

Results and Discussion

To calculate the photon capture in the quantum field of the building curtain wall skin, the motion of photon generation flow has been determined by integrating Eqs. (6.24) and (6.25). Necessarily, the functional unit area $J(\omega)$, the excited quantum field, and the unit area $J(\omega)$ are calculated considering the constant weak coupling point, and the Weisskopf-Wigner approximation mechanism and Markovian unique equation of probability in order to determine the accurate photon generation capture [35, 50].

Consequently, a peak high frequency cutoff Ω_C is calculated to keep away the bifurcation of DOS in a 3D PV cell. Necessarily, a tipped high frequency cutoff at Ω_d which controls the positive DOS in 2D and 1D PV cells has also been calculated. Hence $Li_2(x)$ acts as an algorithm function and $e_{rfc}(x)$ acts as an additional function [42, 51]. Thus, the DOS of various PV cells, here, is represented as $\varrho_{PC}(\omega)$, which is determined by calculating photonic energy frequencies of Maxwell's rules in the PV *nano* structure of the curtain wall skin [12, 16, 35]. For a 1D PV cell, the represented

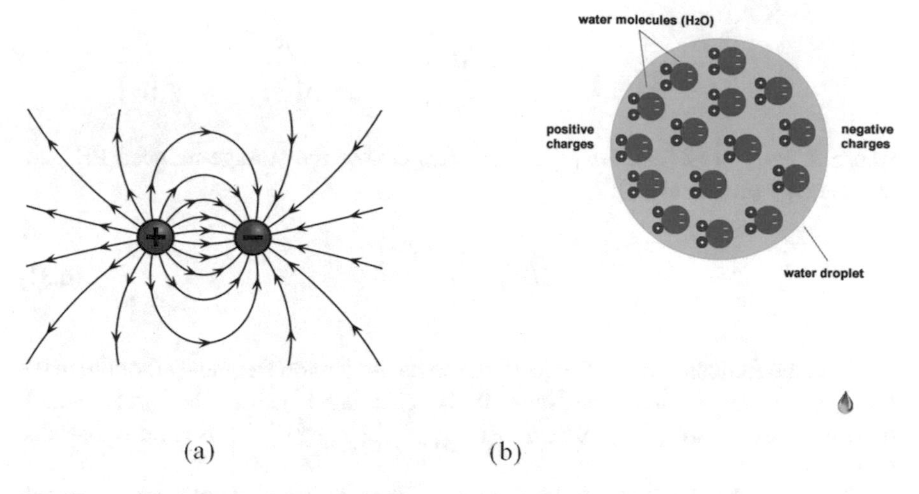

Fig. 6.5 (**a**) The formation of electrostatic force, and (**b**) its functional action to the cloud molecule for the transforming process to convert it into water droplet by static electro force of positive and negative charges of water particles to put it down into the insulator tank

DOS is thus expressed as $\varrho_{PC}(\omega) \propto \frac{1}{\sqrt{\omega - \omega_e}} \Theta(\omega - \omega_e)$, where $\Theta(\omega - \omega_e)$ represents the Heaviside step function and ω_e expresses the frequency of the photonic energy generation [1, 11].

This DOS is thus determined to confirm a 3D isentropic function in the PV cells to acquire an accurate qualitative state of the non-Weisskopf-Wigner mode of photons in the PV cell [11, 13, 38]. Naturally, this 3D PV cell will be the functional DOS in the PBE area of DOS, $\varrho_{PC}(\omega) \propto \frac{1}{\sqrt{\omega - \omega_e}} \Theta(\omega - \omega_e)$, which is then integrated into the electromagnetic field (EMF) vector of PV cell to determine the energy generation accurately [24, 52]. Considering the 2D and 1D of the PV cells, the photonic DOS is exhibited as a pure algorithm divergence which is found close to the PBE and expressed as $\varrho_{PC}(\omega) \propto -[\ln|(\omega - \omega_0)/\omega_0| - 1]\Theta(\omega - \omega_e)$, where ω_e denotes the midpoint of tip algorithm. The functional area $J(\omega)$ is thus clarified as the energy generation field of the DOS of the PV cell where the photonic magnitude of $V(\omega)$ depends on the photonic band (PB) of the PV cell [25, 41]:

$$J(\omega) = \varrho(\omega)|V(\omega)|^2 \tag{6.29}$$

Hence, the PB frequency ω_c and proliferative photonic dynamic are considered as the function $u(t, t_0)$ for photon energy generation in the relation $\langle a(t)\rangle = u(t, t_0)$ $\langle a(t_0)\rangle$. It is therefore determined using the functional integral equation and is expressed as

$$u(t,t_0) = \frac{1}{1-\Sigma'(\omega_b)} e^{-i\omega(t-t_0)} + \int_{\omega_e}^{\infty} d\omega \frac{J(\omega) e^{-i\omega(t-t_0)}}{\left[\omega - \omega_c - \Delta(\omega)\right]^2 + \pi^2 J^2(\omega)} \quad (6.30)$$

where $\Sigma'(\omega_b) = \left[\partial\Sigma(\omega)/\partial\omega\right]_{\omega=\omega_b}$ and $\Sigma(\omega)$ denote the storage-induced PB photonic energy proliferations:

$$\Sigma(\omega) = \int_{\omega_e}^{\infty} d\omega^{\prime \frac{J(\omega')}{\omega-\omega'}} \quad (6.31)$$

Here, the frequency ω_b in Eq. (6.19) denotes the photon frequency module in the PBG ($0 < \omega_b < \omega_e$) and thus it is calculated using the areal condition: $\omega_b - \omega_c - \Delta(\omega_b) = 0$, where $\lesssim\Delta(\omega) = \mathcal{P}\left[\int d\omega' \frac{J(\omega')}{\omega - \omega'}\right]$ is a primary-value integral.

Therefore, the detailed photon dynamics, considering the proliferation magnitude $|u(t, t_0)|$, have been calculated and are clarified as 1D, 2D, and 3D quantum field with respect to various PBG areas [25, 31, 40]. The photon dynamic rate $\kappa(t)$ is clarified, neglecting the function $\delta = 0.1\omega_e$. The result revealed that dynamic photons are generated at a high rate once ω_c crosses from the PBG to PB area. Because the range in $u(t, t_0)$ is $1 \geq |u(t, t_0)| \geq 0$, the crossover area as related to the condition is denoted as $0.9 \gtrsim |u(t \to \infty, t_0)| \geq 0$ where this corresponded to $-0.025\omega_e \lesssim \delta \lesssim 0.025\omega_e$, with a production rate $\kappa(t)$ within the PBG ($\delta < -0.025\omega_e$) and in the area of the PBE ($-0.025\omega_e \lesssim \delta \lesssim 0.025\omega_e$) of the PV cell.

The dynamic photon capture is almost exponential for $\delta \gg 0.025\omega_e$, which is a Markov factor that is clarified as the function of $\delta = 0.1\omega_e$ [17, 35]. In the crossover area ($-0.025\omega_e \lesssim \delta \lesssim 0.025\omega_e$), the PB frequency is small in the vicinity of the PBE, which sharply increases the mode of the dynamic photon generation [14, 53]. Thus, this proliferation of dynamic photon capture confirms the energy-state photonic counts in the vicinity of the PBG in the PV cell where the photons are in a nonequilibrium photonic state [3, 54].

Then, the photon proliferation dynamics in the quantum field are clarified considering thermal variation with respect to the photon concentration function $v(t, t)$ by determining the nonequilibrium photon scattering theorem [4, 55]:

$$v(t,t) = \int_{t_0}^{t} dt_1 \int_{t_0}^{t} dt_2 u^*(t_1,t_0) \tilde{g}(t_1,t_2) u(t_2,t_0) \quad (6.32)$$

Here, the two-time correlation function $\tilde{g}(t_1,t_2) = \int d\omega J(\omega) \bar{n}(\omega,T) e^{-i\omega(t-t')}$ reveals the photonic dynamic variations induced by the thermal relativistic condition, where $\bar{n}(\omega,T) = 1/\left[e^{\hbar\omega/k_B T} - 1\right]$ is the proliferation of the photon generation in the PV cell at the optimum temperature T and is expressed as

$$v(t,t\to\infty)=\int_{\omega_e}^{\infty}d\omega\mathcal{V}(\omega)$$

with

$$\mathcal{V}(\omega)=\bar{n}(\omega,T)\left[\mathcal{D}_1(\omega)+\mathcal{D}_d(\omega)\right] \quad (6.33)$$

Here, Eq. (6.20) is simplified to determine the nonequilibrium condition: $\mathcal{V}(\omega)=\bar{n}(\omega,T)\mathcal{D}_d(\omega)$. Under low-temperature conditions, Einstein's photon fluctuation dissipation is not dynamically viable at the PB, but the connecting photonic dormant structures are measurable (i.e., the field intensity) [5, 9, 50]; $n(t)=\langle a^{\dagger}(t)a(t)\rangle=|u(t,t_0)|^2n(t_0)v(t,t)$, where $n(t_0)$ represents the primary PB. Therefore, the plotted number of dynamic photons versus temperature has been confirmed as the nonequilibrium proliferated photon generation [19, 35]. To be more specific, the first PB has been considered as the Fock-state photon number n_0, i.e., $\rho(t_0)=|n_0\rangle\langle n_0|$, which is obtained mathematically through the quantum dynamics of the photons and then by solving Eq. (6.33), with respect to the state of photon production at time t:

$$\rho(t)=\sum_{n=0}^{\infty}\mathcal{P}_n^{(n_0)}(t)|n_0\ n_0| \quad (6.34)$$

$$\mathcal{P}_n^{(n_0)}(t)=\frac{\left[v(t,t)\right]^n}{\left[1+v(t,t)\right]^{n+1}}\left[1-\Omega(t)\right]^{n_0}\times\sum_{k=0}^{\min\{n_0,n\}}\binom{n_0}{k}\binom{n}{k}\left[\frac{1}{v(t,t)}\frac{\Omega(t)}{1-\Omega(t)}\right]^k \quad (6.35)$$

where $\Omega(t)=\frac{|u(t,t_0)|^2}{1+v(t,t)}$. Therefore, the result reveals that an electron-state photon will evolve into different Fock states of $|n_0\rangle$ is $\mathcal{P}_n^{(n_0)}(t)$. The proliferation of photon dissipation $\mathcal{P}_n^{(n_0)}(t)$ in the primary state $|n_0=5\rangle$ and steady-state limit $\mathcal{P}_n^{(n_0)}(t\to\infty)$ is thus due to the generation of photon emission that will ultimately reach the thermal nonequilibrium state which is expressed as

$$\mathcal{P}_n^{(n_0)}(t\to\infty)=\frac{\left[\bar{n}(\omega_c,T)\right]^n}{\left[1+\bar{n}(\omega_c,T)\right]^{n+1}} \quad (6.36)$$

To probe this huge photon capture, a further calculation of the photon distribution within the quantum field has been conducted through the high-temperature coherent states and solving Eq. (6.19) considering the proliferation state of photons, and it is expressed as

$$\rho(t)=\mathcal{D}\left[\alpha(t)\right]\rho_T\left[v(t,t)\right]\mathcal{D}^{-1}\left[\alpha(t)\right] \quad (6.37)$$

where $\mathcal{D}\left[\alpha(t)\right]$ = exp $\{\alpha(t)\alpha^{\dagger}-\alpha^{*}(t)\alpha\}$ denotes the displacement functions with $\alpha(t) = u(t, t_0)\alpha_0$ and

$$\rho_{\mathrm{T}}\left[v(t,t)\right]=\sum_{n=0}^{\infty}\frac{\left[v(t,t)^{n}\right]}{\left[1+v(t,t)\right]^{n+1}}\# \; nn| \tag{6.38}$$

Here, ρ_T denotes a thermal state with an average particle quantum $v(t, t)$, where Eq. (6.11) suggests that the peak-point photon energy generation state will be evolved into a thermal state [49, 54], which is considered as the functional state of the photon $\mathcal{D}\left[\alpha(t)\right]|n$ [46] in the PV cell. Thus, the photon number representation, Eq. (6.38), is calculated as

$$m\|\rho(t)\|n = J(\omega) = \mathrm{e}^{-\Omega(t)|\alpha_0|^2}\frac{\left[\alpha(t)\right]^{m}\left[\alpha^{*}(t)\right]^{n}}{\left[1+v(t,t)\right]^{m+n+1}}$$
$$= \sum_{k=0}^{\min\{m,n\}}\frac{\sqrt{m!n!}}{(m-k)!(n-k)!k!}\left[\frac{v(t,t)}{\Omega(t)|\alpha_0|^2}\right]^{k} \tag{6.39}$$

where photons are captured in the quantum field ($\langle m|\rho(t)|n\rangle$) into the building curtain wall skin, and then the relativistic thermal state $[1 + v(t, t)]^{m+n+1}$ and non-equilibrium condition $[\alpha(t)]^m[\alpha^*(t)]^n$ of the curtain wall skin will deliver tremendous amount of photon energy within the vicinity.

Electricity Transformation

To transform this tremendous amount of photon energy into electricity, the curtain wall skin PV panel is connected into a series and parallel circuit of a single-diode solar cell. The PV cell is then implemented by the *I–V* relationship of the single-diode circuit and this *I–V* relationship into the PV panel has been clarified as

$$I = I_{\mathrm{L}} - I_{\mathrm{O}}\left\{\exp^{\left[\frac{q(V+I_{\mathrm{Rs}})}{AkT_{\mathrm{c}}}\right]}-1\right\}-\frac{\left(V+I_{\mathrm{Rs}}\right)}{R_{\mathrm{Sh}}} \tag{6.40}$$

Here, I_L denotes the photon formation current, *Io* denotes the ideal current flow into the diode, R_s denotes the resistance in a series, *A* denotes the diode function, *k* (= 1.38×10^{-23} W/m^2 K) denotes the Boltzmann's constant, *q* (=1.6×10^{-19} C) denotes the charge amplitude of the electron, and T_c denotes the cell temperature at optimum condition. Consequently, the *I-q* linked in the PV cells varies in the diode cell which is expressed as the dynamic current as follows [4, 45]:

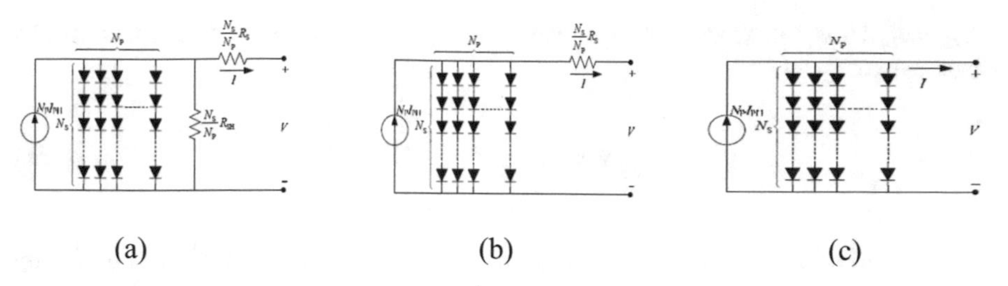

Fig. 6.6 Functional circuit diagram of the acting PV cell at the state of (**a**) normal, (**b**) normalized, and (**c**) perfect modes

$$I_{\text{O}} = \text{I}_{\text{Rs}} \left(\frac{T_{\text{C}}}{T_{\text{ref}}} \right)^3 \exp \left[\frac{qE_{\text{G}} \left(\frac{1}{T_{\text{ref}}} - \frac{1}{T_{\text{C}}} \right)}{KA} \right] \tag{6.41}$$

where IR_{s} denotes the dynamic current representing the functional transformation of solar radiation and qE_{G} denotes the bandgap solar radiation in the acting PV cell at normal, normalized, and perfect modes of electricity generation (Fig. 6.6).

Here, considering this acting PV cell, the *I–V* relationship with the exception of *I–V* curve, a calculative result of linked *I–V* curves among all cells of the PV panel, has been determined [51, 52]. Thus, the equation is rewritten as follows in order to determine the *V–R* relationship much more accurately:

$$V = -IR_{\text{s}} + K \log \left[\frac{I_{\text{L}} - I + I_{\text{o}}}{I_{\text{o}}} \right] \tag{6.42}$$

where *K* denotes the constant $\left(= \frac{AkT}{q} \right)$ and I_{mo} and V_{mo} denote the current and voltage in the acting PV panel. Subsequently, the relationship among I_{mo} and V_{mo} shall remain motional in the PV cell *I–V* which can be written as

$$V_{\text{mo}} = -I_{\text{mo}} R_{\text{Smo}} + K_{\text{mo}} \log \left(\frac{I_{\text{Lmo}} - I_{\text{mo}} + I_{\text{omo}}}{I_{\text{omo}}} \right) \tag{6.43}$$

where I_{Lmo} denotes the photon-induced current, I_{omo} denotes the dynamic current in the diode, R_{smo} denotes the resistance in series, and K_{mo} denotes the factorial constant.

Once all non-series (Ns) cells are interlinked in the series, then the series resistance is calculated as the sum of each cell series resistance $R_{\text{smo}} = N_{\text{s}} \times R_{\text{s}}$ current considering the functional coefficient of the constant factor $K_{\text{mo}} = N_{\text{s}} \times K$. Since the flow of current dynamics into the circuit is lined to the cells in a series connection, the current dynamics in Eq. (6.40) remains the same in each part of $I_{\text{omo}} = I_{\text{o}}$ and

$I_{Lmo} = I_L$. Thus, the mode of I_{mo}–V_{mo} relationship for the N_s series of connected cells can be expressed by

$$V_{mo} = -I_{mo} N_S R_S + N_S K \log\left(\frac{I_L - I_{mo} + I_o}{I_o}\right) \tag{6.44}$$

Naturally, the current–voltage relationship can be further modified considering all parallel connections in N_P cells in all parallel modes and can be described as follows [9, 21]:

$$V_{mo} = -I_{mo}\frac{R_s}{N_p} + K \log\left(\frac{N_{sh} I_L - I_{mo} + N_p I_o}{N_p I_o}\right) \tag{6.45}$$

Since the photon-induced current primarily depends on the solar radiation and optimum temperature configuration of the PV cell, the current dynamic is calculated as

$$I_L = G\left[I_{SC} + K_I\left(T_c - T_{ref}\right)\right] \times V_{mo} \tag{6.46}$$

where I_{sc} denotes the PV current at 25 °C and KW/m², K_I denotes the acting PV panel coefficient factor, T_{ref} denotes the PV panel's optimum temperature, and G denotes the solar energy in mW/m² [44, 53].

Finally, in order to determine a link in the frequency of ν_r to $\nu_r + d\nu_r$ among the density of solar (DOS) radiation, the primarily produced solar energy volts by the acting PV panel are converted into electricity energy by the counting number of light quanta (Fig. 6.7). With the peak solar irradiance, it could be emitted at 1.4 eV with an energy count of 27.77 mW/m² eV considering a mean of 5-h solar radiation harvesting per day at peak levels; thus, the equivalent of 27,770 kWh/year or 7.6 kWh/day energy is calculated as the usable energy for a building [18, 52]. Notwithstanding, there are some losses in the conversion of solar energy which is

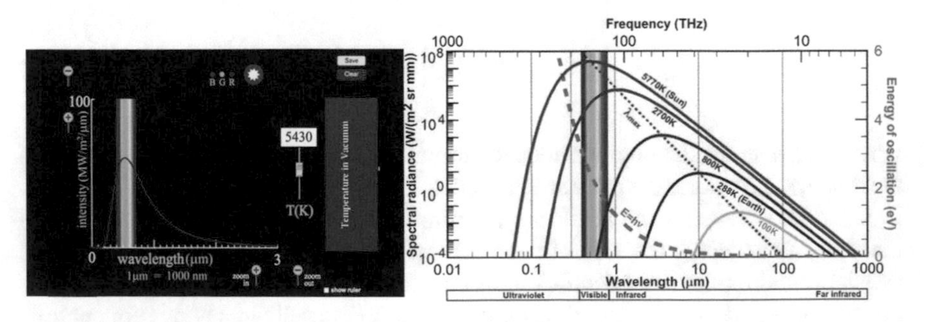

Fig. 6.7 The graph shows the solar irradiance at various frequencies and the peak temperature of 5770 K which suggest that the calculative power is 6.31 × 10⁷ (W/m²) since peak E is 1.410 (eV); tip λ is 0.88 (μm); and tip μ is 2.81 × 10⁷ (W/m² eV)

DC into AC energy that is called "efficiency ratio," which is normally 80% [7, 16]. However, the surface of the acting solar panel is a unique semiconductor in the energy-transforming process where the net current conversion by this curtain wall skin is nearly 125% higher compared to the standard solar panel and thus it will level the net energy production calculated as (27,770 × 1.25 × 0.8) = 27,770 kWh/year or 7.6 kWh/day [3, 23, 27]. Simply electricity generation will remain equal to the solar energy initially calculated which was emitted before into the solar panel. An ideal residential house requires an average 12 kWh/day [7, 12, 15]. Since the acting PV panel-generated solar energy is equivalent to 27,770 kWh/year or 7.6 kWh/day, it will need two solar panels of 1 m^2 each that meet the net energy need for a residential building [23, 36]. The mean energy consumption per month of a commercial office or buildings of 32 m × 31 m with 30 m height approximately is 10,000 kWh/day; thus, the total 1 m^2 PV panels will require 1195 units (945 + 250) with the capacity of 7.6 kWh/unit energy production that can provide a total energy × 1195 = 9082 kWh/day to meet the daily energy demand of about 10,000 kWh/day for a commercial office or building [19, 37, 38].

Electrostatic Force Analysis

The electrostatic force generation around the insulator tank on the roof of a building has been determined in order to confirm the tug down of the cloud water; initially the dynamic photon proliferation is calculated by integrating *HSEF* electric fields; thus, the local U(1) gauge field will allow to add a mass term of the functional particle of $\varnothing' \rightarrow e^{i\alpha(x)}\varnothing$. It is then further clarified by explaining the variable derivative of transformation law of scalar field using the following equation [16, 52]:

$$\begin{aligned} &\partial_\mu \rightarrow D_\mu = \partial_\mu = ieA_\mu \quad [\text{covariant derivatives}] \\ &A'_\mu = A_\mu + \frac{1}{e}\partial_\mu \alpha \quad [A_\mu \text{ derivatives}] \end{aligned} \tag{6.47}$$

Here, the local U(1) gauge denotes the invariant *HSEF* for a complex scalar field which is further expressed as

$$\mathfrak{h} = (D^\mu)^\dagger (D_\mu \varnothing) - \frac{1}{4} F_{\mu v} F^{\mu v} - V(\varnothing) \tag{6.48}$$

The term $\frac{1}{4} F_{\mu v} F^{\mu v}$ is the dynamic term for the gauge field of the acting PV panel and $V(\varnothing)$ denotes the extra term in the *HSEF* which is $V(\varnothing^*\varnothing) = \mu^2(\varnothing^*\varnothing) + \lambda(\varnothing^*\varnothing)^2$.

Therefore, the generation of *HSEF* ($\mathfrak{h}$) under the perturbational function of the quantum field has been confirmed by the calculation of mass scalar particles ϕ_1 and ϕ_2 along with a mass variable of μ. In this condition $\mu^2 < 0$ had an infinite number of quantum which is clarified by $\phi_1^2 + \phi_2^2 = -\mu^2 / \lambda = v^2$ and the $\mathfrak{h}$ through the variable derivatives using further shifted fields η and ξ defined the quantum field as $\phi_0 = \frac{1}{\sqrt{2}}\left[(\upsilon+\eta)+i\xi\right]$:

$$\text{Kinetic term: } \mathfrak{h}(\eta,\xi) = (D^{\mu}\phi)^{\dagger}(D^{\mu}\phi)$$

$$= (\partial^{\mu} + ieA^{\mu})\phi^{*}\left(\partial_{\mu} - ieA_{\mu}\right)\phi \quad (6.49)$$

Thus, this expanding term in the $\mathfrak{h}$ associated to the scalar field suggests that *HSEF* electric field is prepared to initiate the generation of static electricity force into its quantum field to tug down the cloud water [12, 21].

To determine this tug down of water by static electricity force, hereby, a non-variable function of ready dynamics has been implemented for the calculation of $\bar{\varphi}[s_0]$ to confirm the expected value of s_0 for capturing cloud water in cubic meter per second [12, 52]. Thus, the corrective functional asymptotic formulas are used as follows:

$$\bar{\varphi}[s_0] = 2s_0(\ln 4s_0 - 2) + \ln 4s_0(\ln 4s_0 - 2) - \frac{(\pi^2 - 9)}{3} + s_0^{-1}\left(\ln 4s_0 + \frac{9}{8}\right) + \cdots (s_0 \gg 1) \quad (6.50)$$

$$\bar{\varphi}[s_0] = \left(\frac{2}{3}\right)(S_0 - 1)^{\frac{3}{2}} + \left(\frac{5}{3}\right)(S_0 - 1)^{\frac{5}{2}} - \left(\frac{1507}{420}\right)(S_0 - 1)^{\frac{7}{2}}\ (1/2 \text{ instead of } 1). \quad (6.51)$$

Then the final equation can be rewritten as follows where s_0 is the areal value of static electricity force generated in the plastic tank (1 m^2):

$$\bar{\varphi}[s_0] = \left(\frac{2}{3}\right)(S_0 - 1)^{\frac{3}{2}} + \left(\frac{5}{3}\right)(S_0 - 1)^{\frac{5}{2}} - \left(\frac{1507}{420}\right)(S_0 - 1)^{\frac{7}{2}} \text{ gallon water}(s_0 = 1) \quad (6.52)$$

$$\bar{\varphi}[s_0] = 0.8474\ \text{m}^3\ \text{h}$$

The function $\bar{\varphi}[s_0]$ is thus revealed as $0.5 < s_0$; for larger s_0, it contains natural logarithmic which is s_0 to confirm the tug down of the cloud water by the *HSEF* into the plastic tank of the building placed on the roof which is nearly 224 gallons per second once 1 m^2 plastic tank is used to capture the cloud water.

In average 100 gallons of water is required per day per person in a standard daily life for a four-person family [33, 37]. Therefore, a total of (100_{gallons}/day/person × 4_{persons} × 365_{days}) 146,000 gallons of water will be needed yearly for a small family of four persons. In an ideal building of 32 m × 31 m with a height of 30 m which has the standard capacity of 100 units with the total average of four persons each unit will occupy 400 persons for the total building [15, 16]. Thus, the total 400

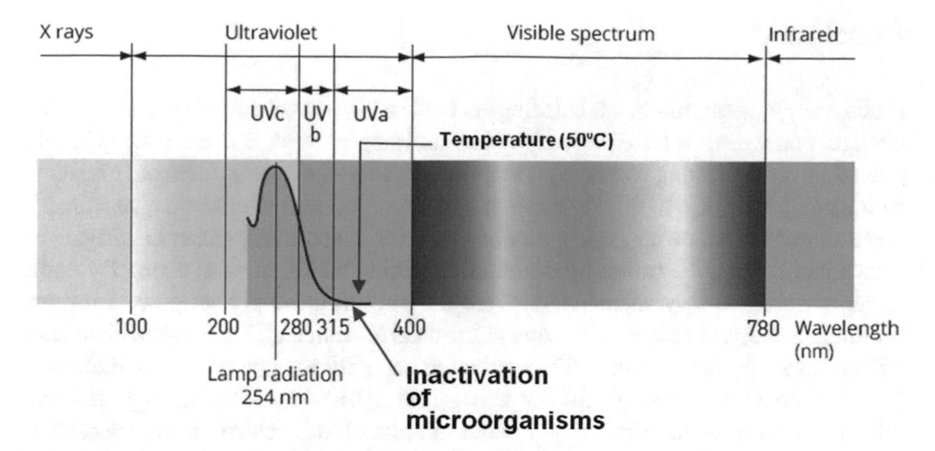

Fig. 6.8 The photoinduced physical irradiance application to purify the water which confirms that if the UV radiation of 320 nm of solar light is implemented into the cloud water with a temperature of only 50 °C, it eliminates all microorganisms immediately

persons require 146,000 × 100 = 14,600,000 gallons of water per year. Since a standard 1 m^2 plastic tank can trap 224 gallons per hour, tugging down of cloud water by HSEF described above shall require only ten (10) ideal 1 m^2 insulator tanks to meet the total water demand for a standard building of 32 m × 31 m with a height of 30 m.

In Situ Water Treatment

Since the gathered water in the insulated tank is simply the liquid state of cloud water, it will not require any conventional process of sedimentation, coagulation, and chlorination to clean the water to make it potable. Just implementation of the UV application and filtration is required to treat the water to meet the US National Primary Drinking Water Standard code [6, 52]. It is the most unadorned approach to treat water by utilizing SODIS framework (SOlar DISinfection), where a simple chamber is filled with water and exposed to full daylight for a few hours (Fig. 6.8). Once the water temperature reaches 50 °C with a UV radiation of 320 nm, the activation process will be accelerated rapidly to eliminate all microorganisms which will be then usable immediately for domestic water supply [15, 52].

This technology is not just an imaginative wellspring of water supply for a building by the usage of electrostatic power to catch the cloud water and treat it in situ by UV application for domestic use, but indeed it will without a doubt be a novel, coordinated, and inventive field in science to fulfill the potable water crisis throughout the world.

Conclusion

Since energy utilization in the building sector is a significant contributor for climate change, converting solar energy by using building exterior curtain wall skin will indeed be an interesting technology to mitigate the global energy demand and environmental perplexity. In this research, it is hence proposed to confirm a sustainable and atmosphere benevolent energy technology for the building sector to mitigate its total energy demand by utilizing the exterior curtain wall skin as an acting PV panel to capture solar energy and convert it into AC current to supply net powers for that building. Simply, the approach in this research is to design all buildings with at least 25% of exterior curtain wall skin to have the capacity to capture solar radiation in order to convert it into electricity energy to fulfill its 100% energy demand. Alongside, to mitigate a building's water supply, cloud water is also proposed to trap by static electricity force creating insulator tank nearby the building and treat in situ by UV technology in order to use as potable water to fulfill the net water demand for a building. Simply, integration of these two mechanisms with a smart building technology to solve the total energy and water crisis for a building will certainly be a new field of science in building and clean energy technology which is benevolent to the environment.

Acknowledgements This research was supported by Green Globe Technology under the grant RD-02017-07 for building a better environment. Any findings, predictions, and conclusions described in this chapter are solely performed by the authors and it is confirmed that there is no conflict of interest for publishing in a suitable journal.

References

1. Chang, D. E., Sørensen, A. S., Demler, E. A. & Lukin, M. D. A single-photon transistor using nanoscale surface plasmons. Nature Physics. 3, 807–812 (2007).
2. Dao Zhou, Frede Blaabjerg, Mogens Lau, Michael Tonnes. "Thermal analysis of multi-MW two-level wind power converter", IECON 2012 - 38th Annual Conference on IEEE Industrial Electronics Society, 2012.
3. Englund, D. et al. Resonant excitation of a quantum dot strongly coupled to a photonic crystal nanocavity. Phys. Rev. Lett. 104, 073904 (2010).
4. Hossain, M. F. "Green science: Independent building technology to mitigate energy, environment, and climate change", Renewable and Sustainable Energy Reviews, 2017.
5. Lang, C. et al. Observation of resonant photon blockade at microwave frequencies using correlation function measurements. Phys. Rev. Lett. 106, 243601 (2011).
6. Leijing Yang, Sheng Wang, Qingsheng Zeng, Zhiyong Zhang, Tian Pei, Yan Li & Lian-Mao Peng (2011). Efficient photovoltage multiplication in carbon nanotubes – Nature Photonics pp 672 – 676.
7. Najjari, A. B. Voitkiv, A. Artemyev, A. Surzhykov. Simultaneous electron capture and bound-free pair production in relativistic collisions of heavy nuclei with atoms, Phys. Rev. A 80, 012701 (2009).
8. Roy, D. Two-photon scattering of a tightly focused weak light beam from a small atomic ensemble: An optical probe to detect atomic level structures. Phys. Rev. A 87, 063819 (2013).

9. Sayrin, C. et al. Real-time quantum feedback prepares and stabilizes photon number states. Nature 477, 73 (2011).
10. Andreas Reinhard. "Strongly correlated photons on a chip", Nature Photonics, 2011.
11. G. Baur, K. Hencken, D. Trautmann. Revisiting unitarity corrections for electromagnetic processes in collisions of relativistic nuclei. Phys. Rep. 453, 1 (2007).
12. Armani, D. K., Kippenberg, T. J., Spillane, S. M. & Vahala, K. J. Ultra-high-Q toroid microcavity on a chip. Nature 421, 925 (2003).
13. Hossain, M. F. "Solar energy integration into advanced building design for meeting energy demand and environment problem: Climate change, photoenergy, solar panel, and clean energy", International Journal of Energy Research, 2016.
14. Gleyzes, S. et al. Quantum jumps of light recording the birth and death of a photon in a cavity. Nature 446, 297 (2007).
15. Hossain, M. F. Solar energy integration into advanced building design for meeting energy demand and environment problem. *Inter. J. Ener. Res.* 40, 1293–1300 (2016).
16. Hossain, M. F. "Breakdown of Bose–Einstein photonic structure to produce sustainable energy", Energy Reports, 2019.
17. Guerlin, C. et al. Progressive field-state collapse and quantum non-demolition photon counting. Nature 448, 889 (2007).
18. Birnbaum, K. M. et al. Photon blockade in an optical cavity with one trapped atom. Nature 436, 87–90 (2005).
19. Josette, M., Scott, R. The ERECTA gene regulates plant transpiration efficiency in Arabidopsis. *Nat.* 436, 866–870 (2005).
20. Dayan, B. et al. A photon turnstile dynamically regulated by one atom. Science 319, 1062–1065 (2008).
21. Lü, Xin-You, Wei-Min Zhang, Sahel Ashhab, Ying Wu, and Franco Nori. "Quantum-criticality-induced strong Kerr nonlinearities in optomechanical systems", Scientific Reports, 2013.
22. Zhu, Y., Xiaoyong, H., Hong, Y., Qihuang, G. On-chip plasmon-induced transparency based on plasmonic coupled nanocavities. *Sci. Rep.* 4, 3752 (2014).
23. Hossain, M. F. Theory of global cooling. *Ener. Sus. Soc.* 7, 6-24 (2016).
24. Hossain, M. F. "Photon application in the design of sustainable buildings to console global energy and environment", Applied Thermal Engineering, 2018.
25. Hossain, M. F. "Theoretical mechanism to breakdown of photonic structure to design a micro PV panel", Energy Reports, 2019.
26. Douglas, J. S., H. Habibian, C.-L. Hung, A. V. Gorshkov, H. J. Kimble, and D. E. Chang. "Quantum many-body models with cold atoms coupled to photonic crystals", Nature Photonics, 2015.
27. Hossain, M. F. "Green Technology: Transformation of Transpiration Vapor to Mitigate Global Water Crisis", Polytechnica, 2019.
28. Tobias, D., Wheeler & Abraham, Stroock, D. The transpiration of water at negative pressures in a synthetic tree. *Nat.* 455, 208–212 (2008).
29. Jaivime, E., Scott, J. McDonnell. Global separation of plant transpiration from groundwater and streamflow. *Nat.* 525, 91–94 (2015).
30. Joannopoulos, J. D., Villeneuve, P. R. & Fan, S. Photonic crystals: putting a new twist on light. Nature 386, 143 (1997).
31. Tame, M. S., K. R. McEnery, Ş. K. Özdemir, J. Lee, S. A. Maier, and M. S. Kim. "Quantum plasmonics", Nature Physics, 2013.
32. Hossain, M. F. "Sustainable technology for energy and environmental benign building design", Journal of Building Engineering, 2019.
33. Hossain, M. F. "Photonic thermal control to naturally cool and heat the building", Applied Thermal Engineering, 2018.
34. Scott, J., Zachary, D. Terrestrial water fluxes dominated by transpiration. *Nat.* 496, 347–350 (2013).

35. Hossain, M. F. "Green science: Advanced building design technology to mitigate energy and environment", Renewable and Sustainable Energy Reviews, 2018.
36. Liao, J. Q. & Law, C. K. Correlated two-photon transport in a one-dimensional waveguide side-coupled to a nonlinear cavity. Phys. Rev. A 82, 053836 (2010).
37. Hossain, M. F. "Transforming dark photons into sustainable energy", International Journal of Energy and Environmental Engineering, 2018.
38. Hossain, M. F. "Photon energy amplification for the design of a micro PV panel", International Journal of Energy Research, 2018.
39. Zhang, W. M., Lo, P. Y., Xiong, H. N., Tu, M. W. Y. & Nori, F. General Non-Markovian Dynamics of Open Quantum Systems. Phys. Rev. Lett. 109, 170402 (2012).
40. Hossain, M. F. "Green science: Decoding dark photon structure to produce clean energy", Energy Reports, 2018.
41. Hossain, M. F. "Design and construction of ultra-relativistic collision PV panel and its application into building sector to mitigate total energy demand", Journal of Building Engineering, 2017.
42. Li, Qiong, D. Z. Xu, C. Y. Cai, and C. P. Sun. "Recoil effects of a motional scatterer on single-photon scattering in one dimension", Scientific Reports, 2013.
43. Shen, J. T. & Fan, S. Strongly correlated two-photon transport in a one-dimensional waveguide coupled to a two-level system. Phys. Rev. Lett. 98, 153003 (2007).
44. Yan, W., Heng, F. Single-photon quantum router with multiple output ports. *Sci. Rep.* 4, 4820 (2014).
45. Langer, L., Poltavtsev, S., Bayer, M. Access to long-term optical memories using photon echoes retrieved from semiconductor spins. *Nat. Phot.* 8, 851–857 (2014).
46. Peter Fratzl. "Biomaterial systems for mechanosensing and actuation", Nature, 11/26/2009.
47. Pregnolato, T., Lee, E., Song, J., Stobbe, D., Lodahl, P. Single-photon non-linear optics with a quantum dot in a waveguide. *Nat. Commun.* 6, 8655 (2015).
48. Yan, Wei-Bin, and Heng Fan. "Single-photon quantum router with multiple output ports", Scientific Reports, 2014.
49. Tu, M. W. Y. & Zhang, W. M. Non-Markovian decoherence theory for a double-dot charge qubit. Phys. Rev. B 78, 235311 (2008).
50. Reed, M., Maxwell, L. Connections between groundwater flow and transpiration partitioning. *Sci.* 353, 377-380 (2015).
51. Shi, T., Fan, S. & Sun, C. P. Two-photon transport in a waveguide coupled to a cavity in a two-level system. Phys. Rev. A 84, 063803(2011).
52. Soto, W., Klein, S. et al. Improvement and validation of a model for photovoltaic array performance. *Sol. Ener.* 80, 78–88 (2006).
53. Xiao, Y. F. et al. Asymmetric Fano resonance analysis in indirectly coupled microresonators. Phys. Rev. A 82, 065804 (2010).
54. Yan, Wei-Bin, Jin-Feng Huang, and Heng Fan. "Tunable single-photon frequency conversion in a Sagnac interferometer", Scientific Reports, 2013.
55. Yuwen, W., Yongyou, Z., Qingyun, Z., Bingsuo, Z., Udo, S. "Dynamics of single photon transport in a one-dimensional waveguide two-point coupled with a Jaynes-Cummings system". *Sci. Rep.* 6, 33867 (2016).

Chapter 7
Advanced Green Building Technology

Abstract An advanced sustainable mechanism is being described to fulfill the complete need of energy for a building that can be created by the building itself. To meet the complete energy demand for a building, the domestic biowaste including human feces is suggested to be executed into converting process in situ where separated sludge is to be collected into an anaerobic shut-tank bioreactor (BR) into the basement, facilitating to form biogas (CH_4) by *methanogenesis* in order to convert biogas into electricity energy to power the entire building. Then, the discharged wastewater is to be stored in another detention tank in situ in order to conduct a complete treatment process of primary, secondary, tertiary, and UV application to utilize the treated wastewater for gardening. Execution of this technology indeed shall be an inventive field of green building science where a building can form electricity by itself to satisfy its total energy need without any connection with the utility companies which is benevolent to environment.

Keywords Domestic biowaste · Bioreactor · In site biowaste treatment technology · Methanogenesis · Bioenergy · Environmental sustainability

Introduction

Environmental vulnerability is correlated much with the building sector since 40% of global fossil energy is consumed by the building sector throughout the world [1–3]. In 2018, the net energy consumption globally accounted for 5.59×10^{20} J = 559 EJ, where 2.236×10^{20} EJ energy is alone engulfed by the building sector [4, 5]. Consequently, the building sector is triggered to release nearly 8.01×10^{11} ton CO_2 (218 gtC by the building sector per year of worldwide total carbon production of 545 gtC; 1 gtC = 10^9 ton C = 3.67 gt CO_2) into the atmosphere [6, 7]. The quickening of fossil fuel consumption by the building sector is getting higher and higher globally and the situation shall remain unchanged until an innovative technology is developed to power the building sector globally. At present, the

M. F. Hossain, *Global Sustainability in Energy, Building, Infrastructure, Transportation, and Water Technology*,
https://doi.org/10.1007/978-3-030-62376-0_7

atmospheric CO_2 level is 400 ppm where the building sector is the major player for creating this high level of CO_2 concentration into the atmosphere and it is accelerating by 2.11% per year which is the clear and present danger to the survival of all living beings in this planet in the near future [8–10]. Necessarily, the atmospheric CO_2 level must be lowered to a clean breathable level of 300 ppm CO_2. Therefore, a sustainable energy mechanism in the building sector is an urgent demand to confirm a clean and green environment on earth.

There are some recent interesting studies showing that a person can produce average feces of 0.4 kg/day that can form 0.4 m^3 biogas/day and this amount of biogas (0.4 m^3/day) production is good enough to cook three meals for a family of four persons in a day [11–13]. However, no one has shown that the mechanism of using the cellar of a building as an acting bioreactor to transform biowaste into electricity energy can satisfy the total energy demand of a building.

Therefore, in this research, a net zero carbon release by a building has been proposed by producing bioenergy by the building itself and transforming it into electricity energy to meet its net energy need. Simply, the domestic biowaste including human stool and wastewater of the building is being chosen to collect it into the sealed separation chamber into the basement. Thereafter, this biowaste is being isolated into (1) wastewater and (2) sludge and transferred into two separation tanks into the cellar. Then the wastewater is being conducted for treatment process in situ by integrating required chemical and physical processes in order to use for landscaping. Consequently, the solid biowaste has been permitted to undergo *methanogenesis* process in the bioreactor to form bioenergy and then convert it into electricity energy. Implementation of this innovative mechanism shall indeed be a promising technology in green building technology to fulfil the net need for a building which is delivered by the building itself.

Materials and Methods

For the conversion of domestic biowaste into bioenergy, a structurally sound long-lasting bioreactor (BR) needs to be designed. Thus, load-resistant factor design (LRFD) bioreactor must be constructed for a structurally sound bioreactor to operate regularly under high water velocity pressure considering the mathematical calculation of water velocity (379 mile/h), water density (1.2 kg/m^3), and friction loss cofactor 1.00/m^2, respectively [12, 14, 15]. As the water dynamic force is 0.5 of half of the density of the water, the equation for water force into the bioreactor can be expressed as $p_w = 0.5\rho C_p v_r^2$, where p_w represents the water force (Pa), ρ considers water density (kg/m^3), C_p denotes water force gradient which is 1, and v_r^2 is the water velocity (m/s) into the bioreactor. Thus, the net resultant force of $P_w = 0.5 \times 1.2$ kg/$m^3 \times 379^2$ m/s is 86,185 Pa of the water pressure resistance capacity of the bioreactor. It can be simplified as force of F = area × drag coefficient (constant = 1.00) × water dynamic force by the following equation, $F = 1$ $m^2 \times 1.0 \times 86{,}185 = 86{,}185$ N (8788 kgf) = 19,375 ibf, to confirm that the bioreactor is structurally sound with water velocity less than 19,000 ibf to operate the bioreactor normally throughout the year.

Once the sophisticated water force resistance of the two-chamber bioreactor has been constructed, the bioreactor is to be connected into biowaste chamber in order to collect the biowaste into the shut separation chamber into the basement. The other chamber is to be connected with the separated wastewater for the process of treatment of primary, secondary, and tertiary mechanism and then implemented into UV application to disinfect wastewater. The UV application and filtration constitute the simplest way of treating wastewater involving *disinfection* (DIS) system in which one fills a detention chamber with water and exposes it to full UV light for a few hours (Fig. 7.1). Once the wastewater temperature hits 50 °C due to the subject of UV light of approximately 320 nm, it functions immediately to kill all bacteria, viruses, and molds and disinfects water completely through bacteriological disinfection process.

This treatment mechanism removes nearly 100% microorganisms and other contaminants from the wastewater effluent which could be used for local gardening.

Then, the other product sludge (human feces including domestic waste) in another chamber of the bioreactor is being conducted for disinfection process in situ into an anaerobic chamber (Fig. 7.2). This is the conversion mechanism performed by electrochemical filters of activated carbon nanotubes (CNT), which has the capability to electrolyze and oxidize pollutants in the anode actively from the sludge [6, 16, 17]. It is an advanced mechanism of biowaste disinfection mechanism that combines both electrolysis and oxidation process into the anode of carbon nanotubes and catalyzes the process of oxidation by H_2O_2 into the cathode of carbon nanotubes. The function here is to accelerate the rate of sludge treatment, and its active oxidation process into the tank is being calculated and demonstrated a pathway that H_2O_2 flow is very much effective to disinfect the biowaste by the electrode and the cathode potential in order to achieve the content of biowaste pH, flow rate, and oxygen dissolved into a normal clean biomaterial form [18–20]. Hence, the maximum flow of H_2O_2 is being accounted for 1.38 mol/L/m^2 C by achieving CNT/L/m^2 with the implementation of cathode

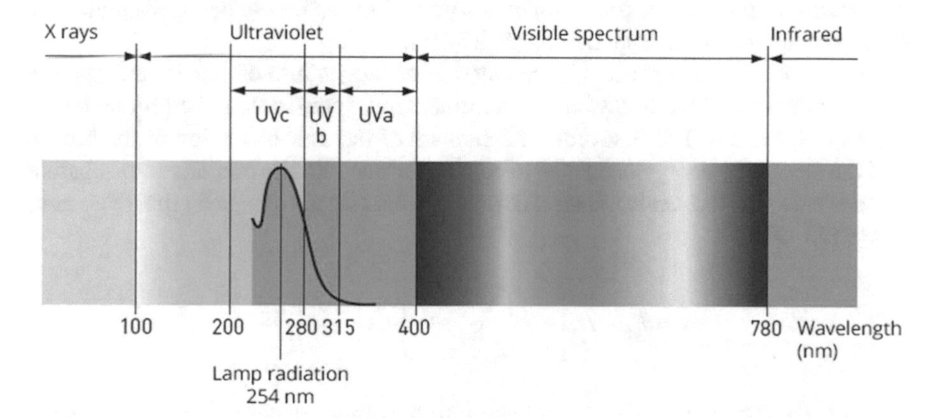

Fig. 7.1 The application of photo-physics radiation in purifying water that illustrates that once one applies UV light of 320 nm into wastewater, it begins to kill the microorganisms once the temperature momentum hits 50 °C

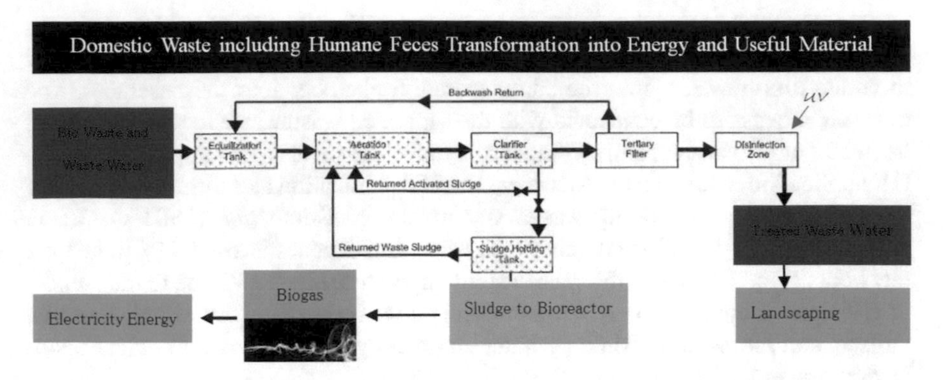

Fig. 7.2 The schematic diagram of wastewater treatment mechanism where treated water could be utilized for landscaping and the sludge is to be used for transforming process to produce energy

potential V −0.4 (vs. Ag/AgCl), with a pH of 6.46, with the flowing rate of 1.5 mL/min and the dissolved oxygen (DO) content of 1.95 mol/L/m^2. Additionally, phenol (C_6H_5OH) is being induced as an aromatic element for addressing the removal efficiency by clarifying the oxidation rate directed to the H_2O_2 flow [21–23]. Consequently, the electrochemical carbon nanotube filters activate the H_2O_2 generation tremendously in order for carbon nanotubes to work most effectively to remove the organic contaminants from the biowaste at nearly 100%.

Once the sludge is disinfected, the product is placed into the closed bioreactor tank to allow for anaerobic co-digestion process [24–26]. Thereafter, the product is heated at 95 °F for 15 days which will stimulate the growth of anaerobic bacteria of *Desulfovibrio* and *Methanococcus*, which engulf the organic material of the sludge and produce biogas through biosynthesis process (Fig. 7.3).

Then the biogas is to be conducted for transforming process to generate electricity energy through the semiconductor diodes of the circuit panel (Fig. 7.4). Hence, the electricity production from biogas into the circuit panel is being examined by detailed mathematical computations [5, 27, 28].

Hence, to achieve a successful conversion of biogas into electricity energy, the first-order perturbation theory has been implemented considering the production of biogas [18, 19, 29]. The first-order mechanism of the transformation of the biogas into the electricity energy needs the adequate surface into the bioreactor to separate the electrons into the semiconductor to produce the electric charge by the given term below [30–32]:

$$I = I_{\mathrm{ph}} - I_{\mathrm{ph}}\left[\exp\left(\frac{V + R_{\mathrm{s}}I}{V_{\mathrm{r}}}\right) - 1\right] - \frac{V + R_{\mathrm{s}}I}{R_{\mathrm{p}}} \tag{7.1}$$

Here I is the current and V is the voltage into the circuit panel. I_{ph} ($=N_pI_{\mathrm{ph,cell}}$) is the electricity energy-created current running inside the circuit module which consists of N_p cells that are connected in parallel. I_0 ($=N_{\mathrm{pI0,\,cell}}$) is called the reverse current

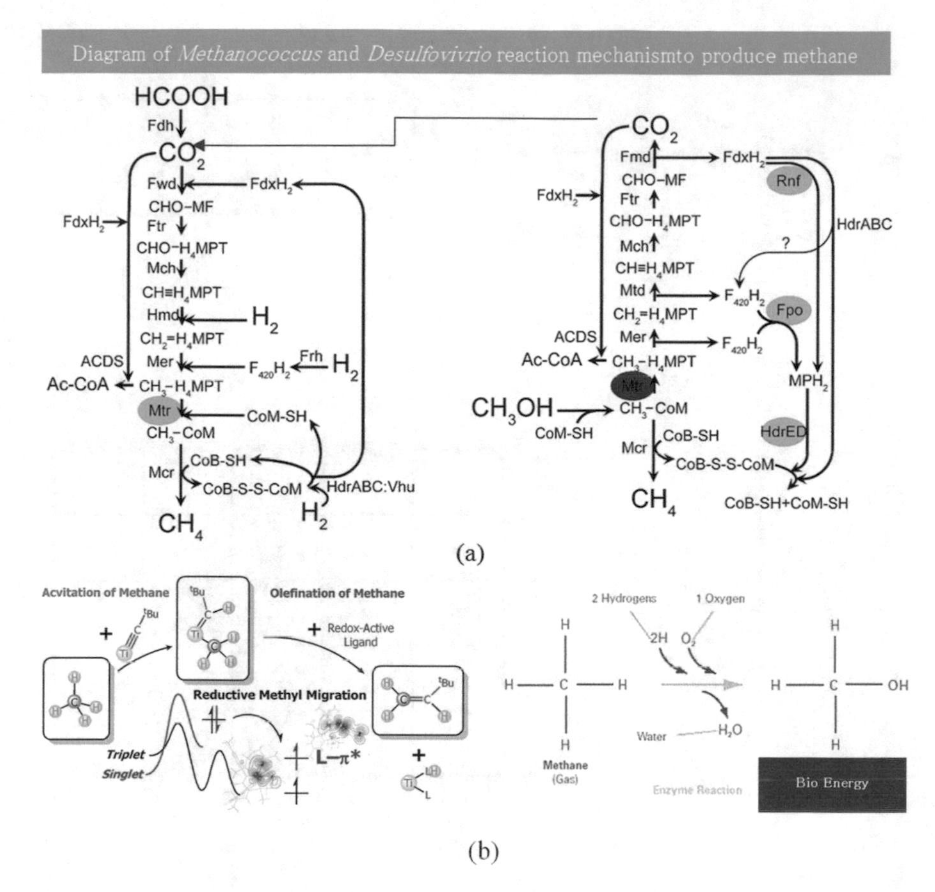

Fig. 7.3 (**a**) Biosynthesis mechanism of methanogenesis shows the conduction of chain reaction to form methane from sludge where two bacteria of Methanococcus and Desulfovibrio are the primary inhibitors to conduct this reaction, (**b**) the process showing the transformation of methane into bioenergy

passing through N_p cells that are connected in parallel, wherein the reverse saturation current $I_{0,\text{cell}}$ passes through each cell. Subsequently, V_T ($=aN_s \cdot kT/q$) is represented as a matrix of thermal stress of N_s cells that are connected in series where ($\sim 1.5 = 1.0$) keeping in mind the diode ideality factor k ($=1.38e^{-23}$ J/K) is a constant, q ($=1.602e^{-19}$ C) is the charge on an electron, and T is the temperature in kelvin. Here R_P is the equivalent resistance in parallel while R_s is the equivalent resistance in series for circuit generator. Depending on the operational point, the circuit device, in practice, operates as a mixed performance of the current source or the voltage source [33, 34]. Practically, for the circuit panel, the effect of R_P parallel resistance will be greater in the operating area having a current source, while the R_s series resistance has a bigger effect on the functioning of the photovoltaic modules when the device works in the area having a voltage source [1, 35, 36]. Based on the studies of various researchers, it can be concluded that for simplifying the model, the value

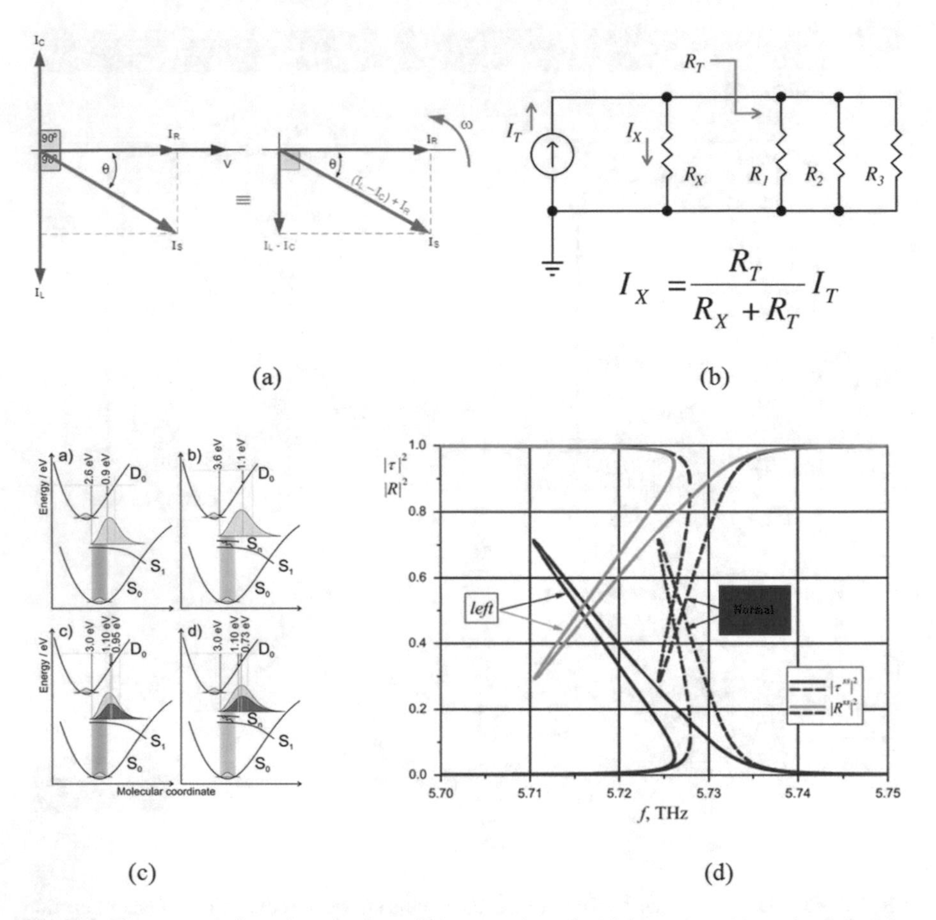

Fig. 7.4 The conversation of mechanism of bioenergy into electricity energy, (**a**) mode of electricity production dynamics with respect to power factor (pf), (**b**) flow of electricity current generation, (**c**) net electric energy production (eV) rate at molecular rate, (**d**) rate of electricity energy generation at normal and normalized circuit parameters, respectively

of R_P can be ignored as it is very high [37–39]. Likewise, the value of R_s, being very low, can be neglected too [40–42], and thus the temperature of the circuit panel can be shown as follows [2, 43, 44]:

$$T = 3.12 + 0.25\frac{S}{S_n} + 0.899T_a - 1.3v_a + 273 \tag{7.2}$$

Here S and S_n (=1000 W/m^2) are the electricity energy available in working condition, respectively; T_a is the surrounding temperature; and v_a is the surrounding energy flow. The I–V features of photovoltaic panel are based on the internal qualities of the device, i.e., R_s and R_P; consequently, electricity energy as well as surrounding temperature affect outer features. The electricity energy that is responsible for producing the electric current is linked linearly to the electricity energy and temperature and can be stated as follows [45–47]:

$$I_{\mathrm{ph}} = \left(I_{\mathrm{ph},n} + \alpha_{\mathrm{I}} \Delta T\right) \frac{S}{S_n} \tag{7.3}$$

Here I_{ph} is the current that is produced because of biogas at STC and $\Delta T = T - T_n$, T is the temperature of the circuit panel because of the electricity energy, whereas T_n is the supposed temperature. For preventing any problems faced by the electricity energy current in deciding the series resistance (very low) as well as the parallel resistance (very high), it has been presumed that $I_{sc} \approx I_{ph}$ so that an explanation can be given for the complex circuit modeling and the open-circuit voltage that is dependent on the temperature can be confirmed [35, 48, 49]. This can be shown as follows:

$$V_{\mathrm{oc}} = V_{\mathrm{oc},n}\left(1 + \alpha_{\mathrm{v}} \Delta T\right) + V_{\mathrm{T}} \ln\left(\frac{S}{S_n}\right) \tag{7.4}$$

Here $V_{oc,n}$ is the open-circuit voltage that is calculated at the given conditions and α_v is the voltage-temperature coefficient. The electrical and thermal features of the electricity energy panels can be achieved from these characteristics which are integrated to achieve the I–V curve to produce much electricity energy Eq. (7.1). The characteristics of the suggested electricity energy panel should consist of the following: the short-circuit current/temperature coefficient (α_I), the open-circuit voltage-temperature coefficient (α_v), the experimental peak power (P_{max}), the insignificant short-circuit current ($I_{sc,n}$), the maximum power point (MPP) voltage (V_{mp}), the MPP current (I_{mpp}), and the insignificant open-circuit voltage ($V_{oc,n}$), to calculate at the supposed conditions or standard test conditions (STC) of temperature $T = 298$ K and electricity energy of $S = 1000$ W [34, 50]. The simple equation at STC can be expressed as follows:

$$I = I_{\mathrm{ph},n} - I_{0,n}\left[\exp\left(\frac{V + R_{\mathrm{s}} I}{V_{\mathrm{T},n}}\right) - 1\right] - \frac{V + R_{\mathrm{s}} I}{R_{\mathrm{p}}} \tag{7.5}$$

Here "n" is evaluated at STC and the values are expected to show that the resistance in series and the resistance in parallel are not dependent on each other. Hence, the modeling in Eq. (7.5) can be simplified as below:

$$I = I_{\mathrm{ph},n} - I_{0,n}\left[\exp\left(\frac{V + R_{\mathrm{s}} I}{V_{\mathrm{T},n}}\right) - 1\right] \tag{7.6}$$

There are three significant points on the I–V curve of electricity energy: maximum power point (V_{mp}, I_{mpp}), open circuit (V_{oc}, 0), and short circuit (0, I_{sc}) that can be shown as

$$I_{\text{sc},n} = I_{\text{ph},n} - I_{0,n}\left[\exp\left(\frac{R_{\text{s}} I_{\text{sc},n}}{V_{\text{T},n}}\right) - 1\right] \tag{7.7}$$

$$0 = I_{\text{ph},n} - I_{0,n}\left[\exp\left(\frac{V_{\text{oc},n}}{V_{\text{T},n}}\right) - 1\right] \tag{7.8}$$

$$I_{\text{mpp},n} = I_{\text{ph},n} - I_{0,n}\left[\exp\left(\frac{V_{\text{mpp},n} + R_{\text{s}} I_{\text{mpp},n}}{V_{\text{T},n}}\right) - 1\right] \tag{7.9}$$

The diode saturation current can thus be shown by its dependence on the temperature of the bioreactor [30]:

$$I_0 = I_{0,n}\left(\frac{T_n}{T}\right)^3 \exp\left[\frac{qE_{\text{G}}}{ak}\left(\frac{1}{T_n} - \frac{1}{T}\right)\right] \tag{7.10}$$

Here E_{G} represents the bandgap energy of the electricity energy. Equation (7.8) shows that the diode saturation current at the STC and the photocurrent at STC are linked:

$$I_{0,n} = \frac{I_{\text{ph},n}}{\left[\exp\left(\frac{V_{\text{oc},n}}{V_{\text{T},n}}\right) - 1\right]} \tag{7.11}$$

The electricity generation model can be further enhanced if Eq. (7.8) is substituted by

$$I_0 = \frac{I_{\text{sc},n} + \alpha_{\text{I}} \Delta T}{\exp\left(\frac{V_{\text{oc},n+\alpha_{\text{v}}\Delta T}}{V_{\text{T}}}\right) - 1} \tag{7.12}$$

By assuming $V_{\text{oc},n}/V_{\text{T},n} \gg 1$, $I_{0,n}$ can be shown as

$$I_{0,n} = I_{\text{ph},n} \exp\left(-\frac{V_{\text{oc},n}}{V_{\text{T},n}}\right) \tag{7.13}$$

Using Eqs. (7.13) and (7.6), it can be shown that

$$V = V_{\text{oc},n} + V_{\text{T},n} \ln\left(1 + \frac{I_{\text{ph},n}^{\ -I}}{I_{0,n}}\right) - R_{\text{s}} I \tag{7.14}$$

Thus, Eq. (7.14) is considered as modest electricity energy generation model that is transformed from the biogas from bioreactor and it can be explained as simply as the following equation:

$$V = V_{\text{oc},n} + V_{\text{T},n} \ln\left(1 + \frac{I}{I_{\text{ph},n}}\right) - R_{\text{s}} I \tag{7.15}$$

Results and Discussion

Since the anaerobic *co-digestion* of domestic biowaste including human feces is done in an anaerobic bioreactor, the *methanogenesis* process begins to produce biogas into the bioreactor right way (Fig. 7.5). Naturally, the formation of biogas from the biowaste is examined by computerized gas chromatograph [51–53].

Therefore, a model of bioreactor module described the generation of maximum bioenergy from domestic waste considering protective anerobic detention chamber (Fig. 7.1a). Naturally, the model of the bioreactor module is being simplified by the determination of accurate form of the current–voltage (*I–V*) curb considering the mode of single-diode electricity circuit [29, 34, 54].

The next step is to calculate the electricity energy generation I_{pv} from biogas production by the calculation of the mode of current flow in the diode panel (Fig. 7.6a), accounting for *I–V–R* relationship (Fig. 7.6b), and biogas received by the diode to convert to alternating current (AC) for the domestic energy demand (Fig. 7.6c).

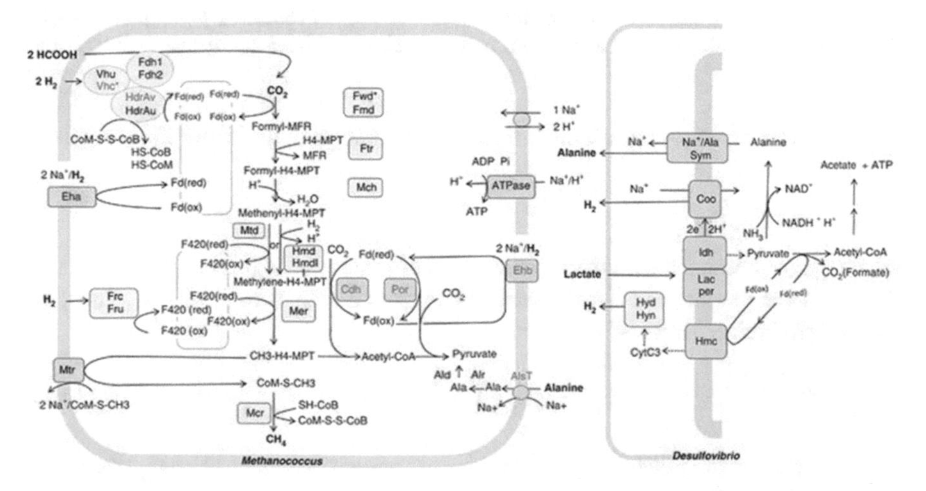

Fig. 7.5 The pathway of the methanogenesis mechanism depicts the biosynthesis of *Methanococcus maripaludis* and *Desulfovibrio vulgaris* to conduct bioenergy generation by consuming sludge

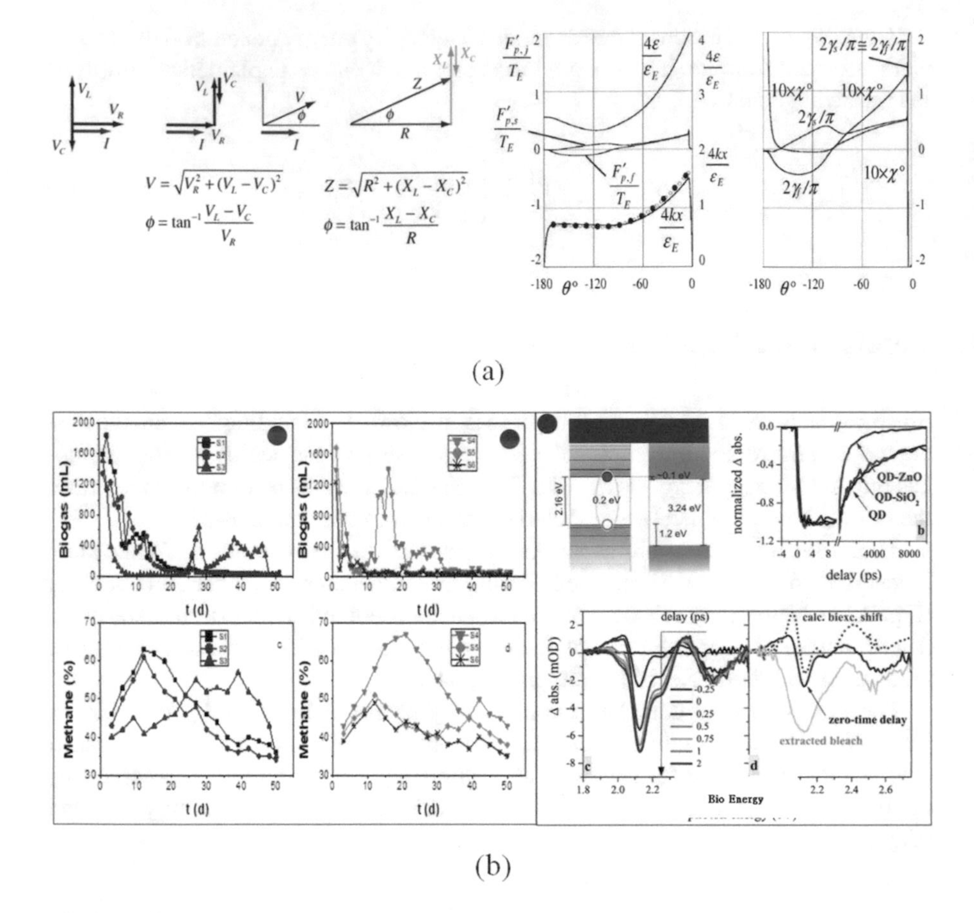

Fig. 7.6 (**a**) The biowaste transformation rate into the bioreactor in different directions and angles, (**b**) the production rate of biogas and the bioenergy considering bioreactor methane content of the biowaste

The below equation represents the electricity energy output from biogas (CH_4):

$$P_{\mathrm{pv}} = \eta_{\mathrm{pvg}} A_{\mathrm{pvg}} G_{\mathrm{t}} \tag{7.16}$$

where η_{pvg} represents the methane generation efficiency, A_{pvg} represents the electricity energy generation, and G_{t} represents the current flow in the circuit cell. Thus, η_{pvg} can be rewritten as follows:

$$\eta_{\mathrm{pvg}} = \eta_{\mathrm{r}} \eta_{\mathrm{pc}} \left[1 - \beta\left(T_{\mathrm{c}} - T_{\mathrm{cref}}\right)\right] \tag{7.17}$$

η_{pc} represents the power factor effectiveness once it is equal to 1; β represents the energy cofactor (0.004–0.006/°C); η_{r} represents the mode of energy production; and T_{cref} is the cell temperature in °C which can be obtained from the following equation:

$$T_c = T_a + \left(\frac{\text{NOCT} - 20}{800}\right) G_t \tag{7.18}$$

Here, T_a represents the ambient temperature in °C, G_t represents the current flow in a circuit cell (W/s), and NOCT represents the standard operating cell temperature in Celsius (°C) degree. The total electricity energy production in the circuit panel is estimated by the following equation:

$$I_t = I_b R_b + I_d R_d + (I_b + I_d) R_r \tag{7.19}$$

The current flow into the circuit cells is determined by the functional mode of its P-N junction that is able to produce electricity by conducting the interconnection of series-parallel configuration of the circuit cell [56–58].

Implementation of the standard single-diode circuit cell and the function of N_s series and N_p parallel connection in relation to current generation can be expressed as

$$I = N_p \left[I_{ph} - I_{rs} \left[\exp\left(\frac{q(V + IR_s)}{\text{AKTN}_s} - 1 \right) \right] \right] \tag{7.20}$$

where

$$I_{rs} = I_{rr} \left(\frac{T}{T_r}\right)^3 \exp\left[\frac{E_G}{\text{AK}} \left(\frac{1}{T_r} - \frac{1}{T} \right) \right] \tag{7.21}$$

Here, in Eqs. (7.20) and (7.21), q represents the generation of electron charge (1.6×10^{-19} C), K is the Boltzmann's constant, A represents the cell standard cofactor, and T represents the cell temperature (K). IR_s represents the cell reverse current at T, T_r represents the cell referred temperature, I_{rr} represents the reverse current at T_r, and E_G represents the bandgap energy flow into the circuit cell (Fig. 7.7). The electric current I_{ph} formation conforming the circuit cell's temperature can be simplified as follows:

$$I_{ph} = \left[I_{SCR} + k_i (T - T_r) \frac{S}{100} \right] \tag{7.22}$$

I_{SCR} represents the cell short-circuit current and electricity energy generation, k_i represents the short-circuit current temperature coefficient, and S represents the electricity energy (kW). Thus, the I–V relationship into the circuit cell can be expressed simply as

$$I = I_{ph} - I_D \tag{7.23}$$

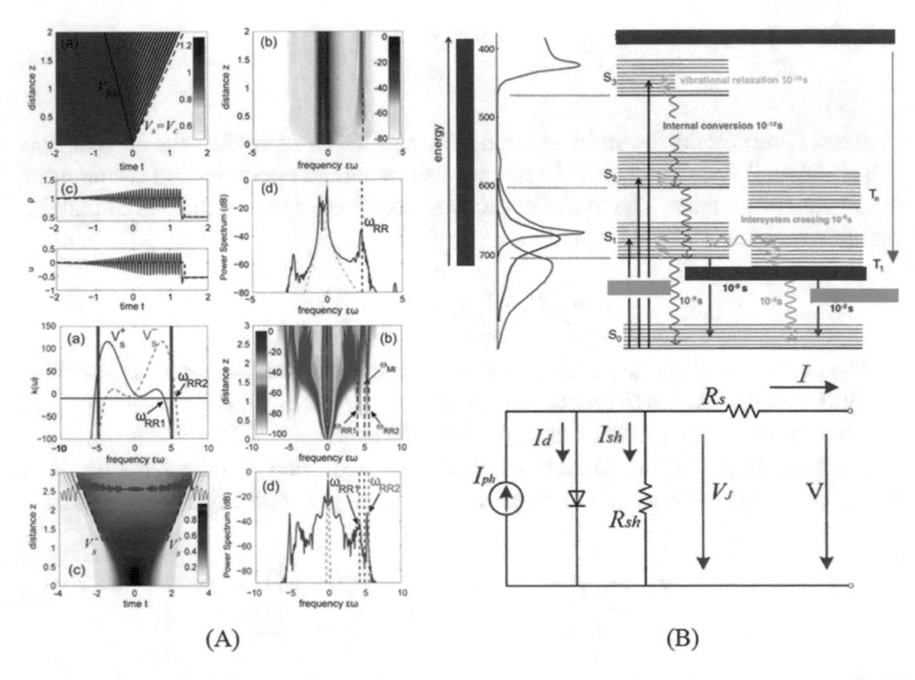

Fig. 7.7 (**a**) MATLAB simulation calculating electricity energy generation from bioenergy showing various frequencies and distances of the electric charges into the single-diode circuit cell, (**b**) shows the conversion mechanism of electricity energy DC into AC for the use as the prime source of power supply for a building

$$I = I_{\mathrm{ph}} - I_0 \left[\exp\left(\frac{q\left(V + R_s I\right)}{\mathrm{AKT}} - 1 \right) - \frac{V + R_s I}{R_{\mathrm{sh}}} \right] \tag{7.24}$$

I_{ph} represents the electricity current (A), I_{D} represents the functional current (A), I_0 represents the inverse current (A), A represents the functional constant, q represents the charge of the electron (1.6×10^{-19} C), K is the Boltzmann's constant, T represents the cell temperature (°C), R_s represents the series resistance (ohm), R_{sh} represents the shunt resistance (Ohm), I represents the cell current (A), and V represents the circuit cell voltage (V). Thus, the output electricity current into the circuit panel is thus described as follows:

$$I = I_{pv} - I_{\mathrm{D1}} - \left(\frac{V + IR_{\mathrm{S}}}{R_{\mathrm{sh}}} \right) \tag{7.25}$$

where

$$I_{D1} = I_{01}\left[\exp\left(\frac{V + IR_s}{a_1 V_{T1}}\right) - 1\right] \tag{7.26}$$

$$I_{D2} = I_{01}\left[\exp\left(\frac{V + IR_s}{a_2 V_{T2}}\right) - 1\right] \tag{7.27}$$

I_{01} and I_{02} represent the reverse currents of cell, respectively, and V_{T1} and V_{T2} represent the thermal voltages of the respective cell. The cell idealist constants are denoted as a_1 and a_2. Then the simplified equation of the cell mode is described as

$$v_{oc} = \frac{V_{oc}}{cKT / q} \tag{7.28}$$

$$P_{max} = \frac{\frac{V_{oc}}{cKT/q} - \ln\left(\frac{V_{oc}}{cKT/q} + 0.72\right)}{\left(1 + \frac{V_{oc}}{KT/q}\right)}\left(1 - \frac{V_{oc}}{\frac{V_{oc}}{I_{SC}}}\right)\left(\frac{V_{oc0}}{1 + \beta \ln \frac{G_0}{G}}\right)\left(\frac{T_0}{T}\right)^{\gamma} I_{sc0}\left(\frac{G}{G_0}\right)^{\alpha} \tag{7.29}$$

where ν_{oc} represents the normal value of the open-circuit voltage, V_{oc} represents the thermal voltage $V_t = nkT/q$, c represents the constant current flow, K is the Boltzmann's constant, T represents the temperature in Kelvin, α represents the nonlinear cofactor, q represents the electron charge, γ represents the factor representing all the nonlinear temperature-voltage function, while β represents the cell module coefficient. Since Eq. (7.29) depicted the tip energy generation by the circuit cell, the equation of total power output for an array with N_s cells connected in series and N_p cells connected in parallel with power P_M for each mode can be expressed as

$$P_{array} = N_s N_p P_M \tag{7.30}$$

Conversely, the derivative of the power with respect to current will equate to peak electricity energy production:

$$\left.\frac{dP}{dI}\right|_{mpp} = \left.\frac{d(VI)}{dI}\right|_{mpp} = V_{mpp} + I_{mpp}\left.\frac{dV}{dI}\right|_{mpp} \tag{7.31}$$

and

$$V = V_{oc,n} + V_{T,n} \ln\left(1 - \frac{I}{I_{ph,n}}\right) - R_s I \tag{7.32}$$

Thus, the net electricity energy production volt (V) from biogas is finally computed as $V = V_{oc,n} + V_{T,n} \ln\left[1-(I/I_{ph,n})\right] - R_s I$, where the total amount of power has been determined using the equation

$$P_{max} = \frac{\frac{V_{oc}}{cKT/q} - \ln\left(\frac{V_{oc}}{cKT/q} + 0.72\right)}{\left(1 + \frac{V_{oc}}{KT/q}\right)} \left(1 - \frac{V_{oc}}{\frac{V_{oc}}{I_{SC}}}\right) \left(\frac{V_{oc0}}{1 + \beta \ln \frac{G_0}{G}}\right) \left(\frac{T_0}{T}\right)^{y} I_{sc0} \left(\frac{G}{G_0}\right)^{a}$$

considering the paraperter of $P_{array} = N_s N_p P_M$. The electricity energy production is therefore accomplished per mole biogas production which is equivalent to 1.4 eV/mol. Since 0.4 kg biowaste can produce 81 mol biogas, the total electricity generation from 0.4 kg biowaste is equivalent to $1.4 \times 81 = 113.4$ eV (cc). Since 1.4 eV is equal to 27.77 kWT, the total electricity energy production would be $(27.77 \text{ kW} \times 81) = 2249.37$ kW·eV/day [29, 59, 60]. If a commercial building and an office consumption is roughly 2200 kWh/day for a building with a 20 m × 20 m footprint and a height of 20 m, 0.4 kg/day biowaste is sufficient enough to meet the total energy demand for this building which is environmentally friendly.

Conclusion

The advancement of building construction in both urban and suburban regions around the globe has been quickening tremendously for the past 50 years. As a result, the adverse environmental impact is expanding rapidly due to the traditional use of fossil fuel by the building sector throughout the world. Subsequently, conventional household waste and wastewater treatment are the cause of serious ecological contamination, which harms human well-being, hindering the kingdom of animals and plants in the terrestrial and aquatic environment. Here, "advanced green building technology," an inventive technology, could be the streamline science to mitigate the complete energy need for a building without using any utility service connection. Simply because this innovative technology can form sustainable energy by using the building cellar as an active bioreactor to deliver biogas from household biowaste, and then convert it into electricity energy to meet the net energy demand of a building which is environmentally friendly.

Acknowledgements This research was conducted by the support of the grant RD-02018-01 provided by Green Globe Technology to build a better environment. It does not have any financial interest by any means. Any discoveries, conclusions, and recommendations expressed in this chapter are exclusively those of the author, who affirms that this research has no conflict of interest for publication in a suitable journal.

References

1. Canadell, J.G. et al: Contributions to accelerating atmospheric CO_2 growth from economic activity, carbon intensity, and efficiency of natural sinks, P. Natl. Acad. Sci. USA, 104, 18866-18870, 2007.
2. Denman, K.L., Brasseur, G., Chdidthaisong, A., Ciais, P., Cox, P.M., Dickinson, et al, Cambridge University Press, Cambridge, UK and New York, USA, 2007.
3. Hossain, M. F. (2018). Bose-Einstein (B-E) Photonic Energy Structure Reformation for Cooling and Heating the Premises Naturally. 142, 100-109 https://doi.org/10.1016/j.applthermaleng.2018.06.057. Advanced Thermal Engineering. (Elsevier).
4. Corinne Le Querel, Robbie M. Andrew et al. Global Carbon Budget 2016. Earth System Science Data, 8, 605-649, 2016. https://doi.org/10.5194/essd-8-605-2016.
5. Earles, J.M. Yeh, S., and Skog, K.E: Timing of carbon emissions from global forest clearance, Nature Climate Change, 2, 682-685, 2012.
6. Erb, K-H, et al: Bias in the attribution of forest carbon sinks, Nature Climate Change, 3, 854-856, 2013.
7. Gonzalez-Gaya, et al: High atmospheric – ocean exchange of semivolatile aromatic hydrocarbons, Nat. Geosci., 9, 438-442, https://doi.org/10.1038/ngeo2714, 2016.
8. Alfredo, M. et al. "In Focus: Biotechnology and chemical technology for biorefineries and biofuel production", Journal of Chemical Technology & Biotechnology, 2017.
9. Achard, F., et al: Determination of tropical deforestation rates and related carbon losses from 1990 to 2010. Glob. Change Biol., 20. 2540-2554, 2014.
10. Yin, Y., Ciais, P., Chevallier, F et al.: Variability of fire carbon emissions in Equatorial Asia and its non-linear sensitivity to El, Nino, Geophys. Res. Lett., 43, 10472-10479, 2016.
11. Ashley, B. et al. "Accelerating net terrestrial carbon uptake during the warming hiatus due to reduced respiration", Nature Climate Change, 2017.
12. Hossain, M. F. (2019). Breakdown of Bose-Einstein Photonic Structure to Produce Sustainable Energy. 5, 202-209. Energy Report. (Elsevier). 2018.
13. Hossain, M. F. (2018). Green Science: Decoding Dark Photon Structure to Produce Clean Energy. Energy Reports. 4; 41-48. (Elsevier).
14. Hossain, M. F. (2019). Sustainable Technology for Energy and Environmental Benign Building Design. 22, 130-139. Journal of Building Engineering. (Elsevier).
15. J.J.: El Nino and a record CO_2 rise, Nature Climate Change, 6, 806-810, 2016.
16. Li, W., Ciais, P., Wang, Y., Peng, S. et al: Reducing uncertainties in decadal variability of the global carbon budget with multiple datasets, P. Natl. Acad. Sci. USA, https://doi.org/10.1073/pnas.1603956113, in press, 2016.
17. J.W.C: Increase in observed net carbon dioxide uptake by land and oceans during the last 50 years. Nature, 488, 70-72, 2012.
18. Hossain, M. F. (2016). Production of Clean Energy from Cyanobacterial Biochemical Products. Strategic Planning for Energy and the Environment. 3; 6-23 (Taylor and Francis).
19. Karl, David M., and Matthew J. Church. "Microbial oceanography and the Hawaii Ocean Time-series programme", Nature Review Microbiology, 2014.
20. Liu, Y., Xie, J., Ong, C., Vecitis, C. and Zhou, Z. (2015). Electrochemical wastewater treatment with carbon nanotube filters coupled with in situ generated H_2O_2. *Environmental Science: Water Research & Technology*, 1(6), pp. 769-778.
21. Anav Alessandro, Prince Friedlingstein et al. "Spatio-temporal patterns of terrestrial gross primary production: A review: GPP spatio-temporal patterns", Reviews of Geophysics, 2015.
22. Hossain, M. F. "Global environmental vulnerability and the survival period of all living beings on earth", International Journal of Environmental Science and Technology, 2018.
23. Hossain, M. F. (2018). Photonic Thermal Energy Control to Naturally Cool and Heat the Building. Applied Thermal Engineering. 131, 576–586. (Elsevier).

24. Andres, R., Boden, T., and Higdon, D: A new evaluation of the uncertainty associated with CDIAC estimates of fossil fuel carbon dioxide emission, Tellus B, 66, 23616, https://doi.org/10.3402/tellusb.v66.23616, 2014.
25. Boden, T.A and Andres, R.J.: Global Regional, and National Fossil-Fuel CO_2 Emissions, available at: http://cdiac.ornl.gov/trends/emis/overview_2013.html (last access: April 2016), Oak Ridge National Laboratory, US Department of Energy, Oak Ridge, Tenn., USA, 2016.
26. Yanbiao Liu, Peng Wu, Fuqiang Liu, Fang Li, Xiaoqiang An, Jianshe Liu, Zhiwei Wang, Chensi Shen, Wolfgang Sand. "Electroactive Modified Carbon Nanotube Filter for Simultaneous Detoxification and Sequestration of Sb(III)", Environmental Science & Technology, 2019.
27. Feely, R. A. "Global nitrogen deposition and carbon sinks", Nature Geoscience, 07/2008.
28. Kenneth, G. et al. "Unravelling the link between global rubber price and tropical deforestation in Cambodia", Nature Plants, 2018.
29. Houghton, R.A. "Balancing the Global Carbon Budget", Annual Review of Earth and Planetary Sciences, 2017.
30. Ciais, P. et al: Chapter 6: Carbon and other Biogeochemical Cycles, in: Climate Change 2013 The Physical Science Basis, edited by: Stocker, T., Qin. D., and Platner, G-K., Cambridge University Press, Cambridge, 2013.
31. Coherent spin–photon coupling using a resonant exchange qubit. A. J. Landig, J. V. Koski, P. Scarlino, U. C. Mendes, A. Blais, C. Reichl, W. Wegscheider, A. Wallraff, K. Ensslin & T. Ihn. Nature 560, pages 179–184 (2018).
32. Weiland, P. (2009). Biogas production: current state and perspectives. *Applied Microbiology and Biotechnology*, 85(4), pp. 849-860.
33. Kane, M. (2003). Small hybrid solar power system. *Energy*, 28(14), pp. 1427-1443.
34. Prietzel, Jorg, Lothar Zimmermann, Alferd Schubert, and Dominik Christophel. "Organic matter losses in German Alps forest soils since the 1970s most likely caused by warming", Nature Geoscience, 2016.
35. Dietzenbacher, E., Pei J.S., and Yang, C.H.: Trade, Production fragmentation, and China's carbon dioxide emissions, J. Environ. Econ, Manag., 64, 88-101-2012.
36. Van der Werf, G.R., Dempewolf, J. et al.: Climate regulation of fire emissions and deforestation in equatorial Asia, P. Natl. Acad. Sci. USA, 15, 20350-20355, 2008.
37. Colonna, Piero, Emiliano Casati, Carsten Trapp, Tiemo Mathijssen, Jaakko Larjola, Teemu Turunen-Saaresti, and Antti Uusitalo. "Organic Rankine Cycle Power Systems: from the Concept to Current Technology, Applications and an Outlook to the Future", Journal of Engineering for Gas Turbines and Power, 2015.
38. Gelfand, I., Sahajpal, R., Zhang, X., Izaurralde, R., Gross, K. and Robertson, G. (2013). Sustainable bioenergy production from marginal lands in the US Midwest. *Nature*, 493(7433), pp. 514-517.
39. Izadyar, N., Ong, H., Chong, W. and Leong, K. (2016). Resource assessment of the renewable energy potential for a remote area: A review. *Renewable and Sustainable Energy Reviews*, 62, pp. 908-923.
40. Chen, J., Dimitar Dimitrov, Tatiana Dimitrova, Paul Timans et al. "Carrier density profiling of ultra-shallow junction layer through corrected C-V plotting", Extended Abstracts - 2008 8th International Workshop on Junction Technology (IWJT '08), 2008.
41. Chen, Y. et al. "A pan-tropical cascade of fire driven by El Niño/Southern Oscillation", Nature Climate Change, 2017.
42. J. Milliman and R. Mei-e, in Climate Change: Impact on Coastal Habitation, D. Eisma, Rd. (CRC Press, Boca Raton, FL, 1995), pp. 57-83.
43. Davis, S.J and Calderia, K: Consumption-based accounting of CO_2 emissions, P. Natl. Acad. Sci. USA, 107, 5687-5692, 2010.
44. J.C. van Dam, Ed., Impacts of Climate Change and Climate Variability on Hydrological Regimes (Cambridge Univ, Press, Cambridge, 1999).
45. Arnell, N. et al, in Climate Change 1995: Impacts, Adaptations, and Mitigation of Climate Change, R.T. Watson et al, Eds, (Cambridge Univ. Press, Cambridge, 1996), pp. 325-363.

46. Chevallier, F: On the statistical optimality of CO_2 atmospheric inversions assimilating CO_2 column retrievals, Atmos, Chem, Phys., 15, 11133-11145, doi: https://doi.org/10.5194/acp-15-11133-2015, 2015.
47. Schwietzke, S. et al.: Upward revision of global fossil fuel methane emissions based on isotope database, Nature, 538, 88-91, 2016.
48. C.S., and Regnier, P.A.G.: The changing carbon cycle of the coastal ocean, Nature, 504, 61-70, 2013.
49. Pierre, R., Ronny Lauerwald, Philippe Ciais. "Carbon Leakage through the Terrestrial-aquatic Interface: Implications for the Anthropogenic CO_2 Budget", Procedia Earth and Planetary Science, 2014.
50. Romero-García, J., Sanchez, A., Rendón-Acosta, G., Martínez-Patiño, J., Ruiz, E., Magaña, G. and Castro, E. (2016). An Olive Tree Pruning Biorefinery for Co-Producing High Value-Added Bioproducts and Biofuels: Economic and Energy Efficiency Analysis. *BioEnergy Research*, 9(4), pp. 1070-1086.
51. Duce, R. A.et al: Impacts of atmospheric anthropogenic nitrogen on the open ocean, Science, 320, 893-897, 2008.
52. Hossain, M. F. (2018). Green Science: Advanced Building Design Technology to Mitigate Energy and Environment. Renewable and Sustainable Energy Reviews. 81 (2), 3051-3060. (Elsevier).
53. Liu, Zhu, Dabo Guan et al. "Reduced carbon emission estimates from fossil fuel combustion and cement production in China", Nature, 2015.
54. Hossain, M. F. (2017). Green Science: Independent Building Technology to Mitigate Energy, Environment, and Climate Change. Renewable and Sustainable Energy Reviews. 73; 695-705. (Elsevier).
55. Ruiz, H., Martínez, A. and Vermerris, W. (2016). Bioenergy Potential, Energy Crops, and Biofuel Production in Mexico. *BioEnergy Research*, 9(4), pp. 981-984.
56. Hossain, M. F. (2018). Photon energy amplification for the design of a micro PV panel. International Journal of Energy Research. https://doi.org/10.1002/er.4118. (Wiley). 2017.
57. Grätzel, M. (2001). Photoelectrochemical cells. *Nature*, 414(6861), pp. 338-344.
58. P.P., Miller, et al: Audit of the global carbon budget: estimate errors and their impact on uptake uncertainty, Biogeosciences, 12, 2565-2584, https://doi.org/10.5194/bg-12-2565-2015, 2015.
59. Hossain, M. F. (2016). Solar Energy Integration into Advanced Building Design for Meeting Energy Demand. International Journal of Energy Research. 40, 1293-1300. (Wiley).
60. S.L. Postel et al., Science 271, 785 (1996).

Part IV
Innovative Infrastructures and Transportation Engineering

Chapter 8
Green Infrastructure

Abstract Since the 1970s there has been increased massive development of urban infrastructure due to the construction of hardscapes that are eventually causing severe environmental crisis. We certainly need advanced technology for the urban infrastructure to mitigate this environmental perplexity. Therefore, in this chapter a green *infrastructure* system has been proposed by implementing green alley, stormwater collection, green roof, and eventually urban forest management by all possible advanced engineering applications to confirm an environmentally friendly infrastructure to develop a resilient urban which can make environment green. Simply the green infrastructure construction will support to mitigate local transformation of climate change that will help to deal with, recover, and manage the stability because of the environmental crises by its adaptive capacity of implementation of sophisticated technologies in stormwater management, green roof construction, and urban forestry management which are the better way to boost the urban environmental sustainability.

Keywords Sustainable infrastructure · Green alleys · Stormwater management · Green roof construction · Urban forest management · Environmental sustainability

Introduction

The recent threat of climate change is due to the challenges of increasing conventional infrastructure system in urban area causing severe losses of the ability of the urban ecological adaptability, flexibility, and sustainability [1, 2]. Hence, the conventional infrastructure system impacts heavily the built environment locally and thus this cost deeply to mitigate global greenhouse gas (GHG) emissions in order to meet the Paris Protocol for establishing long-term GHG mitigation to zero level by 2050 [3–5]. Therefore, there is an urgent need to upgrade the conventional infrastructure system into green infrastructure system in order to keep its functional role to mitigate climate change. Stormwater management is an important part in the

M. F. Hossain, *Global Sustainability in Energy, Building, Infrastructure, Transportation, and Water Technology*,
https://doi.org/10.1007/978-3-030-62376-0_8

infrastructure system because stormwater falls onto the roofs, streets, parking lots, and concrete or other impermeable materials that cannot reach the soil due to the conventional system of infrastructure. It is because the conventional stormwater drainage system uses gutters, storm sewers, and other engineering collection methods to collect the water and discharge to the nearest water body, which causes stormwater runoff to flooding which becomes one of the major causes of water pollution and leads to a disaster for the infrastructure in urban areas because of the trash, heavy metals, and other pollutants it carries [6–8]. To mitigate these problems, green infrastructure development could be an excellent option just because when stormwater falls onto green infrastructure, water will be able to penetrate into the soil easily due to its excellent permeable ability. Thus, in this study the construction of modern green alley including stormwater storage tunnels, green roof, and urban forest management is proposed to confirm the development of green infrastructure to change the mechanism on conventional infrastructure system to mitigate the climate change eventually.

Methods and Materials

To achieve rapid stormwater runoff into sewers and avoid flood, permeable construction of urban alleys for green infrastructure such as green alley, rainwater collection, green roof, and urban forest management has been proposed. Thus, the proposition for permeable pavement construction in this study has been conducted to confirm the construction of green alleys in order to permeate stormwater into the ground in a most efficient way. Hence, the construction of proper subsoiling is being proposed for the construction of permeable pavement, which refers to the maintenance of a porous layer of soil underneath; the permeable pavement is expected to penetrate 3 in. of stormwater from a 1-h storm in its 30–35-year life which means a 70–90% drainage volume reduction [9–11]. Subsequently, here, another method of rainwater collection by controlling stormwater connection to the downspouts from homes and commercial buildings and reconnecting them to a collection or slow dispersion system, of a cistern for storage and then rain garden for slow dispersion, or a rainwater harvesting design system for both collecting and storing stormwater for later use has been proposed. Simply rain garden is proposed to install at unpaved space because of its versatile features to mimic natural hydrology by permeation, evaporation, transpiration, and evapotranspiration of runoff; rain gardens that collect and absorb stormwater from rooftops, sidewalks, and streets should further be developed as bioretention or bioinfiltration cells to avoid flooding.

Subsequently, green roof system, another green infrastructure technology, is construction to a roof covered with waterproof membrane with plants or trees suitable for the local climate for 3–15 in. of soil, sand, or gravel by providing shadows, and cutting down carbon dioxide indirectly through lowering cooling demand for electricity [12–14]. Simply, green roofs can reduce albedo, the heat-trapping element, tremendously as well as it detains nearly 30% runoff while for storms less than 1 in., the number rises to 90% [15, 16]. In addition, seasonal and physiological evapotranspiration rates of plants of the roof also affect the effectiveness of runoff control, with summer growing season being better than winter. As the green roof is made up of plants and trees, it has similar benefits as trees including the transformation of water and air pollutants mentioned above [17, 18]. With the help of a 1000-square-foot green roof, around 40 pounds of particulate matter can be removed annually, which equals to the annual emission of 15 passenger cars [19, 20]. In addition to the benefits of removing pollutants, green roofs can also provide climate change mitigation through managing temperature.

Finally the proper urban forestry management has been proposed since it is the significant contributor to improve urban life quality, including having plenty of benefits including blocking and penetrating rainwater to avoid flooding and improving water quality, absorbing and transforming air pollutants, providing wind breaks to protect buildings from strong wind, and reducing heat island effect [21–23]. Therefore, planting and maintaining trees in urban area shall indeed be an excellent option considering its benefits for resilience, adaptation, and climate mitigation. According to estimation, a typical medium-sized tree can block around 2380 gallons of rainfall per year. When it comes to mitigating urban heat island impacts, trees typically absorb 70–90% of sunlight in summer and 20–90% in winter (because of the seasonal variation between deciduous trees and evergreens) and further reduce the maximum surface temperature of the roofs and walls by 11–25 °C [24, 25]. The new shade trees, planted around houses, can lead to annual cooling energy savings of 1% per tree and annual heating energy use can be decreased by almost 2% per tree [26–28]. Besides, direct energy savings from shading by trees could reduce carbon emissions around 1.5–5% due to decreases in cooling energy use [29, 30]. As mentioned above, trees can absorb air pollutants; these include particulate matter, sulfur dioxide, ground-level ozone, nitrogen oxides, and carbon monoxide. As estimated by a research, urban trees in the USA remove 784,000 tons of pollutant per year, which creates $3.8 billion economic value [31, 32]. By increasing the urban tree coverage rate such as in New York City by 10%, the ground-level ozone can be cut down by 3%; besides, a quarter ton of NOx and over 1 ton of particulate matter will be cut down per day with one million additional trees in the city [33, 34].

Eventually the proposition of better approach for the development of green infrastructure including sustainable urban design and planning and smart growth technology by incorporating the green infrastructure construction shall indeed overcome the problems of stormwater and environmental crisis management [24, 35]. Subsequently, the best management system must be practiced; especially where higher density housing is present in populated area green management should be practiced, providing green open spaces, large-scale urban forestry projects in neighborhoods, and greenbelts around cities in order to enable coastal wetlands to buffer against flooding. Besides, in flood zones, local building codes may be required to follow for raising buildings or bridges above current and future flood levels or setting the first floors at floodable positions with the application of green management. As green infrastructure is a method for creating more resilient metropolitan communities, improving environmental sustainability, smart growth, and climate adaptation in urban areas concurrently, smart growth practice may be the tool of urban design and planning to improve resource efficiencies, increase building density, realize mixed land uses, build more open space, achieve public transit-oriented development, and enhance quality of life.

Results and Discussion

Having a sustainable and healthy rural community with the significance of acquiring green infrastructure is the prime task for the urban authorities to collect and manage stormwater properly. It is the duty of the urban authorities to monitor the stormwater management technology by conducting the development green infrastructure to acquire environmentally friendly stormwater management by constructing green alleys [12, 13]. Thus, it is important to carry out the investigations within the catchment area of water, which accounts for 10–15% of the yearly water distribution from any conventional infrastructure and stormwater management system that must take good care of the removal of prolonged nitrogen which causes environmental pollution. Consequently, to deal with the customary measures to oversee stormwater management, urban authorities must analyze different climate change settings to handle the impacts of climate change due to the conventional infrastructure system. As a result, stormwater effective management shall turn out to be more and more important, and thus it will work actively to separate stormwater from joint system, thereby causing a reduction in the overflow risk, and ensuring a smoother flow which will eventually avert overflow of sewage system facilities, or drainage system. Besides, maximizing its permeable ability of the infrastructure by designing green infrastructure will increase the capacity of the standard management of a 10-year reign event within a 24-h period to make it adapted for projected increases in frequency and intensity of stormwater management properly. Simply, the green infrastructure shall help the management of stormwater by bioswales, and can help for water retention during the transferring of runoff from one place to another by its installation in long narrow spaces, which can be installed between the sidewalk and the curb with vegetated, mulched, or xeriscape channel in the rain garden. Consequently, the planter box can work properly with vertical walls and either open or closed bottoms to fit for space-limited urban areas acting as streetscape to act as additional rain garden to mitigate the stormwater for urban area. Nearly four billion people staying within the urban areas, yearly, dispose waste of more than 2 billion tons globally in an unmannered way of stormwater disposal [8, 16]. It is thus necessary for stormwater management principles like "polluter pays principle" to be practiced, and for the Waste Framework Directive to be set out to console the pollution of the environment by constructing green alleys. In order to construct green alleys, the stormwater management approach must be set with waste prevention technology as a preferred option, followed by reuse, then recycling, other recovery forms like energy from waste, and finally improved last disposal that is followed by enforced strict monitoring. Clearly, it is essential to appropriately manage stormwater so as it does not adversely disturb places of spatial interest; be a nuisance via noise or odors; pose risk to animals, plants, soil, or water; cause harm to the environment; or endanger the health of human beings in the urban area [15, 34].

Subsequently, due to the conventional roof in the urban area based on the present information, the worldwide heat trapping albedo reflected back to the atmosphere increased more rapidly ever since 1000 than in the course of past centuries [5, 18]. In 2007, the Intergovernmental Panel on Climate Change (IPCC) reported that this increase in reflection of albedo into the atmosphere is beyond the worst-case scenarios to meet the goal of Paris protocol. Over the last 100 years, the efficiency of natural sinks like the roof surface, which absorb albedo, has declined, indicating that human efforts to lessen these heat-trapping elements will have to be increasingly effective [7, 12]. Consequently, establishing environmental compatibility examination and effect estimates of utilized products crossways the whole life cycle is important and thus the development of green roofing should carefully be measured to attain the best environmental management, and ultimately secure a sustainable urban system. It is thus estimated that compared to traditional roofs, green roofs can cut down heat-trapping element nearly by 2% in winter and 6% in summer in urban areas. Simply, construction of the green roof would be a great way to control climate change and keep urban areas environmentally friendly.

Eventually, urban forests can play an important role in benefitting the environmental conditions of urban system since it controls local climate, slowing wind and stormwater, and filters air and sunlight. Simply, the urban forest indeed plays a critical role in cooling the urban heat island effect, and potentially reducing the number of unhealthy ozone days that plague major cities in peak summer months. As cities struggle to comply with air quality standards, urban forest can help to clean the air by eliminating the most serious pollutants in the urban atmosphere cycle, nitrogen oxides (NOx), sulfuric oxides (SOx), and heat-trapping albedo elements. Ground-level ozone, or smog, is created by chemical reactions between NOx and volatile organic compounds (VOCs) in the presence of sunlight in urban areas. Urban vehicle emissions (especially diesel), and emissions from industrial facilities, are the major sources of NOx. Besides the urban vehicle emissions, industrial emissions, gasoline vapors, chemical solvents, trees, and other plants are the major sources of VOCs that cause a serious impact on the urban environment. Urban particulate pollution, or particulate matter (PM10 and PM25), is made up of microscopic solids or liquid droplets that can be inhaled and retained in lung tissue causing serious health problems and thus most particulate pollution begins as smoke or diesel soot and can cause serious health risk to people with heart and lung diseases and irritation to healthy citizens. Here the urban forestry can play a tremendous role in mitigating these diseases and enhancing the health of urban citizens and thus these urban trees are the most important, cost-effective solution to reducing pollution and improving air quality in the urban area.

Simply it can be said that the construction of green infrastructure, which is a basic system of green alleys, green roofs, and urban forest, serves urban system to facilitate the essential routes to function the infrastructure system properly and ensure comfortable day-to-day lives of the habitants in the urban area [19, 24]. Since the green infrastructure of green alley structures of rainwater collection, green roof, and urban forest management are interconnected systems it plays a critical role along with the physical components of the urban area by offering best sustainable infrastructure services to facilitate or improve the social condition of urban lives.

Conclusion

To conclude, green infrastructure, mentioned in this chapter, contributes to proper dealing of stormwater runoff management, green roof for albedo control, and urban forest to mitigate greenhouse gases and air pollutants, increasing biodiversity, providing less heat stress, managing climate adaptation, realizing sustainability, and improving human life quality in the urban area. It is thus used to provide an ecological framework for social, economic, and environmental health of the urban areas. Therefore, the application of sophisticated system mentioned above needs to be realized for the construction of green infrastructure which indeed shall be a good solution to achieve environmental and sustainability goals and further create a better resilient community of the urban area.

Acknowledgements This research was supported by Green Globe Technology under the grant RD-02019-03. Any findings, conclusions, and recommendations expressed in this chapter are solely those of the author and do not necessarily reflect those of Green Globe Technology.

References

1. Andres, R. J., Boden, T. A. & Higdon, D. A new evaluation of the uncertainty associated with CDIAC estimates of fossil fuel carbon dioxide emission. Tellus B Chem. Phys. Meteorol. 66, 23616 (2014).
2. Hossain, Md. Faruque (2016). Solar Energy Integration into Advanced Building Design for Meeting Energy Demand. International Journal of Energy Research. 40, 1293-1300. (Wiley).
3. Achard, F. et al. Determination of tropical deforestation rates and related carbon losses from 1990 to 2010. Glob. Change Biol. 20, 2540–2554 (2014).
4. Fairhead, James, and Melissa Leach. 1996. Misreading the African Landscape: Society and Ecology in a Forest-Savanna Mosaic. Cambridge: Cambridge University Press.
5. Mason Earles, J., Yeh, S. & Skog, K. E. Timing of carbon emissions from global forest clearance. Nat. Clim. Change 2, 682–685 (2012).
6. Andreozzi, R., Caprio, V., Ciniglia, C., De Champdor_e, M., Lo Giudice, R., Marotta, R., Zuccato, E., 2004. Antibiotics in the environment: occurrence in Italian STPs, fate, and preliminary assessment on algal toxicity of amoxicillin. Environ. Sci. Technol. 38, 6832-6838.
7. Betts, R. A., Jones, C. D., Knight, J. R., Keeling, R. F. & Kennedy, J. J. El Nino and a record CO2 rise. Nat. Clim. Change 6, 806–810 (2016).
8. Hongyan Bao, Jutta Niggemann, Li Luo, Thorsten Dittmar, Shuh-Ji Kao. "Aerosols as a source of dissolved black carbon to the ocean", Nature Communications, 2017.

9. Bagnall, J.P., Malia, L., Lubben, A.T., Kasprzyk-Hordern, B., 2012b. Stereoselective biodegradation of amphetamine and methamphetamine in river microcosms. Water Res. 47, 5708-5718.
10. Duce, R. A. et al. Impacts of atmospheric anthropogenic nitrogen on the open ocean. Science 320, 893–897 (2008).
11. Hossain, Md. Faruque, and Mukai, Hiroshi (2000). Importance of Nutrients (N, P, and NO_3+NO_2) in Growth of the Surfgrass, Phyllospadix iwatensis Makino.
12. Ballantyne, A. P., Alden, C. B., Miller, J. B., Tans, P. P. & White, J. W. C. Increase in observed net carbon dioxide uptake by land and oceans during the past 50 years. Nature 488, 70–72 (2012).
13. Ballantyne, A. P. et al. Audit of the global carbon budget: Estimate errors and their impact on uptake uncertainty. Biogeosciences 12, 2565–2584 (2015).
14. de Sherbinin, A. 2002. Land-Use and Land-Cover Change. In A CIESIN Thematic Guide. Palisades, NY: Center for International Earth Science Information Network of Columbia University.
15. Hossain, Md. Faruque (2018). Green Science: Advanced Building Design Technology to Mitigate Energy and Environment. Renewable and Sustainable Energy Reviews. 81 (2), 3051-3060. (Elsevier).
16. Hossain, Md. Faruque (2017). Green Science: Independent Building Technology to Mitigate Energy, Environment, and Climate Change. Renewable and Sustainable Energy Reviews. 73; 695-705. (Elsevier).
17. Box, P. 2002. Spatial Units as Agents: Making the Landscape an Equal Player in Agent-Based Simulations. In Integrating Geographic Information Systems and Agent-Based Modeling Techniques for Simulating Social and Ecological Processes, edited by R. H. Gimblett. New York: Methuen.
18. Schwietzke, S. et al. Upward revision of global fossil fuel methane emissions based on isotope database. Nature 538, 88–91 (2016).
19. Gibson, C.C., E. Ostrom, and T.K. Ahn. 2000. The concept of scale and the human dimensions of global change: A survey. Ecological Economics 32:217-239.
20. Hossain, Md. Faruque (2017). Invisible Transportation Infrastructure Technology to Mitigate Energy and Environment. Energy, Sustainability, and Society. 7:27. (Springer).
21. Bridges, E.M., and L.R. Oldeman. 1999. Global assessment of human-induced soil degradation. Arid Soil Research and Rehabilitation 13 (4):319-325.
22. Bauer, J. E. et al. The changing carbon cycle of the coastal ocean. Nature 504, 61–70 (2013).
23. Diouf, A., and E. F. Lambin. 2001. Monitoring land-cover changes in semi-arid regions: Remote sensing data and field observations in the Ferlo, Senegal. Journal of Arid Environments 48:129–148.
24. Davis, S. J. & Caldeira, K. Consumption-based accounting of CO_2 emissions. Proc. Natl. Acad. Sci. U.S.A. 107, 5687–5692 (2010).
25. Donald, Carey E., Marc R. Elie, Brian W. Smith, Peter D. Hoffman, and Kim A. Anderson. "Transport stability of pesticides and PAHs sequestered in polyethylene passive sampling devices", Environmental Science and Pollution Research, 2016.
26. Denman, K. L. et al. Couplings Between Changes in the Climate System and Biogeochemistry (Cambridge University Press, 2007).
27. Emilio Garcia-Robledo, Cory C. Padilla, Montserrat Aldunate, Frank J. Stewart, Osvaldo Ulloa, Aurélien Paulmier, Gerald Gregori and Niels Peter Revsbech. Cryptic oxygen cycling in anoxic marine zones. PNAS (2017) 114 (31) 8319-8324.
28. Postel, S. L., Daily, G. C. & Ehrlich, P. R. Human appropriation of renewable fresh water. Science 271, 785 (1996).
29. Li, W. et al. Reducing uncertainties in decadal variability of the global carbon budget with multiple datasets. Proc. Natl. Acad. Sci. U.S.A. 113, 13104–13108 (2016).
30. Liu, Z. et al. Reduced carbon emission estimates from fossil fuel combustion and cement production in China. Nature 524, 335–338 (2015).

31. Erb, K.-H. et al. Bias in the attribution of forest carbon sinks. Nat. Clim. Change 3, 854–856 (2013).
32. Parazoo et al. "Contrasting carbon cycle responses of the tropical continents to the 2015–2016 El Niño", Science, 2017.
33. Gonzalez-Gaya, B. et al. High atmosphere-ocean exchange of semivolatile aromatic hydrocarbons. Nat. Geosci. 9, 438–442 (2016).
34. Hossain, Md. Faruque (2016). Theory of Global Cooling. Energy, Sustainability, and Society. 6:24. (Springer).
35. Houghton, R. Balancing the global carbon budget. Annu. Rev. Earth Planet. Sci. 35, 313–347 (2007).

Chapter 9
Invisible Roads and Transportation Engineering

Abstract A new technology ***invisible road*** has been proposed in this chapter to confirm zero emission of greenhouse gas from transportation infrastructure system to green infrastructure and transportation system. Simply underground ***maglev*** system is to be constructed for all transportation systems to run the vehicle smoothly just over 2 ft over the earth surface by propulsive and impulsive force at flying stage. Vehicles will not require any energy since it will run by superconducting electromagnetic force created by the maglev technology. A wind energy modeling has also been added to meet the vehicle's energy demand when it will run on non-maglev area. Naturally all maglev infrastructure network will be covered by evergreen herb except pedestrian walkways to absorb CO_2, ambient heat, and moisture (vapor) from the surrounding environment to make it cool. Indeed, the proposed maglev transportation infrastructure technology will be an innovative one in modern engineering science which will also reduce the climate change dramatically.

Keywords Electromagnetic force · Invisible infrastructure · Wind energy · Energy conversion · Cost reduction · Transportation innovation

Introduction

Urban and suburban area massively depends on transportation infrastructure networks which are primarily constructional with concrete and asphalt and it does not have enough vegetation to absorb heat caused by these asphalt and concrete [1, 2]. Recent research found that transportation infrastructure on earth is approximately 0.9% of the total planetary surface area of 196.9 million mi^2 which is equivalent to 1.77 million miles square infrastructure on earth which is causing nearly 6% of global warming by reflecting heat (albedo) back to the space [3, 4]. On the other hand, conventional energy utilization for the transportation sectors is not only costly but also causing adverse environmental impact [5, 6]. A variety of studies have been performed to understand long-term climate variations by conventional energy

M. F. Hossain, *Global Sustainability in Energy, Building, Infrastructure, Transportation, and Water Technology*,
https://doi.org/10.1007/978-3-030-62376-0_9

utilization by the transportation sectors that is casing nearly 28% global energy consumption which is equivalent to megaton CO_2 and is responsible for 28% of global warming and thus infrastructure and transportation fuel cause a total 34% of global warming [7, 8]. To mitigate transportation infrastructure crisis and its adverse environmental impact, I therefore proposed a new technology of maglev transportation infrastructure system for building better transportation infrastructure system.

A recent study by Cai and Chen described the dynamic characteristics, magnetic suspension systems, vehicle stability, and suspension control laws of maglev/guideway coupling systems about the maglev transportation system, but showed that commercial application of this research modeling considering life cycle cost analysis, technology implementation, and infrastructure development did not show any possibility. Therefore, the approach of this research is to apply the maglev transportation infrastructure commercially for confirming a greener and cleaner transportation infrastructure system where all vehicles shall run just over 2 ft above the earth surface at flying stage by the act of propulsive and impulsive superconducting force. Since the vehicles will run by electromagnetic force it will not require any energy while running over the maglev. To mitigate energy consumption when vehicle needs to run on maglev area, additional technology has also been proposed to implement wind energy into the vehicles while it is in a motion as backup energy source. Thus, a detailed mathematical modeling using MATLAB Simulink software has been implemented for this wind energy utilization for the vehicles by performing turbine and drivetrain modeling. A concerted research effort has been performed recently on climate science and found that currently 402 ppm CO_2 is present in atmosphere that is causing global warming which needs to be cut down to 300 ppm CO_2 to confirm global cooling at a comfortable stage. Once maglev transportation infrastructure system is implemented throughout the world it will reduce 34% CO_2 per year. Thus, it will take only $\{ \int_{300}^{402} (1-0.34)dx \} = 67.32$ years to cool the atmosphere, resulting in no more climate change after 68 years. Simply it will be the most innovative technology in modern science to mitigate the cost and global warming dramatically.

Methods and Materials

In order to present maglev transportation infrastructure modeling, I have formulated the following calculation by using MATLAB software in terms of (1) guideway model system by adopting Bernoulli-Euler beam equation of a series of simply supported beams and (2) calculation of magnetic forces for uplift levitation and lateral guidance with allowable levitation and guidance distance considering lateral vibration control LQR algorithm, tuning parameters, and maglev dynamics.

To prepare the guideway modeling considering free body diagram (Fig. 9.1), I have considered multiple magnets with equal intervals (d) that are to be traveling at a various level of speed v, where m = beam weight, c = damping coefficient,

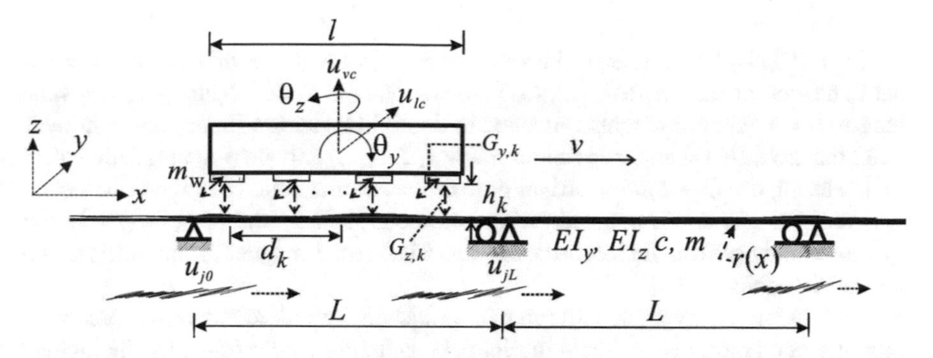

Fig. 9.1 A free body diagram shows the maglev guideway vs. vehicle force considering weight and motion where the superconducting guideway is below the vehicle body. It is functioned by a series of equal-distant concentrated masses to levitate the vehicle up to the superconducting guideway beam; the maglev bar gets stimulated by the lateral multi-support motion which is induced by the superconducting force to allow traveling on longitudinal direction

EI_y = flexural rigidity in the y direction, EI_z = flexural rigidity in the z direction, l = car length, m_w = lumped mass of magnetic wheel, m_v = distributed mass of the rigid car body, and $\theta_{i=x,\, y,z}$ = midpoint rotation components of the rigid car body. Considering these, I have formulated the equations of motion for the jth guideway girder carrying a moving maglev vehicle suspended by multiple magnetic forces as follows:

$$m\ddot{u}_{y,j}+c_y\dot{u}_{y,j}+\mathrm{EI}_y\ddot{u}_{y,j}=\sum_{k=1}^{K}\left[G_{y,k}\left(i_k,h_{y,k}\right)\varphi_j\left(x_k,t\right)\right] \tag{9.1}$$

$$m\ddot{u}_{z,j}+c_z\dot{u}_{z,j}+\mathrm{EI}_z\ddot{u}_{z,j}=p_0-\sum_{k=1}^{K}\left[G_{z,k}\left(i_k,h_{z,k}\right)\varphi_j\left(x_k,t\right)\right] \tag{9.2}$$

and

$$\varphi_j\left(x_k,t\right)=\delta\left(x-x_k\right)\left[H\left(t-t_k-\frac{(j-1)L}{v}\right)-H\left(t-t_k-\frac{jL}{v}\right)\right] \tag{9.3}$$

together with the following boundary conditions with lateral (y direction) support movements:

$$u_{y,j}\left(0,t\right)=u_{yj0}\left(t\right),u_{y,j}\left(L,t\right)=u_{yjL}\left(t\right) \tag{9.4}$$

$$\mathrm{EI}_z\dot{u}_{z,j}\left(0,t\right)=\mathrm{EI}_z\dot{u}_{z,j}\left(L,t\right)=0,$$

$$u_{z,j}\left(0,t\right)=u_{z,j}\left(L,t\right)=0 \tag{9.5}$$

$\mathrm{EI}_y \dot{u}_{y,j}(0,t) = \mathrm{EI}_y \dot{u}_{y,j}(L,t) = 0$ where $(y)' = \partial(y)/\partial x$, $(y) = \partial(y)/\partial t$, $u_{z,j}(x,t)$ = vertical deflection of the jth span, $u_{y,j}(x,t)$ = lateral deflection of the jth span, L = span length, K = number of magnets attached to the rigid levitation frame, $\delta(y)$ = Dirac's delta function, $H(t)$ = unit step function, $k = 1, 2, 3, \ldots, K$th moving magnetic wheel on the beam, $t_k = (k-1)d/v$ = arrival time of the kth magnetic wheel into the beam, x_k = position of the k-th magnetic wheel on the guideway, and $(G_{y,k}, G_{z,k})$ = lateral guidance and uplift levitation forces of the kth lumped magnet in the vertical and lateral directions [9, 10].

Since the maglev vehicle will run over guideway by superconducting force with lateral ground motion (as shown in Fig. 9.1), guidance forces tuned by the maglev system need to be controlled by the lateral motion of the moving maglev vehicle. Therefore, this study adopts the lateral guidance force ($G_{y,k}$) and the uplift levitation force ($G_{z,k}$) to keep and guide the k-th magnet of the vehicle that could be expressed as

$$G_{y,k} = K_0 \left(\frac{i_k(t)}{h_{z,k(t)}} \right)^2 K_{k,z} \tag{9.6}$$

$$G_{y,k} = K_0 \left(\frac{i_k(t)}{h_{z,k(t)}} \right)^2 \left(1 - K_{y,k}\right) \tag{9.7}$$

where $K_{y,k}$ and $K_{z,k}$ represent induced guidance factors and they are given by

$$K_{y,k} = \frac{\chi_k \times h_{y,k}}{W(1+\chi_k)}, K_{z,k} = \frac{\chi_k \times h_{y,k}}{W(1+\chi_k)} \tag{9.8}$$

In Eqs. (9.6) and (9.7), $K_0 = \mu_0 N_0^2 A_0/4$ = coupling factor, $\chi_k = \pi h_{y,k\,z,k}/4h$, W = pole width, μ = vacuum permeability, N_0 = number of turns of the magnet windings, A_0 = pole face area, $i_n(t) = i_0 + \imath_n(t)$ = electric current, $\imath_n(t)$ = deviation of current, and (i_0, h_{y0}, h_{z0}) = desired current and air gaps around a specified nominal operating point of the maglev wheels at *static* equilibrium. And the uplift levitation ($h_{y,k}$) and lateral guidance ($h_{z,k}$) gaps are, respectively, given by

$$h_{y,k}(t) = h_{y0} + u_{l,k}(t) - u_{y,j}(x_k), u_{l,k}(t) = u_{lc}(t) + d_k\theta_z \tag{9.9}$$

$$h_{z,k}(t) = h_{z0} + u_{v,k}(t) - u_{z,j}(x_k) + r(x_k), u_{v,k}(t) = u_{vc}(t) + d_k\theta_y \tag{9.10}$$

where $(u_{l,k}, u_{v,k})$ = displacements of the kth magnetic wheel in the y and z directions, (u_{lc}, u_{vc}) = midpoint displacements of the rigid car, (θ_y, θ_z) = midpoint rotations of the rigid car, $r(x)$ = irregularity of guideway, and d_k = location of the kth magnetic wheel to the midpoint of the rigid beam. As indicated in Eqs. (9.6)–(9.8), the motion-dependent nature and guidance factors ($K_{y,k}$,$K_{z,k}$) dominate the control forces of the maglev vehicle-guideway system. Next, the equations of motion of the 4-DOF rigid maglev vehicle (see Fig. 9.1) are written as

$$M_0 \ddot{u}_{lc} = g(t) + \sum_{k=1}^{K} G_{y,k},\ I_T \ddot{\theta}_Z = g(t) \times l + \sum_{k=1}^{K} \left[G_{y,k} d_k \right] \tag{9.11}$$

$$M_0 \ddot{u}_{vc} = p_0 + \sum_{k=1}^{K} G_{z,k}, I_T \ddot{\theta}_y = -\sum_{k=1}^{K} \left[G_{z,k} d_k \right] \tag{9.12}$$

in which $M_0 = m_v l + K m_w$ = lumped mass of the vehicle, $g(t)$ = control force to tune the lateral response of the maglev vehicle, I_T = total mass moment of inertia of the rigid car, and $p_0 = M_0 g$ = lumped weight of the maglev vehicle.

Though the vehicle will run by electromagnetic force, a wind turbine generator is to be used for powering the vehicle as the additional source of energy to exit the vehicle from road and park where maglev system is not available. Thus, the model is developed by doubly fed induction generator (DFIG) for producing electricity for transportation vehicles [11–13]. The fundamental equation governing the mechanical power of the wind turbine is

$$P_w = \frac{1}{2} C_p (\lambda, \beta) \rho A V^3 \tag{9.13}$$

where ρ is the air density (kg/m^3), C_p is the power coefficient, A is the intercepting area of the rotor blades (m^2), V is the average wind speed (m/s), and λ is the tip speed ratio [14]. The theoretical maximum value of the power coefficient C_p is 0.593; C_p is also known as Betz's coefficient. Mathematically,

$$\lambda = \frac{R\omega}{V} \tag{9.14}$$

R is the radius of the turbine (m), ω is the angular speed (rad/s), and V is the average wind speed (m/s). The energy generated by wind can be obtained by

$$Q_w = P \times (\text{Time})[\text{kWh}] \tag{9.15}$$

It is well known that wind velocity cannot be obtained by a direct measurement from any particular motion [9, 15]. In data taken from any reference, the motion needs to be determined for that particular motion; then, the velocity needs to be measured at a lower motion:

$$v(z) \ln\left(\frac{Z_r}{Z_0}\right) = v(Z_r) \ln\left(\frac{Z}{Z_0}\right) \tag{9.16}$$

where Z_r is the reference height (m), Z is the height at which the wind speed is to be determined, Z_0 is the measure of surface roughness (0.1–0.25 for crop land), $v(Z)$ is the wind speed at height Z (m/s), and $v(Z_r)$ is the wind speed at the reference height z (m/s). The power output in terms of the wind speed shall be estimated using the following equation:

$$P_w(v) = \begin{cases} \dfrac{v^k - v_C^k}{v_R^k - v_C^k} \cdot P_R & v_C \le v \le v_R \\ P_R & v_R \le v \le v_F \\ 0 & v \le v_C \text{ and } v \ge v_F \end{cases} \tag{9.17}$$

where P_R is the rated power, v_C is the cut-in wind speed, v_R is the rated wind speed, v_F is the rated cutout speed, and k is the Weibull shape factor [16]. When the blade pitch angle is zero, the power coefficient is maximized for an optimal TSR [17]. The optimal rotor speed is to be calculated by

$$\omega_{opt} = \frac{\lambda_{opt}}{R} V_{wn} \tag{9.18}$$

which will give

$$V_{wn} = \frac{R\omega_{opt}}{\lambda_{opt}} \tag{9.19}$$

where ω_{opt} is the optimal rotor angular speed in rad/s, λ_{opt} is the optimal tip speed ratio, R is the radius of the turbine in m, and V_{wn} is the wind speed in m/s.

The turbine speed and mechanical powers are depicted in Fig. 9.2 with increasing and decreasing rates of wind speed while the vehicle is in motion. When the wind is steady, the persistence forecasts yield good results. When the wind speed is increased rapidly, sudden "ramps" in power output are generated, which is a tremendous benefit for capturing the energy.

Standard Simulink/Sim Power Systems have been calculated by using Matlab-Simulink for the wind energy conversion that is to be stored in circuit-implemented inverter as a storage buffer, and all the electricity is to be supplied through the battery according to Peukert's law to start the engine and to be used when the vehicle is not in motion.

Though underground maglev system has the capability to allow to run up to 580 kph, the vehicles' high speed shall be calculated based on traffic flow, composition, volume, number and location of access points, and local environment importantly allotting sufficient number of lanes considering Greenshield's model following road and highway capacity analysis (Fig. 9.3).

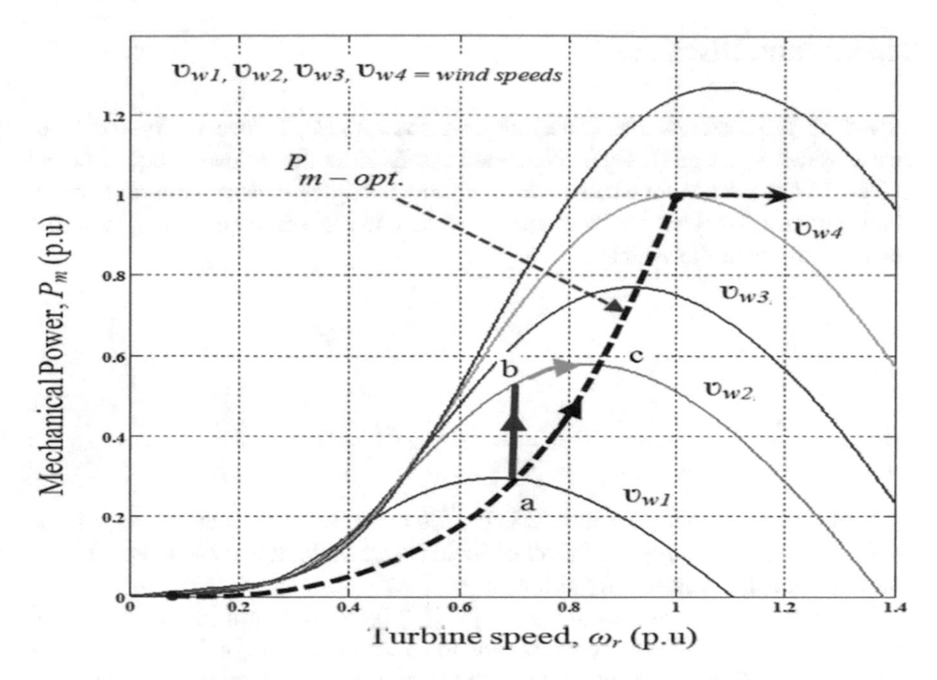

Fig. 9.2 Relationship between mechanical power generation and turbine speeds at different wind speeds for an implementation in a car

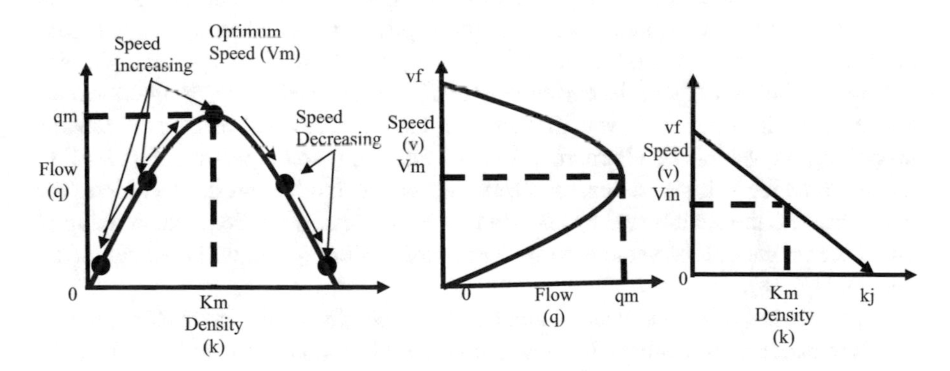

Fig. 9.3 The Greenshield's fundamental diagrams: (**a**) speed vs. vehicle density, (**b**) flow vs. vehicle density, and (**c**) speed vs. flow analysis

Since the maglev technology is invisible, to alert the drivers and pedestrian, the maglev roads, highways, and its exits have to be constructed by landscaping by covering the guideway by herb (green grass) and in between lanes of at least 2 ft should be left blank (no landscaping) in order to differentiate the lanes.

Results and Discussion

Based on the mathematical modeling described above, I have performed load-resistant factor design (LRFD) calculation considering the following equation and selected W24 × 84 beam which is the continuous maglev underground runs (metal track guideway) that need to be structurally sound to carry enough current, load, and levitation force of the vehicles:

$$Fy \propto \frac{nl^2}{h} \tag{9.20}$$

$$Fx \propto \frac{-1}{ktvx} \quad Fx \propto \frac{-nl^2}{h} \tag{9.21}$$

where Fy is the vehicle weight, n is the total number of coils in maglev, l is the current on each coil, h is the height of levitation, t is the thickness of conduction track, and k is the conductivity of track.

To construct under-maglev guideway just 2 ft below the earth surface, there will be a need to have a U-shaped cross section to fix the pole position. Naturally heavy-duty waterproofing membrane is to be used to protect the maglev underground runs for avoiding floods and moisture. It is well researched that the propulsion coils run in elliptical loops along both walls of the guideway, generating magnetic force when electricity runs through them. So, levitation and guidance coils will be formed that will create their own magnetic force once the applied superconducting magnets pass on it where propulsion and levitation are the key factors to run the vehicle. In propulsion, as the direction of the current charges back and forth in the propulsion coils above the wall of the guideway, the north and south poles will reverse repeatedly, propelling the vehicle by alternating force of attracting and repulsion (Fig. 9.4). In levitation as the vehicle passes, an electric current is induced in the coil along the guideway and the vehicle will be levitated by the force of attraction, which will pull up on the magnet in the vehicle, as well as by repulsion, which will push up on the magnet [10, 18].

To create levitation and lateral balance in the vehicle, an electromagnetic induction is to be used. To confirm the most efficient and economical way to produce the powerful magnetic field by using the superconducting coils I have assumed the permanent currents of about 700,000 A to go through these superconducting coils [9], hence creating a strong magnetic field of almost 5 T, i.e., 100,000 times stronger than the earth magnetic field by implementing the following block diagram (Fig. 9.5).

Simply it can be explained that when an electric current flows through the propulsion coils, a magnetic field is produced. The forces of attraction and repulsion between the coils and the superconducting magnets on the vehicle propel the vehicle forward in a flying stage up to 4 ft height where 2 ft shall be considered underground cover and other 2 ft just over the earth surface (Fig. 9.6). The vehicle's speed is to be

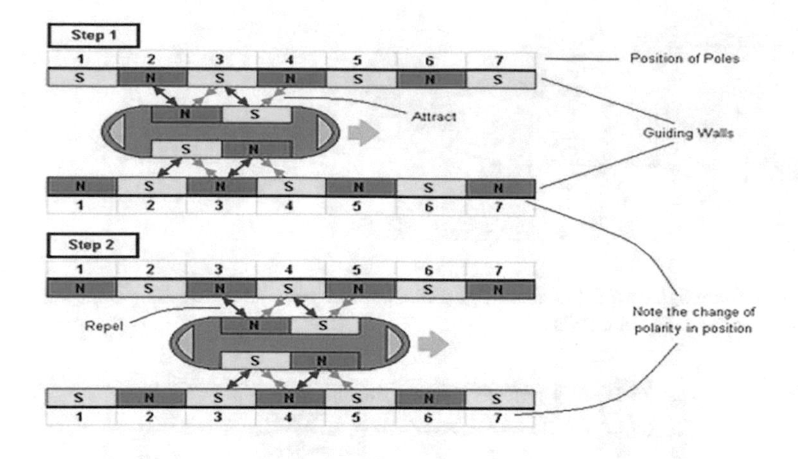

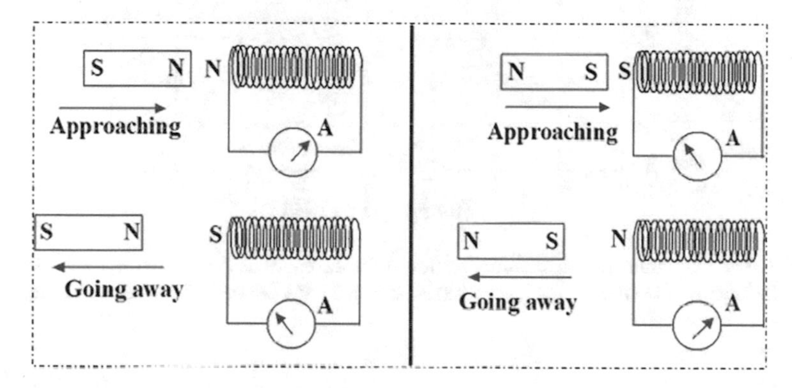

Fig. 9.4 The above figures indicate the polarization of the coil in different cases: (**a**) Schematic diagram of the director of the running vehicle (must be construction with magnet as shown in this diagram) on maglev propulsion via propulsion coils. (**b**) Near the receding S-pole becomes an N-pole to oppose the going away of the bar magnet's S-pole

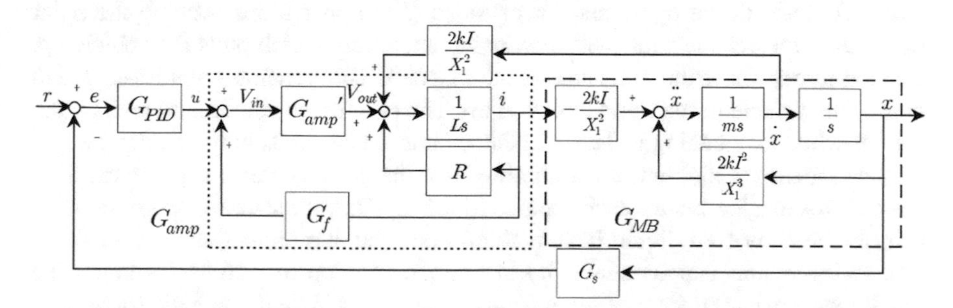

Fig. 9.5 Block diagram to control the mathematically modeled magnetic bearing system, a process to design the driver to operate the electromagnet. Here, the method is to determine the peripheral device values of the linear amplifier circuit that has the desired output by applying a generic algorithm and to identify the magnetic bearing system

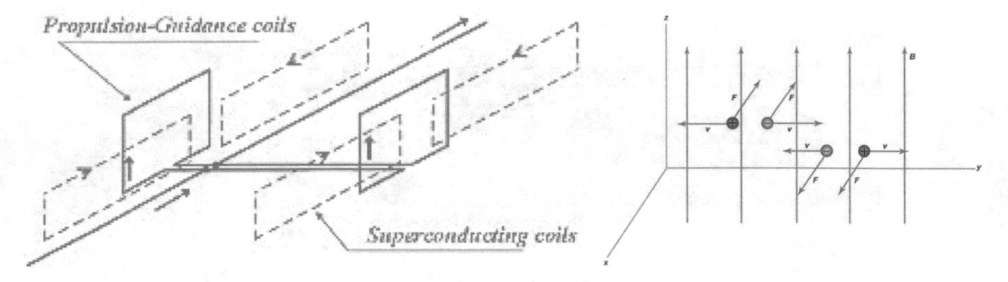

Fig. 9.6 The maglev vehicle's force and directional diagram as shown by propulsion guidance coils and superconducting coils

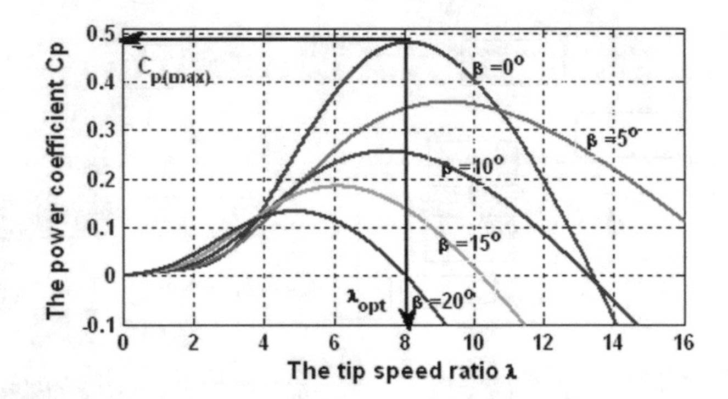

Fig. 9.7 The maximum values of C_p are achieved for the curve associated with $\beta = 2°$. From this curve, the maximum value of $C_p(C_{p,max} = 0.5)$ is obtained for $\lambda_{opt} = 0.91$. This value (λ_{opt}) represents the optimal speed ratio

adjusted by altering the timing of the polarity shift in the propulsion coils' magnetic field between north and south with the possibility of maximum speed of 580 kph [20, 21]. As the vehicle passes just 2 ft above the guideway (1 ft from the earth surface), an electric current is induced in the levitation and guidance coils, creating opposite magnetic poles in the upper and lower loops. The upper loops become the polar opposite of the vehicle's magnets, producing attraction, which pulls the vehicle up. The lower loops have the same pole as the magnets. This generates repulsion, which pushes the vehicle in the same direction up. The two forces combine to levitate the vehicle while maintaining its lateral balance between the walls of the guideway.

Subsequently a niobium-titanium alloy is to be used to create superconducting magnets for maglev, but to reach superconductivity, they must be kept cold. In order to keep the alloy cool, liquid helium should be used at a temperature of −269 °C since alloy retains superconductivity at temperatures up to −263 °C, though the maglev system can operate better at 6 °C to produce sufficient magnetic force.

In addition to underground maglev, the wind turbine generation system for the backup energy source is to be implemented for the optimal operation of the whole system and a robustness test should be performed by adding a wind speed signal and power coefficient.

These conditions permit application of the wind profile considered to be a wind speed signal with a mean value of 8 m/s and a rated wind speed of 10 m/s; the whole system is tested under standard conditions with a stator voltage of approximately 50% for 0.5 s between 4 and 4.5 s, approximately 25% between 6 and 6.5 s, and 50% between 8 and 8.5 s (Fig. 9.7). Thus, the machine is considered to be functioning in ideal conditions (no perturbations and no parameter variations). Moreover, to guarantee a unity power factor at the stator side, the reference for the reactive power has to be set to zero [22]. As a result of increasing wind speed, the generator shaft speed achieves maximum angular speed by tracking the maximum power point speed. Thus, the wind turbine always works optimally since the pole placement technique is to be used to design the tracking control [23]. Consequently, decoupling among the components of the rotor current was also performed to confirm that the control system worked effectively. The bidirectional active and reactive power transfer between the rotor and power system is exchanged by the generator according to the supersynchronous operation, achieving the nominal stator power, and the reactive power can be controlled by the load-side converter to obtain the unit's power factor to generate energy for powering vehicles [24, 25].

Construction Cost Estimate Comparison

Order of magnitude cost estimate was performed by using HCSS (Heavy Bid) software standard union rate of New York State locals with a project of 10% general condition, 10% overhead and profit, and 3% contingency over the hard cost of labor, materials, and equipment comparing between maglev infrastructure and traditional infrastructure system for a sample of 100 miles long and 128 ft wide (12 ft wide of 4 lanes in each direction, two-sided 10 ft service space, and 6 ft median in the center of the road). In order to determine that the underground guideway (w24 × 84) can last long I have calculated again the LRFD to provide the shoring to both sides for the entire 100 miles long and 128 ft wide (12 ft wide of 4 lanes in each direction, two-sided 10 ft service space, and 6 ft median in the center of the road) construction cost considering standard excavation up to 6 ft deep, with appropriate shoring with minimum embedment depth L_4 being 5 ft, standard soil pressure $\Upsilon_s = 120$ lbf/ft^3, angle of pressure $\Phi = 21°$, and soil pressure coefficient $c = 800$ lbf/ft^2. To prepare the conceptual estimate we need to determine the length of soldier piles. I have counted 6′ OC (on center) soldier piles on both sides by illustration and using the following LRFD method the soldier piles must be set to support the necessary excavation and/or earth pressure against collapse:

$$\text{Active earth pressure} \quad K_a = \tan^2\left(45^\circ - \frac{\Phi}{2}\right) \tag{9.22}$$

$$\text{Passive earth pressure} \quad K_b = \tan^2\left(45^\circ + \frac{\Phi}{2}\right) \tag{9.23}$$

Use Eqs. (9.22) and (9.23) to find the lateral earth pressure the solid piles must support:

$$\begin{aligned} P_{EM} &= \Upsilon_s h k_{a,\text{piles}} \\ &= \left(120\text{lbf}/\text{ft}^3\right)(6.0)\tan^2\left(45^\circ - \frac{21^\circ}{2}\right) \\ &= 340.128\text{lbf}/\text{ft}^2 \end{aligned}$$

To determine the type of steel beams required for the soldier piles, we have taken the bending moments about the tributary area of the piles:

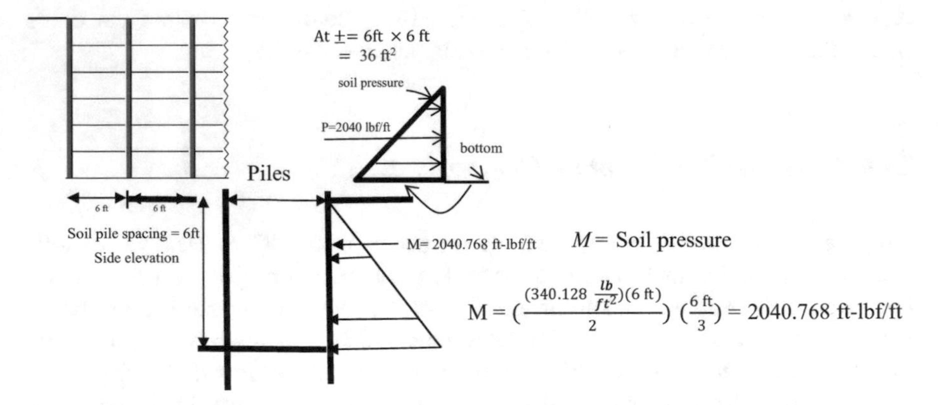

$$M = \left(\frac{\left(340.128\,\text{lb}/\text{ft}^2\right)(6\,\text{ft})}{2}\right)\left(\frac{6\,\text{ft}}{3}\right) = 2040.768\,\text{ft lbf}/\text{ft}$$

The moment is a distributed moment applied to the base of the tributary area of each soldier pile. Therefore, the moment is 2040.768 ft lbf per foot. The total moment on the soldier pile (at the base) is

$$\begin{aligned} M_0 &= M(6\text{ft}) \\ &= (2040.768\ \text{ftlbf}/\text{ft})(6\text{ft}) \\ &= 12{,}244.61\text{ftlbf} \end{aligned}$$

Now,

$$Z_{req} = \frac{M_0}{\Phi b\, Fy} = \frac{(12\text{in.}/\text{ft})(12{,}244.61\text{ft lbf})}{(0.9)(50{,}000\ \text{lbf}/\text{in.}^2)}$$
$$= 3.27\text{in.}^3$$

From AISC tables, the soldier piles have been selected as to be W12 × 26, and the perpendicular support w8 × 12 members 6 ft long.

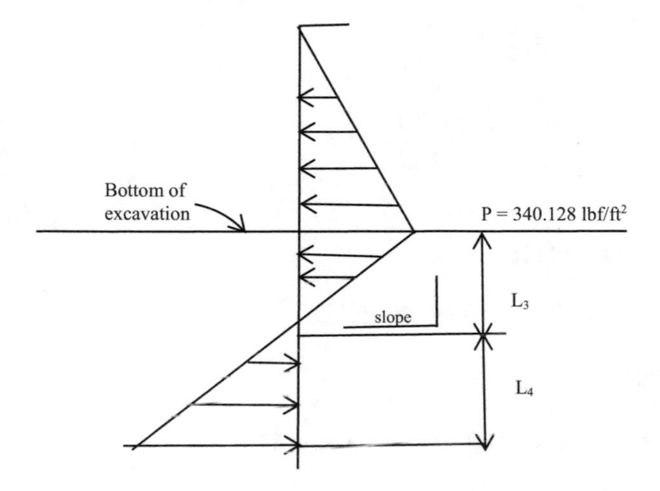

Then we have determined the depth required below subgrade by calculating the passive earth pressure coefficient using Eq. (9.23):

$$K_p = \tan^2\left(45^\circ + \frac{\Phi}{2}\right)$$
$$= \tan^2\left(45^\circ + \frac{21^\circ}{2}\right)$$
$$= 2.12$$

Then we have calculated the active earth pressure coefficient using Eq. (9.22):

$$K_a = \tan^2\left(45^\circ - \frac{\Phi}{2}\right)$$
$$= \tan^2\left(45^\circ - \frac{21^\circ}{2}\right)$$
$$= 0.4724$$

In order to determine the slopes of the excavation, depth is required. Since below the bottom of the excavation, pressure is both considered to be passive and have the same slope. The slope of the pressure profile above the reversal point is calculated from the standard equation for the slope, using L_3 as the rise and Υhk_a as the run (a value equal to the lateral earth pressure, expressed this way for the purpose of cancelation). Thus, the slope of the pressure profile below the reversal point can be calculated similarly, using L_4 as the rise and the product of $\Upsilon L_4 k_p$ as the run. Because the slopes are the same, the two equations can be equated. Rearranging to solve for L_3,

$$\frac{L_3}{\Upsilon hk_a} = \frac{L_4}{\Upsilon L_4 k_p}$$

$$L_3 = \frac{hk_a}{k_p} = \frac{(6\text{ft})(0.4724)}{2.12}$$
$$= 1.337\text{ft}$$

The necessary embedment depth is

$$1.337\text{ ft} + 5\text{ ft} = 6.337\text{ ft}$$

The total required soldier pile length is

$$6.337\text{ ft} + 6\text{ ft} = 12.337\text{ ft}\ (13\text{ ft assumed})$$

So, I have determined that the soldier pile (W12 × 26) should be 13 ft long, and the perpendicular support (w8 × 12) should be 6 ft long as the support for structurally sound maglev construction.

Construction Cost Estimate Comparison

Order of magnitude cost estimate was performed by using HCSS (Heavy Bid) software standard union rate of New York State locals with a project of 10% general condition, 10% overhead and profit, and 3% contingency over the hard cost of labor, materials, and equipment comparing between maglev infrastructure and traditional infrastructure system for a sample of 100 miles long and 128 ft wide (12 ft wide of 4 lanes in each direction, two-sided 10 ft service space, and 6 ft median in the center of the road). To construct the long-lasting and sophisticated underground maglev, I have performed load-resistant factor design (LRFD) calculation and selected W24 × 84 beam wherein the continuous maglev underground runs (structural beam) are structurally sound. Then I have calculated the required shoring concept for 100 miles long and 128 ft wide construction cost considering standard excavation up to 6 ft deep, with appropriate shoring with minimum embedment depth L_4 being

5 ft, standard soil pressure $\Upsilon_s = 120$ lbf/ft^3, angle of pressure $\Phi = 21°$, and soil pressure coefficient $c = 800$ lbf/ft^2 in order to determine the length of soldier piles. So, I have calculated by using LRFD methods again that selected the soldier pile (W12 × 26) that should be 13 ft long, and the perpendicular support (w8 × 12) should be 6 ft long as the support maglev construction.

Cost of Maglev Infrastructure

The proposed maglev infrastructure, therefore, requires shoring, excavation, structural steel, and concrete operation and thus I have calculated the estimate considering the following components:

Shoring at 13′ deep with w24 × 26 steel soldier piles at 6′ OC both sides $2/lf; top rail w8 × 12 both sides $2/lf; 6′ length w8 × 12 perpendicular support 20 OC $2.lf; protection board 1,372,800 ft^2 both sides at $4/ft^2; and thus the total cost would be $23,724,800.

Excavation ($52{,}800'_{\text{length}} \times 128_{\text{width}} \times 6_{\text{deep}} \times 1.3_{\text{fluff factor}}$)/27 is 19,524,266.67 yd^3 at $56/yd^3 cost for digging, stock piling, and backfilling and the total cost would be $1,093,358,933.

Cost of materials: 100 Miles maglev system with structural steel (w24 × 84) support for 8 lanes is $354,816,000; 2 × 2 structural concrete strip footing at $150/yd^3 is $93,866,666; reinforcement bar at 100 lb/yd^3 and cost is $62,577,778; concrete form at $2/ft^2 is $16,896,000; and thus the total cost of material is $528,156,445.

Cost of labor: 200 Iron workers for 2704 working days at $100/h; 100 concrete cement workers for 2704 working days at $90/h; 100 laborers for 2704 working days at $70/h; 50 equipment operators for 2704 working days at $100/h; and thus the total labor cost is $886,912,000 considering standard 8 h a day.

Equipment cost: Ten small renting at $1000/day; 10 small tool renting at $250/day; 271 concrete pumps at $2000/each; and thus the total equipment cost is $34,342,000.

Other cost: Engineering service at $5/ft^2; survey team at $4400/day for each working day; and thus the total cost is $349,817,600. The net construction cost by adding 10% general condition, 10% overhead and profit, and 3% contingency into the excavation, material, labor, equipment, and other cost would be $3,587,063,487.

Cost of Traditional Road Infrastructure

A typical highway consists of 8″ asphalt surface course, 4″ binder course, 4″ base course, and 12″ aggregate with standard wire mesh or framing and thus we have calculated the estimate considering the following components:

Excavation ($52{,}800_{length} \times 128_{width} \times 2.33_{deep} \times 1.3_{fluff\ factor}$)/27 is 7,581,924 yd^3 at $56/$yd^3$ cost for digging, stock piling, and backfilling and the total cost would be $424,587,744.

Cost of materials: $50/$yd^3$; 4″ base course is 834,370 yd^3 at $50/$yd^3$; wire mesh or framing is (528,000 × 128) at $1/$ft^2$; 12″ subbase aggregate is 2,503,111 yd^3 at $25/$yd^3$; and thus the total cost of material is $380,472,775.

Cost of labor: 200 Asphalt cement workers for 2704 working days at $100/h; 200 labor foremen for 2704 working days at $100/h; 200 laborers for 2704 working days at $70/h; 200 equipment operators for 2704 working days at $100/h; 100 truck drivers for 2704 working days at $100/h; 200 small roller engineers for 2704 working days at $100/h; and thus the total cost is $2,249,728,000.

Equipment cost: 200 Roller renting at $1000/week; 200 milling renting at $10,000/week; 100 truck renting at $500/week; and thus the total cost is $502,171,429.

Other cost: Detailing and shop drawing at $10/$ft^2$; engineering service at $5/$ft^2$; survey team at $4400/day for each working day; banking service of 301,037 yd^3 at $1000/$yd^3$; maiden concrete divider is 106,468 yd^3 at $818/$yd^3$; and thus the total cost is $1,326,694,600. The net construction cost by adding 10% general condition, 10% overhead and profit, and 3% contingency into the excavation, material, labor, equipment, and other cost would be $6,805,115,863.

Cost Saving

In this chapter I have calculated cost saving by using standard 100 miles highway of 128 ft wide (12 ft wide of 4 lanes in each direction, two-sided 10 ft service space, and 6 ft median in the center of the road) as an experimental tool to compare construction cost in between conventional and maglev infrastructure system. Total cost estimate for traditional infrastructure is $6,805,115,863 and the maglev infrastructure system cost is only $3,587,063,487 for the same 100 miles highways and the net cost saving is $3,218,052,377 (Table 9.1). Consequently it will reduce neatly 50% of cost once maglev infrastructure system is used for the construction of invisible infrastructure which is also benign to the environment.

Conclusions

Traditional transportation infrastructure construction and maintenance throughout the world are not only expensive, but also consuming 5.6×10^{20} J/year (560 EJ/year) fossil fuel each year which is indeed dangerous of a cliché when discussing about climate [13, 26]. In order to mitigate this issue, better infrastructure transportation planning needs to be achieved where environmental sustainability and climate adaptation have been confirmed to create more resilient and vibrant communities.

Table 9.1 This cost comparison is prepared by using HCSS cost data 2016 for material by utilizing selective manufacturers and labor rate in accordance with international union wage of each specified trade worker considering US location. The equipment rental cost is estimated as current rental market in conjunction with the standard practice of construction of the production

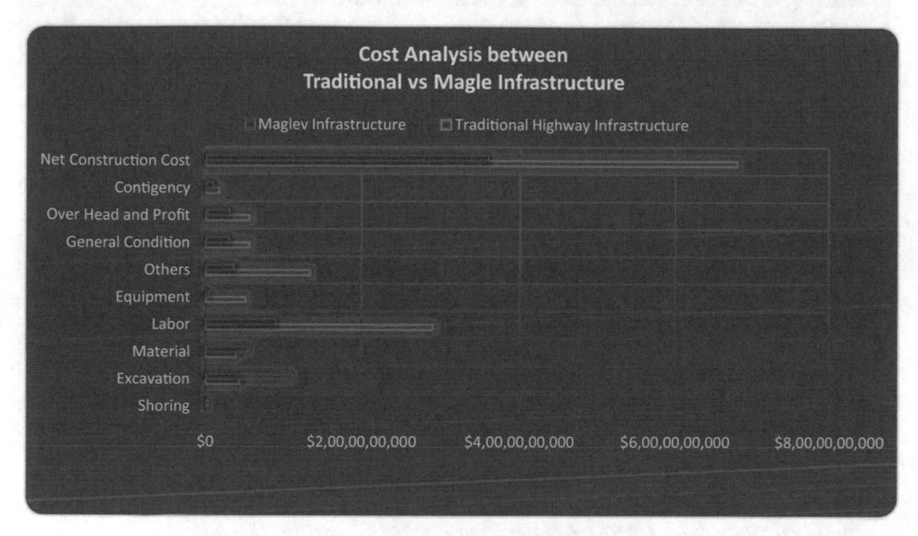

Interestingly, ***invisible infrastructure transportation*** technology proposed in this chapter, for urban infrastructure transportation system, implicated by electromagnetic and superconducting magnets will, thus, be the emergent technology in modern science. This is because the technology is cheaper, and it will run by repulsive force and attractive force at the levitated (flying) stage while it will run on maglev system and will run by air (wind energy) while it is on non-levitated area without consuming fossil fuel. Indeed, the maglev infrastructure transportation system would be the innovative technology ever to console infrastructure, transportation, energy, and global warming crisis.

Acknowledgements This research was supported by Green Globe Technology under the grant RD-02016-06. Any findings, conclusions, and recommendations expressed in this chapter are solely those of the author and do not necessarily reflect those of Green Globe Technology.

References

1. Eltamaly, A.M., Alolah, A.I., Abdel-Rahman, M.H., Improved simulation strategy for DFIG in wind energy applications, (2011) *International Review on Modelling and Simulations (IREMOS),* 4 (2), pp. 525-532.
2. E.W.E Association. Wind directions - the European wind industry magazine Feb. 2012; 31(1).
3. Ghennam, T., Berkouk, E. M., Francois, B., A vector hysteresis current control applied on three-level inverter. Application to the active and reactive power control of doubly fed induc-

tion generator based wind turbine, (2007) *International Review of Electrical Engineering (IREE)*, 2 (2), pp. 250-259.

4. Manfred stieber, *Wind Energy System for Electric Power Generation* (Springer Verlag, Berlin Heidelberg, 2008).
5. Grätzel, M. Photoelectrochemical cells. *Nature*, 414:338–344, 2001.
6. K. D. Kerrouche, A. Mezouar, L. Boumediene, K. Belgacem Modeling and Optimum Power Control Based DFIG Wind Energy Conversion System, *International Review of Electrical Engineering (I.R.E.E.), 2014 Vol. 9, N. 1*
7. T.K.A. Brekken, N. Mohan, Control of a doubly fed induction wind generator under unbalanced grid voltage conditions, *IEEE Trans. On Energy Conversion, vol. 22 n. 1*, Mar 2007, pp. 129-135.
8. Wang, L, Truong, D.-N, Stability Enhancement of DFIG-Based Offshore Wind Farm Fed to a Multi-Machine System Using a STATCOM, *IEEE Trans. on Power Sys, vol. 28 n. 3*, 2013, pp. 2882 – 2889.
9. Mohammad Pichan, Hasan Rastegar, Mohammad Monfared, Two fuzzy-based direct power control strategies for doubly-fed induction generators in wind energy conversion systems, *Energy, vol. 51*, 2013, pp. 154–162.
10. Organisation for Economic Co-operation and Development. World energy outlook. Paris: Organisation for Economic Co-operation and Development; 2010.
11. Gopal Sharma, K., Bhargava, A., Gajrani, K., Stability analysis of DFIG based wind turbines connected to electric grid, (2013) *International Review on Modelling and Simulations (IREMOS)*, 6 (3), pp. 879-887.
12. Gupta, N., Singh, S.P., Dubey, S.P., Palwalia, D.K., Fuzzy logic controlled three-phase three-wired shunt active power filter for power quality improvement, (2011) *International Review of Electrical Engineering (IREE)*, 6 (3), pp. 1118-1129.
13. Heier, Wind energy conversion systems (John Wiley & Sons Inc., New York, 1998).
14. Dürr, M., Cruden, A., Gair, S., and McDonald, J. R., "Dynamic Model of a Lead Acid Battery for Use in a Domestic Fuel Cell System," Journal of Power Sources, Vol. 161, No. 2, pp. 1400-1411, 2006.
15. M. Boutotbat, L. Mokrani, M. Machmoum, Control of a wind energy conversion system equipped by a DFIG for active power generation and power quality improvement, *Renewable Energy, vol. 50*, 2013, pp. 378-386.
16. E. Kamal, M. Koutb, A. A. Sobaih, and B. Abozalam, An intelligent maximum power extraction algorithm for hybrid wind-diesel-storage system, *Int. J. Electr. Power Energy Syst, vol. 32 n. 3*, 2010, pp. 170–177.
17. Abdullah, M. A., Yatim, A. H. M., and Chee Wei, T., "A Study of Maximum Power Point Tracking Algorithms for Wind Energy system," Proc. of IEEE First Conference on Clean Energy and Technology(CET), pp. 321-326, 2011.
18. G. Tsourakisa, B. M. Nomikosb, C.D. Vournasa. Effect of wind parks with doubly fed asynchronous generators on small-signal stability, *Electric Power Systems Research, vol. 79*, 2009, pp. 190-200.
19. Faida, H., Saadi, J., Modelling, control strategy of DFIG in a wind energy system and feasibility study of a wind farm in Morocco, (2010) *International Review on Modelling and Simulations (IREMOS)*, 3 (6), pp. 1350-1362.
20. Tazil M, Kumar V, Bansal RC, Kong S, Dong ZY, Freitas W, et al, Three-phase Doubly Fed Induction Generators; an overview. *IET Journal on Electric Power Applications, vol. 4*, 2010, pp. 75-89.
21. T. Takagi and M. Sugeno, Fuzzy identification of systems and its applications to modelling and control, *IEEE Trans. Syst Man Cybern, vol. 15 n. 1*, 1985, pp.116–132.
22. E. H. Mamdani and S. Assilina, An experiment in linguistic synthesises with a fuzzy logic controller, *International Journal of Man Machine Studies, n. 7*, 1975, pp. 1-13.
23. A. Gaillard, P. Poure, S. Saadate, M. Machmoum. Variable Speed DFIG Wind Energy System for Power Generation and Harmonic Current Mitigation, *Renewable Energy, vol. 34*, 2009, pp. 1545-1553.

24. Vieira. J.P.A, Alves Nunes. Marcus Vinicius, Bezerra. U. H, Barra. W. Jr, New Fuzzy Control Strategies Applied to the DFIG Converter in Wind Generation Systems, *Trans. America Latina, IEEE, vol. 5 n. 3*, 2007, pp.
25. Wang L, Truong D.-N, Stability Enhancement of a Power System with a PMSG-Based and a DFIG-Based Off shore Wind Farm Using a SVC with an Adaptive-Network-Based Fuzzy Inference System, *IEEE Trans. on Power Sys, vol. 60 n. 7*, 2013, pp. 2799 – 2807.
26. E. Kamal, M. Oueidat, A. Aitouch and R. Ghorbani, Robust Scheduler Fuzzy Controller of DFIG Wind Energy Systems, *IEEE Trans. Sustain. Energy, vol. 4 n. 3*, 2013.

Chapter 10
Zero-Emission Vehicles

Abstract Indeed, massive development of transportation sectors accelerating the consumption of fossil fuel has caused nearly 30% of climate change. Necessarily, renewable energy technology needs to be utilized in the transportation sector globally which should be clean in order to develop zero-emission vehicles. In this research, therefore, a modeling of wind energy technology has been developed to capture the wind power by running vehicles to meet its required energy to power. Hence, both theoretical modeling and experimental calculation have been conducted in order to confirm the utilization of wind energy in the transportation vehicle turbine to generate electricity to power the transportation vehicle while the vehicle is in motion as a self-producing energy source. Interestingly both calculative and experimental results suggest that implementation of wind energy in the running vehicle in large scale would be an innovative science to satisfy the global energy need for the transportation sector which is 100% clean and abundant everywhere on earth.

Keywords Wind power · Turbine modeling · Control design · Energy conversion · Powering the transportation vehicles

Introduction

The magnitude of the wind turbine installation for meeting electricity demand of the building and industry sectors seriously impacts the global habitat and the ecosystem [1–3]. Prominent concerns include visibility and misuse of the landscape and the various nuisances affect the harmonium of lifestyle. In addition, the possible radar interference is also a concern for the construction of wind turbine to produce electricity [4, 5]. Besides, the technology application for the conversion of wind power into electricity supply involves cost concerns and engineering challenges along with transmission and operational mechanisms which more likely restrict the use of wind energy globally [1, 3, 6]. Therefore, an appropriate wind energy technology devel-

M. F. Hossain, *Global Sustainability in Energy, Building, Infrastructure, Transportation, and Water Technology*,
https://doi.org/10.1007/978-3-030-62376-0_10

opment and implementation can be interesting to utilize wind power to reduce the impact on energy, ecosystems, and local communities. Simply, wind power has the tremendous potential to meet the near-term (2020) and long-term (2050) energy mitigation goal and greenhouse gas (GHG) emission reductions once appropriate technology has been developed. Though a series of wind power technologies are available across the world which meet nearly 1.8% of global electricity demand, the functional use of wind power in relevance to transportation vehicles to generate electricity is yet limited [7]. Since the wind power interestingly has a great potential to be converted into energy for running transportation vehicles, this research has implemented the wind turbine placement in the running vehicle to satisfy its required energy demand. Thus, the wind turbine modeling is conducted by MATLAB Simulink in this research where the drivetrain model, wind energy conversion chain analysis, kinetic energy conversion mechanism, two interacting main subsystems, and the detailed process of energy conversion system have been calculated mathematically. Subsequently, the control structure, designing, and generator modeling are also analyzed by a set of mathematical calculations for total process of wind energy capture and utilization of it for the transportation vehicle. In order to prove this mechanism for commercial utilization, this technology is introduced into a sedan automobile as an experiment. Interesting experimental results suggested that wind energy implementation in the running vehicle is very much functional which indeed could be an innovative technology for the transportation sector to meet their complete energy demand.

Methodology and Materials

Turbine Modeling for Transportation Sector

Since the differential speed of wind turbine is governed by its electronic equipment, the wind turbine has been proposed to be installed into the vehicle to power the running vehicles through wind [8, 9]. Naturally a mathematical mechanism has been computed in order to simulate the DFIG for producing electricity energy from wind to run the transportation vehicle [4, 10]. Subsequently, the mechanical portion of the wind turbine has been modeled and then the electrical portion analyzed by a standard DC/AC converter [11, 12]. Then the detailed regulation of the power mechanism of wind has been controlled by the state active and co-reactive power within the DFIG and energy storage box of the RST regulator, which is a vector control strategy, calculated by the following equation:

$$P_{\mathrm{w}} = \frac{1}{2} C_{\mathrm{p}} \left(\lambda, \beta \right) \rho A V^{3} \tag{10.1}$$

Here ρ represents the air density (kg/m^3), C_p represents the power coefficient, A is considered as the intercepting area of the wind rotor blades (m^2), V represents the average wind speed (m/s), and λ represents the tip speed ratio [11, 13]. Here, the peak value of the energy coefficient C_p is 0.593 [8, 14]. Thus, the average tip speed ratio (TSR) of wind turbine is defined from the rate of its rotational velocity, which is mathematically calculated as

$$\lambda = \frac{R\omega}{V} \tag{10.2}$$

where R denotes the radius of turbine (m), ω denotes the angular velocity (rad/s), and V is the mean wind velocity (m/s). The power governed by wind turbine is thus calculated as the following equation:

$$Q_w = P \times (\text{Time})\ [\text{kWh}] \tag{10.3}$$

Since various factors can interfere the wind, the wind velocity is measured by dynamic wind speed with respect to factorial error proneness due to accountable obstacles surrounded by the transportation vehicles (Fig. 10.1):

$$v(Z)\ln\left(\frac{Z_r}{Z_0}\right) = v(Z_r)\ln\left(\frac{Z}{Z_0}\right) \tag{10.4}$$

where Z_r denotes the height (m), Z is the wind speed, Z_0 denotes the measuring surface roughness (0.1–0.25), $v(Z)$ denotes the wind velocity in height Z (m/s), and $v(Z_r)$ is the wind velocity at height Z (m/s).

Subsequently, the wind velocity has also been calculated considering the approach of turbine rotation in order to confirm the steady equilibrium wind velocity input into wind turbine which can be expressed by the following equation:

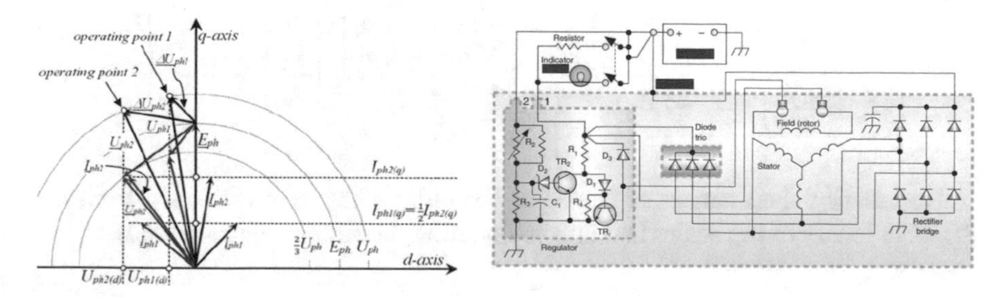

Fig. 10.1 Different functional areas for a standard wind turbine at its maximum rotor speed, while the wind speed is controlled by the stator functional electrical equipment

$$P_w(v) = \begin{cases} \frac{v^k - v_C^k}{v_R^k - v_C^k} \cdot P_R & v_C \le v \le v_R \\ P_R & v_R \le v \le v_F \\ 0 & v \le v_C \text{ and } v \ge v_F \end{cases} \tag{10.5}$$

where P_R is the rated power, v_C represents the cut-in wind velocity, v_R represents the rated wind velocity, v_F represents the rated cutout velocity, and k is considered as the Weibull shape cofactor.

Since the angular velocity of the turbine is related to extracting of the maximum wind power, the maximum power point tracking (MPPT) has been calculated considering the rotor optimum velocity by using the following equation:

$$\omega_{opt} = \frac{\lambda_{opt}}{R} V_{wn} \tag{10.6}$$

which can be converted as

$$V_{wn} = \frac{R\omega_{opt}}{\lambda_{opt}} \tag{10.7}$$

where ω_{opt} is the maximum rotor angular velocity in rad/s, λ_{opt} is the maximum tip velocity factor, R represents the radius of turbine (m), and V_{wn} represents the wind velocity in m/s.

Since the turbine drive force is transferred into the generator gearbox, the drive force is modeled using the torsional multibody dynamics, expressed by the following matrix [15, 16:

$$\begin{bmatrix} \dot{\omega}_t \\ \dot{\omega}_g \\ \dot{T}_{ls} \end{bmatrix} = \begin{bmatrix} \frac{K_t}{J_t} & 0 & \frac{1}{J_t} \\ 0 & -\frac{K_g}{J_g} & \frac{1}{n_g J_g} \\ \left(B_{ls} - \frac{K_{ls}K_r}{J_r}\right) & \frac{1}{n_g}\left(\frac{K_{ls}K_r}{J_g} - B_{ls}\right) & -K_{ls}\left(\frac{J_r + n_g^2 J_g}{n_g^2 J_g J_r}\right) \end{bmatrix} \begin{bmatrix} \omega_t \\ \omega_g \\ T_{ls} \end{bmatrix} + \begin{bmatrix} \frac{1}{J_r} \\ 0 \\ \frac{K_{ls}}{J_r} \end{bmatrix} T_m + \begin{bmatrix} 0 \\ -\frac{1}{J_g} \\ \frac{K_{ls}}{n_g J_g} \end{bmatrix} T_g \tag{10.8}$$

Necessarily, the moment of inertia of the turbine has been estimated from the combined mass of the blades and hub considering the simple equation below:

$$J_t \dot{\omega}_t = T_a - K_t \omega_t - T_g \tag{10.9}$$

and

$$J_t = J_r + n_g^2 J_g$$

$$K_t = K_r + n_g^2 K_g$$

$$T_g = n_g T_{em} \tag{10.10}$$

where J_t represents the turbine rotor moment of inertia in [kg m^2], ω_t represents the rotor angular velocity [rad/s^2], K_t represents the turbine cofactor [Nm/rad/s], and K_g represents the generator damping cofactor [Nm/rad/s] for this turbine model.

Once the turbine modeling has been done, the wind energy modeling is conducted mathematically to transform the wind force into the vehicle turbine in order to form electoral energy to power the running vehicle.

Wind Energy Modeling

Once the maximum achievable wind speed considering the air mass flow by using a turbine has been analyzed, the mechanism of wind energy conversion into the turbine is calculated [17, 18]. Since this wind energy eventually delivers electric power by the kinetic force of DFIG of the turbine, the mechanism of energy conversion modeling has been described by the chain of two interacting subsystems: (a) aerodynamic system (wind speed, wind turbine, and gearbox) and (b) electrical system (DFIG) (Fig. 10.1) [19]. Here, (a) aerodynamic subsystem governs the wind velocity signal on simulations which is analyzed by using the real log determinations of the velocity on the DFIG. Since the wind turbine generates an equivalent wind speed of V into the rotor where both deterministic effects and stochastic variations are turbulence, the deterministic and stochastic parts are determined by the equivalent wind velocity which is denoted by

$$V(t) = V_0 + \sum_{i=1}^{n} A_i \sin(\omega_i t + \varphi_i) \tag{10.11}$$

where V_0 is the average component, and A_i, ω_i, and ψ_i are, respectively, magnitude, pulsation, and initial phase on every turbulence.

Subsequently, the turbulence function of the rotational wind turbine blades is taken considering the WTGS convert energy from the kinetic energy of the wind (Fig. 10.2). Hence, the kinetic power in the stream of wind turbine is a function of rotor speed; thus, it has been multiplied by C_p factor called power coefficient and is expressed by the following equation as an aerodynamic power [20, 21]:

$$P_{aer} = \frac{1}{2} C_p(\lambda, \beta) \rho R \pi^2 V^3 \tag{10.12}$$

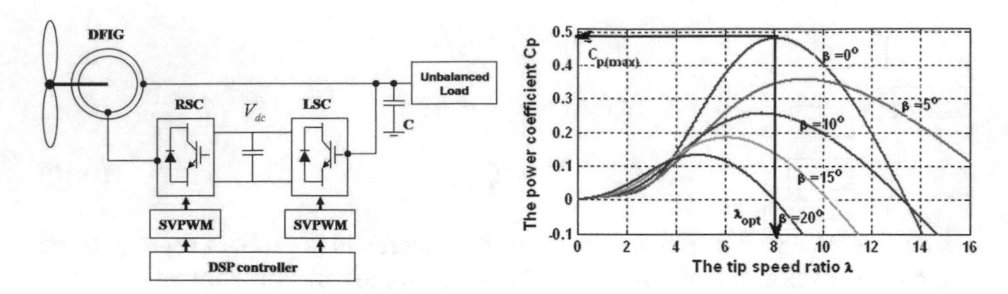

Fig. 10.2 (**a**) Typical wind energy conversion chain and (**b**) power factor variance considering the tip speed ratio and pitch angle

where ρ represents the air density, R represents the blade length, and V represents the wind speed. Hence, the captured wind is denoted by a coefficient $C_p(\lambda)$ considering the function of wind speed (Ω_t) and the turbine pitch angle [14, 22] which is simplified by the following equation:

$$\lambda = \frac{\Omega_T * R}{V} \tag{10.13}$$

Consequently, the turbine torque is calculated from the rate of the aerodynamic power into the turbine shaft speed [23, 24]. Since the turbine is functionally coupled with the turbine shaft, the gearbox ratio G is determined to clarify the aerodynamic power, which is expressed by

$$\begin{cases} T_g = \dfrac{T_{aer}}{G} \\ \Omega_t = \dfrac{\Omega_g}{G} \end{cases} \tag{10.14}$$

where T_g is the driving torque of the generator and Ω_g is the generator shaft speed.

Electrical Subsystem Modeling

To use the electrical subsystem to transform the wind energy into electric energy for powering the rotor induction generator of the transportation vehicle, the variable wind velocity has been modeled [25, 26]. It is simply to produce electrical energy where the DFIG is being actively working to feed wind into the generator consistently [27–29]. Hence, current fluxes across electrical subsystem are thus expressed by [21, 30]

$$\begin{aligned}
V_{ds} &= \frac{d\phi_{ds}}{dt} + \frac{R_s}{L_s}\phi_{ds} - \omega_s * \phi_{qs} - M\frac{R_s}{L_s}i_{qr} \\
V_{qs} &= \frac{d\phi_{qs}}{dt} + \omega_s * \phi_{ds} + \frac{R_s}{L_s}\phi_{qs} - M\frac{R_s}{L_s} \\
i_{qr}\left(V_{dr} - \frac{M}{L_s}V_{ds}\right) &= \sigma L_r \frac{di_{dr}}{dt} - M\frac{R_s}{L_s^2}\phi_{ds} + \frac{M}{L_s}\omega\phi_{qs} + \left(R_r + \frac{M^2}{L_s^2 R_s}\right)i_{dr} - \sigma L_r \omega_r i_{qr} \\
\left(v_{qr} - \frac{M}{L_s}v_{ds}\right) &= \sigma L_r \frac{di_{qr}}{dt} - \frac{M}{L_s}\omega\phi_{ds} - M\frac{R_s}{L_s^2}\phi_{qs} + \sigma L_r \omega_r i_{dr} + \left(R_r + \frac{M^2}{L_s^2 R_s}\right)i_{dr}
\end{aligned} \tag{10.15}$$

where

R_s and R_r represent the stator and rotor resistances
L_s and L_t represent the stator and rotor inductances
M and σ represent the mutual inductance and leakage cofactors
$\Omega = p\Omega_g$ represents the electrical velocity and p represents the pair pole number

In order to simplify the stator and rotor flux for detailing the equation it is further explained as

$$\begin{cases}
\phi_{ds} = L_s i_{ds} + M i_{dr} \\
\phi_{sq} = L_s i_{sq} + M i_{qr} \\
\phi_{rd} = L_r i_{dr} + M i_{ds} \\
\phi_{rd} = L_r i_{qr} + M i_{qs}
\end{cases} \tag{10.16}$$

where i_{ds}, i_{qs}, i_{dr}, and i_{qr} are the direct and quadratic currents transformed from the wind which are defined further as

$$\begin{cases}
P_s = v_{ds} * i_{ds} + v_{qs} * i_{qs} \\
Q_s = v_{qs} * i_{ds} - v_{ds} * i_{qs}
\end{cases} \tag{10.17}$$

where the electromagnetic torque is clarified as below which is the primary force to conduct electrical subsystem operation:

$$T_{em} = p\left(i_{qs}\phi_{ds} - i_{ds}\phi_{qs}\right) \tag{10.18}$$

Hence, the wind turbine electric system works much efficiently; thus, the wind turbine DFIG control system has been described as a cascade control structure around two subsystem controls: (a) the wind turbine control subsystem and (b) the DFIG control electric power converter system (Fig. 10.3).

Subsequently, the clarification of electrical voltage formations is used with a fixed switching frequency for the load-side converter from the connection of DFIG [31, 32].

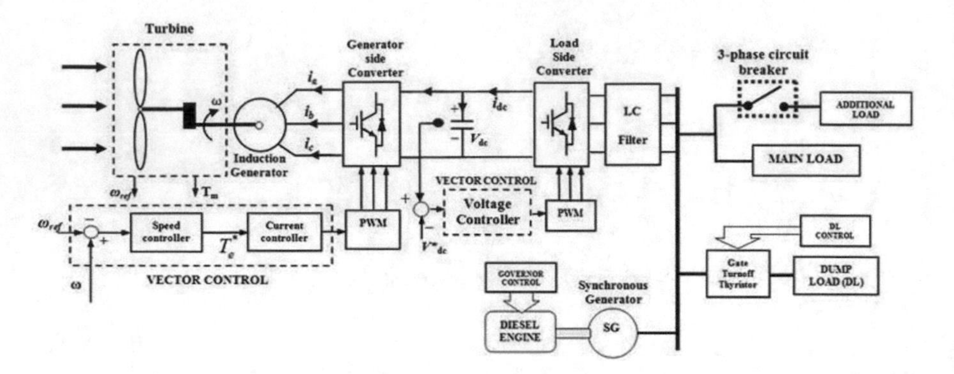

Fig. 10.3 Block diagram of the whole system where these control levels are modeled considering turbine, generator-side converter, and load-side converter in order to transform electricity energy into the three-phase circuit breaker

Since the DFIG is interconnected to the power rotor of the transportation vehicle, the stator flux vector at the *d*-axis directional has been calculated and expressed as the following equation (Fig. 10.4):

$$\phi_s = \phi_{ds} \Rightarrow \phi_{qs} = 0 \tag{10.19}$$

Conducting the above equation, the vehicle power system, the functional voltage V_s, and the stator's constant flux ϕ_s are calculated as

$$\begin{cases} v_{ds} = 0 \\ v_{qs} = \omega_s * \phi_s = V_s \end{cases} \tag{10.20}$$

Hence, the stator voltage vectors are in the direction of quadrate advances which is further clarified by using Eqs. (10.11) and (10.5) to obtain voltages for rotor:

$$\begin{cases} v_{dr} = \sigma L_r \dfrac{di_{dr}}{dt} + R_r i_{dr} - \sigma L_r \omega_r i_{qr} + \dfrac{M}{L_s}\dfrac{d\phi_{ds}}{dt} \\ v_{qr} = \sigma L_r \dfrac{di_{qr}}{dt} + R_r i_{qr} + \sigma L_r \omega_r i_{dr} + g\dfrac{M}{L_s} V_s \end{cases} \tag{10.21}$$

where V_s represents the stator voltage and g represents the slip range and thus it can be rewritten as follows:

$$\begin{cases} v_{dr} = \sigma L_r \dfrac{di_{dr}}{dt} + R_r i_{dr} + \text{fem}_d \\ v_{qr} = \sigma L_r \dfrac{di_{qr}}{dt} + R_r i_{qr} + \text{fem}_q \end{cases} \tag{10.22}$$

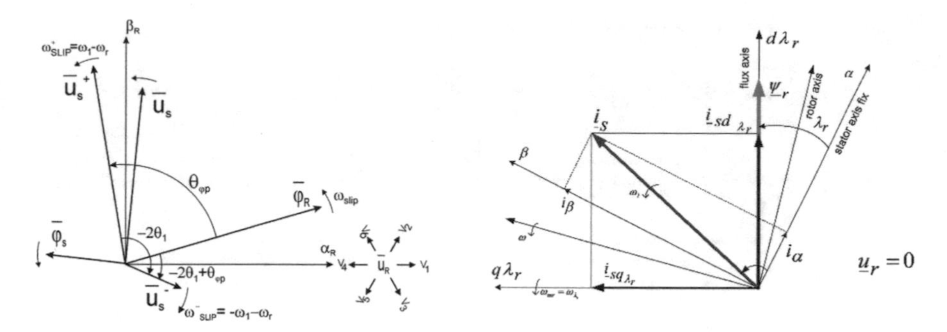

Fig. 10.4 Stator flux orientation leads to stator's constant flux ϕ_s which is calculated as the voltage power

with fem$_d$ and fem$_q$ being the crossed coupling terms between the d-axis and q-axis:

$$\begin{cases} \text{fem}_d = -\sigma L_r \omega_r i_{qr} \\ \text{fem}_q = \sigma L_r \omega_r i_{dr} + s \dfrac{M}{L_s} V_s \end{cases} \tag{10.23}$$

Further considering Eqs. (10.10) and (10.12), the current fluxes are simplified as

$$\begin{cases} \phi_{ds} = L_s i_{ds} + M i_{dr} \\ 0 = L_s i_{qs} + M i_{qr} \end{cases} \tag{10.24}$$

From (10.15), the determination of the deduced currents is expressed as

$$\begin{cases} i_{ds} = \dfrac{\phi_{ds} - M i_{dr}}{L_s} \\ i_{qs} = -\dfrac{M}{L_s} i_{qr} \end{cases} \tag{10.25}$$

Finally, the stator is linked to these rotor currents precisely as

$$\begin{cases} P_s = -V_s * \dfrac{M}{L_s} i_{qr} \\ Q_s = -V_s * \dfrac{M}{L_s} \left(i_{dr} - \dfrac{\phi_{ds}}{M} \right) \end{cases} \tag{10.26}$$

where the stator active and reactive powers are controlled by means of i_{qr} and i_{dr}, respectively (Fig. 10.5).

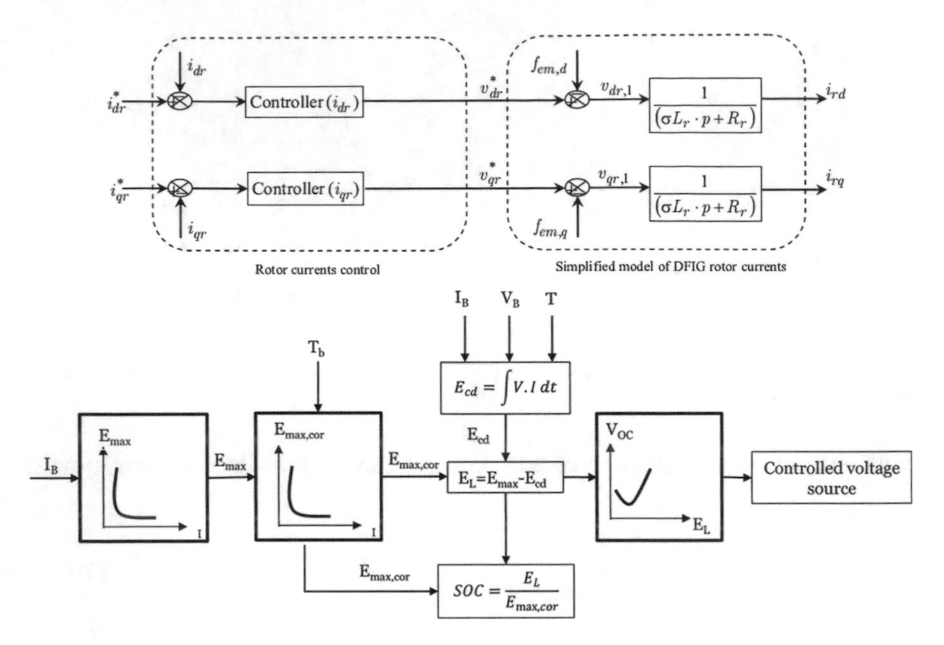

Fig. 10.5 Functional diagram of control system of the oriented DFIG to power the running vehicle by delivering controlled voltage

Simply, the oriented control of the DFIG can then be applied to power the transportation vehicle by using the precedent variable control of power system to run the vehicles.

Results and Discussions

Wind Turbine Modeling

The calculation of wind turbine generation considering the cascade control algorithm reveals that the optimal operation of the whole system at stator flux-oriented and MPPT mechanisms has functioned properly. Therefore, this mechanism has been successfully tracked to get the wind force rate in the pitch control functions by the calculation of the rate of the power coefficient of the wind velocity [33, 34]. Hence, the robustness test of the wind speed signal and voltage dips demonstrates the inherent ability of the fuzzy logic controller to deal with the wind turbine rotor. Thus, the analysis reveals that the tip values of C_p are achieved for the curve in relation to $\beta = 2°$ (Fig. 10.2) where the maximum value of C_p ($C_{p,max} = 0.5$) is $\lambda_{opt} = 0.91$ and this value (λ_{opt}) represents the optimum speed of the rotor. Hence, the wind power is denoted as wind velocity signal of 8 m/s mean value; thus, the total mechanism is calculated under the condition of the stator dynamic of around 50% of 0.5 s between 4 and 4.5 s, around 25% between 6 and 6.5 s, and 50% between 8 and 8.5 s [35, 36]. Since the wind turbine is considered as working over ideal conditions to guarantee a unique wind intake factor into the rotor, the stator active and reactive winds are calculated in accordance to MPPT and fuzzy logic controller. This is because this condition permits to control the DFIG mechanism which plays an active role in wind transformation in order to clarify the maximum angular speed of the generator shaft [37, 38].

Since the generator shaft velocity reaches the maximum angular velocity, the wind regulation is increased at peak rated wind speed in order to control the maximum power point. Therefore, it is a very good decoupling mechanism between the bidirectional components of rotor's active and reactive speed to determine wind energy intake rate. With the functional wind transfer into the rotor, the nominal stator force will confirm the unit wind intake power into the wind turbine. Hence, the bidirectional control involves cross coupling between the two axes (Eq. (10.14)) in order to determine the precise wind energy balance as $P_s + P_r = P_m$, where $P_s = T_{em}\omega_s$ and $P_m = T_q\omega$. To confirm this stator active power, the calculation of MPPT control by analyzing the electromagnetic torque suggested that a predefined turbine power-speed plays a vital role in tracking the maximum power point in the rotor [39, 40].

The turbine shaft speed is then determined to achieve the maximum power cofactor which shows that $P_s = T_{em}\omega_s$ where T^*_{em} is the referenced electromagnetic torque deduced from MPPT control strategy. Since the MPPT control reveals the efficiency of the wind turbine, the calculation of rotor dynamics using Eq. (10.17) can be determined by the referenced active and reactive powers, as follows:

$$\begin{cases} i^*_{qr} = -\dfrac{L_s}{MV_s} P^*_s \\ i^*_{dr} = -\dfrac{L_s}{MV_s}\left(Q^*_s - \dfrac{V_s^2}{\omega_s L_s}\right) \end{cases} \tag{10.27}$$

This whole mechanism suggests an accurate speed of the wind intake by the wind turbine where MPPT control tracks the wind velocity continuously and adjusts the imposed electromagnetic torque of the DFIG to track its aerodynamics accordingly (Fig. 10.6). Thus, the aerodynamic subsystem has been clarified as shown in the block diagram to confirm the rate of wind intake and processed by this mechanism.

Simply, this MPPT control and DFIG block diagram reveals wind harvesting by the turbine by controlling the generator speed; thus, the optimal wind intake rate is determined by the stator flux calculation of wind turbine.

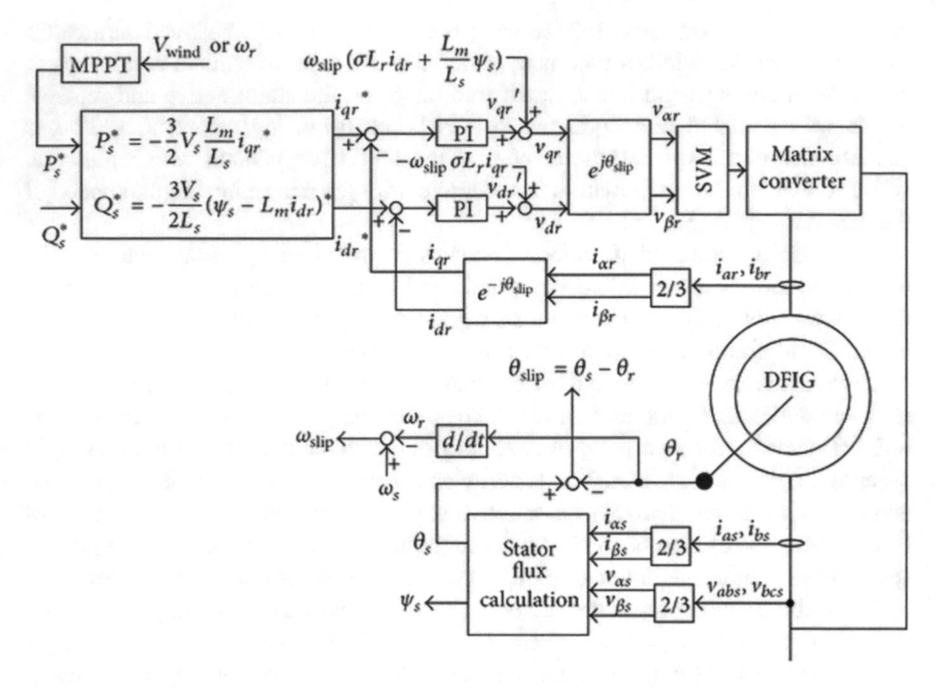

Fig. 10.6 The block diagram of the MPPT control considering the enslaved DFIG velocity of the aerodynamic subsystem to determine the rate of wind intake and process

Wind Energy Modeling

The fuzzy logic controller (FLC) robustness has been calculated considering its nonlinear procedures and characteristics in order to model wind energy to convert it into electricity energy with respect to the Mamdani fuzzy inference system [41, 42]. The basic configuration of this fuzzy logic controller includes fuzzifications which determines the net wind energy conversion rate [18, 43]. Thus, the fuzzy inference system (FIS) evaluates the FLC mechanisms in order to determine the wind energy conversation process (modeling) by presenting the control rules of the fuzzy block diagram (Fig. 10.7). The fuzzification block simply reveals the calculated risk output of wind energy which is controlled by the rules of the defuzzification of wind turbine [44–46].

This FLC is utilized to determine the wind turbine and its DFIG subsystem to deal with the wind energy transformation mechanism to produce electric energy which is expressed by

$$\begin{cases} e_{\Omega_g(n)} = \Omega_g^*(n) - \Omega_g(n) \\ \Delta e_{\Omega_g(n)} = \Omega_g^*(n) - \Omega_g(n-1) \end{cases} \tag{10.28}$$

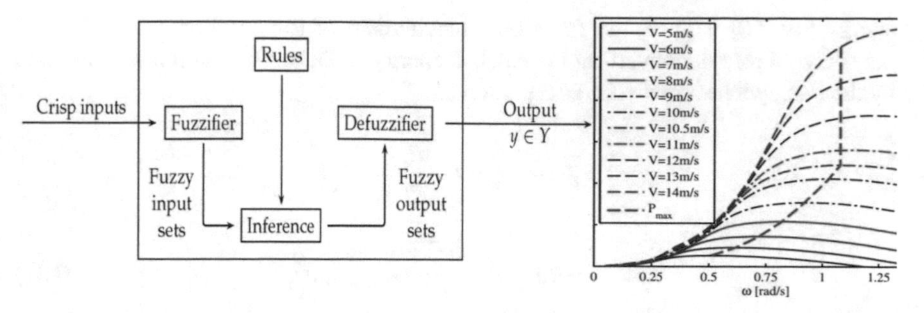

Fig. 10.7 Block diagram for fuzzy controller calculated the robustness of nonlinear characteristics of wind velocity to transform it into electricity energy

The net realistic energy variants in the fuzzy controller are subsequently calculated from this fuzzy controller. Therefore, the net wind energy production is calculated by triangular, trapezoidal, and symmetrical speed functions of the wind turbine generator. Simply the FLC of DFIG has been calculated by analyzing the net current formation of the wind turbine and is expressed by

$$\begin{cases} e_{i_{dr}}(n) = i_{dr}^{*}(n) - i_{dr}(n) \\ e_{i_{qr}}(n) = i_{qr}^{*}(n) - i_{qr}(n) \end{cases} \tag{10.29}$$

And again, the equation is simplified as

$$\begin{cases} \Delta e_{i_{dr}}(n) - i_{dr}^{*}(n) - i_{dr}(n-1) \\ \Delta e_{i_{qr}}(n) = i_{qr}^{*}(n) - i_{qr}(n-1) \end{cases} \tag{10.30}$$

Finally, the input and output variables of wind energy conversion mechanism by the fuzzy controller have been quantized and thus the wind energy transformation rate is determined by the effect of wind speed in changing the output power into the rotor of the wind turbine [2, 19]. Therefore, the changes in the output power in the rotor while the vehicle is running are computed in order to have a trade-off between accuracy and complexity of wind energy conversion through the rigorous simulation to produce desirable enteric energy for running the vehicle.

Electrical Subsystem Modeling

Once the wind energy modeling has been analyzed, the implementation of this wind energy into the electrical subsystem is calculated to run the vehicle. Therefore, both induction and synchronous wind energy driven by the rotor is captured by permanent magnet synchronous generator (PMSG) in order to convert it into electric energy [49, 50]. Hence, the functional mechanism of the gearbox of this PMSG plays a vital role in converting the electric energy to DC current and it is quantified by the *d–q* synchronous voltage equation as

$$V_d = -R_s i_d - L_d \frac{di_q}{dt} + \omega L_q i_q$$

$$V_q = -R_s i_q - L_q \frac{di_q}{dt} - \omega L_d i_d + \omega \lambda_{\mathrm{m}} \tag{10.31}$$

The electronic torque is given by

$$T_{\mathrm{e}} = 1.5\rho\left[\lambda i_q + \left(L_d - L_q\right) i_d i_q\right] \tag{10.32}$$

where L_q represents the q axis inductance, L_d is considered as the d axis inductor, i_q represents the q axis current, i_d represents the d axis current, V_q represents the q axis voltage, V_d denotes the d axis voltage, ω_{r} represents the angular speed of rotor, λ represents the amplitude of flux induced, and p is the number of pairs of poles.

Then the electronic torque is simplified by the following equation which demonstrates the dynamic modeling of wind energy as

$$\begin{bmatrix} V_{qs} \\ V_{ds} \\ V_{qr} \\ V_{dr} \end{bmatrix} = \begin{bmatrix} R_{\mathrm{s}} + pL_{\mathrm{s}} & 0 & pL_{\mathrm{m}} & 0 \\ 0 & R_{\mathrm{s}} + pL_{\mathrm{s}} & 0 & pL_{\mathrm{m}} \\ pL_{\mathrm{m}} & -\omega_{\mathrm{r}} L_{\mathrm{m}} & R_{\mathrm{r}} + pL_{\mathrm{r}} & -\omega_{\mathrm{r}} L_{\mathrm{r}} \\ \omega_{\mathrm{r}} L_{\mathrm{m}} & pL_{\mathrm{m}} & \omega_{\mathrm{r}} L_{\mathrm{r}} & R_{\mathrm{r}} + pL_{\mathrm{r}} \end{bmatrix} \begin{bmatrix} i_{qs} \\ i_{ds} \\ i_{qr} \\ i_{dr} \end{bmatrix} \tag{10.33}$$

where the stator side is expressed as

$$\begin{aligned} \lambda_{ds} &= L_{\mathrm{s}} i_{ds} + L_{\mathrm{m}} i_{dr} \\ \lambda_{qs} &= L_{\mathrm{s}} i_{qs} + L_{\mathrm{m}} i_{dr} \\ L_{\mathrm{s}} &= L_{\mathrm{ls}} + L_{\mathrm{m}} \\ L_{\mathrm{r}} &= L_{\mathrm{lr}} + L_{\mathrm{m}} \\ V_{ds} &= R_{\mathrm{s}} i_{ds} + \frac{d}{dt}\lambda_{ds} \\ V_{qs} &= R_{\mathrm{s}} i_{qs} + \frac{d}{dt}\lambda_{qs} \end{aligned} \tag{10.34}$$

It is further simplified as

$$\begin{aligned}
\lambda_{dr} &= L_{\mathrm{r}} i_{dr} + L_{\mathrm{m}} i_{ds} \\
\lambda_{qr} &= L_{\mathrm{r}} i_{qr} + L_{\mathrm{m}} i_{qs} \\
V_{dr} &= R_{\mathrm{r}} i_{dr} + \frac{d}{dt}\lambda_{dr} + \omega_{\mathrm{r}}\lambda_{qr} \\
V_{qr} &= R_{\mathrm{r}} i_{qr} + \frac{d}{dt}\lambda_{qr} - \omega_{\mathrm{r}}\lambda_{dr}
\end{aligned} \tag{10.35}$$

In case of air gap flux leakage in the generation, the equations are further corrected as

$$\begin{aligned}
\lambda_{dm} &= L_{\mathrm{m}}\left(i_{ds} + i_{dr}\right) \\
\lambda_{qr} &= L_{\mathrm{m}}\left(i_{qr} + i_{qs}\right)
\end{aligned} \tag{10.36}$$

where R_{s}, R_{r}, L_{m}, L_{ls}, L_{lr}, ω_{r}, i_d, i_q, V_d, V_q, λ_d, and λ_q represent the stator winding resistance, motor winding resistance, magnetizing inductance, stator leakage inductance, rotor leakage inductance, electrical rotor angular speed, current, voltage, and fluxes, respectively, of the *d–q* model, respectively [32]. Then the net output power and torque of turbine (T_{t}) in terms of rotational speed can be obtained by the following equations:

$$P_{\mathrm{w}} = \frac{1}{2}\rho A C_{\mathrm{p}}\left(\lambda,\beta\right)\left(\frac{R\omega_{\mathrm{opt}}}{\lambda_{\mathrm{opt}}}\right)^{3} \tag{10.37}$$

$$T_{\mathrm{t}} = \frac{1}{2}\rho A C_{\mathrm{p}}\left(\lambda,\beta\right)\left(\frac{R}{\lambda_{\mathrm{opt}}}\right)^{3}\omega_{\mathrm{opt}} \tag{10.38}$$

The power coefficient (C_{p}) is a nonlinear function expressed by the fitting equation [51] in the form

$$C_{\mathrm{p}}\left(\lambda,\beta\right) = c_1\left(c_2\frac{1}{\lambda_i} - c_3\beta - c_4\right)\mathrm{e}^{-c_5\frac{1}{\lambda_i}} + c_6\lambda \tag{10.39}$$

with

$$\frac{1}{\lambda_i} = \frac{1}{\lambda + 0.08\beta} - \frac{0.035}{\beta^3 + 1} \tag{10.40}$$

Finally, the output of the wind energy generator module is processed by an energy conversion circuit diagram by implementing the converter of SimPower Systems. The result of the circuit model generation electricity is calculated as volt to power the vehicle (Fig. 10.8).

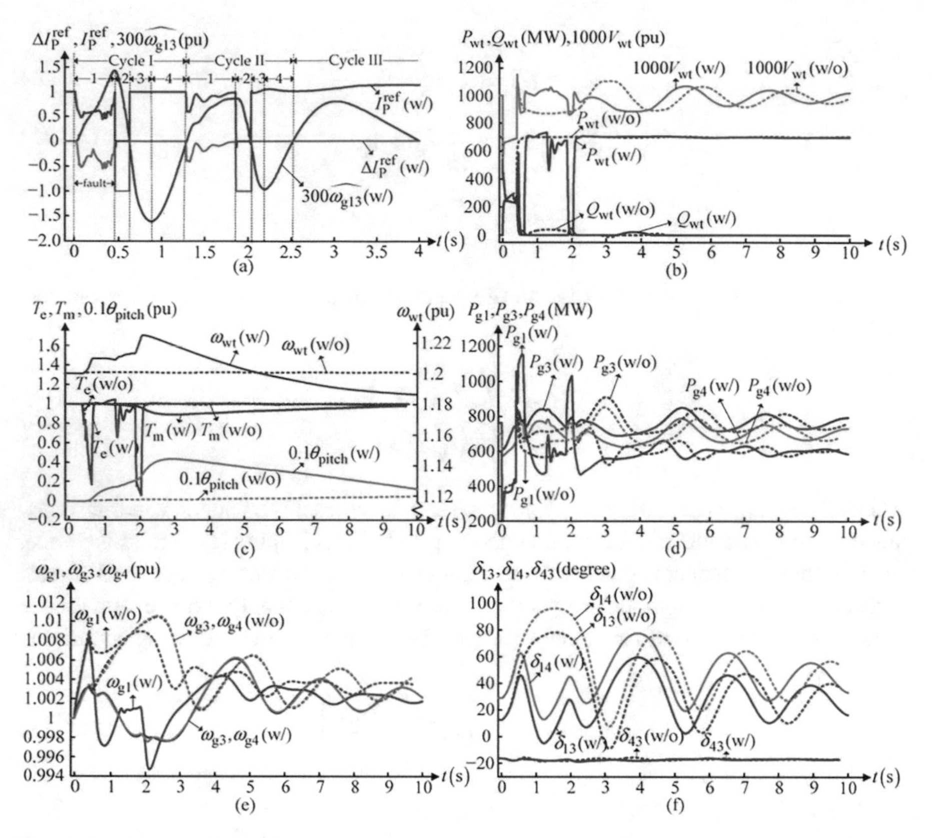

Fig. 10.8 The aerodynamic power generation by illustrating the circuit diagram of wind generator current at various degrees

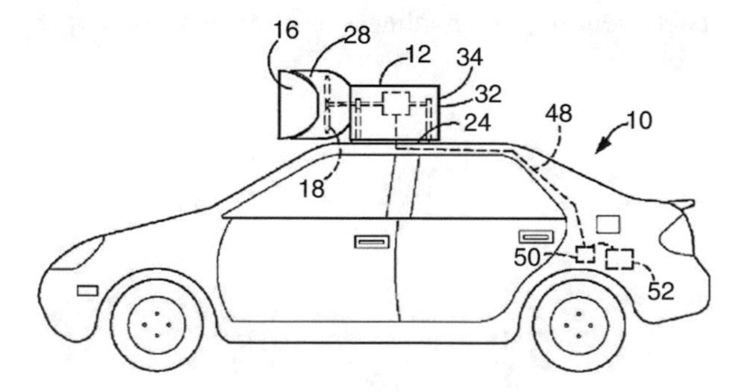

Fig. 10.9 A conceptual model of wind turbine to produce energy for powering the car while it is in motion

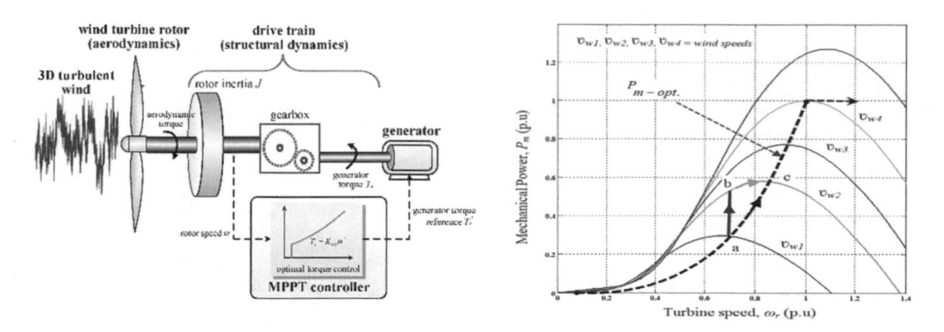

Fig. 10.10 The relationship among mechanical power formation and turbine speeds at different turbine speeds once it is implemented into a car

Mathematic Experiment on Car

The calculative application of this wind power in the car has been implemented to use the wind energy eventually for all transportation vehicles. Interesting, the design of wind turbine for the car and its mechanical and electrical system encountered including all operational systems, strength of the wind, shears, and intensity and frequency of the turbulent fluctuations have been calculated which suggested that application of wind energy in the car is very much functional (Fig. 10.9).

The results of the turbine velocity and mechanical function have been illustrated as per the following graph as per the rate of wind speed increase or decrease during the car in motion, and they show that the persistence of electric energy generation for running depends on the wind velocity where the increment of the wind speed rapidly causes sudden "ramps" in power output which are a tremendous source of capturing wind to convert it into electricity energy to run the car (Fig. 10.10).

Here, the amount of power output from a wind energy electric system (WEES) depends upon the peak power points; thus, this experimental car is utilized for extracting maximum power from the WEES. The main MPPT control methods are presented in this experiment, and the MPPT controllers are used for extracting maximum possible power in WEES [49, 51]. The power generation is interestingly related to the speed of wind which is coming from the speed of running car. The analysis is therefore clarified as per the following figure which depicts the relationship between wind speed (due to the motion of a car) miles per hour (mph) and kWh power production. The results show at the tip a mean 8 kWh at mph for 10 mph average where energy starts to produce 2 mph wind speed just after the engine gets started by battery. Since a standard car requires 20 kWh to get fully energized, if only the car runs at 10 mph for 2 h it will get fully charged to run for an average 200 miles; consequently if the car runs at 60 mph it will take only 20 min to get fully charged to run for the same mileage (Fig. 10.11).

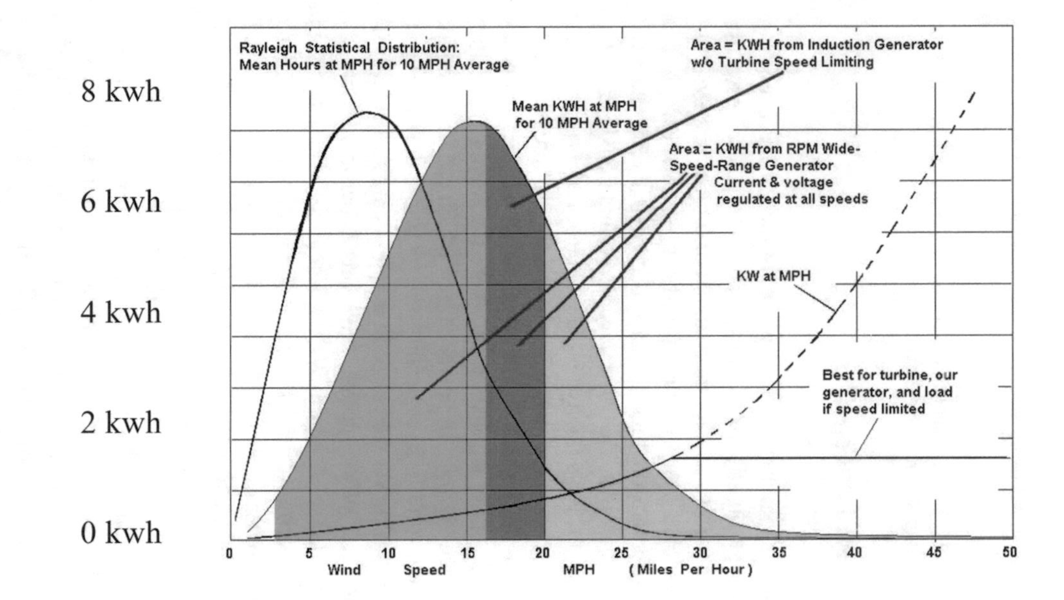

Fig. 10.11 The transformation of wind power into the car in relation to the wind speed vs. energy production including the mean hours (10 mph average), kWh from induction generator, and kWh from RPM speed range generator current and voltage

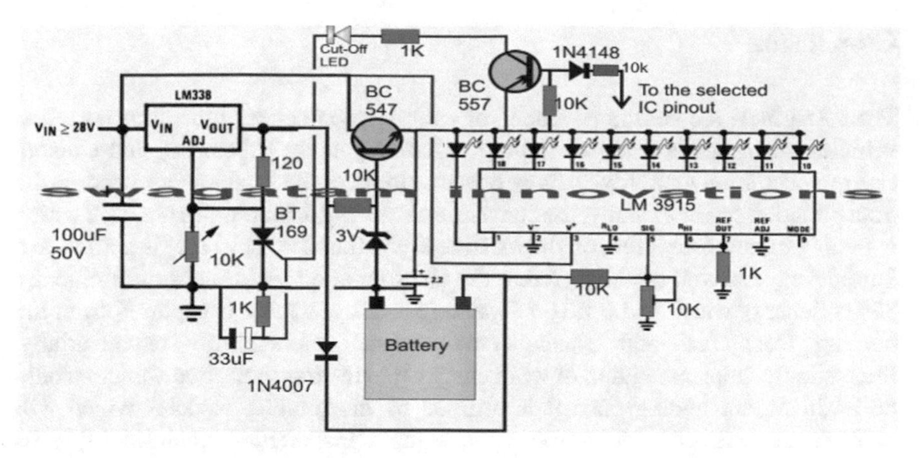

Fig. 10.12 Battery modeling of a car to ignite the engine and the car when not in motion. A block diagram of a car also shown for determining V_{oc} and SOC to control its voltage source

Battery Modeling

Finally, battery is to be utilized as the starting power of the ignition of the vehicle and can also be used (Fig. 10.12) by the implication of Peukert's law of battery charge which is

$$t_{\text{discharge}} = H\left(\frac{C}{IH}\right)^{k}$$

Here t represents the battery charge time, C represents the battery capacity, I represents the current flow, H represents the rated discharge time, and k is Peukert's coefficient which is calculated by the following formula:

$$k = \frac{\log T_2 - \log T_1}{\log I_1 - \log I_2}$$

where I_1 and I_2 represent the variance of the charge current flow rates, and T_1 and T_2 represent the corresponding time for completely charged battery acquired by

$$t_{\text{charging}} = \frac{\text{Ampere hour of battery}}{\text{Charging current}}$$

Conclusion

The fossil fuels are yet the backbone of energy resources for most transportation vehicles which are already dwindling, becoming more expensive, and causing environmental vulnerability. This is because to meet the total energy need in the transportation sector globally, the conventional energy technologies are used which is putting tremendous stress on the fossil fuel and thus the fossil fuel is getting to a finite level. It is well established that the transportation sector consumes globally 30% of energy which is 5.6×10^{20} J/years (560 EJ/year) that is captured from the burning fossil fuel and causing environmental vulnerability exponentially. Interestingly, implementation of wind energy into the transportation vehicle would be an important finding once it is utilized by the running vehicles which will increase the use of wind power to run the transportation vehicles globally. Concretely it can be said that once the vehicles start to run by air it indeed would be the gateway of new technology to eradicate the energy cost for the transportation sectors which also will play a key factor in mitigating climate change exponentially.

Acknowledgements This research was supported by Green Globe Technology under the grant RD-02019-06 for building a better environment. Any findings, predictions, and conclusions described in this chapter are solely performed by the authors and it is confirmed that there is no conflict of interest for publishing this research in a suitable journal.

References

1. Bilel Touaiti, Hechmi Ben Azza, Mohamed Jemli. "A MRAS observer for sensorless control of wind-driven doubly fed induction generator in remote areas", 2016 17th International Conference on Sciences and Techniques of Automatic Control and Computer Engineering (STA), 2016.
2. Eunice Ribeiro, Ana Monteiro, Antonio J. Marques Cardoso, Chiara Boccaletti. "Fault tolerant small wind power system for telecommunications with maximum power extraction", 2014 IEEE 36th International Telecommunications Energy Conference (INTELEC), 2014.
3. Getachew Bekele, Getnet Tadesse. "Feasibility study of small hydro/PV/wind hybrid system for off-grid rural electrification in Ethiopia", Applied Energy, 2012.
4. Gunasekaran Nallappan, Young-Hoon Joo. "Robust Sampled-data Fuzzy Control for Nonlinear Systems and Its Applications: Free-weight Matrix Method", IEEE Transactions on Fuzzy Systems, 2019.
5. Hossain, Md. Faruque. "Photonic thermal control to naturally cool and heat the building", Applied Thermal Engineering, 2018.
6. Hossain, Md. Faruque. "Bose-Einstein (B-E) Photon Energy Reformation for Cooling and Heating the Premises Naturally", Applied Thermal Engineering.
7. K. Ghedamsi, D. Aouzellag. "Improvement of the performances for wind energy conversions systems", International Journal of Electrical Power & Energy Systems, 2010.
8. K. Kerrouche, A. Mezouar, Kh. Belgacem. "Decoupled Control of Doubly Fed Induction Generator by Vector Control for Wind Energy Conversion System", Energy Procedia, 2013.

9. K. Kerrouche, A. Mezouar, L. Boumedien. "A simple and efficient maximized power control of DFIG variable speed wind turbine", 3rd International Conference.
10. D Julius Mwaniki, Hui Lin, Zhiyong Dai. "A Condensed Introduction to the Doubly Fed Induction Generator Wind Energy Conversion Systems", Journal of Engineering, 2017.
11. Hossain, Md. Faruque. "Design and construction of ultra-relativistic collision PV panel and its application into building sector to mitigate total energy demand", Journal of Building Engineering, 2017.
12. Kheira Belgacem, Abelkader Mezouar, Najib Essounbouli. "Design and Analysis of Adaptive Sliding Mode with Exponential Reaching Law Control for Double-Fed Induction Generator Based Wind Turbine", International Journal of Power Electronics and Drive Systems (IJPEDS), 2018.
13. Hossain, Md. Faruque. "Advanced Building Design", Elsevier BV, 2019.
14. Hossain, Md. Faruque. "Theoretical mechanism to breakdown of photonic structure to design a micro PV panel", Energy Reports 2019.
15. H. Amimeur, D. Aouzellag, R. Abdessemed, K. Ghedamsi. "Sliding mode control of a dual-stator induction generator for wind energy conversion systems", International Journal of Electrical Power & Energy Systems, 2012.
16. Phan, Dinh-Chung, and Shigeru Yamamoto. "Rotor speed control of doubly fed induction generator wind turbines using adaptive maximum power point tracking", Energy, 2016.
17. Hossain, Md. Faruque. "Invisible transportation infrastructure technology to mitigate energy and environment", Energy, Sustainability and Society, 2017.
18. Mario Di Nardo, Mosè Gallo, Teresa Murino, Liberatina Carmela Santillo. "System Dynamics Simulation for Fire and Explosion Risk Analysis in Home Environment", International Review on Modelling and Simulations (IREMOS), 2017.
19. Mohammed Ouassaid, Kamal Elyaalaoui, Mohammed Cherkaoui. "Reactive power capability of squirrel cage asynchronous generator connected to the grid", 2015 3rd International Renewable and Sustainable Energy Conference (IRSEC), 2015.
20. Farhad Ilahi Bakhsh, Dheeraj Kumar Khatod. "A new synchronous generator based wind energy conversion system feeding an isolated load through variable frequency transformer", Renewable Energy.
21. Huang Ligang, Wang Xiangdong, Yan Kang. "Optimal speed tracking for double fed wind generator via switching control", The 27th Chinese Control and Decision Conference (2015 CCDC), 2015.
22. Vipin Chandra Upadhyay, K. S. Sandhu. "Reactive Power Management of Wind Farm Using STATCOM", 2018 International Conference on Emerging Trends and Innovations In Engineering And Technological Research (ICETIETR), 2018.
23. Mezouar, A.. "Adaptive sliding-mode-observer for sensorless induction motor drive using two-timescale approach", Simulation Modelling Practice and Theory, 2008.
24. Nabil Taib, Brahim Metidji, Toufik Rekioua. "Performance and efficiency control enhancement of wind power generation system based on DFIG using three-level sparse matrix converter", International Journal of Electrical Power & Energy Systems, 2013.
25. "Modeling, Identification and Control Methods in Renewable Energy Systems", Springer Science and Business Media LLC, 2019.
26. Riouch, Tariq, and Rachid El-Bachtiri. "Advanced Control Strategy of Doubly Fed Induction Generator Based Wind-Turbine during Symmetrical Grid Fault", International Review of Electrical Engineering (IREE), 2014.
27. Hossain, Md. Faruque, Nowhin Fara. "Integration of wind into running vehicles to meet its total energy demand", Energy, Ecology and Environment, 2016.
28. Hossain, Md. Faruque. "Photon application in the design of sustainable buildings to console global energy and environment", Applied Thermal Engineering, 2018.
29. Jogendra Singh, Mohand Ouhrouche. "Chapter 15 MPPT Control Methods in Wind Energy Conversion Systems", InTech, 2011.

30. Kamal Elyaalaoui, Mohammed Ouassaid, Mohamed Cherkaoui. "Supervision system of a wind farm based on squirrel cage asynchronous generator", 2016 International Renewable and Sustainable Energy Conference (IRSEC), 2016.
31. Manfred Stieber, *Wind Energy System for Electric Power Generation* (Springer, Verlag Berlin Heidelberg, 2008).
32. S. Abdeddaim, A. Betka. "Optimal tracking and robust power control of the DFIG wind turbine", International Journal of Electrical Power & Energy Systems, 2013.
33. Hossain, Md. Faruque. "Solar Energy Integration into Advanced Building Design for Meeting Energy Demand and Environment Problem". *Journal of Energy Research, 17: 49–55,* 2016.
34. Hossain, Md. Faruque. "Infrastructure and Transportation", Elsevier BV, 2019.
35. Kamal, E., M. Oueidat, A. Aitouche, and R. Ghorbani. "Robust Scheduler Fuzzy Controller of DFIG Wind Energy Systems", IEEE Transactions on Sustainable Energy, 2013.
36. Loucif, Mourad, and Abdelmadjid Boumediene. "Modeling and direct power control for a DFIG under wind speed variation", 2015 3rd International Conference on Control Engineering & Information Technology (CEIT), 2015.
37. Hossain, Md. Faruque. "Solar energy integration into advanced building design for meeting energy demand and environment problem". International Journal of Energy Research, 2016.
38. Junyent-Ferre, A.. "Modeling and control of the doubly fed induction generator wind turbine", Simulation Modelling Practice and Theory, 201010.
39. Maimaitireyimu Abulizi, Ling Peng, Bruno Francois, Yongdong Li. "Performance Analysis of a Controller for Doubly-Fed Induction Generators Based Wind Turbines Against Parameter Variations", International Review of Electrical.
40. Ouled Amor, Walid, A. Ltifi, and M. Ghariani. "Study of a Wind Energy Conversion Systems Based on Doubly-Fed Induction Generator", International Review on Modelling and Simulations (IREMOS), 2014.
41. M. Deepthi Rani, M. Satyendra Kumar. "Development of Doubly Fed Induction Generator Equivalent Circuit and Stability Analysis Applicable for Wind Energy Conversion System", 2017 International Conference on Recent Advances in Electronics and Communication Technology (ICRAECT), 2017.
42. Zohoori, Alireza, Abolfazl Vahedi, Mohammad Ali Noroozi, and Santolo Meo. "A new outer-rotor flux switching permanent magnet generator for wind farm applications: Flux switching permanent magnet generator for wind farm applications", Wind Energy, 2016.
43. Taveiros, F.E.V., L.S. Barros, and F.B. Costa. "Back-to-back converter state-feedback control of DFIG (doubly-fed induction generator)-based wind turbines", Energy, 2015.
44. Abdullah Asuhaimi B. Mohd Zin, Mahmoud Pesaran H. A, Azhar B. Khairuddin, Leila Jahanshaloo, Omid Shariatu, An overview on doubly fed induction generators' controls and contributions to wind based electricity generation, *Renewable and Sustainable Energy Reviews, vol. 27*, 2013, pp. 692-708.
45. Bakhsh, Farhad Ilahi, and Dheeraj Kumar Khatod. "A novel method for grid integration of synchronous generator based wind energy generation system", 2014 IEEE International Conference on Power Electronics Drives and Energy Systems (PEDES), 2014.
46. R. Karthikeyan, A. K. Parvathy. "Real time energy optimization using cyber physical controller for microsmart grid applications", 2016 International Conference on Recent Trends in Information Technology (ICRTIT), 2016.
47. Ali Darvish Falehi. "Augment dynamic and transient capability of DFIG using optimal design of NIOPID based DPC strategy", Environmental Progress & Sustainable Energy, 2017.
48. Hossain, Md. Faruque. "Bose-Einstein (B-E) photon energy reformation for cooling and heating the premises naturally", Applied Thermal Engineering, 2018.
49. Kiflom Gebrehiwot, Md. Alam Hossain Mondal, Claudia Ringler, Abiti Getaneh Gebremeskel. "Optimization and cost-benefit assessment of hybrid power systems for off-grid rural electrification in Ethiopia", Energy, 2019.

50. Xizheng Zhang, Yaonan Wang. "Robust Fuzzy Control for Doubly Fed Wind Power Systems with Variable Speed Based on Variable Structure Control Technique", Mathematical Problems in Engineering, 2014.
51. Youcef Saidi, Abdelkader Mezouar, Yahia Miloud, Mohammed Amine Benmahdjoub. "A Robust Control Strategy for Three Phase Voltage t Source PWM Rectifier Connected to a PMSG Wind Energy Conversion System", 2018 International Conference on Electrical Sciences and Technologies in Maghreb (CISTEM), 2018.

Chapter 11
Flying Transportation Technology

Abstract To have pleasant road trips, avoid commuting, and spend less time on journey to arrive at the desired destination much faster, a model of ***flying transportation technology*** has been proposed. The vision behind this technology is to design an economical, safe, and environmentally friendly mode of transportation. This study seeks to present a 3D numerical simulation of external flow for a flying automobile with well-designed rectangular NACA 9618 wings. To enhance its airborne capabilities, this car's aerodynamic traits have been professionally measured and adjusted, such that it utilizes minimal takeoff velocity. Besides, the vehicle will have an integrated 3D *k-omega* turbulence model, which captures a fundamental flow physics enhancing the performance during takeoff. This forms the theoretical basis of the flying car. The numerical aspect comprises a limited-edition Reynolds-averaged Navier–Stokes equation (RANS) comprehensible schemes. Generally, the vehicle is designed with highly functional wings that allow divergent deployments during takeoff to maximize its air performance. Because of the utilization of wind during the flying process, the model has integrated wind turbines that enables wind recodification to propel the car in the air. Considering the combination of technologies involved in the design of the flying car, it is one of the most sophisticated inventions that will not only facilitate safe transportation and save trillions of dollars annually but also significantly help in saving the ecosystem.

Keywords 3D numerical simulation · 3D k-omega turbulence modeling · Flying car design · Wind energy modeling · Cleaner energy implementation · Smart transportation technology

Introduction

The construction of roads and other transportation infrastructure consumes 0.9% of the earth surface. In addition, 1.77 million dollars is spent on constructing every mile of the infrastructure. As a result, several forest covers are destroyed to pave the

M. F. Hossain, *Global Sustainability in Energy, Building, Infrastructure, Transportation, and Water Technology*,
https://doi.org/10.1007/978-3-030-62376-0_11

way for the infrastructure, thus contributing to an approximated 6% of climate change. This costs about \$5575 trillion (100,000,000 mi^2 × 5280 sf × 5280 sf × \$200 per SF cost) as the heat is reflected back into space. The current statistics indicate that 2% of the existing infrastructure undergoes annual repairs totaling to \$12 trillion, with another \$55.75 trillion spent in new infrastructure development annually. The rate of infrastructure development is estimated at 1% annually. Other than the infrastructure development and rehabilitation processes, road transport results in frequent traffic jams that cost trillions annually [1, 2]. The fact that the current traditional transport system heavily relies on fossil fuels, which results in undesirable emissions in the atmosphere, also makes it unfriendly to the environment. Compared to the flying transportation technology, the new model is much more safe, convenient, and economical. A lot of research has been conducted in designing the flying car, and it might help save both the environment and the economy upon its completion.

The engineering disciplines have of late seen diverse integration of computational fluid dynamic (CFD) models on a large-scale basis. This is a commendable step that will significantly improve the affordability of the models and enhance their power. The CFD model can also be adopted in the development of flying cars. It, however, requires much more precision as the flying cars are much more complicated than the regular race cars and aircraft currently in use primarily due to the additional features of its wings that support its deployment during takeoffs. This study, therefore, aims to introduce a propulsive and levitative force model that can be used in addressing the aerodynamics takeoff velocity and drag force computational fluid problems to enhance the performance of the flying cars. The research will adopt the use of MATLAB software to illustrate the model.

Materials, Methods, and Simulation

The flying vehicle's performance in the air is mainly determined by takeoff velocity, propelling forces, drag force, stability, and proper control capabilities. These are the areas that require significant attention, especially during the wing designing phase. When creating the flying car's flanks, the shape, aspect ratio, cross sections, and surface areas must be considered as they will determine the stability, control, and takeoff forces required for the maximum performance of the car. This research involves an illustration of the Mach number contours and the physical model of a complete 3D CFD, a *k-omega* turbulence mode required in a flying vehicle (Fig. 11.1). To validate the reliability of a CFD design in designing flying cars, the study will review various deployment histories involving its use.

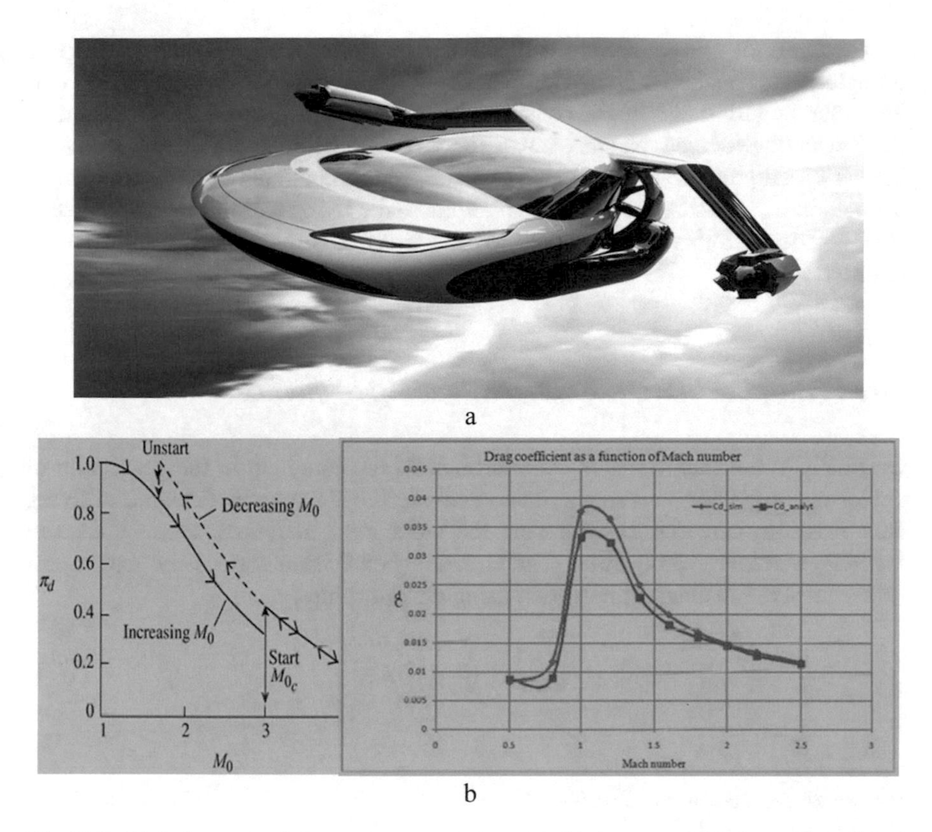

Fig. 11.1 (**a**) Indicates the flying car's conceptual model, (**b**) the proposed flying vehicle's Mach number contours

Solving the Concept Numerically

As discussed earlier, the 3D standard *k-omega* turbulence model is based on the transport velocity model as an empirical model in assuring accurate dissipation rates and for turbulence kinetic energy and is indicated in Eq. (11.2):

$$\psi = \left[-\left(\frac{v_\infty r_o^2}{4} \right) \frac{r_o}{r} + \left(\frac{3 v_\infty r_o}{4} \right) r - \left(\frac{v_\infty}{2} \right) r^2 \right] \sin^2 \theta \tag{11.1}$$

$$v_r = v_\infty \left[1 - \frac{3}{2} \left(\frac{r_o}{r} \right) + \frac{1}{2} \left(\frac{r_o}{r} \right)^3 \right] \cos\theta \tag{11.2}$$

$$v_\theta = -v_\infty \left[1 - \frac{3}{4} \left(\frac{r_o}{r} \right) - \frac{1}{4} \left(\frac{r_o}{r} \right)^3 \right] \sin\theta \tag{11.3}$$

The equation utilizes second coupled hierarchy formulation to adequately find the solution to standard *k-omega* turbulence with shear flow corrections. The numerical solution will, therefore, be considered as an entirely limited volume scheme of the Reynolds-averaged, Navier–Stokes force compressible that neutralizes the flying car's total pressure and wall temperatures. To achieve reliable codes that can be adopted in invalidating the baseline solutions to the flying car's wing designs 9618 NACA series airfoils can be incorporated to achieve maximum aerodynamic performance (Fig. 11.2).

The Flying Car's Wind Energy Modeling Sequence

Originally, the installed single wind turbines in the cars help in the generation of energy necessary for powering it, thus satisfying the high energy demands of flying cars. It is also essential that a doubly fed induction generator is also installed to facilitate electricity production. The equation below summarizes the entire processes involved in energy production using wind turbines:

$$P_{\mathrm{w}} = \frac{1}{2} C_{\mathrm{p}} (\lambda, \beta) \rho A V^3 \tag{11.4}$$

where

V = average speed of wind (m/s)
P = air density (kg/m^3)
C = Betz's coefficient (maximum value of 0.593)
λ = tip speed ratio
A = rotor blades' intercepting areas (m^2)

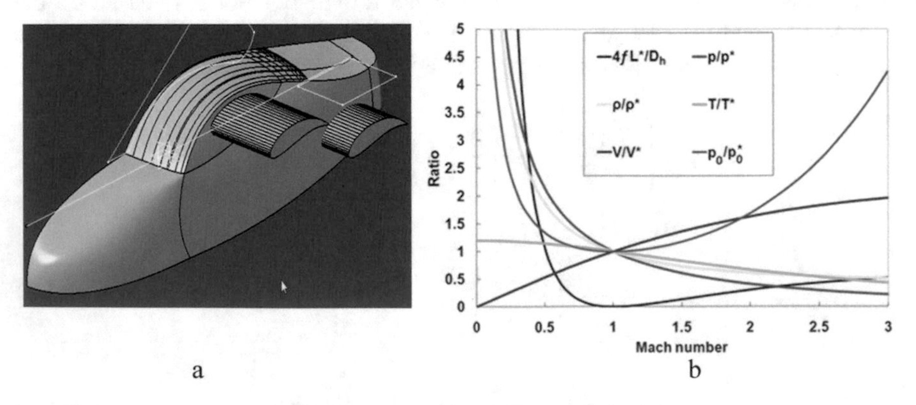

Fig. 11.2 (**a**) Indicates half of the flying car's 3D idealized model. (**b**) Aerodynamic traits in relation to the velocity and the Mach number

Here the, tip speed ratio equation is rewritten as

$$\lambda = \frac{R\omega}{V} \tag{11.5}$$

In the above equation,

ω = angular speed in rad/s
V = wind speed average (m/s)
R = turbine radius (m)

To calculate the wind-generated energy the equation below is utilized:

$$Q_{\mathrm{w}} = P \times (\text{Time}) \; [\text{kWh}] \tag{11.6}$$

The direct measurements from a given motion cannot be used in obtaining velocity; thus, a lower motion should be first established using the equation below:

$$v(Z)\ln\left(\frac{Z_{\mathrm{r}}}{Z_0}\right) = v(Z_{\mathrm{r}})\ln\left(\frac{Z}{Z_0}\right) \tag{11.7}$$

with

Z = reference height (m)
Z_0 = surface roughness measurement (in crop lands = 0.1–0.25)
$v(Z)$ = speed of wind at Z height (m/s)
$v(Z_r)$ = speed of wind at Z (m/s)

The equation below represents the estimation of power output based on the wind speed:

$$P_{\mathrm{w}}(v) = \begin{cases} \dfrac{v^k - v_{\mathrm{C}}^k}{v_{\mathrm{R}}^k - v_{\mathrm{C}}^k} \cdot P_{\mathrm{R}} & v_{\mathrm{C}} \le v \le v_{\mathrm{R}} \\ P_{\mathrm{R}} & v_{\mathrm{R}} \le v \le v_{\mathrm{F}} \\ 0 & v \le v_{\mathrm{C}} \text{ and } v \ge v_{\mathrm{F}} \end{cases} \tag{11.8}$$

In the above equation,

P_{R} = represents the rated power
v_{C} = wind speed cut-in
v_{R} = total wind rated
v_{F} = cutout speed rated
k = shape factor of Weibull

The primary purpose of altering the angular speed is to enable the extraction of optimum power levels, a technique commonly known as maximum power point tracking (MPPT). In scenarios where the blade pitch is at a zero angle, the optimum TRS is obtained by maximizing the power coefficients:

$$\omega_{\text{opt}} = \frac{\lambda_{\text{opt}}}{R} V_{\text{wn}} \tag{11.9}$$

This results in

$$V_{\text{wn}} = \frac{R\omega_{\text{opt}}}{\lambda_{\text{opt}}} \tag{11.10}$$

with

ω_{opt} as the optimum speed of the rotor angle measured in rad/s
λ_{opt} representing the ratio of optimum tip speed
R denoting the turbine radius (m)
V indicating the speed of wind (m/s)

Figure 11.3 indicates the changes in the rates of wind speed and power when the car is active [3, 4]. From the analysis of the figure, it is clear that quality results in the car's performance can only be achieved in steady winds. On the other hand, a rapid increment in the speed of wind results in the rise in energy levels.

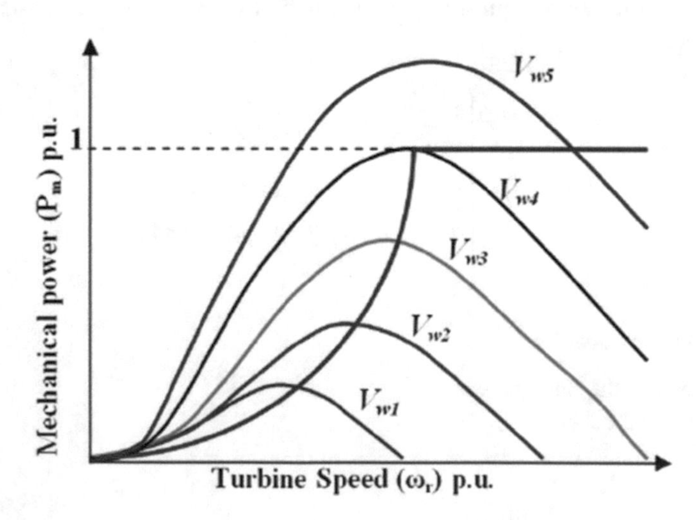

Fig. 11.3 Correlation between turbine speed and generated power

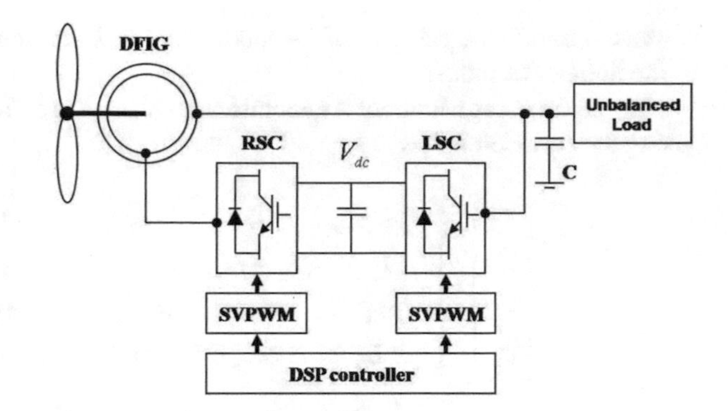

Fig. 11.4 The process of wind conversion to usable energy through the use of LSC, RSC, DFIG, DSP, and SVPWM

Conversion of Wind Energy

The conditions above require adequate airflow monitoring and regulation at specific kinetic energy levels. This is usually indicated by a wind energy conversion system (WECS) used in DFIG, such as the speed of the wind, gearbox, and wind turbine, among others (Fig. 11.4).

Modeling the Generator

Two options are always available when energy is required for the wind turbines. This includes the use of induction generators or synchronous models. The use of an asynchronous model drastically increases reliability. However, it results in a significant reduction of the nacelle weight [5, 6]. The equation below indicates the synchronous model principles based on a *d*–*q* referencing framework:

$$V_q = -R_s i_q - L_q \frac{di_q}{dt} - \omega L_d i_d + \omega \lambda_m \tag{11.11}$$

$$V_d = -R_s i_d - L_d \frac{di_q}{dt} + \omega L_q i_q \tag{11.12}$$

in which electronic torque is indicated by

$$T_e = 1.5\rho\left[\lambda i_q + \left(L_d - L_q\right) i_d i_q\right] \tag{11.13}$$

In the equation, q axis inductance is represented by L_q, while L_d indicates d axis inductance. i_q is the current from the q axis while i_d current from d axis.

V_q represents voltage from the q axis, and the energy from the d axis is indicated by V_d.

ω_r represents the velocity of the angular rotor, λ the induced flux amplitude, and p the number of poles.

The squirrel cage induction generator can also be used in modeling the generator using the equation below:

$$\begin{bmatrix} V_{qs} \\ V_{ds} \\ V_{qr} \\ V_{dr} \end{bmatrix} = \begin{bmatrix} R_s + pL_s & 0 & pL_m & 0 \\ 0 & R_s + pL_s & 0 & pL_m \\ pL_m & -\omega_r L_m & R_r + pL_r & -\omega_r L_r \\ \omega_r L_m & pL_m & \omega_r L_r & R_r + pL_r \end{bmatrix} \begin{bmatrix} i_{qs} \\ i_{ds} \\ i_{qr} \\ i_{dr} \end{bmatrix} \tag{11.14}$$

Beginning from the stator position

$$\begin{aligned} \lambda_{ds} &= L_s i_{ds} + L_m i_{dr} \\ \lambda_{qs} &= L_s i_{qs} + L_m i_{dr} \\ L_s &= L_{ls} + L_m \\ L_r &= L_{lr} + L_m \\ V_{ds} &= R_s i_{ds} + \frac{d}{dt}\lambda_{ds} \\ V_{qs} &= R_s i_{qs} + \frac{d}{dt}\lambda_{qs} \end{aligned} \tag{11.15}$$

Beginning from the rotor position

$$\begin{aligned} \lambda_{dr} &= L_r i_{dr} + L_m i_{ds} \\ \lambda_{qr} &= L_r i_{qr} + L_m i_{qs} \\ V_{dr} &= R_r i_{dr} + \frac{d}{dt}\lambda_{dr} + \omega_r \lambda_{qr} \\ V_{qr} &= R_r i_{qr} + \frac{d}{dt}\lambda_{qr} - \omega_r \lambda_{dr} \end{aligned} \tag{11.16}$$

To determine the air flux gaps

$$\begin{aligned} \lambda_{dm} &= L_m \left(i_{ds} + i_{dr} \right) \\ \lambda_{qr} &= L_m \left(i_{qr} + i_{qs} \right) \end{aligned} \tag{11.17}$$

In the above equation, R_s, R_r, ω_r, i_d, i_q, V_d, V_q, λ_d, λ_q, L_m, L_{ls}, and L_{lr} indicate the resistance from wind by the stator and the fluxes. Energy conversion diagram-implemented inverter is often used in the preparation of the wind energy output. This is done through the calculation in Simulink-MATLAB.

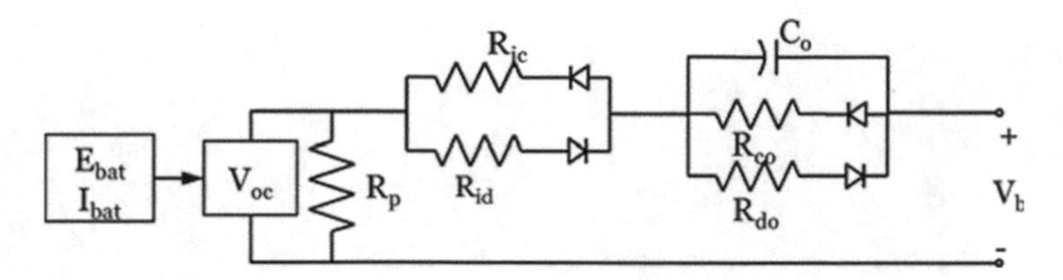

Fig. 11.5 The power backup model in a motionless car's battery for the ignition of the vehicle and to provide required energy while the transport is not in motion

Battery Modeling

In situations where the car is nonfunctional or where the engine power is exceeded by the power produced, battery modeling is adopted in order to conserve the extra power [7, 8]. This process can be calculated by using the formula below:

$$t_{\text{discharge}} = H\left(\frac{C}{IH}\right)^{k} \tag{11.18}$$

In the formula, the discharge time of the battery is represented by t. C shows the capacity of the battery in ampere hour value (Fig. 11.5). The drawn current is indicated by I, while H represents the time taken to discharge the power. The Peukert's coefficient is shown by k and can be calculated by using the equation below:

$$k = \frac{\log T_2 - \log T_1}{\log I_1 - \log I_2} \tag{11.19}$$

in which I_1 and I_2 represent the rates of current discharge. T_1 and T_2 show the duration of discharges. Because of the changes that take place after particular recharge cycles, including a decrement in the capacity of the battery, the k value must be redefined by adopting 1.3–1.4. To calculate the exact time taken to discharge a battery fully, the equation below is used:

$$t_{\text{charging}} = \frac{\text{Ampere hour of battery}}{-\text{Charging current}} \tag{11.20}$$

Results, Optimization, and Discussion

In relation to proposed models, the fly car has been identified to be able to produce a satisfactory lift with a proper takeoff velocity within various wing positions and attack angles. Also, an in-depth analysis of the study indicates that the car has the

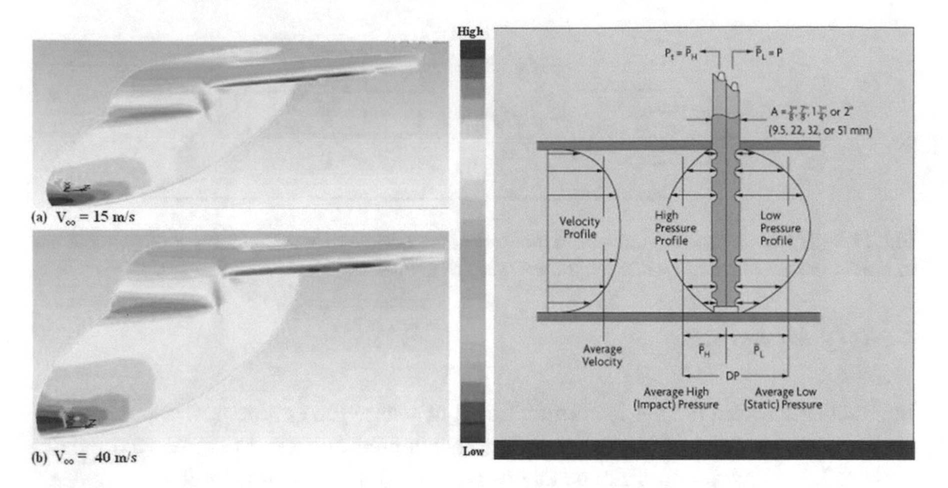

Fig. 11.6 (**a**) A flying vehicle's contours at different speeds (15 and 40 m/s), (**b**) a short high-wing flying car's model static pressure contours at a different velocity

ability to attain an appropriate lift even in low-takeoff-velocity situations. This is because of the presence of the deployable high wings shown in Fig. 11.6. Mathematically, it is evident that the flying vehicle's stream velocity ranges and the various geometrical options are able to produce sufficient upward force, thus lifting the car even in low-takeoff-velocity conditions and also across multiple wing positions.

The findings from the assessment of the aerodynamic characteristics and the features of the external flow of the flying car indicate that the car's speed heavily relies on the aerodynamic traits and the steadiness of the free stream velocity. This property enables the vehicle to initiate a takeoff from the ground, its flying ability, and proper control at different altitudes, as indicated in Fig. 11.7.

Through the adoption of these principles, this study adequately depicts the required standards and conditions necessary to facilitate the flying vehicle's takeoff and flying pattern. In order to achieve better performance and to keep the car on air, the body of the vehicle must be streamlined, and the wings must be made suitable for enhancing the lifting abilities of the vehicle (Fig. 11.7). One way to achieve this is through the installation of NACA 9816 wings at various points of the vehicle. This will enable the vehicle to adequately launch attacks at various angles in the air, facilitating a smooth flying experience.

From the robustness tests conducted in the simulation phase, the findings indicated that the flying car's capability of handling the conditions is remarkable. This test was conducted by adding voltage dips and wind speed signals. Other suitable controls that can be added to enhance the performance of the flying car are DFIG and MPPT control. These controls can also be used in controlling the reactive powers and the stator active powers to facilitate a united power on the stator position. Also, after an increment in the wind speed, the generator shaft is able to detect an increased performance in angular momentum. In such situations, the reactive and

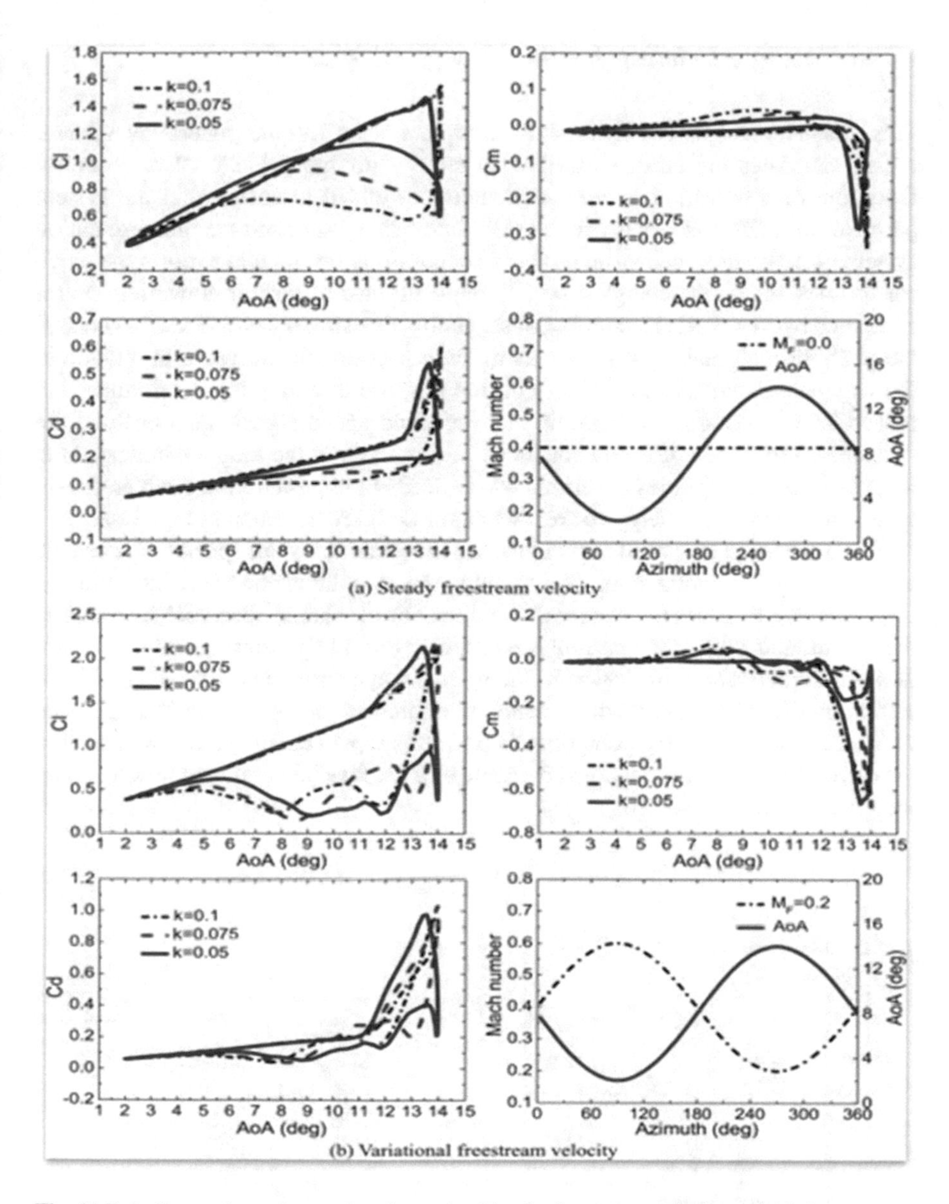

Fig. 11.7 Indicates the car's speed as determined by the freestream velocity variable in respect to the aerodynamic traits to facilitate the takeoff, fly, and optimization of the velocity, steady freestream velocity

the active bidirectional power transfer between the power and the rotor system can be calculated through the realization of the nominal potentials of the stator and the supersynchronous operation. The stator power is reactive and is controlled by the converter on the load side to produce sufficient energy to power the car.

Wind Energy Modeling for the Flying Vehicles

This chapter presents the flying car's complete wind turbine generating system, which facilitates the production of wind energy for the vehicle. Also, a control algorithm of a cascade has been meticulously molded to ensure that the system performs optimally, especially in the MPPT control system and the flux orientation system [9, 10]. This mechanism has been successfully installed in a prototype sedan car to track the active energy power through the use of a rotor converter control computed by MATLAB Simulink. Subsequently, this strategy is also used to control the pitch through making adjustments in the power coefficient value in relation to the variation in the speed of wind with the purpose of extracting maximum wind power. The addition of voltage dip and the wind speed signals also enabled the robustness test to be efficiently conducted. From the test, the findings indicated the wind turbine's capabilities at various wind speeds in relation to the pitch angle, tip speed ratio, and wind energy conversion chain DFIG represented in Fig. 11.8.

The increase in the speed of wind causes the generator shaft speed to achieve its extreme angular momentum by trailing the absolute power point velocity. Subsequently, to affirm the workability of the control scheme, disassociation among the constituents of the rotor current is measured (Fig. 11.8). Consequently, the function of the generator with regard to the splendid synchronous action, insignificant station power, and volatile power which is organized through a consignment-side transformer to acquire the component's power aspect to create energy was used in the calculation of the bidirectional dynamic and reactive power transmission among the rotor and electric scheme.

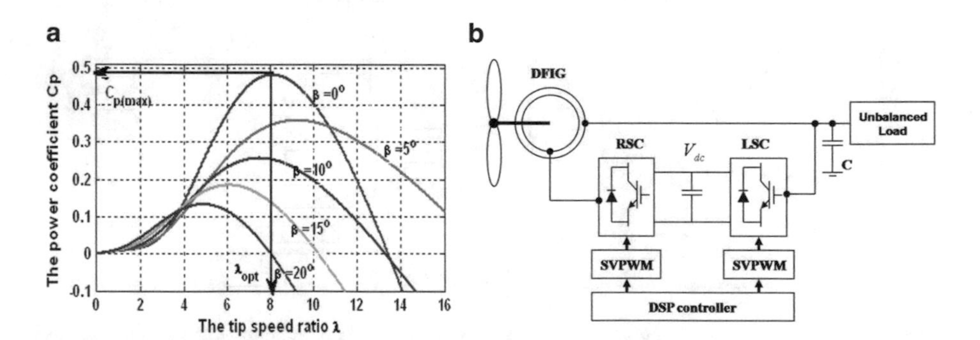

Fig. 11.8 (**a**) Indicates how to achieve the C_p's maximum values (0.5) in relation to the $\beta = 2°$. The value $\lambda_{opt} = 0.91$ indicates the maximum speed ratio (8 m/s) with a 10 m/s wind speed rating. The entire system testing was conducted under tight conditions with an approximated 50% (0.5–4.5), 25% (6–6.5), and 50% (8–8.5); (**b**) for the DFIG control, the profile of the wind was considered as the wind speed signal, which permits the application of DSP control of the wind turbine to form the V_{dc} energy in the turbine. Thus, the wind turbine is considered to be working and in proper condition. In addition, to confirm a unity power factor at the stator position, the turbine is given the reactive power as zero, where the stator active and reactive powers are regulated by the MPPT system [11, 12]

An analysis of the electromagnetic torque implies that a predefined power speed plays a crucial role in trailing the extreme power point into the rotor to convert the wind into energy computation of MPPT control. The calculation of MPPT, which was done, had the aim of confirming this stator active power. Thus, the achieving of the maximum power cofactor depicted that the optimum electromagnetic torque deduced from MPPT is by determination of the rapidity of the shaft of the wind-driven turbine. Since the control of the MPPT revealed the effectiveness of the wind-driven turbine, consequently the computation of the rotor dynamics is determined by the referenced reactive and active powers that are expressed by the equation below:

$$\begin{cases} i_{qr}^{*} = -\dfrac{L_{s}}{MV_{s}} P_{s}^{*} \\ i_{dr}^{*} = -\dfrac{L_{s}}{MV_{s}} \left(Q_{s}^{*} - \dfrac{V_{s}^{2}}{\omega_{s} L_{s}} \right) \end{cases} \tag{11.21}$$

The whole of this mechanism tries to imply an accurate wind speed intake by the wind-driven turbine where the MPPT control tracked the velocity of the wind uninterruptedly and attuned the imposed electromagnetic torque of the DFIG to follow its aerodynamics accordingly to produce the energy. Therefore, the block diagram of the clarified subsystem confirms the rate of wind intake processed by this mechanism (Fig. 11.9).

What is being revealed in the MPPT control and DFIG block diagram is that a double rate of harvesting wind by the turbine is attained by controlling the speed of the generator. Thus the stator flux computation of the wind turbine, which is passed through the drivetrain to convert it into electric energy for operating the vehicle, determines the optimal wind intake rate.

Both induction and synchronous variable speed are produced by the drivetrain at the turbine, and they both had a direct impact on the synchronous gearbox. Consequently, the application of the gearbox in variable-speed wind turbine shaft is controlled to produce the maximum rate of energy being produced and given out. Therefore, the drivetrain model is carried out with regard to the *d–q* synchronous and is expressed as in the equation below:

The electronic torque is given by

In the above equations,

L_q represents the q axis mutual induction
L_d represents the d axis mutual induction
i_q represents the q axis flow of electricity
i_d represents the d axis flow of electricity
V_q represents the q axis energy
V_d represents the d axis energy
ω_r represents the rate of change of angular rotor position
λ refers to the maximum extent of oscillation of flux made

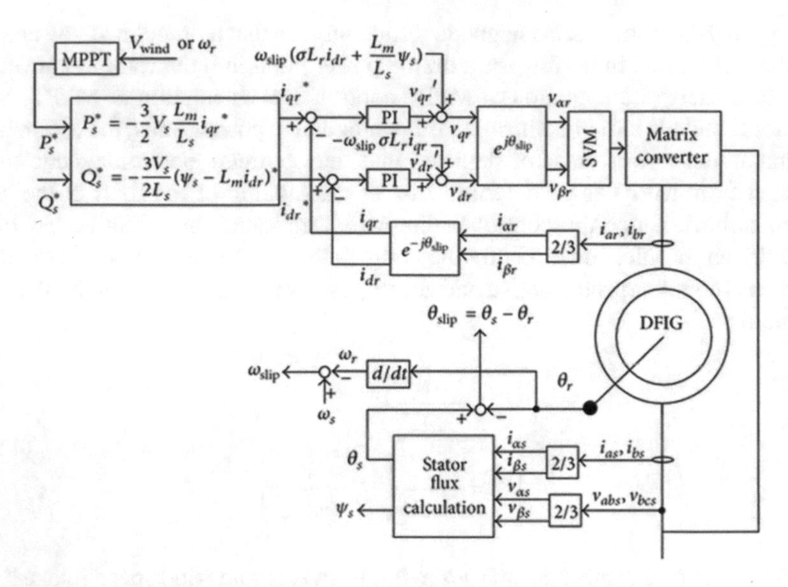

Fig. 11.9 The block diagram of the MPPT (V_{wind}) control bearing in mind the incarcerated DFIG velocity of the aerodynamic subsystem to find out the rate of wind intake and the transformation process of the wind energy into electric energy for operating the running vehicle by means of stator flux rotor initiation of wind turbine of the shipping vehicle

p represents the number of pairs of poles

The equation below can be applicable in case of dynamic modeling and is also pertinent in the instance of squirrel-confined initiation generator (SCIG):

From the stator side, the equivalence is

From the rotor side, the equivalence is

For the inflight gap flux connection, the equivalences are

In the equation

R_s = stator curving resistance
R_r = motor sneaking resistance
L_m = magnetizing inductance
L_{ls} = stator seepage inductance
L_{lr} = dynamic rotor angular velocity
ω_r = current, energy
λ_q = fluidities

A substitution for the above equations can be expressed as

$$P_{\mathrm{w}} = \frac{1}{2}\rho A C_{\mathrm{p}}(\lambda,\beta)\left(\frac{R\omega_{\mathrm{opt}}}{\lambda_{\mathrm{opt}}}\right)^3 \tag{11.22}$$

And it can be used to obtain the productivity power and rotating force of the turbine (T_t) in terms of gyratory rapidity:

$$T_{\mathrm{t}} = \frac{1}{2}\rho A C_{\mathrm{p}}(\lambda,\beta)\left(\frac{R}{\lambda_{\mathrm{opt}}}\right)^3 \omega_{\mathrm{opt}} \tag{11.23}$$

The power constant (C_{p}) is a counterclockwise principle articulated by the appropriate equivalence in the method:

$$C_{\mathrm{p}}(\lambda,\beta) = c_1\left(c_2\frac{1}{\lambda_i} - c_3\beta - c_4\right)\mathrm{e}^{-c_5\frac{1}{\lambda_i}} + c_6\lambda \tag{11.24}$$

with

$$\frac{1}{\lambda_i} = \frac{1}{\lambda + 0.08\beta} - \frac{0.035}{\beta^3 + 1} \tag{11.25}$$

The value of constants c_1–c_6 has been explained in later section. Consequently, the energy conversion circuit diagram implemented in Simulink systems by solicitation of wind energy alteration mechanism processes the productivity of the airstream energy generation module.

Wind Energy Conversion

Determination of the sturdiness of the fuzzy lucidity checker (FLC) is done in consideration of its nonlinear processes and physiognomies to wind energy contrivance to convert it into electrical energy. To determine the operation of the conversion of wind by the presentation of the control rules of the fuzzy block diagram, the evaluation of the FLC mechanism by the fuzzy inference system is applicable. The fuzzification block reveals the calculated risk output of wind energy that is controlled by the rules of the defuzzification of the airstream turbine [13, 14].

The determination of the airstream turbine and its DFIG subsystem to deal with the wind energy transformation mechanism to come up with electric energy is by utilization of this FLC as expressed below:

$$\begin{cases} e_{\Omega_g(n)} = \Omega_g^*(n) - \Omega_g(n) \\ \Delta e_{\Omega_g(n)} = \Omega_g^*(n) - \Omega_g(n-1) \end{cases} \tag{11.26}$$

The subsequent determination of the net realistic energy variants into the fuzzy controller is usually from the fuzzy controller. With regard to the above statement, it is evident that the net wind energy production affirms the triangle, trapezoidal, and symmetrical speed purposes of the wind-driven turbine. Analyzation of the formation of the net current of the turbines driven by wind is simply by the precise calculation of the FLC of DFIG by expression of the equation below:

$$\begin{cases} e_{i_{dr}(n)} = i_{dr}^*(n) - i_{dr}(n) \\ e_{i_{dr}}(n) = i_{qr}^*(n) - i_{qr}(n) \end{cases} \tag{11.27}$$

The above equation can be further simplified as

$$\begin{cases} \Delta e_{i_{dr}}(n) = i_{dr}^*(n) - i_{dr}(n-1) \\ \Delta e_{i_{qr}}(n) = i_{qr}^*(n) - i_{qr}(n-1) \end{cases} \tag{11.28}$$

In conclusion, the input and output of the conversion of wind energy using the fuzzy controller have been quantized. Consequently, the effect of wind speed in altering the output power into the rotor of the airstream-driven turbine is used to determine the rate at which the wind is transforming. Consequently, there is the computation of changes in the output power in the rotor with regard to the trade-off between exactitude and intricacy of the conversion of wind energy in the electrical subsystem via simulation that is extremely thorough and careful to produce enteric energy that is useful and desirable for the operation of a vehicle.

Electrical Subsystem

Once the wind energy modeling has been analyzed, the implementation of this wind energy into the electrical subsystem has been calculated to run the vehicle. Therefore, both induction and synchronous wind energy driven by the rotor are captured by a permanent synchronous generator (PMSG) to convert it into electric power [15, 16]. Hence, the functional mechanism of the gearbox of this PMSG plays a vital role in converting the electrical energy to DC current, and the *d–q* synchronous voltage equation quantifies it.

Analysis of the wind energy modeling implies that the enactment of this wind energy into the electrical subsystem has been premeditated to operate the vehicle. It is thus clear that both induction and synchronous wind energy driven by the rotor are captured by a permanent synchronous generator (PMSG) to convert it into electric energy. Hence, the efficient mechanism of the gearbox of this PMSG plays a vivacious role in converting the electrical energy to DC current, and it is quantified by the *d–q* synchronous voltage equation as

The electronic torque is given by

In the above equation, L_q represents the q axis mutual induction, L_d is considered as the d axis inductor, i_q represents the q axis flow of electricity, i_d represents the d axis flow of electricity, V_q represents the q axis energy, V_d denotes the d axis energy, ω_r represents the range of speed of rotor, and λ represents the maximum extent of the oscillation of the induced flux. P is the quantity of pairs of poles. The demonstration of the dynamic modeling of wind simplifies the electronic torque, as illustrated in the equation below:

The equation that expresses the stator side is

A simplified form of the above equation is

If an air gas flux leaks into the generation, the equation is further simplified to where R_s, R_r, L_m, L_{ls}, L_{lr}, ω_r, i_d, i_q, V_d, V_q, λ_d, and λ_q represent the stator resistance from one end to another, motor resistance from one end to another, self-inductance of an inductor with magnetic core, stator leakage inductance, rotor outflow inductance, electrical rotor angular velocity, electric flow, energy, and fluxes correspondingly of the *d–q* model [17]. Then the net productivity power and rotating force of turbine (T_t) in terms of gyratory velocity can be attained by the following equations:

The fitting equation below expresses the power constant (C_p), which is a discipline that is not linear:

with

In conclusion, an energy transformation circuit diagram processes the output of the wind energy generator module through the implementation of the converter of Simulink power systems. Volt that assists in powering the vehicle is the result of what is calculated as the result of the circuit mode electricity generation (Fig. 11.10).

Generator Modeling

One of the two between initiation and asynchronous dynamo can be applicable in the WT schemes. Adjustable velocity straight-driven multipole perpetual magnet synchronous dynamos (PMSGs) are similarly expansively applicable in airstream-driven energy schemes since they have higher competence, subordinate weight, cheap upkeep, and stress-free workability and since they do not necessitate a responsive and attracting flow of electricity. The existence of a gearbox in variable-velocity WTs results in an additional encumbrance of fee and upkeep. Exhausting a straight-driven PMSG does both the increasing of consistency and decreasing of the mass of the nacelle.

The classical for a PMSG is its basis on a *d–q* synchronous situation frame. The equation of PMSG energy is given as

$$V_d = -R_s i_d - L_d \frac{di_q}{dt} + \omega L_q i_q$$

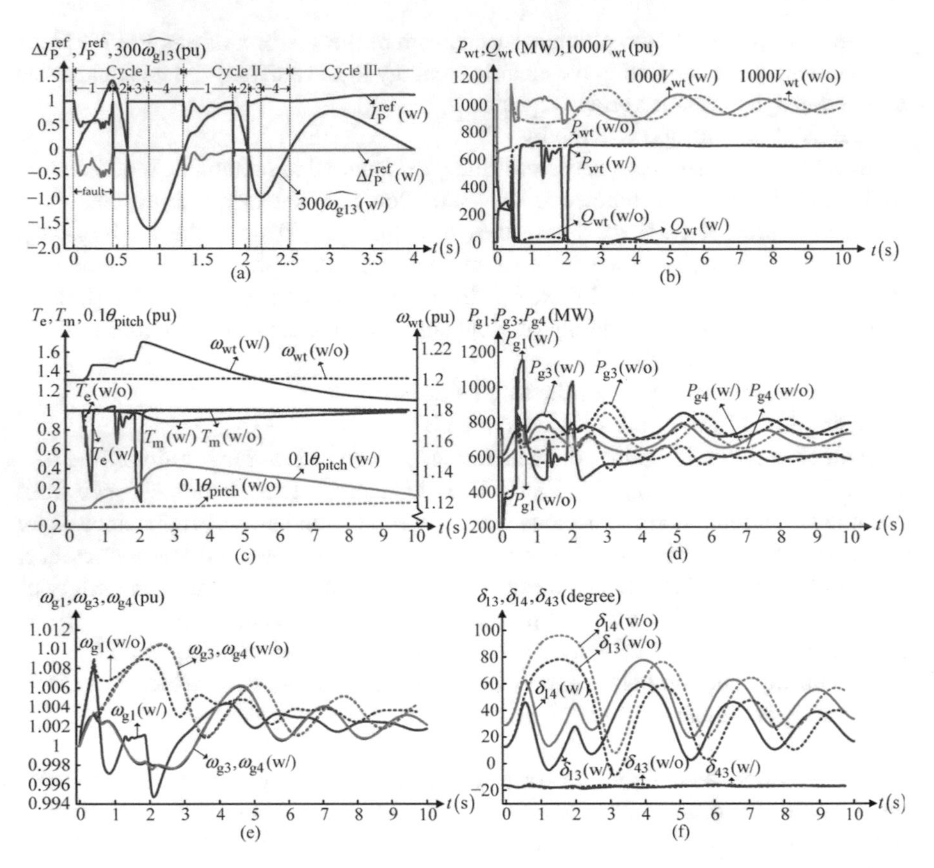

Fig. 11.10 Aerodynamic power generation by an installed wind turbine, which is linked to the electrical subsystem circuit to determine the current generation (MW) at various degrees of the angle of the pitch of the wind turbine rotation

The automated torque is given by

A WT system can accommodate both an induction and a synchronous generator. Because of their unique features and characteristics such as high efficiency, reduced cost of maintenance, and ease in controlling and since they can function without the use of reactive and magnetizing current, the variable-velocity direct-driven multi-pole perpetual magnet synchronous dynamos (PMSGs) are also expansively being used in airstream-driven power schemes. An extra cost of maintenance and burden is created by the existence of a gearbox in variable-speed WT generators. The decreasing of the weight of the nacelle is not the only impact created by the use of a direct-driven gearbox in variable-speed WTs but also the generation of an extra burden of cost. A *d–q* synchronous reference frame provides the basis for the model for a PMSG. The equivalence of PMSG is

The electronic torque is given by

In the above equation, the *q* axis mutual induction is L_q, *d* axis mutual induction is L_d, *q* axis flow of electricity is i_q, i_d is the *d* axis flow of electricity, V_q is the *q* axis energy, V_d is the d axis energy, ω_r is the rate of change of angular position of the rotor, λ is the amplitude of flux prompted, and p is the quantity of pairs of poles. The equivalence that follows, which can be applicable in conjunction with a static d–q frame of reference for dynamic modeling, is relevant with regard to a squirrel cage induction dynamo (SCIG):

The equations that are applicable for the stator side are

The equations that are applicable for the rotor side are

The equations that apply to the air gap flux connotation are

where R_s, R_r, L_m, L_{ls}, L_{lr}, ω_r, i_d, i_q, V_d, V_q, λ_d, and λ_q are the stator winding resistance, motor winding resistance, fascinating inductance, stator leakage inductance, rotor leakage inductance, electrical rotor angular velocity, flow of electricity, energy, and fluxes, correspondingly, of the d–q model. Obtaining of the torque turbine (T_t) and production of power in terms of consistent speed are done through substitution of the formulas below:

$$P_w = \frac{1}{2}\rho A C_p(\lambda,\beta)\left(\frac{R\omega_{opt}}{\lambda_{opt}}\right)^3$$

Battery Modeling

The application of battery modeling plays an essential role in the functioning of a vehicle. One of the uses of application of battery modeling in a car is functioning as the starting power that ignites the car and operation as the source that provides backup whenever a car is not moving (Fig. 11.11). Battery modeling functions as the source that provides backup whenever a car is not moving by integrating the Peukert's law of battery charge which is

$$t_{discharge} = H\left(\frac{C}{IH}\right)^k \tag{11.29}$$

In the above formula, the time for battery charge is represented by t, C represents the capacity of the battery, the time that is rated discharge is represented by H, and the constant for Peukert is represented by k. Peukert's constant is calculated using the formula

$$k = \frac{\log T_2 - \log T_1}{\log I_1 - \log I_2} \tag{11.30}$$

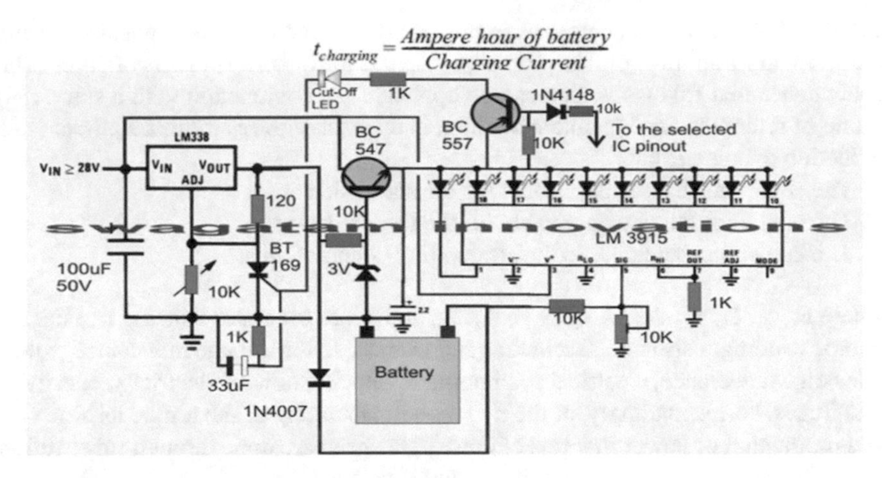

Fig. 11.11 A block diagram of the battery modeling of a car is shown to determine the V_{oc} and SOC to control its voltage source to ignite the engine and run the car when it is not in motion

In the above formula, the variance of the charge current flow rates is represented by I_1 and I_2, and T_1 and T_2 represent the corresponding time to charge the battery completely. The formula used to acquire the corresponding time to charge the battery entirely is

$$t_{\text{charging}} = \frac{\text{Ampere hour of battery}}{\text{Charging current}}$$

Conclusion

Coextensively, conventional transport infrastructure systems all over the world cause adverse environmental impacts, and also result in a waste of funds annually. The solution for balancing the link amid traffic, price, energy, and atmosphere in this research may come from the introduction of flying vehicle technology. Subsequenlty, the installation of a wind turbine into the flying vehicle helps in the production of energy by itself when the vehicle is functioning in consideration of the sophisticated energy demand and the necessity for a more conducive ecosystem. In recent days, the construction of infrastructure has been costing trillions of dollars; the amount of fossil fuel consumed by the transport sector annually is 5.6×10^{20} J/year (560 EJ/year). Besides substructure building costing trillions of dollars annually and the transport segment consuming 5.6×10^{20} J/year (560 EJ/year) fossil fuel annually, the two are also accountable for approximately 34% of the total yearly change in climatic conditions. If the sectors involved in this technology take it incumbent upon themselves to develop this technology appropriately, a new epoch of science to dispatch unembellished transport, substructure, and energy glitches can be seen with this technology.

Acknowledgements This research was supported by Green Globe Technology under the grant RD-02018-03. Any findings, conclusions, and recommendations expressed in this chapter are solely those of the author and do not necessarily reflect those of Green Globe Technology.

References

1. X.J. Zheng, J.J. Wu, and Y.H. Zhou, Effect of spring non-linearity on dynamic stability of a controlled maglev vehicle and its guideway system, J. Sound Vib., 279:201–215, 2005.
2. X.J. Zheng, J.J. Wu, and Y.H. Zhou, Numerical analyses on dynamic control of five-degree-of-freedom maglev vehicle moving on flexible guideways, J. Sound Vib., 235:43–61, 1997.
3. Tien, H., Scherer, C., Scherpen, J. and Muller, V. (2016). Linear Parameter Varying Control of Doubly Fed Induction Machines. *IEEE Transactions on Industrial Electronics*, 63(1), pp. 216-224.
4. Vieira. J.P.A, Alves Nunes. Marcus Vinicius, Bezerra. U. H, Barra. W. Jr, New Fuzzy Control Strategies Applied to the DFIG Converter in Wind Generation Systems, *Trans. America Latina, IEEE, vol. 5 n. 3*, 2007, pp.
5. Page, Lewis, "Terrafugia flying car gets road-safety exemptions", The Register, 4 July 2011; retrieved 11 July 2011.
6. Sivasankar, G. and Suresh Kumar, V. (2014). Improving Low Voltage Ride Through Capability of Wind Generators Using Dynamic Voltage Restorer. *Journal of Electrical Engineering*, 65(4).
7. S.J. Thompson, Congressional Research Service, "High Speed Ground Transportation (HGST): Prospects and Public Policy," Apr. 6, 1989, p. 5.
8. Xu, Hailiang, Xiaojun Ma, and Dan Sun. "Reactive Current Assignment and Control for DFIG Based Wind Turbines during Grid Voltage Sag and Swell Conditions", Journal of Power Electronics, 2015.
9. Eltamaly, A.M., Alolah, A.I., Abdel-Rahman, M.H., Improved simulation strategy for DFIG in wind energy applications, (2011) *International Review on Modelling and Simulations (IREMOS)*, 4 (2), pp. 525-532.
10. G. Tsourakisa, B. M.Nomikosb, C.D. Vournasa. Effect of wind parks with doubly fed asynchronous generators on small-signal stability, *Electric Power Systems Research, vol. 79*, 2009, pp. 190-200.
11. Haines, Thomas B. (19 March 2009). "First roadable airplane takes flight". Aircraft Owners and Pilots Association (AOPA). Retrieved 2009-03-19.
12. Hossain, Md. Faruque. "Solar energy integration into advanced building design for meeting energy demand and environment problem, International Journal of Energy Research, 2016.
13. Durbin, Dee-Ann (2012-04-02). "Flying car gets closer to reality with test flight". boston.com. Associated Press. Retrieved April 20, 2012.
14. Ghennam, T., Berkouk, E. M., Francois, B., A vector hysteresis current control applied on three-level inverter. Application to the active and reactive power control of doubly fed induction generator based wind turbine, (2007) *International Review of Electrical Engineering (IREE)*, 2 (2), pp. 250-259.
15. E. Kamal, M. Koutb, A. A. Sobaih, and B. Abozalam, An intelligent maximum power extraction algorithm for hybrid wind-diesel-storage system, *Int. J. Electr. Power Energy Syst, vol. 32 n. 3*, 2010, pp. 170–177.
16. Khodakarami, J. and Ghobadi, P. (2016). Urban pollution and solar radiation impacts. *Renewable and Sustainable Energy Reviews*, 57, pp. 965-976.
17. Kazemi MV, Moradi M, Kazemi RV. Minimization of powers ripple of direct power controlled DFIG by fuzzy controller and improved discrete space vector modulation, *Electr Power Syst Res, vol. 89*, 2012, pp. 23-30.

Part V
Clean Water and Sanitation System

Chapter 12
Potable Water

Abstract Water is a natural resource which is used globally for domestic, agricultural, industrial, recreational, and environmental activities. All living things require water to survive and thus it is said that *water means life and life means water*. The earth surface consists of 25% of land and 75% of water, and all creatures need water to survive. Among the surface water, 97% is saline while only 3% is freshwater where two-thirds of the freshwater exist in the form of glaciers or ice caps. The rest of the freshwater is unfrozen groundwater, and aboveground water. Freshwater is a renewable resource, yet the world's supply of groundwater is steadily decreasing, with depletion occurring most prominently in the whole world which is threatening our survival in the near future. In this chapter, therefore, a detailed strategic investigation of natural water resources has been done to conserve and recycle water more efficiently for ultimate global sustainability in water perplexity.

Keywords Natural water resources · Transpiration · Water vapor · Static electricity force · Potable water · Wastewater · Photocatalytic · Nanotechnology · UV technology · Renewable water engineering

Introduction

The conventional water (H_2O) harvesting by groundwater extraction and digging of ponds, lakes, and canals is clarified in order to conserve water naturally for our daily lives to run our domestic, agricultural, industrial, and recreational operation properly. This is the most indispensable natural element on earth which is an inorganic compound, represented as liquid with attributes of tastelessness, odorlessness, and nearly colorlessness with a little bit blue under normal condition. It is the simplest hydrogen chalcogenide molecule which is described as a universal solvent because of its ability to dissolve many substances, e.g., many salts, sugar, simple alcohols, proteins, polysaccharides, DNA, oxygen, and carbon dioxide [1–3]. Therefore, water molecule is called a polar molecule with an electrical dipole because of its

M. F. Hossain, *Global Sustainability in Energy, Building, Infrastructure, Transportation, and Water Technology*,
https://doi.org/10.1007/978-3-030-62376-0_12

oxygen atom having the higher electronegativity compared to the hydrogen atoms. Simply, the oxygen atom carries a slightly negative charge, while hydrogen atoms are slightly positive and thus each water molecule contains one oxygen atom and two hydrogen atoms connected by covalent bonds. Interestingly, it is the only substance on earth that can exist as solid, liquid, and gas in normal conditions whereas its liquid state exists at standard ambient temperature and pressure, its solid state exists as ice, and its gaseous state exists as steam or water vapor [1, 4, 5]. It also exists in the subforms of ice packs and icebergs, fog, glaciers, aquifers, atmospheric humidity, and dew, which are the main components of earth's streams situated in different locations [1, 6]. Under different circumstances, water can have different attributes; when standard pressure is 1 atm, water is in the form of liquid between 0 °C (32 °F) and 100 °C (212 °F). However, the melting point will change to −5 °C at 600 atm and −22 °C at 2100 atm, which is used as the principle to explain the movement of glaciers, ice skating, and buried lakes of Antarctica. What is more, the melting point will rapidly increase again, leading to several exotic forms of ice that do not exist at lower pressures when pressure is higher than 2100 atm [7–9]. The boiling point of water also changes due to the changes of pressure. For instance, the point is 374 °C at 220 atm but 68 °C (154 °F) at the top of Mount Everest with the atmospheric pressure of 0.34 atm. The effect of this change is applied on pressure cooking, steam engine design, deep-sea hydrothermal vents, etc. Simply, water's attributes change under extreme conditions and thus when the pressure is extremely low (lower than 0.006 atm), water cannot exist in liquid form and passes directly from solid to gas by sublimation, which is used for freeze-drying of food. When the pressure is extremely high (higher than 221 atm), the liquid and gas states are no more distinguishable, and a new state called supercritical steam replaces [10–12]. Another characteristic of water is that it becomes less dense as it freezes compared to most liquids. The maximum density of water in its liquid form (at 1 atm) is 1000 kg/m^3 (62.43 lb./ft^3), which occurs at 3.98 °C (39.16 °F) while the density of ice is 917 kg/m^3 (57.25 lb./ft^3). Thus, through calculation, water expands 9% in volume as it freezes, which reflects the fact that ice floats on liquid water [4, 13].

Simply this natural resource, H_2O, is unique, which needs to be utilized calculatively to meet our daily lives and thus this research has been conducted to clarify all the possible natural sources of water collection in order to conserve this valuable element to protect the modern civilization and future generation.

Methods and Materials

The global water dimension has been clarified by custodial analysis to identify the net water resource on earth by using MATLAB software. Therefore, the calculation of barotropic (depth-averaged) water modeling considers Boussinesq approximation, hydrostatic momentum, and mass balances, and water material tracer conservation of earth topography coordinate is determined as

$$z = z(x,y,\sigma) \tag{12.1}$$

where z is the cartesian height and σ is the vertical distance from the earth surface measured as the fraction of water column thickness (i.e., $-1 \leq \sigma \leq 0$, $\sigma = 0$ corresponds to the water surface, $z = \zeta$, and $\sigma = -1$ corresponds to the water bottom, $z = -h(x,y)$) and thus the resulting system of coordinates is clarified as nonorthogonal chain rules for derivatives in order to accurately measure the net water reserve on earth as

$$\left.\frac{\partial}{\partial x}\right|_z = \left.\frac{\partial}{\partial x}\right|_\sigma - \left.\frac{\partial z}{\partial x}\right|_\sigma \cdot \frac{\partial}{\partial z} \tag{12.2}$$

Here, the classical σ-coordinate is rewritten as

$$z = \sigma \cdot h(x,y). \tag{12.3}$$

It is then combined with nonlinear stretching, $S(\sigma)$:

$$z(x,y,\sigma) = S(\sigma) \cdot h(x,y) \tag{12.4}$$

And further generalization of the water surface S-coordinate is also clarified to be more specific to determine the water source on earth from σ- to z-coordinates. Thus, it is chosen to implement a set of z-levels $\{z^*_{(k+½)} | k = 0, 1, \ldots, N\}$ of earth water topography, where $z* \ 1 \ 2 = -h_{max}$ is chosen to be the maximum depth of water and $z* \ N \,\text{I}\ 1 \ 2 = 0$ is the unperturbed free surface 2 of water where the starting bottom of water is referred to as $k = 0$ and the water topography equation is expressed as

$$z_{\frac{1}{2}}(x,y) = -h(x,y) \tag{12.5}$$

and for each $k = 1, \ldots, N - 1$ set

$$z_{k+\frac{1}{2}}(x,y) = \max\left(z^*_{k+\frac{1}{2}}, z_{k-\frac{1}{2}}(x,y) + \Delta z_{\min}\right) \tag{12.6}$$

where $\Delta z_{\min}$ is the chosen minimal vertical water range on earth (*n.b.*, surfacing of coordinate isolines, $\Delta z_{\min} \leq h_{\min}/N$, where $h_{\min}$ is the minimal depth). In horizontal water supply, $\Delta z_{\min}$ is chosen as infinite in order to get the resultant system to be equivalent to a z-coordinate of earth surface and clarify the source of water on earth (Fig. 12.1).

Therefore, the surface water analysis, here, refers to water location in river, lake, or freshwater wetland on earth. It is naturally refilled and lost by precipitation and discharge to the oceans, evapotranspiration, groundwater recharge, and evaporation, respectively [5, 14]. Though precipitation is the only natural input to any surface

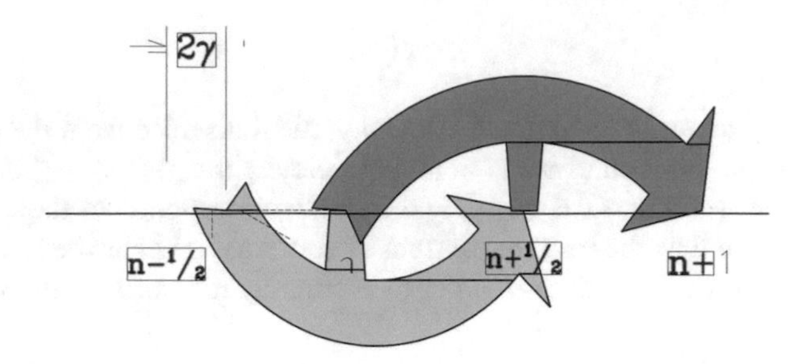

Fig. 12.1 Geometrical interpretation of earth water presence formulated as primary data ˀ-1 and secondary data ˀ that are being linearly interpolated to $n - 1/2 + 2Y$ in order to getting the resultant water resources on earth. Here, $n + 1/2$ is the advance prognostic variables of groundwater sources from n to $n + 1$, and $2Y$ is the net water source on the earth surface

water system, many other factors also influence the total quantity of water available at any given time and proportions of water loss in the system and thus, in this model, calculation of water of wetlands and artificial reservoirs, runoff characteristics of the land in the watershed, storage capacity in lakes, timing of precipitation, permeability of the soil beneath these storage bodies, and local evaporation rates are also clarified computationally. Additionally, the groundwater analysis has been conducted which refers to the part of surface water located in the subsurface pore space of soil and rocks and the water flows within aquifers below the water table which is seepage from surface water and springs, respectively [7, 10].

Results and Discussion

The results of net water resources on earth suggested that split-explicit hydrostatic and hydrodynamic is temporal at an average barotropic mode to account for the exact conservation and constancy preservation of natural water which can be expressed as

$$\frac{\partial D\bar{u}}{\partial t} + \ldots = -gD\nabla_x\zeta + \{gD\nabla_x\zeta + \mathcal{F}\} \tag{12.7}$$

Here, ζ refers to the earth surface, and ρ_0 refers to a constant water density, g is the acceleration of gravity, $D = h + \zeta$ is the total depth, u is the depth-averaged velocity, $\nabla_x\zeta$ is a shorthand for $\partial\zeta/\partial x$, and F is the vertically integrated pressure gradient:

$$\mathcal{F} = -\frac{1}{\rho_0}\int_{-h}^{\zeta}\frac{\partial P}{\partial x}dz \tag{12.8}$$

The latter F is a functional of the topography, free surface gradient, and free surface itself, as well as the vertical distribution of density and its gradient, and is expressed as

$$\mathcal{F} = \mathcal{F}\left[\nabla_x \zeta, \zeta, \nabla_x \rho(z), \rho(z)\right] \tag{12.9}$$

Thus, the mathematical analysis suggests that water reserve on earth is quite scary to meet the total water demand in the near future which is generated by the simple relationships of $S(\sigma)$ and the independent coordinates involving a full three-dimensional (3D) transformation water sources on earth described as

$$\left\{ z_{k+\frac{1}{2}} = z_{k+\frac{1}{2}}(x,y), \quad k = 0,1,\ldots,N \right\} \tag{12.10}$$

Here, if the water is at rest, the surface water elevation is $\zeta = 0$; hence $z_{N+1/2} = 0$, and the whole set corresponding to zero surface water $\{z^{(0)}_{k+1/2}\}$ is referred to as an unperturbed coordinate system (Fig. 12.2). In the case of a nonzero ζ, all $z^{(0)}_{k+1/2}$ are displaced by a distance proportional to ζ and the distance from the bottom as the fraction of unperturbed water depth:

$$z_{k+\frac{1}{2}} = z^{(0)}_{k+\frac{1}{2}} + \zeta\left(1 + \frac{z^{(0)}_{k+\frac{1}{2}}}{h}\right) \tag{12.11}$$

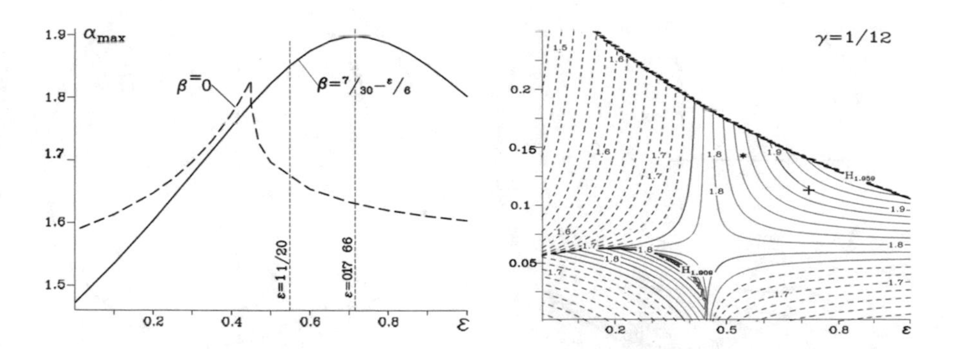

Fig. 12.2 Top: water stability limit on earth $\alpha_{\max}$ as a function of ε for algorithm with $\gamma = 1/12$ for two different settings of β: (solid) along the line of vanishing O (α^5) term (2.36) and (dashed) $\beta = 0$. Bottom: $\alpha_{\max}$ as function of γ, β with fixed $\gamma = 1/12$. Contours below $\alpha = 1.75$ are shown in dashed lines. The appearance of two maxima of water stability on earth, at $(\varepsilon, \beta) = (0.83, 0.126)$ just on the edge of asymptotic instability and $(0.39, 0.044)$. The straight dashed line approximately parallel to the edge corresponds to a zero O (α^5) truncation term. The asterisk (*) and cross (+) on this line denote locations of the minimal truncation error and maximum stability limit among the fourth-order algorithms

Here $z_{1/2} \equiv z^{(0)}{}_{1/2} \equiv -h$ and $z_{N+1/2} \equiv \zeta$ and thus the surface water height is $\Delta z_k \equiv z_{k+1/2} - z_{k-1/2}$, which is related to the unperturbed height of water depth $\Delta z^{(0)}{}_k \equiv z^{(0)}{}_{k+1/2} - z^{(0)}{}_{k-1/2}$ according to

$$\Delta z_k = \Delta z_k^{(0)}\left(1+\frac{\zeta}{h}\right) \tag{12.12}$$

where the multiplier $(1 + \zeta/h)$ is independent of the vertical coordinate of earth surface where water surface elevation affects only the temperature increment (hence $\Delta z_k = \Delta z^{(0)}{}_k + \zeta/N$) regardless of its unperturbed size $\Delta z^{(0)}{}_k$ in order to calculate the net water reserve on earth properly (Fig. 12.3).

Therefore, the surface distributed water analysis, here, breaks down global water location in river, lake, or freshwater wetland on earth and has been presented in Fig. 12.4.

Necessarily surface water dynamics, considering the pressure gradient force, vertical system of coordinates of time-stepping algorithms for the barotropic and baroclinic momentum, and their mutual interaction, confirm weighted averaging barotropic motions and baroclinic time step tide calculation confirming that the total quantity of water available on earth surface at any given time is perturbed by misuse of the water by the following factors.

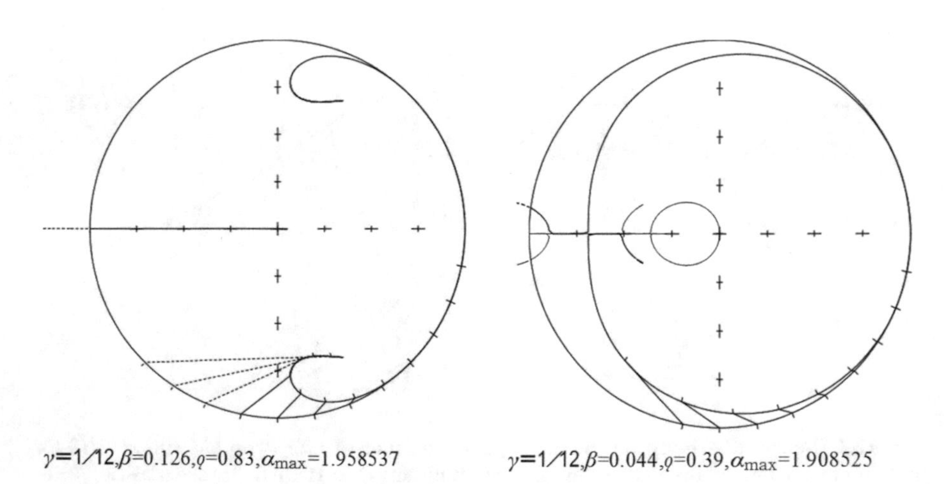

Fig. 12.3 Characteristic roots of algorithms corresponding to the primary (top) and secondary (bottom) maxima of water stability on earth limit

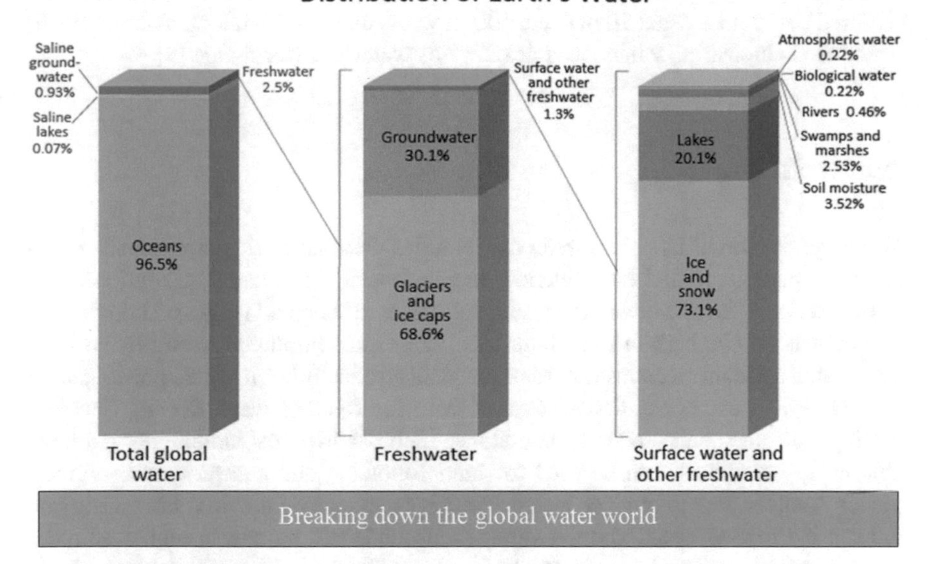

Fig. 12.4 A graphical distribution of the locations of water on earth. Only 3% of the earth's water is freshwater. Most of it is in icecaps and glaciers (69%) and groundwater (30%), while all lakes, rivers, and swamps combined only account for a small fraction (0.3%) of the earth's total freshwater reserves

Domestic Use

The clarification of water use conducted revealed that approximately 8% of water use worldwide is for domestic purposes, which include cooking, laundry, toilet flushing, drinking water, bathing, gardening, and cleaning [5]. Excluding water for gardening, basic domestic water requirement is at around 50 L per person per day according to estimation [15, 16]. Drinking water, more special than other uses, refers to water that is of sufficiently high quality so that it can be consumed or used without risk of harm, which can also be called potable water. In most developed countries, the water provided for domestic, commerce, or industry is at drinking water standard though only a small proportion is consumed for food preparation.

Agricultural Use

Compared to domestic use, agricultural use of water occupies a large amount of water use. According to estimation, 70% of water use worldwide is for irrigation, with 15–35% of irrigation withdrawals being unsustainable [15]. To satisfy one person's daily food need, 2000–3000 L of water will be consumed, which is a considerable amount compared to drinking 2–5 L [13, 17]. Therefore, according to calculation, if enough food is provided for over seven billion people now on the earth, the water required could fill a canal 10 m deep, 100 m wide, and 2100 km long. Another major water use is industrial, which occupies 22% of water use worldwide [5].

Industrial Use

Major application of industrial water use includes thermoelectric power plants, manufacturing plants, ore and oil refineries, and hydroelectric dams. Water in these processes can play the role of cooling agent, solvent, and chemical reagent. Though water withdrawal may be high in some industries, these consumptions are still much lower than that in agriculture. Industrial water uses can also include renewable power generation. Hydroelectric power derives energy from the force of water flowing downhill, driving a turbine connected to a generator, which is a low-cost, nonpolluting, renewable energy source. It can be used for load following and can provide continuous power. Another way to produce hydroelectricity depends on the Sun. First, heat from the sun evaporates water, and the vapor then condenses as rain in higher altitudes, flows downhill, and produces electricity. Except for the two plants above, pumped storage hydroelectric plants use grid electricity to pump water uphill when energy demand is low, and then use the stored water to produce electricity when demand is high. Hydroelectric power plants require the creation of artificial lake. Due to the larger surface area exposed to the elements, evaporation from this lake is higher compared to a natural river, as well as the water consumption. The process of driving water through the turbine and tunnels or pipes briefly removes the water from natural environment, creating water withdrawal, which impact on wildlife varies largely depending on the design of the plant. Another industrial water use is pressurized water, which is often used in water blasting, water jet cutters, and high-pressure water guns. These uses not only work well and safe, but also have no harm to the environment. Cooling of machinery from overheating is also an industrial water use, which only occupies a small proportion of water consumption. In many large-scale industrial processes like oil refining, fertilizer production, other chemical plant uses, thermoelectric power production, and natural gas extraction from shale rock, water is also used. Discharge of untreated water coming from these processes is pollution, which includes discharged solutes (chemical pollution) and increased water temperature (thermal pollution). These industrial water uses mentioned above normally require pure water and apply a variety of purification techniques both in water supply and discharge. Most of the pure water is produced onsite from natural freshwater or municipal gray water. Industrial consumption of water is generally much lower than withdrawal because laws require industrial gray water to be treated and disposed to the environment. One exception is the thermoelectric power plant using cooling towers with high consumption of water equal to their withdrawal because most of the withdrawn water is evaporated as part of the cooling process. In addition, business activities including entertainment expand quickly nowadays, which increases the demand of water supply and sanitation, and further leads to higher pressure on water resources and natural ecosystem.

Recreational Use

Recreational water use accounts for generally a very small but increasing percentage in total water use. Normally, recreational water use is tied to reservoirs. If a reservoir is kept fuller than it would otherwise be for recreation, then the water retained could be categorized as recreational usage. Common recreational usage includes whitewater boating, angling, water skiing, and swimming. Recreational water use is usually nonconsumptive. Golf courses are often accused of using excessive amount of water, especially in drier regions. In order to avert environmentalist's thoughts of wasting water, some governments categorize golf course usage as agricultural. However, it is not clear whether recreational irrigation has an obvious influence on water resources till now because of the inapplicability of reliable data. Besides, many golf courses use either primarily or exclusively treated effluent water, which will not affect potable water. In Arizona, an organized lobby concentrated on educating the public on how golf affects the environment has been established in the form of the Golf Industry Association. One disadvantage of recreational water use is that it may reduce the availability of water for other users at specific times and places. For example, water retained in a reservoir to allow boating in the late summer is not available to farmers during the spring planting season. Water released for whitewater rafting may not be available for hydroelectric generation during the time of peak electrical demand.

Environmental Use

Environmental water use also accounts for a small but increasing percentage of total water use. The source of environmental water includes water stored in retention tank and released for environmental purposes or water retained in waterways through regulatory limits of abstraction. Then the usages of environmental water include fish ladders, water releasing from reservoirs timed to help fish spawn, restoring of more natural flow regimes, watering of natural or artificial wetlands, and creating of wildlife habitat. Similar to recreational usage, environmental usage is nonconsumptive but may influence other users. For example, water released from a reservoir to help fish spawn may not be available to farms upstream, and water retained in a river to maintain waterway health would not be available to water abstractors downstream.

Irrigation Use

Irrigation is necessary to grow and increase the productivity of crops. Various irrigation methods involve different relationships between water consumption, capital cost of equipment and structures, and crop yield. Some types of irrigation methods which are simple, cheap, and less efficient include furrow and overhead sprinkler, during which much water evaporates, runs off, or drains below the root zone. Other irrigation methods which are more efficient and more expensive include drip or trickle irrigation, surge irrigation, and some types of sprinkler systems where the sprinklers are operated near ground level. Issues often insufficiently considered when applying irrigation methods are salinization of groundwater and contaminant accumulation leading to water quality declines, which is the reason of the sharp rising of water scarcity.

Consequently, the misuse of natural resource is leading to its runout in the near future due to the rapid increase in world population. In 2000, the world population was 6.2 billion. This number will increase to 9.7 billion by 2050 according to the estimation of the UN which growth mainly comes from developing countries suffering water stress [18]. Therefore, the water demand will also increase unless water conservation or recycling increases [18]. The World Bank explained the result of the UN's research that access to water will be one of the main challenges in the near future [3]. Simultaneously, urbanization is also increasing rapidly. In low-density communities, small private wells and septic tanks are commonly used to provide water, which is not so useful in high-density urban areas. To deliver water to individuals and concentrate wastewater, urbanization requires large investment from individuals and businesses in water infrastructure. In 60% of European cities with more than 100,000 people, the replenishing rate of groundwater is less than consuming rate. Even if some water can be utilized, a large amount of cost will be needed to capture it. Recent research revealed that in 2025, water shortage problems will be more obvious among poorer areas with limited resources and rapid population growth, such as the Middle East, parts of Asia, and Africa. Besides, at that time, to provide clean water and adequate sanitation, large urban and peri-urban areas will need to construct new infrastructure, which may lead to growing conflicts between government and agricultural water users, who currently consume most of the water. Compared to these poorer areas, more developed areas of North America, Europe, and Russia will not suffer the threat, not only because of their relative wealth, but also because their population can better align with available water resources. Subsequently, one of the major concerns today is water pollution and many countries have tried hard to find solutions to solve this problem. Among all pollutants threatening water supplies, the most widespread one is the discharge of raw sewage, sludge, garbage, or even toxic pollutants into natural water, which is most common in underdeveloped countries and some quasi-developed countries like India, Iran, China, and Nepal. Treated sewage can form sludge, which may be placed in landfills, spread out on land, incinerated, or dumped at sea. Except for sewage, nonpoint source pollution such as agricultural runoff, urban storm water runoff, and chemical wastes is also a major source of pollution in some countries. Simply, the misuse of water resources worldwide significantly impacts the climate change because of the close connections between the climate and hydrological cycle. Increased hydrologic variability and change in climate have and will continue to have a profound impact on the water sector through the hydrologic cycle, water availability, water demand, and water allocation at the global, regional, basin, and local levels.

Conclusion

Once it was thought that water was an infinite resource since earth has plenty of places for rivers, reservoirs, and canals to have abundant water for everybody. It may be because there were less than half of the current number of people on the planet who consumed less sources and did not need much water to produce these sources. Unfortunately, nowadays, the condition of water resource is much more intense because of the explosive increase of people and food and comfortableness they need. Even more water will be needed in the future because the population worldwide is forecasted to rise to 9 billion by 2050. Thus, this study suggested that the current availability of water on a global scale will suffer severe water scarcity since one-third of the world's population will not have clean drinking water, and a fifth of the world's people, which is more than 1.2 billion, will live in areas of physical water scarcity, which means a lack of water to meet all demands by the year of 2050 if we cannot control the misuses of water. Simply, this research suggested that water must be used wisely for the benefit of future generation, but unfortunately, the continuation of misuse of today's water consumption shall certainly lead to crises in the whole world in the near future.

Acknowledgements This research was supported by Green Globe Technology under the grant RD-02017-07 for building a better environment. Any findings, predictions, and conclusions described in this chapter are solely performed by the author.

References

1. Mann, M.E., R.S. Bradley, and M.K. Hughes, 1998: "Global-scale temperature patterns and climate forcing over the past six centuries", Nature, 392: 779-787.
2. Werner, Johannes P., Juerg Luterbacher, and Jason E. Smerdon. "A Pseudoproxy Evaluation of Bayesian Hierarchical Modelling and Canonical Correlation Analysis for Climate Field Reconstructions over Europe", Journal of Climate, 2012.
3. P. R. Mahaffy et al., Science 341, 263–266 (2013).
4. *Staff (November 22, 2016).* "Scalloped Terrain Led to Finding of Buried Ice on Mars". NASA. Retrieved November 23, 2016.
5. Martín-Torres, F. Javier; Zorzano, María-Paz; Valentín-Serrano, Patricia; Harri, Ari-Matti; Genzer, Maria (April 13, 2015). "Transient liquid water and water activity at Gale crater on Mars". Nature Geocience. 8: 357–361. https://doi.org/10.1038/ngeo2412. Retrieved April 14, 2015.
6. Hossain, Md. Faruque (2016). "Theory of Global Cooling: Energy, Sustainability, and Society".
7. Baker, V.R.; Strom, R.G.; Gulick, V.C.; Kargel, J.S.; Komatsu, G.; Kale, V.S. (1991). "Ancient oceans, ice sheets and the hydrological cycle on Mars". Nature. 352 (6348): 589–594. https://doi.org/10.1038/352589a0.
8. A. G. Suits, Acc. Chem. Res. 41, 873–881 (2008).
9. D. L. Huestis, T. G. Slanger, B. D. Sharpee, J. L. Fox, Faraday Discuss. 147, 307–322 (2010).
10. Hossain, Md. Faruque. "Solar energy integration into advanced building design for meeting energy demand and environment problem": International Journal of Energy Research, 2016.

11. H. Gao, Y. Song, W. M. Jackson, C. Y. Ng, J. Chem. Phys. 138, 191102 (2013).
12. M. A. Gharaibeh, D. J. Clouthier, J. Chem. Phys. 132, 114307. (2010).
13. Bibring, J.-P.; Langevin, Yves; Poulet, François; Gendrin, Aline; Gondet, Brigitte; Berthé, Michel; Soufflot, Alain; Drossart, Pierre; Combes, Michel; Bellucci, Giancarlo; Moroz, Vassili; Mangold, Nicolas; Schmitt, Bernard; Omega Team, the; Erard, S.; Forni, O.; Manaud, N.; Poulleau, G.; Encrenaz, T.; Fouchet, T.; Melchiorri, R.; Altieri, F.; Formisano, V.; Bonello, G.; Fonti, S.; Capaccioni, F.; Cerroni, P.; Coradini, A.; Kottsov, V.; et al. (2004). "Perennial Water Ice Identified in the South Polar Cap of Mars". Nature. 428(6983).
14. Ojha, L.; Wilhelm, M. B.; Murchie, S. L.; McEwen, A. S.; Wray, J. J.; Hanley, J.; Massé, M.; Chojnacki, M. (2015). "Spectral evidence for hydrated salts in recurring slope lineae on Mars". Nature Geoscience. 8: 829–832. https://doi.org/10.1038/ngeo2546.
15. "Mars Ice Deposit Holds as Much Water as Lake Superior". *NASA. November 22, 2016.* Retrieved November 23, 2016.
16. Howard, A.; Moore, Jeffrey M.; Irwin, Rossman P. (2005). "An intense terminal epoch of widespread fluvial activity on early Mars: 1. Valley network incision and associated deposits". Journal of Geophysical Research. 110:2005JGRE.11012S14H. https://doi.org/10.1029/2005JE002459.
17. L. Archer et al., J. Quant. Spectrosc. Radiat. Transf. 117, 88–92. (2013).
18. M. P. Grubb et al., Science 335, 1075–1078 (2012).

Chapter 13
Wastewater

Abstract Wastewater mitigation is a major problem in the world. Water treatment refers to processes that improve the water quality by removing contaminants and undesirable components or reducing their concentration to make it more suitable for specific end use including river flow maintenance, irrigation, water recreation, drinking, industrial water supply, or other uses. It is used to convert wastewater into an effluent that can be further returned to the water cycle with minimal impact on the environment or that can be directly reused, which can also be called as water reclamation. The treatment process takes place in a wastewater treatment plant (WWTP), often referred to as a Water Resource Recovery Facility (WRRF) or a sewage treatment plant. In this chapter, both water treatment and wastewater treatment are discussed where most advanced technological applications are applied. Providing clean and affordable water to meet human needs is a great challenge of the twenty-first century. Worldwide, water supply struggles to keep up with the fast-growing demand, which is exacerbated by population growth, global climate change, and water quality deterioration. The need for technological innovation to enable integrated water management cannot be overstated. Nanotechnology holds great potential in advancing water and wastewater treatment to improve treatment efficiency as well as to augment water supply through safe use of unconventional water sources. Here we review recent developments in nanotechnology for water and wastewater treatment. The discussion covers candidate nanomaterials, properties and mechanisms that enable the applications, advantages and limitations as compared to existing processes, and barriers and research needs for commercialization. By tracing these technological advances to the physicochemical properties of nanomaterials, this review outlines the opportunities and limitations to further capitalize on these unique properties for sustainable water management.

Keywords Water supply demand · Conventional wastewater treatment · Environmental pollution · Nanotechnology · Environmental sustainability

M. F. Hossain, *Global Sustainability in Energy, Building, Infrastructure, Transportation, and Water Technology*,
https://doi.org/10.1007/978-3-030-62376-0_13

Introduction

Water is the most essential substance for all life on earth and a precious resource for human civilization. Reliable access to clean and affordable water is considered one of the most basic humanitarian goals, and remains a major global challenge for the twenty-first century. Our current water supply faces enormous challenges, both old and new. Worldwide, some 780 million people still lack access to improved drinking water sources [1, 2]. It is urgent to implement basic water treatment in the affected areas (mainly in developing countries) where water and wastewater infrastructure are often nonexistent. In both developing and industrialized countries, human activities play an ever-greater role in exacerbating water scarcity by contaminating natural water sources. The increasingly stringent water quality standards, compounded by emerging contaminants, have brought new scrutiny to the existing water treatment and distribution systems widely established in developed countries. The rapidly growing global population and the improvement of living standard continuously drive up the demand. Moreover, global climate change accentuates the already uneven distribution of fresh water, destabilizing the supply. Growing pressure on water supplies makes using unconventional water sources (e.g., stormwater, contaminated freshwater, brackish water, wastewater, and seawater) a new norm, especially in historically water-stressed regions. Furthermore, current water and wastewater treatment technologies and infrastructure are reaching their limit for providing adequate water quality to meet human and environmental needs. Recent advances in nanotechnology offer leapfrogging opportunities to develop next-generation water supply systems. Our current water treatment, distribution, and discharge practices, which heavily rely on conveyance and centralized systems, are no longer sustainable. The highly efficient, modular, and multifunctional processes enabled by nanotechnology are envisaged to provide high-performance, affordable water and wastewater treatment solutions that rely less on large infrastructure [3, 4]. Nanotechnology-enabled water and wastewater treatment promises to not only overcome major challenges faced by existing treatment technologies, but also provide new treatment capabilities that could allow economic utilization of unconventional water sources to expand the water supply. Here, we provide an overview of recent advances in nanotechnologies for water and wastewater treatment. The major applications of nanomaterials are critically reviewed based on their functions in unit operation processes. The barriers for their full-scale application and the research needs for overcoming these barriers are also discussed. The potential impact of nanomaterials on human health and ecosystem as well as any potential interference with treatment processes are beyond the scope of this review and thus will not be addressed here.

Current and Potential Applications for Water and Wastewater Treatment

Nanomaterials are typically defined as materials smaller than 100 nm in at least one dimension. At this scale, materials often possess novel size-dependent properties different from their large counterparts, many of which have been explored for applications in water and wastewater treatment. Some of these applications utilize the smoothly scalable size-dependent properties of nanomaterials which relate to the high specific surface area, such as fast dissolution, high reactivity, and strong sorption. Others take advantage of their discontinuous properties, such as superparamagnetism, localized surface plasmon resonance, and quantum confinement effect. These applications are discussed below based on nanomaterial functions in unit operation processes (Table 13.1). Most applications discussed below are still in the stage of laboratory research. The pilot-tested or field-tested exceptions will be noted in the text.

Adsorption

Adsorption is commonly employed as a polishing step to remove organic and inorganic contaminants in water and wastewater treatment. Efficiency of conventional adsorbents is usually limited by the surface area or active sites, the lack of selectivity, and the adsorption kinetics. Nano-adsorbents offer significant improvement with their extremely high specific surface area and associated sorption sites, short intraparticle diffusion distance, and tunable pore size and surface chemistry.

Carbon-Based Nano-Adsorbents

Organic Removal

CNTs have shown higher efficiency than activated carbon on adsorption of various organic chemicals [5, 6]. Its high adsorption capacity mainly stems from the large specific surface area and the diverse contaminant-CNT interactions. The available surface area for adsorption on individual CNTs is their external surfaces [7, 8]. In the aqueous phase, CNTs form loose bundles/aggregates due to the hydrophobicity of their graphitic surface, reducing the effective surface area. On the other hand, CNT aggregates contain interstitial spaces and grooves, which are high-adsorption energy sites for organic molecules [9, 10]. Although activated carbon possesses comparable measured specific surface area as CNT bundles, it contains a significant number of micropores inaccessible to bulky organic molecules such as many antibiotics and pharmaceuticals [11, 12]. Thus, CNTs have much higher adsorption capacity for some bulky organic molecules because of their larger pores in bundles and more accessible sorption sites.

Table 13.1 Current and potential applications of nanotechnology in water and wastewater treatment

Applications	Representative nanomaterials	Desirable nanomaterial properties	Enabled technologies
Adsorption	Carbon nanotubes	High specific surface area, highly assessable adsorption sites, diverse contaminant-CNT interactions, tunable surface chemistry, easy reuse	Contaminant preconcentration/ detection, adsorption of recalcitrant contaminants
	Nanoscale metal oxide	High specific surface area, short intraparticle diffusion distance, more adsorption sites, compressible without significant surface area reduction, easy reuse, some are superparamagnetic	Adsorptive media filters, slurry reactors
	Nanofibers with core-shell structure	Tailored shell surface chemistry for selective adsorption, reactive core for degradation, short internal diffusion distance	Reactive nano-adsorbents
Membranes and membrane processes	Nano-zeolites	Molecular sieve, hydrophilicity	High-permeability thin-film nanocomposite membranes
	Nano-Ag	Strong and wide-spectrum antimicrobial activity, low toxicity to humans	Anti-biofouling membranes
	Carbon nanotubes	Antimicrobial activity (unaligned carbon nanotubes)	Anti-biofouling membranes
		Small diameter, atomic smoothness of inner surface, tunable opening chemistry, high mechanical and chemical stability	Aligned carbon nanotube membranes
	Aquaporin	High permeability and selectivity	Aquaporin membranes
	Nano-TiO_2	Photocatalytic activity, hydrophilicity, high chemical stability	Reactive membranes, high-performance thin-film nanocomposite membranes
	Nano-magnetite	Tunable surface chemistry, superparamagnetic	Forward osmosis
Photocatalysis	Nano-TiO_2	Photocatalytic activity in UV and possibly visible light range, low human toxicity, high stability, low cost	Photocatalytic reactors, solar disinfection systems
	Fullerene derivatives	Photocatalytic activity in solar spectrum, high selectivity	Photocatalytic reactors, solar disinfection systems

(continued)

Table 13.1 (continued)

Applications	Representative nanomaterials	Desirable nanomaterial properties	Enabled technologies
Disinfection and microbial control	Nano-Ag	Strong and wide-spectrum antimicrobial activity, low toxicity to humans, ease of use	POU water disinfection, anti-biofouling surface
	Carbon nanotubes	Antimicrobial activity, fiber shape, conductivity	POU water disinfection, anti-biofouling surface
	Nano-TiO_2	Photocatalytic ROS generation, high chemical stability, low human toxicity and cost	POU to full-scale disinfection and decontamination
Sensing and monitoring	Quantum dots	Broad absorption spectrum, narrow, bright, and stable emission which scales with the particle size and chemical component	Optical detection
	Noble metal nanoparticles	Enhanced localized surface plasmon resonances, high conductivity	Optical and electrochemical detection
	Dye-doped silica nanoparticles	High sensitivity and stability, rich silica chemistry for easy conjugation	Optical detection
	Carbon nanotubes	Large surface area, high mechanical strength and chemical stability, excellent electronic properties	Electrochemical detection, sample preconcentration
	Magnetic nanoparticles	Tunable surface chemistry, superparamagnetism	Sample preconcentration and purification

A major drawback of activated carbon is its low adsorption affinity for low-molecular-weight polar organic compounds. CNTs strongly adsorb many of these polar organic compounds due to the diverse contaminant-CNT interactions including hydrophobic effect, π-π interactions, hydrogen bonding, covalent bonding, and electrostatic interactions [8, 13]. The π electron-rich CNT surface allows π-π interactions with organic molecules with C-C bonds or benzene rings, such as polycyclic aromatic hydrocarbons (PAHs) and polar aromatic compounds [14, 15]. Organic compounds which have -COOH, -OH, and -NH_2 functional groups could also form hydrogen bond with the graphitic CNT surface which donates electrons [16, 17]. Electrostatic attraction facilitates the adsorption of positively charged organic chemicals such as some antibiotics at suitable pH [12, 18].

Heavy Metal Removal

Oxidized CNTs have high adsorption capacity for metal ions with fast kinetics. The surface functional groups (e.g., carboxyl, hydroxyl, and phenol) of CNTs are the major adsorption sites for metal ions, mainly through electrostatic attraction and chemical bonding [19, 20]. As a result, surface oxidation can significantly enhance the adsorption capacity of CNTs. Several studies show that CNTs are better adsorbents than activated carbon for heavy metals (e.g., Cu^{2+}, Pb^{2+}, Cd^{2+}, and Zn^{2+}) and the adsorption kinetics is fast on CNTs due to the highly accessible adsorption sites and the short intraparticle diffusion distance [16, 21].

Overall, CNTs may not be a good alternative for activated carbon as wide-spectrum adsorbents. Rather, as their surface chemistry can be tuned to target specific contaminants, they may have unique applications in polishing steps to remove recalcitrant compounds or in pre-concentration of trace organic contaminants for analytical purposes. These applications require small quantity of materials and hence are less sensitive to the material cost.

Produced by exfoliating graphite with strong acids and oxidizers, graphite oxide is a potentially low-cost adsorbent. It was recently reported that sand granules coated with graphite oxide were efficient in removing Hg^{2+} and a bulky dye molecule (rhodamine B); its performance was comparable to commercial activated carbon [22, 23].

Regeneration and Reuse

Regeneration is an important factor that determines the cost-effectiveness of adsorbents. Adsorption of metal ions on CNTs can be easily reversed by reducing the solution pH. The metal recovery rate is usually above 90% and often close to 100% at pH <2 [16, 24]. Moreover, the adsorption capacity remains relatively stable after regeneration. Lu et al. reported that Zn^{2+} adsorption capacity of SWNT and MWNT decreased less than 25% after ten regeneration and reuse cycles, while that of activated carbon was reduced by more than 50% after one regeneration [16, 25]. A statistical analysis based on the best-fit regression of Zn^{2+} adsorption capacity and the number of regeneration and reuse cycles suggested that CNT nano-adsorbents can be regenerated and reused up to several hundred times for Zn^{2+} removal while maintaining reasonable adsorption capacity [26, 27].

Metal-Based Nano-Adsorbents

Metal oxides such as iron oxide, titanium dioxide, and alumina are effective, low-cost adsorbents for heavy metals and radionuclides. The sorption is mainly controlled by complexation between dissolved metals and oxygen in metal oxides [28, 29]. It is a two-step process: fast adsorption of metal ions on the external surface, followed by the rate-limiting intraparticle diffusion along the micropore walls [30, 31]. Their nanoscale counterparts have higher adsorption capacity and faster kinetics because of the higher specific surface area, shorter intraparticle diffusion distance, and larger number of surface reaction sites (i.e., corners, edges, vacancies). For instance, as the particle size of nano-magnetite decreased from 300 to 11 nm, its arsenic adsorption capacity increased more than 100 times [32, 33]. Much of this observed increase in adsorption was attributed to the increase in specific surface area as the 300 and 20 nm magnetite particles have similar surface area-normalized arsenic adsorption capacity (~6 μmol/m^2 or 3.6 atoms/nm^2). However, when particle size was reduced to below 20 nm, the specific surface area-normalized adsorption capacity increased, with 11 nm magnetite nanoparticles absorbing three times more arsenic (~18 μmol/m^2 or 11 atoms/nm^2), suggesting a "nanoscale effect." This "nanoscale effect" was attributed to the change of magnetite surface structure which creates new adsorption site vacancies [34, 35].

In addition to high adsorption capacity, some iron oxide nanoparticles, e.g., nano-maghemite and nano-magnetite, can be superparamagnetic. Magnetism is highly volume dependent as it stems from the collective interaction of atomic magnetic dipoles. If the size of a ferro- or ferri-magnet decreases to the critical value (~40 nm), the magnet changes from multiple domains to single domain with higher magnetic susceptibility [17, 36]. As the size further decreases, magnetic particles become superparamagnetic, losing permanent magnetic moments while responding to an external magnetic field, which allows easy separation and recovery by a low-gradient magnetic field. These magnetic nanoparticles can be used either directly as adsorbents or as core material in a core-shell nanoparticle structure where the shell provides the desired function while the magnetic core realizes magnetic separation (Fig. 13.1).

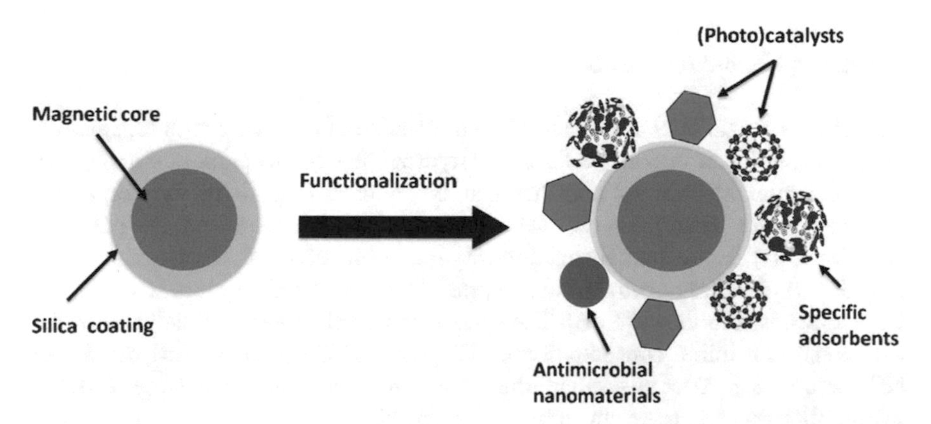

Fig. 13.1 Multifunctional magnetic nanoparticles. Magnetic nanoparticles are used as the core material in a core-shell nanoparticle structure where the shell provides the desired function while the magnetic core realizes magnetic separation. Silica coating helps functionalization due to the rich silica chemistry

Metal oxide nanocrystals can be compressed into porous pellets without significantly compromising their surface area when moderate pressure is applied [16, 37]. The pore volume and pore size can be controlled by adjusting the consolidation pressure. Thus, they can be applied in forms of both fine powders and porous pellets, which are the likely forms to be used in industry.

Metal-based nanomaterials have been explored to remove a variety of heavy metals such as arsenic, lead, mercury, copper, cadmium, chromium, and nickel, and have shown great potential to outcompete activated carbon [38, 39]. Among them, the application for arsenic removal has attracted much attention. Despite being a good adsorbent for many organic and inorganic contaminants, activated carbon has limited capacity for arsenic, especially for As(V). Several metal oxide nanomaterials including nanosized magnetite and TiO_2 have shown arsenic adsorption performance superior to activated carbon [40–42]. Metal (hydr)oxide nanoparticles can also be impregnated onto the skeleton of activated carbon or other porous materials to achieve simultaneous removal of arsenic and organic co-contaminants, which favors point-of-use (POU) applications [43, 44].

Regeneration and Reuse

Metal oxide nano-adsorbents can be easily regenerated by changing solution pH [39, 45]. In many cases, the adsorption capacity of metal oxide nano-adsorbents is well maintained after several regeneration and reuse cycles [46, 47]. However, reduced adsorption capacity after regeneration has also been reported [48, 49].

Above all, metal-based nano-adsorbents can be produced at relatively low cost. The high adsorption capacity, low cost, easy separation, and regeneration make metal-based nano-adsorbents technologically and economically advantageous.

Polymeric Nano-Adsorbents

Dendrimers are tailored adsorbents that are capable of removing both organics and heavy metals. Their interior shells can be hydrophobic for sorption of organic compounds while the exterior branches can be tailored (e.g., hydroxyl- or amine-terminated) for adsorption of heavy metals. The sorption can be based on complexation, electrostatic interactions, hydrophobic effect, and hydrogen bonding [50, 51]. A dendrimer-ultrafiltration system was designed to recover metal ions from aqueous solutions [52, 53]. The system achieved almost complete removal of Cu^{2+} ions with initial concentration of 10 ppm and Cu^{2+}-to-PAMAM dendrimer-NH_2 ratio of 0.2. After adsorption, the metal ion-laden dendrimers were recovered by ultrafiltration and regenerated by decreasing pH to 4.

Potential Application in Water Treatment

Nano-adsorbents can be readily integrated into existing treatment processes in slurry reactors or adsorbers. Applied in the powder form, nano-adsorbents in slurry reactors can be highly efficient since all surfaces of the adsorbents are utilized and the mixing greatly facilitates the mass transfer. However, an additional separation unit is required to recover the nanoparticles. Nano-adsorbents can also be used in fixed or fluidized adsorbers in the form of pellets/beads or porous granules loaded with nano-adsorbents. Fixed-bed reactors are usually associated with mass transfer limitations and head loss; but it does not need future separation process. Applications of nano-adsorbents for arsenic removal have been commercialized, and their performance and cost have been compared to other commercial adsorbents in pilot tests [54]. ArsenXnp is a commercial hybrid ion-exchange medium comprising iron oxide nanoparticles and polymers. ADSORBSIA™ is a nanocrystalline titanium dioxide medium in the form of beads from 0.25 to 1.2 mm in diameter. Both nano-adsorbents were highly efficient in removing arsenic and ArsenXnp required little backwash [54, 55]. The estimated treatment cost for ArsenXnp is $0.25–$0.35/1000 gal if the medium is regenerated, similar to $0.37/1000 gal of Bayoxide E33, a high-performance granular iron oxide adsorbent [54, 56]. ArsenXnp and ADSORBSIA™ have been employed in small- to medium-scale drinking water treatment systems and were proven to be cost competitive.

Membranes and Membrane Processes

The basic goal of water treatment is to remove undesired constituents from water. Membranes provide a physical barrier for such constituents based on their size, allowing the use of unconventional water sources. As the key component of water treatment and reuse, they provide high level of automation and require less land and chemical use, and the modular configuration allows flexible design [4]. A major challenge of the membrane technology is the inherent trade-off between membrane selectivity and permeability. The high energy consumption is an important barrier to the wide application of pressure-driven membrane processes. Membrane fouling adds to the energy consumption and the complexity of the process design and operation. Furthermore, it reduces the lifetime of membranes and membrane modules.

The performance of membrane systems is largely decided by the membrane material. Incorporation of functional nanomaterials into membranes offers a great opportunity to improve the membrane permeability, fouling resistance, and mechanical and thermal stability, as well as to render new functions for contaminant degradation and self-cleaning.

Nanofiber Membranes

Electrospinning is a simple, efficient, and inexpensive way to make ultrafine fibers using various materials (e.g., polymers, ceramics, or even metals) [64, 176]. The resulting nanofibers have high specific surface area and porosity and form nanofiber mats with complex pore structures. The diameter, morphology, composition, secondary structure, and spatial alignment of electrospun nanofibers can be easily manipulated for specific applications [176]. Although nanofiber membranes have been commercially employed for air filtration applications, their potential in water treatment is still largely unexploited. Nanofiber membranes can remove micron-sized particles from aqueous phase at a high rejection rate without significant fouling [270]. Thus, they have been proposed to be used as pretreatment prior to ultrafiltration or reverse osmosis (RO). Functional nanomaterials can be easily doped into the spinning solutions to fabricate nanoparticle-impregnated nanofibers or those formed in situ [176]. The outstanding features and tunable properties make electrospun nanofibers an ideal platform for constructing multifunctional media/membrane filters by either directly using intrinsically multifunctional materials such as TiO_2 or introducing functional materials on the nanofibers. For example, by incorporating ceramic nanomaterials or specific capture agents on the nanofiber scaffold, affinity nanofiber membranes can be designed to remove heavy metals and organic pollutants during filtration.

Nanocomposite Membranes

A significant number of studies on membrane nanotechnology have focused on creating synergism or multifunction by adding nanomaterials into polymeric or inorganic membranes. Nanomaterials used for such applications include hydrophilic metal oxide nanoparticles (e.g., Al_2O_3, TiO_2, and zeolite), antimicrobial nanoparticles (e.g., nano-Ag and CNTs), and (photo)catalytic nanomaterials (e.g., bimetallic nanoparticles, TiO_2).

The main goal of adding hydrophilic metal oxide nanoparticles is to reduce fouling by increasing the hydrophilicity of the membrane. The addition of metal oxide nanoparticles including alumina, silica, zeolite, and TiO_2 to polymeric ultrafiltration membranes has been shown to increase membrane surface hydrophilicity, water permeability, or fouling resistance [18, 30, 217, 258]. These inorganic nanoparticles also help enhance the mechanical and thermal stability of polymeric membranes, reducing the negative impact of compaction and heat on membrane permeability [84, 258].

Antimicrobial nanomaterials such as nano-Ag and CNTs can reduce membrane biofouling. Nano-Ag has been doped or surface grafted on polymeric membranes to inhibit bacterial attachment and biofilm formation on the membrane surface as well as inactivate viruses [74, 216, 360]. However, its long-term efficacy against membrane biofouling has not been reported. Appropriate replenishment of nano-Ag needs to be addressed for practical application of this technology. CNTs inactivate bacteria upon direct contact [33]. High bacterial inactivation (>90%) has been achieved using polyvinyl-*N*-carbazole-SWNT nanocomposite at 3 wt% of SWNT [2]. As CNTs are insoluble in water and not consumed, there is no need for replenishment. However, as direct contact is required for inactivation, long-term filtration experiments are needed to determine the impact of fouling on the antimicrobial activity of CNTs. Addition of oxidized MWNT at low weight percentage (up to 1.5 wt%) also increases the hydrophilicity and permeability of polysulfone membranes [57].

Photocatalytic nanoparticle-incorporated membranes (a.k.a. reactive membranes) combine their physical separation function and the reactivity of a catalyst toward contaminant degradation. Much effort has been devoted to develop photocatalytic inorganic membranes consisting of nanophotocatalysts (normally nano-TiO_2 or modified nano-TiO_2) [55]. Metallic/bimetallic catalyst nanoparticles such as nano zerovalent iron (nZVI) and noble metals supported on nZVI have been incorporated into polymeric membranes for reductive degradation of contaminants, particularly chlorinated compounds [330, 331]. nZVI serves as the electron donor and the noble metals catalyze the reaction.

Thin-Film Nanocomposite (TFN) Membranes

Development of TFN membranes mainly focuses on incorporating nanomaterials into the active layer of thin-film composite (TFC) membranes via doping in the casting solutions or surface modification. Nanomaterials that have been researched for such applications include nano-zeolites, nano-Ag, nano-TiO_2, and CNTs. The impact of nanoparticles on membrane permeability and selectivity depends on the type, size, and amount of nanoparticles added.

Nano-zeolites are the most frequently used dopants in TFN and have shown potential in enhancing membrane permeability. The addition of nano-zeolites leads to more permeable, negatively charged, and thicker polyamide active layer [184]. One study reported that water permeability increased up to 80% over the TFC membrane, with the salt rejection largely maintained (93.9 ± 0.3%) [141]. TFN membranes doped with 250 nm nano-zeolites at 0.2 wt% achieved moderately higher permeability and better salt rejection (>99.4%) than commercial RO membranes [186]. It was hypothesized that the small, hydrophilic pores of nano-zeolites create preferential paths for water. However, water permeability increased even with pore-filled zeolites, although less than the pore-open ones, which could be attributed to defects at the zeolite-polymer interface. Nano-zeolites were also used as carriers for antimicrobial agents such as Ag, which imparts antifouling property to the membrane [185]. The zeolite TFN technology has reached the early stage of commercialization. QuantumFlux, a seawater TFN RO membrane, is now commercially available (www.nanoH2O.com).

Incorporation of nano-TiO_2 (up to 5 wt%) into the TFC active layer slightly increased the membrane rejection while maintaining the permeability [91]. When the concentration of nano-TiO_2 exceeded 5 wt%, the water flux increased at the cost of reducing rejection, suggesting defect formation in the active layer. Upon UV irradiation, TiO_2 can degrade organic contaminants and inactivate microorganisms. This helps reduce organic and biological fouling as well as remove contaminants that are not retained by the membrane. However, the close adjacency between the photocatalyst and the membrane may also lead to detrimental effects on polymeric membrane materials, which needs to be addressed for long-term efficacy [48].

CNTs (unaligned) also found their application in TFN membranes due to their antimicrobial activities. Tiraferri et al. covalently bonded SWNTs to a TFC membrane surface [304]. This approach is advantageous as it uses relatively small amount of the nanomaterial and minimizes perturbation of the active layer. The resulting TFN membrane exhibited moderate antibacterial properties (60% inactivation of bacteria attached on the membrane surface in 1-h contact time), potentially reducing or delaying membrane biofouling.

Biologically Inspired Membranes

Many biological membranes are highly selective and permeable. Aquaporins are protein channels that regulate water flux across cell membranes. Their high selectivity and water permeability make their use in polymeric membranes an attractive approach to improve membrane performance. Aquaporin-Z from *Escherichia coli* has been incorporated into amphiphilic triblock-polymer vesicles, which exhibit water permeability at least an order of magnitude over the original vesicles with full rejection to glucose, glycerol, salt, and urea [163]. One potential design is to coat aquaporin-incorporated lipid bilayers on commercial nanofiltration membranes. On this front, limited success was achieved [150]. Aligned CNTs have been shown both experimentally and theoretically to provide water permeation much faster than what the Hagen-Poiseuille equation predicts, owing to the atomic smoothness of the nano-sized channel, and the one-dimensional single-file ordering of water molecules while passing through the nanotubes [125, 133]. It was predicted that a membrane containing only 0.03% surface area of aligned CNTs will have flux exceeding current commercial seawater RO membranes [257]. However, high rejection for salt and small molecules is challenging for aligned CNT membranes due to the lack of CNTs with uniform sub-nanometer diameter. Functional group gating at the nanotube opening has been proposed to enhance the selectivity of aligned CNT membranes [214]. By grafting carboxyl functional groups on sub-2 nm CNT openings, 98% rejection of $Fe(CN_6)^{3-}$ was achieved at low ionic strength by Donnan exclusion [94]. However, KCl rejection was only 50% at 0.3 mM, and decreased to almost zero at 10 mM. Grafting bulky functional groups at the tube opening could physically exclude salts. However, steric exclusion will significantly reduce membrane permeability [242]. Thus at the current stage, aligned CNT membranes are not capable of desalination. To achieve reliable salt rejection, the CNT diameter must be uniformly smaller than 0.8 nm [124]. A key barrier for both aquaporin and aligned CNT membranes is the scale-up of the nanomaterial production and membrane fabrication. Large-scale production and purification of aquaporins are very challenging. To date, chemical vapor deposition (CVD) is the most common way to make aligned nanotubes. A continuous high-yield CVD prototype has been designed for producing vertically aligned CNT, paving the way for large-scale production [76]. A post-manufacturing alignment method using magnetic field was also developed [215].

Nanocomposite and TFN membranes have good scalability as they can be fabricated using current industrial manufacturing processes. The high water permeability can reduce the applied pressure or required membrane area and consequently cut cost. This strategy may greatly improve the energy efficiency for treatment of waters with low osmosis pressure, but it may have limited advantage in seawater RO, whose energy consumption is already close to the thermodynamic limit [85]. A recent review ranked current membrane nanotechnologies based on their potential performance enhancement and state of commercial readiness [257].

Forward Osmosis

Forward osmosis (FO) utilizes the osmotic gradient to draw water from a low-osmotic-pressure solution to a high-osmotic-pressure one (i.e., the draw solution). The diluted draw solution is then treated by reverse osmosis or thermal processes to generate pure water. FO has two major advantages over the pressure-driven reverse osmosis: it does not require high pressure, and the membrane is less prone to fouling.

The key to FO is to have a draw solute with high osmolality and that is easily separable from water. Chemicals currently employed for draw solutions include NaCl and ammonia bicarbonate. Therefore, RO or thermal treatment, both energy intensive, is required to recover water from the draw solution. Magnetic nanoparticles were recently explored as a new type of draw solute for its easy separation and reuse. Hydrophilic coating was employed to aid dissolution and increase osmotic pressure. An FO permeate flux higher than 10 L/m^2/h was achieved using 0.065 M poly(ethylene glycol) diacid-coated magnetic nanoparticles when deionized water was used as the feed solution [104]. Magnetic nanoparticles were also applied to recover draw solutes. In a recent study, magnetic nanoparticles ($Fe_3O_4@SiO_2$) were used to recover $Al_2(SO_4)_3$ (the draw solute) through flocculation [193].

Photocatalysis

Photocatalytic oxidation is an advanced oxidation process for removal of trace contaminants and microbial pathogens. It is a useful pretreatment for hazardous and nonbiodegradable contaminants to enhance their biodegradability. Photocatalysis can also be used as a polishing step to treat recalcitrant organic compounds. The major barrier for its wide application is the slow kinetics due to limited light fluence and photocatalytic activity. Current research focuses on increasing photocatalytic reaction kinetics and photoactivity range (Table 13.2).

Nano-Photocatalyst Optimization

TiO_2 is the most widely used semiconductor photocatalyst in water/wastewater treatment owing to its low toxicity, chemical stability, low cost, and abundance as raw material. It generates an electron/hole (e^-/h^+) pair upon absorbing a UV photon, which later either migrates to the surface and forms reactive oxygen species (ROS) or undergoes undesired recombination. The photoactivity of nano-TiO_2 can be improved by optimizing particle size and shape, reducing e^-/h^+ recombination by noble metal doping, maximizing reactive facets, and surface treatment to enhance contaminant adsorption.

The size of TiO_2 plays an important role in its solid-phase transformation, sorption, and e^-/h^+ dynamics. Among the crystalline structures of TiO_2, rutile is the most stable for particles larger than 35 nm, while anatase, which is more efficient in producing ROS, is the most stable for particles smaller than 11 nm [99, 352]. A major cause for the slow reaction kinetics of TiO_2 photocatalysis is the fast recombination of e^- and h^+. Decreasing TiO_2 particle size lowers volume recombination of e^-/h^+, and enhances interfacial charge carrier transfer [356]. However, when particle size is reduced to several nanometers, surface recombination dominates, decreasing photocatalytic activity. Therefore, the photocatalytic activity of TiO_2 is maximum due to the interplay of the aforementioned mechanisms, which lies in the nanometer

Table 13.2 TiO_2 photocatalyst optimization

Optimization objectives	Optimization approaches	Optimization mechanisms	Water treatment applications
Enhance photocatalytic reaction kinetics	Size	More surface-reactive sites, higher reactant adsorption, lower electron/hole recombination	High-performance UV-activated photocatalytic
	Nanotube morphology	Shorter carrier-diffusion paths in the tube walls, higher reactant mass transfer rate toward tube surface	Reactors
	Noble metal doping	Better electron/hole separation, lower electron/hole recombination	
	Reactive crystallographic facets	Higher reactant sorption, better electron/hole separation, lower electron/hole recombination	
Expand photoactivity range	Metal impurity doping Anion doping Dye sensitizer doping Narrow bandgap Semiconductor doping	Impurity energy levels Bandgap narrowing Electron injection Electron injection	Low-energy-cost solar/visible light-activated photocatalytic reactors

range. TiO_2 nanotubes were found to be more efficient than TiO_2 nanoparticles in the decomposition of organic compounds [201]. The higher photocatalytic activity was attributed to the shorter carrier-diffusion paths in the tube walls and faster mass transfer of reactants toward the nanotube surface.

Noble metal doping can reduce the e^-/h^+ recombination because the photoexcited electrons tend to migrate to the noble metals with lower Fermi levels while the holes stay in TiO_2 [244]. The photocatalytic activity of TiO_2 can also be promoted by creating highly reactive crystallographic facets. Because high-energy {001} facets diminish quickly during crystal growth, anatase TiO_2 is usually dominated by the low-energy {101} facets. Using specific capping agent (usually fluoride), the percentage of {001} facets can be increased from less than 10% to up to 89%, substantially enhancing hydroxyl radical production and organic compound decomposition [117, 237]. The enhanced activity stems from the strong adsorption of reactants on high-energy facets and the spatial separation of electrons and holes on specific crystal facets [192, 237]. The optimal percentage of {001} facets for photocatalysis is still debated [192]. Improving contaminant adsorption by modifying photocatalyst surface is another way to enhance photocatalytic activity due to the short lifetime of ROS. However, little has been done in this area.

Another actively pursued research area is to extend the excitation spectrum of TiO_2 to include visible light. The general strategy is doping metal impurities, dye sensitizers, narrow bandgap semiconductors, or anions into nano-TiO_2 to form hybrid nanoparticles or nanocomposites [99, 245]. Metals and anions create impurity energy levels or narrow the bandgap; upon visible light excitation, dye sensitizers and narrow bandgap semiconductors inject electrons into TiO_2 to initiate the catalytic reactions. Among these methods, anion (especially nitrogen) doping was considered the most cost effective and feasible for industrial applications, although their stability and long-term efficacy have not been tested. Decreased nitrogen concentration during photocatalysis has been reported [99, 156].

Other than TiO_2, WO_3 and some fullerene derivatives also have the potential to be used in photocatalytic water treatment. WO_3 has a narrower bandgap than TiO_2, allowing it to be activated by visible light (<450 nm) [159]. Pt doping further enhances WO_3 reactivity by facilitating multielectron reduction of O_2 and improving e^-/h^+ separation [154]. Aminofullerenes generate $^1O_2$ under visible light irradiation (<550 nm) and have been known to degrade pharmaceutical compounds and inactivate viruses [172, 195]. Fullerol and C_{60} encapsulated with poly(*N*-vinylpyrrolidone) can produce $^1O_2$ and superoxide under UVA light [35]. Aminofullerenes are more amenable to immobilization than fullerol and are more effective for disinfection purposes due to their positive charge. $^1O_2$ has lower oxidation potential than hydroxyl radicals produced by TiO_2, while it is a more selective ROS and consequently less susceptible to quenching by nontarget background organic matter. Fullerenes are currently much more expensive and not as readily available as TiO_2.

Potential Applications in Water Treatment

The overall efficiency of a photocatalytic water treatment process strongly depends on the configuration and operation parameters of the photoreactor. Two configurations are commonly used: slurry reactors and reactors using immobilized TiO_2. Various dispersion/recovery or catalyst immobilization techniques are being pursued to maximize its efficiency. Extensive investigation on operating parameters has been carried out with these lab- or pilot-scale systems. A recent critical review outlines the effects of water quality and a wide range of operating parameters including TiO_2 loading, pH, temperature, dissolved oxygen, contaminant type and concentration, light wavelength, and intensity [59]. Readers are referred to this review for details regarding process optimization. A commercial product, Purifics Photo-Cat™ system, has a treatment capacity as high as two million gallons per day with a small footprint of 678 ft^2. Pilot tests showed that the Photo-Cat™ system is highly efficient for removing organics without producing waste streams and it operates with relatively low specific power consumption of about 4 kWh/m^3 [3, 25, 324]. Nano-TiO_2-facilitated solar disinfection (SODIS) has been extensively tested and appears to be a feasible option to produce safe drinking water in remote areas of developing countries. The SODIS system can be small scale for one person or scaled up to medium-size solar compound parabolic collectors.

Photocatalysis has shown great potential as a low-cost, environmental friendly, and sustainable water treatment technology. However, there are several technical challenges for its large-scale application, including (1) catalyst optimization to improve quantum yield or to utilize visible light; (2) efficient photocatalytic reactor design and catalyst recovery/immobilization techniques; and (3) better reaction selectivity.

Metal oxide nanomaterials such as TiO_2 and CeO_2 as well as carbon nanotubes have been studied as catalysts in heterogeneous catalytic ozonation processes that provide fast and comparatively complete degradation of organic pollutants. Both radical-mediated and non-radical-mediated reaction pathways have been proposed [241]. The adsorption of ozone and/or pollutants on the catalyst surface plays a critical role in both mechanisms. Nanomaterials have large specific surface area and an easily accessible surface, leading to high catalytic activity. Some nanomaterials were also reported to promote decomposition of ozone into hydroxyl radicals, facilitating degradation process through radical-mediated routes [250]. For future industrial scale applications, a better understanding of the mechanism of nanomaterial-enabled catalytic ozonation is in critical need.

Disinfection and Microbial Control

The dilemma between effective disinfection and formation of toxic disinfection by-products (DBPs) poses a great challenge for the water industry. It is now well recognized that conventional disinfectants, such as chlorine disinfectants and ozone, can form toxic DBPs (e.g., halogenated disinfection by-products, carcinogenic nitrosamines, bromate). UV disinfection emerged as an alternative for oxidative disinfection as it produces minimal DBPs, while it requires high dosage for certain viruses (e.g., adenoviruses). These limitations urge the development of alternative methods that can enhance the robustness of disinfection while avoiding DBP formation.

Our previous review on antimicrobial nanomaterials highlighted the potential of nanotechnology in disinfection and microbial control [178]. Many nanomaterials, including nano-Ag, nano-ZnO, nano-TiO_2, nano-Ce_2O_4, CNTs, and fullerenes, exhibit antimicrobial properties without strong oxidation, and hence have lower tendency to form DBPs (Table 13.3). The antimicrobial mechanisms of these nanomaterials, their merits, limitations, and applicability for water treatment and the critical research needs are thoroughly discussed in that review paper [178]. Thus, only a brief update mainly regarding nano-Ag and carbon-based nanomatcrials will be provided here.

Antimicrobial Mechanisms

Nano-Ag is currently the most widely used antimicrobial nanomaterial. Its strong antimicrobial activity, broad antimicrobial spectrum, low human toxicity, and ease of use make it a promising choice for water disinfection and microbial control. It is now well accepted that the antimicrobial activity of nano-silver largely stems from the release of silver ions [334, 335]. Silver ions can bind to thiol groups in vital proteins, resulting in enzyme damage [181]. It has also been reported that silver ions can prevent DNA replication and inducc structural changes in the cell envelope [89]. Thus, the release ratc and bioavailability of silver ions are crucial for the toxicity of nano-Ag. Studies have suggested that physicochemical properties of nano-Ag play an important role in its antimicrobial activity. However, the influence of the size, shape, coating, and crystallographic facet appears to be mainly related to different

Table 13.3 Nanomaterial antimicrobial mechanisms

Nanomaterials	Antimicrobial mechanisms
Nano-Ag	Release of silver ions, protein damage, suppression of DNA replication, membrane damage
Nano-TiO_2	Production of ROS
Nano-ZnO	Release of zinc ions, production of H_2O_2, membrane damage
Nano-MgO	Membrane damage
Nano-Ce_2O_4	Membrane damage
nC_{60}	ROS-independent oxidation
Fullerol and aminofullerene	Production of ROS
Carbon nanotubes	Membrane damage, oxidative stress
Graphene-based nanomaterials	Membrane damage, oxidative stress

release kinetics of silver ions. The presence of common ligands reduces the bioavailability of silver ions and mitigates its toxicity [334]. A recent study found that low concentration (sublethal) of silver ions or nano-Ag enhances *E. coli* growth, suggesting a hermetic response that could be counterproductive to its antimicrobial applications [335].

CNTs kill bacteria by causing physical perturbation of the cell membrane, oxidative stress, or disruption of a specific microbial process via disturbing/oxidizing a vital cellular structure/component upon direct contact with bacterial cells. Graphene and graphite materials exhibit antimicrobial properties through similar mechanisms [191, 313]. The cytotoxicity of CNTs strongly depends on their physicochemical properties. Short, dispersed, and metallic CNTs with small diameters are more toxic [147, 148, 313].

Potential Applications in Water Treatment

Antimicrobial nanomaterials are envisaged to find their applications in three critical challenges in water/wastewater systems: disinfection, membrane biofouling control, and biofilm control on other relevant surfaces. Nano-Ag has good potential for application in POU treatment. It can improve water quality for high-end use, or provide another barrier against waterborne pathogens for vulnerable population. Commercial devices utilizing nano-Ag are already available, e.g., MARATHON® and Aquapure® systems. Nano-Ag has also been incorporated into ceramic microfilters as a barrier for pathogens, which can be employed in remote areas in developing countries [260].

The antimicrobial properties, fibrous shape, and high conductivity of CNTs enable novel CNT filters for both bacteria and virus removal: The thin layer of CNTs effectively removes bacteria by size exclusion and viruses by depth filtration; the retained bacteria are largely inactivated by CNTs within hours [34]. With a small intermittent voltage (2–3 V), MWNTs can directly oxidize attached bacteria and viruses and lead to inactivation in seconds [269, 312]. The applied electric potential also enhances viral transport to the anodic CNTs [269]. Such CNT filters can be used as high-performance POU devices for water disinfection with minimal to no power requirement.

The application of nanomaterials in membrane biofouling control is detailed in section "Membranes and Membrane Processes." They can also be used in other water treatment-related surfaces such as storage tanks and distribution pipes to control pathogen contamination, biofilm formation, and microbial influenced corrosion. Affordable coating techniques that can economize nanomaterial use and maximize its efficacy while allowing for regeneration are in critical need. An alternative approach is to employ nanoscale biofouling-resistant surface structures, a strategy used by marine organisms (dolphins and sharks) and plants (lotus leafs). A common disadvantage of many nanomaterial-enabled disinfection approaches is the lack of disinfection residue, which is crucial for controlling microbial growth during water storage and distribution. Nevertheless, nanotechnology-enabled disinfection can reduce DBP formation as chlorine or other chemical disinfectants are only needed as secondary disinfectants. Long-term efficacy is another major uncertainty for all the aforementioned technologies. Antimicrobial nanomaterials that rely on the release of biocidal ions will be eventually depleted. Controlled release and replenish strategies are thus needed. A potential "on-demand" release strategy is to encapsulate antimicrobial agents into a matrix gated by materials responsive to the presence of microorganisms or biofilms. This "on-demand" mechanism can be further coupled with recognition mechanisms for targeted release (Fig. 13.2). For nanomaterials relying on direct contact, fouling may largely suppress or even eliminate their antimicrobial activity.

Sensing and Monitoring

A major challenge for water/wastewater treatment is water quality monitoring due to the extremely low concentration of certain contaminants, the lack of fast pathogen detection, as well as the high complexity of the water/wastewater matrices. Innovative sensors with high sensitivity and selectivity, and fast response, are in great need.

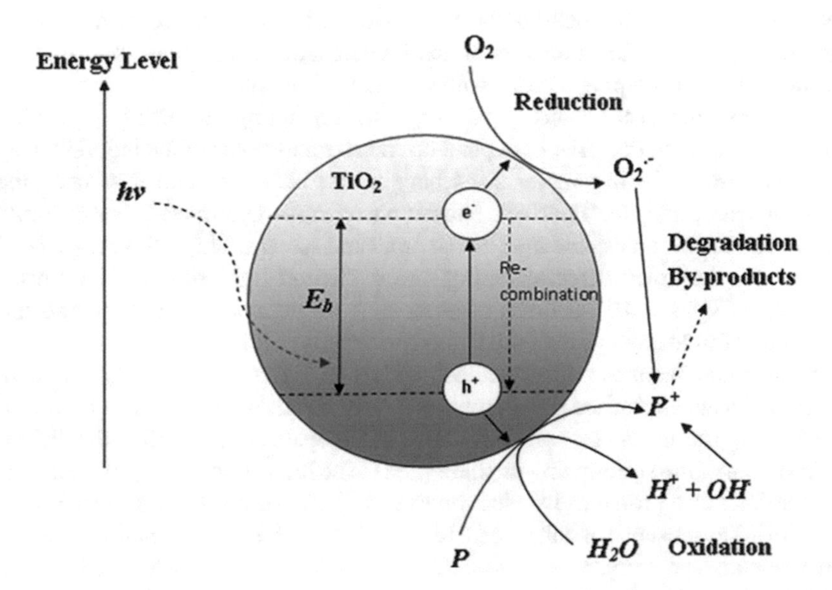

Fig. 13.2 Schematic mechanism for "on-demand" microbial control. "On-demand" microbial control can be achieved by using recognition agents that target a specific microorganism. The responsive gating material is designed to release the antimicrobial agent after the recognition event

Pathogen Detection

Pathogen detection is of critical importance as it is directly related to public health. Conventional indicator systems such as coliform bacteria are slow and fail to monitor the presence of some important or emerging pathogens including viruses (hepatitis A and E, coxsackieviruses, echoviruses, adenoviruses, and Norwalk viruses), bacteria (*Legionella* and *Helicobacter*), and protozoan (*Cryptosporidium* and *Giardia*) [302]. Many of these pathogens are etiologic agents in outbreaks associated with drinking water. Additionally, pathogen detection is the key component of diagnosis-based water disinfection approach, in which disinfection is triggered by the detection of target microorganisms. Active research is going on for developing nanomaterial-enabled pathogen sensors. These sensors usually consist of three major components: recognition agents, nanomaterials, and a signal transduction mechanism [315]. Recognition agents that specifically interact with antigens or other epitopes on the pathogen surface provide the selectivity. Sensitivity and fast response are achieved by the nanomaterial-related signal transduction upon the recognition event. A wide range of recognition agents have been utilized, including antibodies, aptamers, carbohydrates, and antimicrobial peptides [315]. Nanomaterials improve the sensitivity and speed of detection and achieve multiplex target detection owing to their unique physicochemical properties, especially electrochemical, optical, and magnetic properties. These sensors can be employed to detect whole cells as well as biomolecules [302, 315].

The most commonly used nanomaterials in pathogen detection are magnetic nanoparticles, quantum dots (QDs), noble metals, dye-doped nanoparticles, and CNTs. Magnetic nanoparticles and CNTs have been extensively studied for sample concentration and purification. A commercial magnetic nanocomposite, Dynabead®, is available for developing various pathogen detection kits.

QDs are fluorescent nanocrystals of semiconducting materials (e.g., CdSe) whose electronic characteristics depend on the size and shape of the individual crystals. QD particles with smaller sizes have wider bandgaps and thus need more energy to excite and emit light with shorter wavelength. QDs have broad absorption spectra but narrow and stable fluorescent emission spectra. Thus QDs are especially suitable for multiplex detection using one excitation light source. The emission spectrum of QDs is 10–20 times brighter than an organic fluorophore and up to thousands of times more stable than conventional dyes [297, 337].

Noble metal nanomaterials have been widely used in sensors mainly due to their enhanced localized surface plasmon resonance (LSPR), which depends on the size, shape, composition, and separation distance of nanoparticles, as well as the dielectric environment of the surrounding medium [261]. The high conductivity of noble metal nanoparticles also promotes the electron transfer between electrode surface and analyst [174]. The presence of enhanced LSPR leads to high molar extinction coefficient and Rayleigh scattering, as well as enhanced local electromagnetic fields near the nanoparticle surface. Based on theoretical calculation, nano-Au spheres of 40 nm in diameter have an absorption cross section 5 orders of magnitude higher than conventional dyes, while 80 nm nano-Au spheres scatter light 5 orders of magnitude more than fluorescence dyes [139]. Noble metals were used mainly in colorimetric and surface-enhanced Raman spectroscopy (SERS) sensing. Colorimetric assays are fast and simple. The signal transduction relies on the color change of the nanoparticle suspension due to different interparticle distances or aggregation states [152]. It has been studied for the detection of DNA, diagnosis of pathogen infection, and pathogen monitoring in water samples. However, the aggregation state of nanoparticles is sensitive to the solution chemistry and difficult to control. The coexisting water/wastewater constituents will greatly affect results, reducing reproducibility. The SERS phenomenon is attributed to both electromagnetic effect and chemical mechanisms related to the charge transfer between the noble metals and the target molecules [231]. As a result, the efficiency of Raman scattering can be enhanced more than 10^{14}-fold, which is even capable of detecting a single molecule [246].

Silica nanoparticles doped with either organic or inorganic luminescent dyes have been developed for ultrasensitive sensors. The large number of dye molecules confined in a single silica particle guarantees huge improvement in sensitivity. Moreover, the silica matrix protects the dye molecules from the external environment, largely suppressing photobleaching and photodegradation. The outstanding photostability makes dye-doped silica nanoparticles especially advantageous for applications that require high-intensity or prolonged excitations [337]. The rich silica chemistry (e.g., silane chemistry) also helps future surface modification and conjugation.

The high conductivity along the length makes CNTs outstanding electrode materials. As a result, CNTs can greatly facilitate electrochemical detection by promoting electron transfer and electrode-analyst interactions [219]. They have been incorporated into electrodes via random or aligned coating, or used as a single CNT electrode [343]. Semiconducting CNTs can be used in nanoscale field-effect transistor [118]. Besides their excellent electronic properties, the high adsorption capacity of CNTs increases detection sensitivity [66]. The major challenge for CNT-based sensors is the heterogeneity of CNTs. Separation of metallic and semiconducting SWNT has been extensively studied but is still far from perfect. The production and purification processes of CNTs often introduce impurities, contaminants, and even degradation of the CNT structure. Therefore, better synthesis, purification, and separation are required to produce more homogeneous CNTs. Although most of these nanosensors possess excellent photostability and sensitivity, nonspecific binding is still a major challenge for their application in water and wastewater. Strategies to reduce nonspecific binding and prevent undesired nanoparticle aggregation are in critical need.

Trace Contaminant Detection

In trace organic or inorganic contaminant detection, nanomaterials can be used in both concentration and detection. CNTs have great potential for environmental analysis of trace metal or organic pollutants as they offer high adsorption capacity and recovery rate as well as fast kinetics as discussed above. The pre-concentration factors for metal ions were found to be between 20 and 300 with fast adsorption kinetics [83]. CNTs have also been extensively studied for pre-concentrating a variety of organic compounds, many of which were done in real water samples [39]. Adsorption of charged species to CNTs results in changes of conductance, providing the basis for the correlation between analyte concentration and current fluctuation [214]. Other nanomaterials such as nano-Au and QDs have also been used. Nano-Au was used to detect pesticides at ppb levels in a colorimetric assay; modified nano-Au was shown to detect Hg^{2+} and CH_3Hg^+ rapidly with high sensitivity and selectivity [183, 188]. QD-modified TiO_2 nanotubes lowered the detection limits of PAHs to the level of pica-mole per liter based on fluorescence resonance energy transfer [342]. A nanosensor based on CoTe QDs immobilized on a glassy carbon electrode surface was reported to detect bisphenol A in water at concentrations as low as 10 nM within 5 s [347].

Multifunctional Devices

The advance in functional nanomaterials and their convergence with conventional technologies bring opportunities in designing a new family of nanotechnology-enabled multifunctional water treatment devices which are capable of performing multiple tasks in one device. Such multifunctional systems can enhance the overall performance and avoid excessive redundancy, miniaturizing the footprint. Therefore, the multifunctional concept is especially advantageous in decentralized and small-scale applications. Different functional nanomaterials can be integrated onto a common platform based on treatment requirement. Besides magnetic nanoparticles, membranes are a good and extensively studied platform to construct multifunctional devices. Notably, electrospun nanofibers have drawn much attention as an excellent nanomaterial carrier. Owing to the high performance, small footprint, and modular design of nanotechnology-enabled devices, it is envisaged that different functionalities can be assembled in layers of cartridges or as modules arranged in series, allowing optimization/regeneration of each functionality separately [267]. The capacity and functionality of such nanotechnology-enabled systems can be easily manipulated by plugging in or pulling out modules.

Retention and Reuse of Nanomaterials

The retention and reuse of nanomaterials are key aspects of nanotechnology-enabled device design due to both cost and public health concerns. It can be usually achieved by applying a separation device or immobilizing nanomaterials in the treatment system. A promising separation process is membrane filtration which allows continuous operation with small footprint and chemical use. Ceramic membranes are more advantageous than polymeric membranes in photocatalytic or catalytic ozonation applications as they are more resistant to UV and chemical oxidants [48]. The suspended particles in the receiving water are detrimental to reactor membrane hybrid systems as they can be retained by the membrane and can significantly reduce the reaction efficiency. Thus, raw water pretreatment is usually required to reduce the turbidity. Nanomaterials can also be immobilized on various platforms such as resins and membranes to avoid further separation. However, current immobilization techniques usually result in significant loss of treatment efficiency. Research is needed to develop simple, low-cost methods to immobilize nanomaterials without significantly impacting its performance. For magnetic nanoparticles/nanocomposites, low-field magnetic separation is a possible energy-efficient option.

Little is known about the release of nanomaterials from nanotechnology-enabled devices. However, the potential release is expected to be largely dependent on the immobilization technique and the separation process employed. If no downstream separation is applied, nanomaterials coated on treatment system surfaces are more likely to be released in a relatively fast and complete manner, while nanomaterials embedded in a solid matrix will have minimum release until they are disposed of. For nanomaterials that release metal ions, their dissolution needs to be carefully controlled (e.g., by coating or optimizing size and shape). The detection of nanomaterial release is a major technical hurdle for risk assessment and remains challenging. Details regarding detection techniques are beyond the scope of this chapter, and readers are referred to several recent reviews on this topic [72, 303]. Few techniques can detect nanomaterials in complex aqueous matrices and they are usually sophisticated, expensive, and with many limitations. Fast, sensitive, and selective nanomaterial analytical techniques are in great need.

Barriers and Research Needs

Although nanotechnology-enabled water/wastewater treatment processes have shown great promise in laboratory studies, their readiness for commercialization varies widely. Some are already on the market, while others require significant research before they can be considered for full-scale applications. Their future development and commercialization face a variety of challenges including technical hurdles, cost-effectiveness, and potential environmental and human risk. There are two major research needs for full-scale applications of nanotechnology in water/wastewater treatment. First, the performance of various nanotechnologies in treating real natural and wastewaters needs to be tested. Future studies need to be done under more realistic conditions to assess the applicability and efficiency of different nanotechnologies as well as to validate nanomaterial-enabled sensing technologies. Secondly, the long-term efficacy of these nanotechnologies is largely unknown as most lab studies were conducted for relatively short period of time. Research addressing the long-term performance of water and wastewater treatment nanotechnologies is in great need. As a result, side-by-side comparison of nanotechnology-enabled systems and existing technologies is challenging.

Despite the superior performance, the adoption of innovative technologies strongly depends on the cost-effectiveness and the potential risk involved. The current cost of nanomaterials is prohibitively high with few exceptions such as nano-TiO_2, nanoscale ion oxide, and polymeric nanofibers. There are currently two approaches to address the cost issue. One proposed approach is to use low-purity nanomaterials without significantly compromising efficiency as much of the production cost is related to separation and purification [267]. Alternatively, the cost-effectiveness can be improved by retaining and reusing nanomaterials. Nanomaterials possess unique challenges for risk assessment and management as they are small particles instead of molecules or ions for which risk assessment framework and protocols are already in place. Better understanding and mitigating of potential hazards associated with the use of nanomaterials in water and wastewater treatment will lead to broader public acceptance, which is crucial for new technology adoption. The compatibility between aforementioned nanotechnologies and current water and wastewater treatment processes and infrastructure also needs to be addressed. Most treatment plants and distribution systems in developed countries are expected to remain in place for decades to come. As a result, it is important to be able to implement nanotechnology with minimal changes to existing infrastructure in the near term. In the meantime, nanotechnology-enabled treatment processes can be employed in places where water treatment infrastructure does not exist or in POU devices.

Conclusions

Nanotechnology for water and wastewater treatment is gaining momentum globally. The unique properties of nanomaterials and their convergence with current treatment technologies present great opportunities to revolutionize water and wastewater treatment. Although many nanotechnologies highlighted in this review are still in the laboratory research stage, some have made their way to pilot testing or even commercialization. Among them, three categories show most promise in full-scale application in the near future based on their stages in research and development, commercial availability and cost of nanomaterials involved, and compatibility with the existing infrastructure: nano-adsorbents, nanotechnology-enabled membranes, and nanophotocatalysts. All three categories have commercial products, although they have not been applied in large-scale water or wastewater treatment. Several other water treatment nanotechnologies have found their niche applications in POU systems.

The challenges faced by water/wastewater treatment nanotechnologies are important, but many of these challenges are perhaps only temporary, including technical hurdles, high cost, and potential environmental and human risk. To overcome these barriers, collaboration between research institutions, industry, government, and other stakeholders is essential. It is our belief that advancing nanotechnology by carefully steering its direction while avoiding unintended consequences can continuously provide robust solutions to our water/wastewater treatment challenges, both incremental and revolutionary.

Recent Developments in Photocatalytic Water Treatment Technology

In recent years, semiconductor photocatalytic process has shown a great potential as a low-cost, environmental friendly, and sustainable treatment technology to align with the "zero" waste scheme in the water/wastewater industry. The ability of this advanced oxidation technology has been widely demonstrated to remove persistent organic compounds and microorganisms in water. At present, the main technical barriers that impede its commercialization remained on the post-recovery of the catalyst particles after water treatment. This chapter reviews the recent R&D progresses of engineered photocatalysts, photoreactor systems, and process optimizations and modeling of the photooxidation processes for water treatment. A number of potential and commercial photocatalytic reactor configurations are discussed, in particular the photocatalytic membrane reactors. The effects of key photoreactor operation parameters and water quality on the photo-process performances in terms of mineralization and disinfection are assessed. For the first time, we describe how to utilize a multivariable optimization approach to determine the optimum operation parameter so as to enhance process performance and photooxidation efficiency. Both photomineralization and photo-disinfection kinetics and their modeling associated with the photocatalytic water treatment process are detailed. A brief discussion on the life cycle assessment for retrofitting the photocatalytic technology as an alternative waste treatment process is presented. This chapter delivers a scientific and technical overview and useful information to scientists and engineers who work in this field.

Introduction

Increasing demand and shortage of clean water sources due to the rapid development of industrialization, population growth, and long-term droughts have become an issue worldwide. With this growing demand, various practical strategies and solutions have been adopted to yield more viable water resources. The storage of rainwater for daily activities and increasing the catchment capacity for stormwater are just a few examples that could resolve the problems in short term. Water industries and governments in some arid areas with abundant sunlight, less rainfall, and long-term drought have a challenge to seek viable water resources. It is estimated that around four billion people worldwide experience to have no or little access to clean and sanitized water supply, and millions of people died of severe waterborne diseases annually [205]. These statistical figures are expected to grow in the short future, as increasing water contamination due to overwhelming discharge of micropollutants and contaminants into the natural water cycle [272, 296, 327]. In view to suppress the worsening of clean water shortage, development of advanced low-cost and high-efficiency water treatment technologies to treat wastewater is desirable.

One of a few attractive options is the possible reuse of onsite rural wastewater or treated municipal wastewater from treatment plants for agricultural and industrial activities [32, 168]. Since these wastewaters constitute one of the largest possible water resources, its reuse is anticipated to offset more clean water resource. Recycling wastewaters are usually associated with the presence of suspended solids, health-threat coliforms, and soluble refractory organic compounds that are both tedious and expensive to treat [314]. Currently available water treatment technologies such as adsorption or coagulation merely concentrate the pollutants present by transferring them to other phases, but they will still remain and will not be completely "eliminated" or "destroyed" [252]. Other conventional water treatment methods such as sedimentation, filtration, and chemical and membrane technologies involve high operating costs and could generate toxic secondary pollutants into the ecosystem [103]. These concentrated toxic contaminants are highly redundant and have been concerned worldwide due to the increasing environmental awareness and legislation. Chlorination has been the most commonly and widely used disinfection process. The disinfection by-products generated from chlorination are mutagenic and carcinogenic to human health [65, 199, 339]. These have led to the rapid R&D in the field of "advanced oxidation processes (AOPs)" as the innovative water treatment technologies. The rationales of these AOPs are based on the in situ generation of highly reactive transitory species (i.e., H_2O_2, $OH^{\bullet}$, $O_2^{\bullet-}$, O_3) for mineralization of refractory organic compounds, water pathogens, and disinfection by-products [86, 259]. Among these AOPs, heterogeneous photocatalysis employing semiconductor catalysts (TiO_2, ZnO, Fe_2O_3, CdS, GaP, and ZnS) has demonstrated its efficiency in degrading a wide range of ambiguous refractory organics into readily biodegradable compounds, and eventually mineralizing them to innocuous carbon dioxide and water. Among the semiconductor catalysts, titanium dioxide (TiO_2) has received the greatest interest in R&D of photocatalysis technology. The TiO_2 is the most active photocatalyst under the photon energy of 300 nm $< l <$ 390 nm and remains stable after the repeated catalytic cycles, whereas Cds or GaP is degraded along to produce toxic products [205]. Other than these, the multifaceted functional properties of TiO_2 catalyst, such as their chemical and thermal stability or resistance to chemical breakdown and their strong mechanical properties, have promoted its wide application in photocatalytic water treatment. A number of important features for the heterogeneous photocatalysis have extended their feasible applications in water treatment, such as (1) ambient operating temperature and pressure, (2) complete mineralization of parents and their intermediate compounds without secondary pollution, and (3) low operating costs. The fact that the highly reactive oxygen species (ROS) are generated as a result of the photoinduced charge separation on TiO_2 surfaces for microbial inactivation and organic mineralization without creating any secondary pollution is well documented. So far, the application of such TiO_2 catalysts for water treatment is still experiencing a series of technical challenges. The post-separation of the semiconductor TiO_2 catalyst after water treatment remains as the major obstacle toward the practicality as an industrial process. The fine particle size of the TiO_2, together with their large surface area-to-volume ratio and surface energy, creates a strong tendency for catalyst agglomeration during the operation.

Such particle agglomeration is highly detrimental in terms of particle size preservation, surface-area reduction, and its reusable life span. Other technical challenges include the catalyst development with broader photoactivity range and its integration with feasible photocatalytic reactor system. In addition, the understanding of the theory behind the common reactor operational parameters and their interactions is also inadequate and presents a difficult task for process optimization. A number of commonly made mistakes in studying kinetic modeling on either the photomineralization or the photo-disinfection have also been seen over the years.

This review chapter aims to give an overview of the understanding and development of photocatalytic water treatment technology, from fundamentals of catalyst and photoreactor development to process optimization and kinetics modeling, and eventually the water parameters that affect the process efficiency. A short outline of the feasible application of photocatalytic water technology via life cycle interpretation and the possible future challenges are also given.

Fundamentals and Mechanism of TiO_2 Photocatalysis

Heterogeneous TiO_2 Photocatalysis

The fundamentals of photo-physics and photochemistry underlying the heterogeneous photocatalysis employing the semiconductor TiO_2 catalyst have been intensively reported in many literatures [98, 103]. The semiconductor TiO_2 has been widely utilized as a photocatalyst for inducing a series of reductive and oxidative reactions on its surface. This is solely contributed by the distinct lone electron characteristic in its outer orbital. When photon energy (hv) of greater than or equal to the bandgap energy of TiO_2 is illuminated onto its surface, usually 3.2 eV (anatase) or 3.0 eV (rutile), the lone electron will be photoexcited to the empty conduction band in femtoseconds. Figure 13.3 depicts the mechanism of the electron/hole pair formation when the TiO_2 particle is irradiated with adequate hv. The light wavelength for such photon energy usually corresponds to $l < 400$ nm. The photonic excitation leaves behind an empty unfilled valence band, and thus creates the electron/hole pair (e^-/h^+). The series of chain oxidative-reductive reactions (Eqs. (13.1)–(13.11)) that occur at the photon-activated surface was widely postulated as follows:

$$\text{Photoexcitation}: \text{TiO}_2 + hv \rightarrow \text{e}^- + \text{h}^+ \tag{13.1}$$

$$\text{Charge-carrier trapping of e-}: \text{e}^-_{\text{CB}} \rightarrow \text{e}^-_{\text{TR}} \tag{13.2}$$

$$\text{Charge-carrier trapping of h+}: \text{h}^+_{\text{VB}} \rightarrow \text{h}^+_{\text{TR}} \tag{13.3}$$

$$\text{Electron / hole recombination}: \text{e}^-_{\text{TR}} + \text{h}^+_{\text{VB}}\left(\text{h}^+_{\text{TR}}\right) \rightarrow \text{e}^-_{\text{CB}} + \text{heat} \tag{13.4}$$

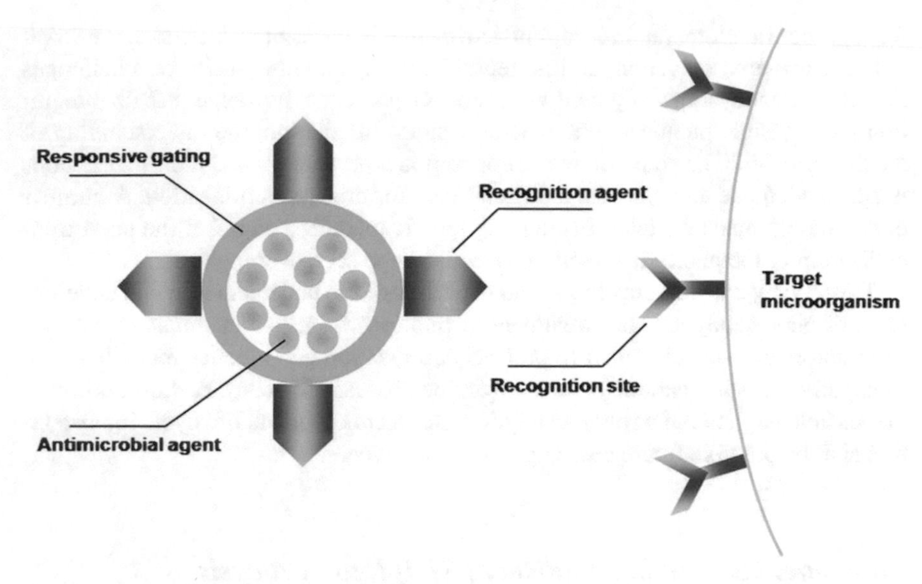

Fig. 13.3 Photoinduced formation mechanism of electron/hole pair in a semiconductor TiO_2 particle with the presence of water pollutant (P)

$$\text{Photoexcited e}^- \text{ scavenging}: (O_2)_{ads} + e^- \rightarrow O_2^{\cdot -} \tag{13.5}$$

$$\text{Oxidation of hydroxyls}: OH^- + h^+ \rightarrow OH^{\cdot} \tag{13.6}$$

$$\text{Photodegradation by OH}^{\cdot}: R-H + OH^{\cdot} \rightarrow R^{'\cdot} + H_2O \tag{13.7}$$

Direct photoholes:

$$R + h^+ \rightarrow R^{+\cdot} \rightarrow \text{Intermediate}(s) / \text{Final degradation products} \tag{13.8}$$

$$\text{Protonation of superoxides}: O_2^{\cdot -} + OH^{\cdot} \rightarrow HOO^{\cdot} \tag{13.9}$$

$$\text{Co-scavenging of e}^-: HOO^{\cdot} + e^- \rightarrow HO_2^- \tag{13.10}$$

$$\text{Formation of } H_2O_2: HOO^- + H^+ \rightarrow H_2O_2 \tag{13.11}$$

The e_{TR}^- and h_{TR}^+ in Eq. (13.4) represent the surface-trapped valence-band electron and conduction-band hole, respectively. It was reported that these trapped carriers are usually TiO_2 surface bounded and do not recombine immediately after photon excitation [101]. In the absence of electron scavengers (Eq. 13.4), the photoexcited electron recombines with the valence-band hole in nanoseconds with simultaneous dissipation of heat energy. Thus, the presence of electron scavengers is vital for prolonging the recombination and successful functioning of photocatalysis. Equation (13.5) depicts how the presence of oxygen prevents the recombination of electron/hole pair while allowing the formation of superoxide radical ($O_2^{\cdot-}$). This $O_2^{\cdot-}$ radical can be further protonated to form the hydroperoxyl radical ($HO_2^{\cdot}$) and subsequently H_2O_2 as shown in Eqs. (13.9) and (13.10), respectively. The $HO_2^{\cdot}$ radical formed was also reported to have scavenging property and thus the coexistence of these radical species can doubly prolong the recombination time of the h_{TR}^{+} in the entire photocatalysis reaction. However it should be noted that all these occurrences in photocatalysis were attributed to the presence of both dissolved oxygen (DO) and water molecules. Without the presence of water molecules, the highly reactive hydroxyl radicals ($OH^{\cdot}$) could not be formed and impede the photodegradation of liquid-phase organics. This was evidenced from a few reports that the photocatalysis reaction did not proceed in the absence of water molecules. Some simple organic compounds (e.g., oxalate and formic acid) can be mineralized by direct electrochemical oxidation where the e_{TR} is scavenged by metal ions in the system without water being present [37]. Although the h_{TR}^{+} has been widely regarded for its ability to oxidize organic species directly, this possibility has remained inconclusive. The h_{TR}^{+} is a powerful oxidant (+1.0 to +3.5 V against NHE), while e_{TR}^{-} is a good redundant (+0.5 to −1.5 V against NHE), depending on the type of catalysts and oxidation conditions.

Many elementary mechanistic studies on different surrogate organic compounds (e.g., phenol, chlorophenol, oxalic acid) have been extensively investigated in the photodegradation over TiO_2 surface. Aromatic compounds can be hydroxylated by the reactive $OH^{\cdot}$ radical that leads to successive oxidation/addition and eventually ring opening. The resulting intermediates, mostly aldehydes and carboxylic acids, will be further carboxylated to produce innocuous carbon dioxide and water. Since the photocatalysis reaction occurs on the photon-activated surface of TiO_2, the understanding of the reaction steps that involve photodegradation of organics is essential in the formulation of kinetic expression. For heterogeneous photocatalysis, the liquid-phase organic compounds are degraded to its corresponding intermediates and further mineralized to carbon dioxide and water, if the irradiation time is extended (Eq. (13.12)):

$$\text{Organic contaminants} \xrightarrow{TiO_2/hv} \text{Intermediate(s)} \rightarrow CO_2 + H_2O \qquad (13.12)$$

The overall photocatalysis reaction as portrayed by Eq. (13.12) can be divided into five independent steps, which are shown in Fig. 13.4 [93, 121]:

1. Mass transfer of the organic contaminant(s) (e.g., A) in the liquid phase to the TiO_2 surface
2. Adsorption of the organic contaminant(s) onto the photon-activated TiO_2 surface (i.e., surface activation by photon energy occurs simultaneously in this step)
3. Photocatalysis reaction for the adsorbed phase on the TiO_2 surface (e.g., A → B)
4. Desorption of the intermediate(s) (e.g., B) from the TiO_2 surface
5. Mass transfer of the intermediate(s) (e.g., B) from the interface region to the bulk fluid

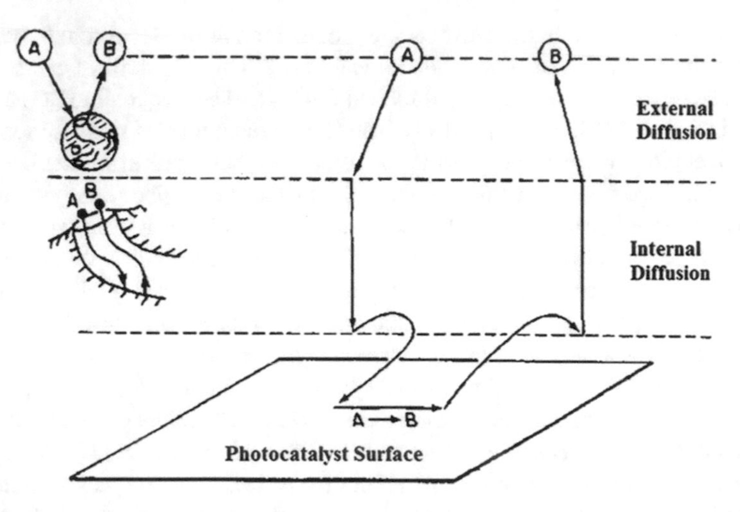

Fig. 13.4 Steps in heterogeneous catalytic reaction [93]

In terms of rate determination, the overall rate of reaction is equal to the slowest step. When the mass transfer steps (1 and 5) are very fast compared with the reaction steps (2, 3, and 4), the organic concentrations in the immediate vicinity of the active sites are indistinguishable from those in the bulk liquid phase. In this scene, the mass transfer steps are not rate limiting and do not affect the overall rate of photocatalytic reaction. Vinodgopal and Kamat [317] reported the dependence of the photodegradation rate of the organic surrogate on surface coverage of the photocatalysts used. This outlines the importance of molecule adsorption or surface contact with the catalyst during the photocatalytic degradation. If the mass transfer steps are rate limiting, a change in the aeration or liquid flow conditions past the TiO_2 photocatalyst may alter the overall photocatalytic reaction rate.

Similarly, the surface interaction of microorganisms with the catalyst used during the photo-disinfection is essential for enhancing the inactivation rate. When the generated ROS contacts closely with the microorganisms, the cell wall will be the initial site of attack [207]. The lipopolysaccharide layer of the cell external wall is the initial site attacked by the photoinduced ROS. This is followed by the site attack on the peptidoglycan layer, peroxidation of the lipid membrane, and eventual oxidation on the protein membrane. All these will cause a rapid leakage of potassium ions from the bacterial cells, resulting in direct reduction of cell viability. The decrease in cell viability is usually linked to the peroxidation of polyunsaturated phospholipid components of the cell membrane (i.e., loss of essential cell functions) and eventually leads to cell death. The formation of oxidative stress and its effects on the cell membrane can be observed using advanced atomic force microscopy or attenuated total reflection Fourier transform infrared spectroscopy. The rate of adsorption and the eventual photoinactivation is known to positively correlate to the bactericidal effect of TiO_2 catalyst. In this instance, the transfer of bacterial cell to the close vicinity of the surface-generated ROS site remains as the rate-limiting step in the photo-disinfection reaction.

Homogeneous Photo-Fenton Reaction

The Fenton reaction is a process that does not involve any light irradiation as compared with the heterogeneous TiO_2 photocatalysis reaction, whereas the photo-Fenton does react up to a light wavelength of 600 nm. It was first recognized in the 1960s and remains one of the most applied AOPs for its ability to degrade high loading of organic compounds in highly saline conditions [17, 202, 243]. Numerous studies on the photo-Fenton degradation of water pollutants such as chlorophenol, pesticides, and phenolic or aromatic compounds with organic loading of up to 25 g/L have been investigated [88, 107, 108, 134, 259]. A number of literatures have provided a comprehensive review of the basic understanding and clarity of the principles underlying the Fenton reaction [111, 243, 262].

In the absence of a light source, hydrogen peroxide (H_2O_2) will be decomposed by Fe^{2+} ions that are present in the aqueous phase, resulting in the formation of hydroxyl radicals. The photo-Fenton reaction is expedited when light source is present, causing rapid H_2O_2 decomposition by ferrous or ferric ions and resulting in the formation of radicals. All these soluble iron hydroxy or iron complexes can absorb not only UV radiation but also visible light. However, the actual oxidizing species responsible for the photo-Fenton reaction is still under discussion [42, 73]. These Fenton and photo-Fenton reactions could occur simultaneously with TiO_2 photocatalysis during UV-Vis irradiation period, post-TiO_2 photocatalysis period, or stand-alone photo-Fenton process. The Fenton reaction is seen to strongly correlate with the post-TiO_2 photocatalysis reaction and thus is described in detail here. The mechanism for the Fenton reaction is shown in Eq. (13.13):

$$Fe^{2+}(aq) + H_2O_2 \rightarrow Fe^{3+}(aq) + OH^- + HO^{\cdot} \tag{13.13}$$

The Fe^{2+} can be reverted back to Fe^{3+} via different mechanisms:

$$Fe^{3+}(aq) + H_2O_2 \rightarrow Fe^{2+}(aq) + HO_2^{\cdot} + H^+ \tag{13.14}$$

$$Fe^{3+}(aq) + HO_2^{\cdot} \rightarrow Fe^{2+}(aq) + O_2 + H^+ \tag{13.15}$$

When a light source is present, the rate of photo-Fenton was reported to be positively enhanced compared to the dark condition. This is mainly due to the regeneration of Fe^{2+} (aq) from the photochemical effect of light and the concurrent generation of the $OH^{\bullet}$ radicals in the system. Such a reversion cycle of $Fe^{2+}(aq) \rightarrow Fe^{3+}(aq) \rightarrow Fe^{2+}(aq)$ continuously generates $HO^{\bullet}$, provided that the concentration of H_2O_2 in the system is substantial. The regeneration of the Fe^{2+} (aq) from Fe^{3+} (aq) is the rate-limiting step in the catalytic iron cycle, if small amount of iron is present. This photoassisted reaction is termed as photo-Fenton reaction, where such reactions could be activated by irradiation wavelengths of up to 600 nm. It was known that this reaction is better functional under longer wavelengths as they are able to overcome the inner filter effects by photolyzing the ferric iron complexes. The inner filter effects refer to the competitive adsorption of photons by other light-absorbing species in the water.

Even if the photo-Fenton has higher photoactivity than the heterogeneous photocatalysis, its feasible operation is largely dependent on several water quality parameters. In the photo-Fenton reaction, the formation of the highly photoactive iron complexes is highly dependent on the water pH and ion content [75]. It was reported that pH 2.8 was the frequent optimum pH for photo-Fenton reaction [42, 85]. This is owing to the fact that at such low pH 2.8, the precipitation does not take place and further promotes the presence of dominant iron species of $[Fe(OH)]^{2+}$ in water. Such a low optimum pH 2.8, however, is not cost effective for operation as it requires high chemical costs for pH rectification. The presence of different ions such as carbonate (CO_3^{2-}), phosphate (PO_4^{3-}), sulfate (SO_4^{2-}), and chlorine (Cl^-) also affects the iron equilibrium in water. These ions have the potential to raise the water pH and effectively lower the photo-Fenton reaction rate. Both CO_3^{2-} and PO_4^{3-} have a double detrimental effect on the reaction, as they precipitate the iron as well as scavenge the OH^- radicals. A higher pH of 4.0–5.0 was determined to be sufficient to sustain the photo-Fenton reaction with 2–6 mM of iron for the initiation of the treatment [107]. To date, the maximal iron loading reported was 450 mg/L [249, 307].

Although H_2O_2 may be generated via the TiO_2 photocatalysis (Eq. 13.11), its relative amount in the system may be inadequate to drive the Fenton reaction. Many researchers have reported the addition of H_2O_2 in enhancing both the photo-Fenton and TiO_2 photocatalysis reactions. The H_2O_2 can inhibit the recombination of electron/hole pair while further providing additional OH^- radicals through the following mechanisms:

$$H_2O_2 + e^- \rightarrow HO^{\cdot} + HO^- \tag{13.16}$$

$$O_2^{\cdot-} + H_2O_2 \rightarrow O_2 + HO^{\cdot} + HO^- \tag{13.17}$$

This combined TiO_2 photocatalysis and photo-dark-Fenton reaction is particularly useful for the disinfection process [81, 210, 211]. The addition of H_2O_2 to the photocatalysis and dark-Fenton system results in a residual disinfection to avoid microbial regrowth. Rincón and Pulgarin [277] performed trials with TiO_2 photocatalysis and photo-Fenton reaction for the disinfection of water contaminated with *Escherichia coli*. They found that the bacterial inactivation rate was higher than the photocatalysis alone and the decrease in bacterial number continued in dark conditions without significant regrowth within the following 60 h. However, it was found that such residual disinfection effect was highly dependent on the light intensity used during the irradiation period, as well as the relative concentrations of Fe^{3+} and H_2O_2. Further addition of H_2O_2 was found to decrease the overall reaction rate in several studies, owing to the formation of less penetrative $HO_2^{\cdot}$ radicals, as described by Eq. (13.18):

$$HO^{\cdot} + H_2O_2 \rightarrow HO_2^{\cdot} + H_2O \tag{13.18}$$

Other combined processes of photo-Fenton and oxidative processes have also been proposed in the literature, such as ozone and ultrasound [23, 308]. However, their significance as compared to the TiO_2/photo-Fenton will not be discussed in detail.

Advancements in Photocatalyst Immobilization and Supports

Since the discovery of photocatalytic effect on water splitting by Fujishima and Honda (1972) using TiO_2 electrode, numerous researches have evolved to synthesize TiO_2 catalyst of different scales, characterize its physical properties, and determine its photooxidation performances to the surface-oriented nature of photocatalysis reaction [97, 127, 144, 160, 321]. The TiO_2 catalyst in nanodimensions allows having a large surface area-to-volume ratio and can further promote the efficient charge separation and trapping at the physical surface [238, 239]. The light opaqueness of this nanoscale TiO_2 catalysts was reported to have an enhanced oxidation capability compared to the bulk TiO_2 catalysts [292]. Although the nanoscale TiO_2 catalysts show considerable improvement in terms of their physical and chemical properties, their particle size and morphology remain the main problems in a large-scale water treatment process [38, 348]. In this section, the current technical challenges that prevent the application of slurry TiO_2 photocatalytic system are discussed together with the possible engineering solutions to resolve the problem. We will have a brief discussion on the modified TiO_2 catalyst with dopants for enhanced photoactivity under solar irradiation.

Challenges in the Development of Photocatalytic Water Treatment Process

To date, the most widely applied photocatalyst in the research of water treatment is the Degussa P-25 TiO_2 catalyst. This catalyst is used as a standard reference for comparisons of photoactivity under different treatment conditions [284]. The fine particles of the Degussa P-25 TiO_2 have always been applied in a slurry form. This is usually associated with a high volumetric generation rate of ROS as proportional to the amount of surface-active sites when the TiO_2 catalyst is in suspension [263]. On the contrary, the fixation of catalysts into a large inert substrate reduces the amount of catalyst-active sites and also enlarges the mass transfer limitations. Immobilization of the catalysts results in increasing the operation difficulty as the photon penetration might not reach every single surface site for photonic activation [263]. Thus, the slurry type of TiO_2 catalyst application is usually preferred.

With the slurry TiO_2 system, an additional process step would need to be entailed for post-separation of the catalysts. This separation process is crucial to avoid the loss of catalyst particles and introduction of the new pollutant of contamination of

TiO_2 in the treated water [338]. The catalyst recovery can be achieved through process hybridization with conventional sedimentation, cross-flow filtration, or various membrane filtrations [80, 90, 353, 357]. Coupled with the pH control strategy close to the isoelectric point for induced coagulation, it was reported that the microfiltration (MF) hybridization can recover the remaining 3% of the catalyst particles for reuse [205]. Several important operating issues with slurry TiO_2 still remain even with a membrane integration process. These include the types of membrane, pore size and blockage, regeneration or back-washing, and fouling [173, 227, 332]. A number of studies have utilized micron-size immobilizers for catalyst fixation that enhance surface contact with contaminants and prevent membrane fouling or pore blocking with rapid back-washing [332, 355]. These immobilizers include catalyst fixation onto activated carbon, mesoporous clays, fibers, or even the membrane itself [63, 165, 170, 358]. The following subsections outline a few catalyst immobilization strategies that are suitable for the use of slurry reactor or membrane reactor or both.

Mesoporous Clays

Natural clays have been used intensively as the support for TiO_2 owing to their high adsorption capacity and cost-effectiveness. Figure 13.5 shows the TiO_2 crystal being deposited on a clay material [63]. Different types of clays have been investigated, which include bentonite, sepiolite, montmorillonite, zeolite, and kaolinite [63, 100, 164, 299, 333]. Although these clays are catalytically inactive, their superior adsorption capacity has been attractive for increasing the surface contact during photocatalysis reaction. It was proposed that the natural clays should not be used directly to immobilize TiO_2. This is owing to the presence of different surface or lattice-bounded impurities that might diffuse and further affect the TiO_2 efficiency of the immobilized layer [63]. In addition, if these impurities are not removed, the polar molecules in the aqueous environment might initiate an internal reaction within the clay structure that results in clay swelling [63]. The swelling will be profound in certain type of clays, where van der Waals forces hold the entire clay in a turbostatic array. This is undesirable, particularly if the photocatalytic reactions take place in a reactor where the hydrodynamics may be strongly affected and consequently leading to the loss of photoactivity. Other factors that might need to be taken into consideration if pillared clays are used as the immobilizer substrate include the density of the clays, particle size distribution range, and complementary photoreactor system used. The use of mesoporous clays as the support for nano-size TiO_2 has been successfully demonstrated in a number of studies, including the slurry or membrane processes [62, 298].

Fig. 13.5 Nanocrystal of TiO_2 deposited on clay materials by SEM imaging. (**a**) 10 μm resolution; (**b**) 3 μm resolution

Nanofibers, Nanowires, or Nanorods

Glass, optical, carbon, titanate, and woven cloth fibers have also been studied as support materials in the photooxidation of various organic contaminants for water purification [263]. Most of these fibers have a protruded rod shape or longitudinal morphology. Using the nanofibers, nanowires, or nanorods, mass transfer limitation can be resolved by their thin longitudinal morphology. The use of less durable immobilizer fibers (e.g., of glass or woven cloths), however, may lead to a low durability as the deposited anatase crystals might wear off, resulting in a loss of photoactivity over reaction time. Such immobilizer fibers also increase pressure drop in the reactor system. On the contrary, a good benefit of nanofibers with commercial success is that they can be fabricated into MF, ultrafiltration (UF), and photocatalytic membranes (PMs). The MF fiber membrane is of particular interest as it shows high pollutant removal rate at low transmembrane pressure (<300 kPa). A commercial success of such fabricated MF and UF membranes has been demonstrated by Zhang and co-workers [353, 354]. Further details on the fabricated nanofibers or nanowire MF membranes can be obtained in the literatures.

Photocatalytic Membrane

Recently the use of PMs has been targeted owing to the photocatalytic reaction that can take place on the membrane surface and the treated water could be continuously discharged without the loss of photocatalyst particles. The PMs can be prepared from different materials and synthesis methods. These include the TiO_2/Al_2O_3 composite membranes [29, 54, 56, 350, 351]. TiO_2 is supported on polymer and metallic membranes or doted polymer membranes containing TiO_2 particles are entrapped within the membrane structure during the membrane fabrication process [9, 21, 22, 155, 157, 229]. Also, the possible TiO_2 organic and inorganic ceramic membranes have been investigated [155, 165, 344]. Figure 13.6 shows different types of PMs for water treatment application [4, 354]. In most studies, however, PMs may encounter various technical problems such as membrane structure deterioration, low photocatalytic activity, and loss of deposited TiO_2 layer over time. To prevent the problems associated with the TiO_2 membrane coating, an approach of using membranes without any deposited TiO_2 layer can be configured into a slurry-membrane hybrid system, which will be outlined in section 4.

Photocatalyst Modification and Doping

As TiO_2 photocatalytic reactions take place under ambient operating conditions, photoactivity is usually constrained by the narrow wavelength spectrum for photonic activation of catalysts. The higher end of UV spectrum required for catalyst activation is usually accompanied by high operating costs. One attractive option is to utilize the vast abundance of outdoor solar irradiation for catalyst activation in a suitably designed photoreactor system. To broaden the photoresponse of TiO_2 catalyst for solar spectrum, various material engineering solutions have been devised, including composite photocatalysts with carbon nanotubes, dyed sensitizers, noble metal or metal ion incorporation, and transition metal and nonmetal doping [99, 189, 244, 318, 349].

The rationale in utilizing these material engineering strategies is to balance both the half-reaction rates of the photocatalytic reaction by adding electron acceptor and modifying the catalyst structure and composition. Figure 13.7 presents the use of

Fig. 13.6 (**a**) FESEM images of the TiO_2 nanowire membrane [354]; (**b**) SEM images of TiO_2 nanotube layer formed in free-standing membrane [4]

different mechanisms to enhance the photoactivity of the catalysts. The presence of electron acceptors could scavenge the excited electrons and altogether prevent the recombination of electron/hole pairs. Recent studies have shown that modified TiO_2 catalysts have an enhanced photoactivity under solar irradiation [138, 177, 287]. CNTs coupling with TiO_2 have shown potential prolongation of electron/hole pairs by capturing the electron within their structure. As for dye-sensitized coupling, the excited dye molecules under solar illumination can provide additional electrons to

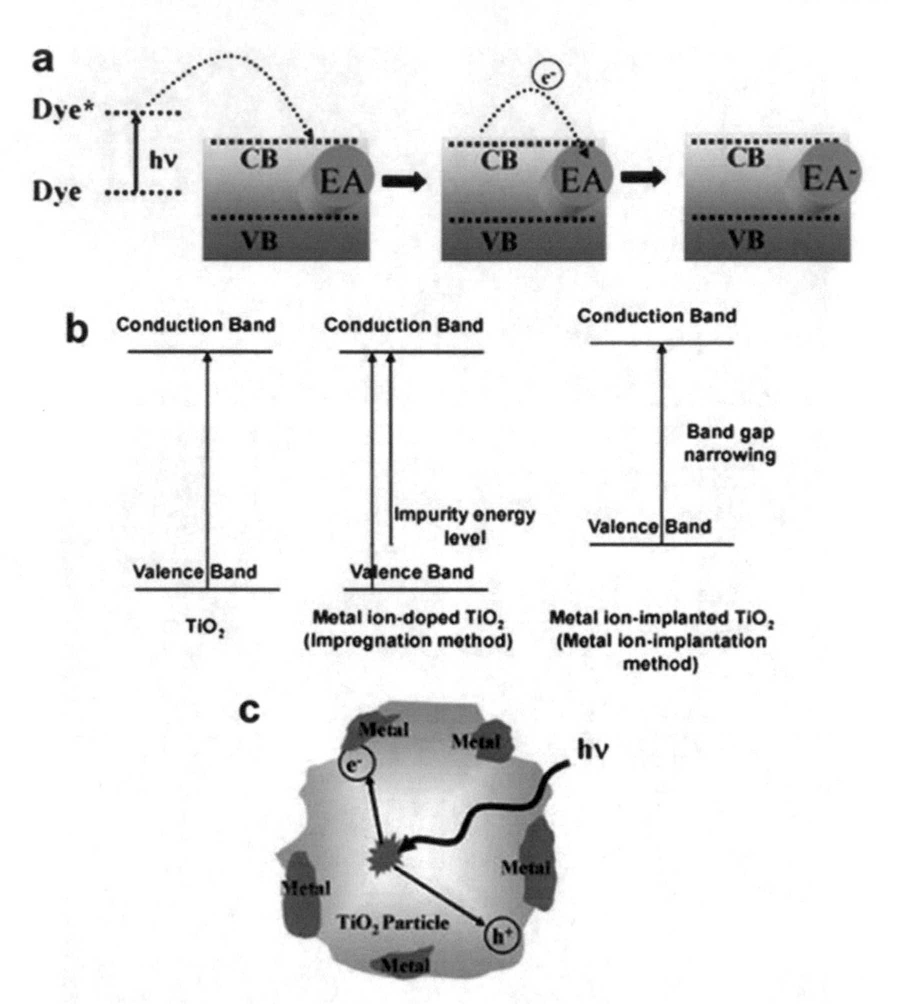

Fig. 13.7 (**a**) Steps of excitation with a sensitizer in the presence of an adsorbed organic electron acceptor (EA); (**b**) scheme of TiO_2 band structures, chemically ion-doped TiO_2 and physically ion-implanted TiO_2; (**c**) electron capture by a metal in contact with a semiconductor surface [205]

the CB for enhancing the formation of electron/hole pairs (Fig. 13.7a) [318]. Dyes such as methylene blue, Azure, erythrosine, rhodamine, and crystal violet have been widely functionalized under solar irradiation [318].

Similarly, noble metals (e.g., Ag, Ni, Cu, Pt, Rh, and Pd) with Fermi level lower than TiO_2 catalyst have also been deposited on the TiO_2 surface for enhanced charge separation (Fig. 13.7b) [244]. These metals were reported to enhance electron transfer, but require good knowledge on the optimal deposited amount needed during the fabrication process. Although noble metal coupling could be efficient in prolonging the surface charge separation, their cost-effectiveness for an industrial application is usually replaced by more economical transition or nonmetal doping. The mechanism of such transition and nonmetal doping, however, is different from the noble metal coupling as the TiO_2 is incorporated into the TiO_2 crystal lattice [10, 136, 137]. Such incorporation introduces impurity in the bandgap of TiO_2 and thus reduces the photonic energy requirements (Fig. 13.7c). More recently, the use of nonmetal dopants (e.g., N, C, F, S) can improve the photoactivity and feasibility of TiO_2 catalysts for industrial application [45, 310]. Further research efforts may be needed to get a better understanding of the photoactivity kinetics, so as to improve the photooxidation efficiency for water treatment.

Photocatalytic Reactor Configuration

Photocatalytic reactors for water treatment can generally be classified into two main configurations, depending on the deployed state of the photocatalysts: (1) reactors with suspended photocatalyst particles and (2) reactors with photocatalyst immobilized onto continuous inert carrier [264]. Various types of reactors have been used in the photocatalytic water treatment, including the annular slurry photoreactor, cascade photoreactor, and downflow contactor reactor [43, 62, 247]. The disparity between these two main configurations is that the first one requires an additional downstream separation unit for the recovery of photocatalyst particles while the latter permits a continuous operation.

Pareek et al. [255] addressed that the most important factors in configuring a photocatalytic reactor are the total irradiated surface area of catalyst per unit volume and light distribution within the reactor. Slurry-type photocatalytic reactor usually performs a high total surface area of photocatalyst per unit volume, while the fixed-bed configuration is often associated with mass transfer limitation over the immobilized layer of photocatalysts. The light photon distribution through either direct or diffuse paths within the reactors needs to be decided [42]. Direct photon utilization means that the photocatalysts are directly activated with light photon, rather with the assistance of various parabolic light deflectors to transfer the photons. To achieve uniformity in photon flux distribution within the reactor, a correct position of light source is essential to ensure maximal and symmetrical light transmission and distribution. The use of photoreactors with assisted parabolic light deflectors nowadays has become unfavorable, owing to the need of special configuration and high operating costs. This type of reactor needs to be specifically designed to ensure the maximal illuminated reactor volume with minimal pressure requirement for good catalyst mixing and dispersion. Until recently, the slurry photocatalytic reactor was still the preferred configuration owing to its high total surface area of photocatalyst per unit volume and ease of photocatalyst reactivation. The photocatalyst particles can be separated by settling tanks or external cross-flow filtration system to enable continuous operation of the slurry reactor. A technically promising solution for solving the downstream separation of photocatalyst particles after treatment is via the application of hybrid photocatalysis-membrane processes. Application of such a hybrid system prevents the use of a coagulation, flocculation, or sedimentation to separate the catalyst particles from the treated water stream. Other benefits include further energy saving and size of process installation and site area required.

The hybrid photocatalytic-membrane reactor system is generally known as the "photocatalytic membrane reactors" (PMRs). This is owing to the nature of the hybrid system wherein the membrane filtration unit could be configured into different positioning with the photocatalytic reactor. Fu et al. [95] designed a submerged membrane reactor (Fig. 13.8) with two different reaction zones: UV slurry TiO_2 zone with a movable baffle that separates the submerged membrane module. These PMRs can be generalized by (1) irradiation of the membrane module and (2) irradiation of a feed tank containing photocatalyst in suspension [226, 227]. For the former configuration, the photocatalyst could be either deposited onto the membrane or suspended in the reaction water. The PMRs allow a continuous operation of the slurry-type reactor without any loss of photocatalyst particles as well as to control the water residence time independently. This enables the treated water to achieve the predefined level before being filtered through the hybrid membrane system. In the PMRs with immobilized PMs, the membrane module functionalized as the support for the photocatalyst particles and barrier against the different organic molecules in the reaction water. Similarly, the membrane also acts as a physical barrier against the photocatalyst particles and organic molecules or intermediate compounds to be degraded in the slurry PMRs.

In the PMRs with immobilized photocatalysts, the photocatalytic reaction takes place on the surface of the membrane or within its pores. The PMs used may be of

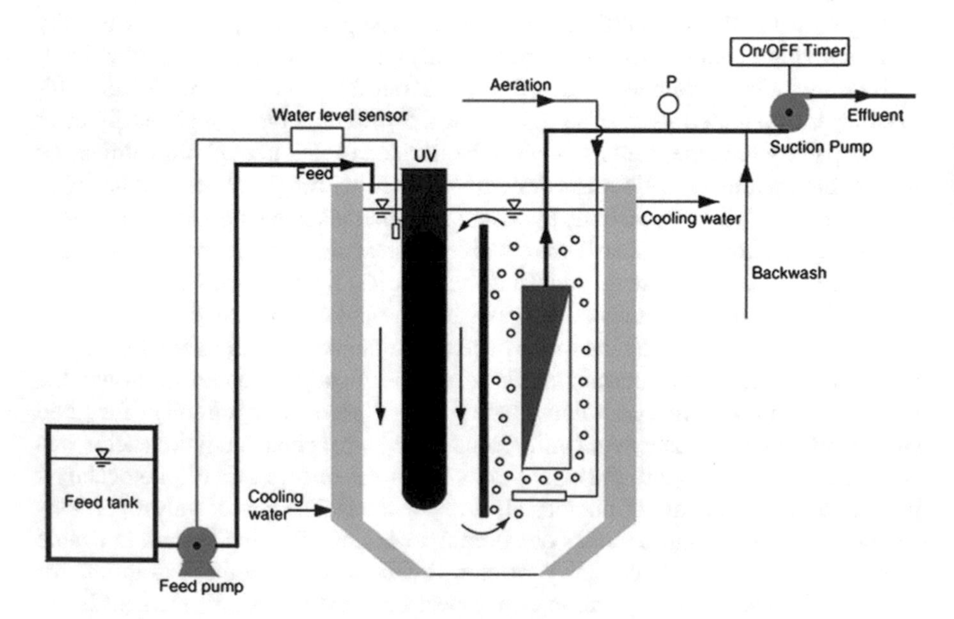

Fig. 13.8 Schematic of submerged membrane photocatalytic reactor [95]

MF [49, 132, 144, 222, 278, 280], UF [227, 293, 294, 298, 310], and nanofiltration (NF) [13, 226–229], depending on the targeted colloidal size and final water quality requirement. The MF membrane is useful when the colloidal size is in the range of 0.1–5 mm, while both UF and NF target a smaller particle size range. The photooxidation efficiency of the contaminants was reported to be higher when an immobilized PM was used, rather than in the case of PMRs with suspended catalyst particles [229]. It was, however, reported that immobilizing the photocatalyst particles might cause severe destruction to the membrane structure owing to their close contact with both UV light and hydroxyl radicals [48]. In view of this, the hybridization configuration of the membrane process using photocatalysts in suspension appears to be the more promising arrangement. This PMR configuration has been well described in the literature for water-phase degradation of humic and fulvic acids, bisphenol A, phenol, 4-nitrophenol, 4-chlorophenol, gray water, *para*-chlorobenzoate, river water, and dyes [48, 95, 132, 145, 222, 226, 227, 229, 278, 280, 293, 294, 298, 310].

With these PMRs, one of the main operational issues is the transmembrane pressure, which determines both the filtration rate and operating costs. It was known that the PMR treatment costs increase if the photocatalysts with small particle and colloidal size are used. With both the MF and UF membrane filtration, the fine photocatalyst particles can cause membrane fouling and subsequently reduce membrane permeate flux. Fu et al. [95] utilized a spherical ball-shaped TiO_2 particle that promotes separation, recovery, and reuse while prolonging the membrane life span as the particles do not cause pore blockage. Besides, the surface charge properties of the photocatalyst particles can also be manipulated to prevent membrane pore blockage. Xi and Geissen [332] integrated a thermoplastic membrane module of cross-flow MF and found that the low permeate flux occurred when the operating pH varied from the isoelectric point of the TiO_2 particles used. This is owing to the pH-induced coagulation–flocculation state of TiO_2 that declines the rate of permeate flux. This was resolved by maintaining the operating pH close to the isoelectric point of TiO_2 by adding certain electrolytes to the TiO_2 slurry. Even with such control strategies, the quality of permeate is low owing to the rapid penetration of small molecules through the membrane used.

Recently, different hybridizations of PMRs with dialysis, pervaporation, and direct-contact membrane distillation (MD) have been used (Fig. 13.9) [16, 41, 232]. Pervaporation is a physical process where usually a selective organophilic membrane was used to act as a selective barrier for the molecules to be degraded. Augugliaro et al. [14] observed that a synergistic effect occurs when pervaporation was used, where the intermediates from the degradation of 4-chlorophenol (i.e., hydroquinone, benzoquinone) could be selectively permeable without competing with 4-chlorophenol for photocatalytic reaction. Other types of organophilic membrane have also been investigated, such as polymeric UF membranes and polyethersulfonate NF membranes [18, 171]. A stronger rejection impact on the membrane is usually associated with the use of the UF or NF membranes. The choice of membrane for an efficient hybridization depends on the organic molecular size, pH, or electrostatic interaction and Donnan exclusion phenomenon [221].

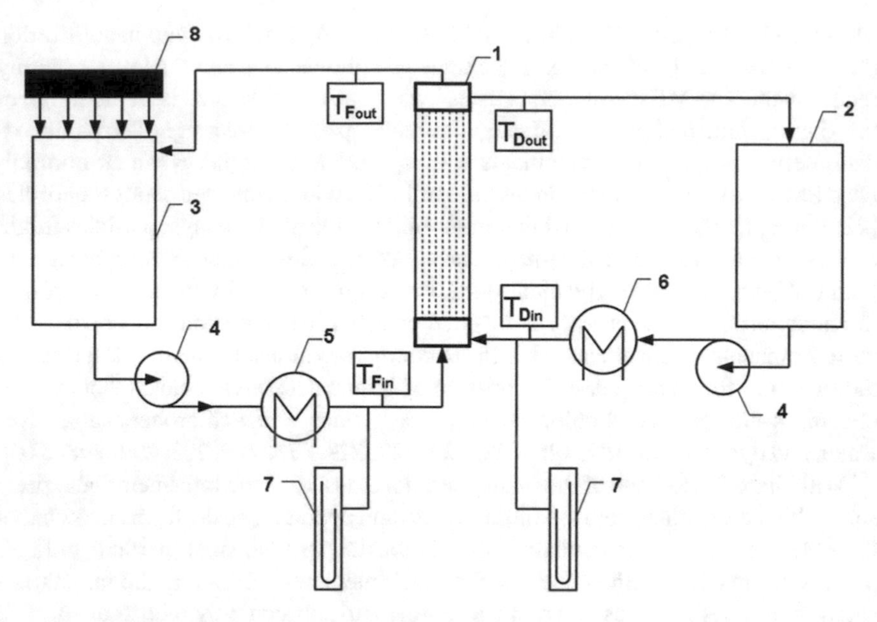

Fig. 13.9 Schematic diagram of the apparatus for hybrid photocatalysis-MD process: (1) membrane module; (2) distillate tank; (3) feed tank (*V* [2.9 dm^3]); (4) pump; (5) and (6) heat exchangers; (7) manometers; (8) UV lamp; T_{Fin}, T_{Din}, T_{Fout}, T_{Dout}—inlet and outlet temperatures of feed and distillate, respectively [232]

The MD is a process where the feed volatile components in water are evaporated through a porous hydrophobic membrane to produce high-quality distillate products [232]. Its main advantage is that there is no membrane fouling when TiO_2 is present. During the process, the volatile stream is maintained inside the membrane pores. The difference in vapor pressure on both sides of the porous membrane remains as a driving force for the process. This force, however, largely depends on the temperatures and solution composition in the layer adjacent to the membrane [113, 305]. It was reported that the feeding temperature in MD can range from 303 to 363 K. Similar scaling-up operational constraints of low permeate flux and high energy demand have been redundant in its current full-scale development. However, some efforts to utilize alternative energy source such as solar energy to enable MD application are found in the literature [31, 169].

Among all the hybrid PMR systems, the pilot Photo-Cat™ system (Fig. 13.10) (manufactured by Purifics Inc., Ontario, London) has shown the potential application. Benotti et al. [25] have evaluated its ability in the removal of 32 pharmaceuticals, endocrine-disrupting compounds, and estrogenic activity from water. They found that 29 targeted compounds and estrogenic activity of greater than 70% were removed while only 3 compounds were less than 50% removed at the highest number of UV passes. In the Photo-Cat™ system, the water stream passes through a pre-filter bag and a cartridge filter before being mixed with a nanoparticle TiO_2

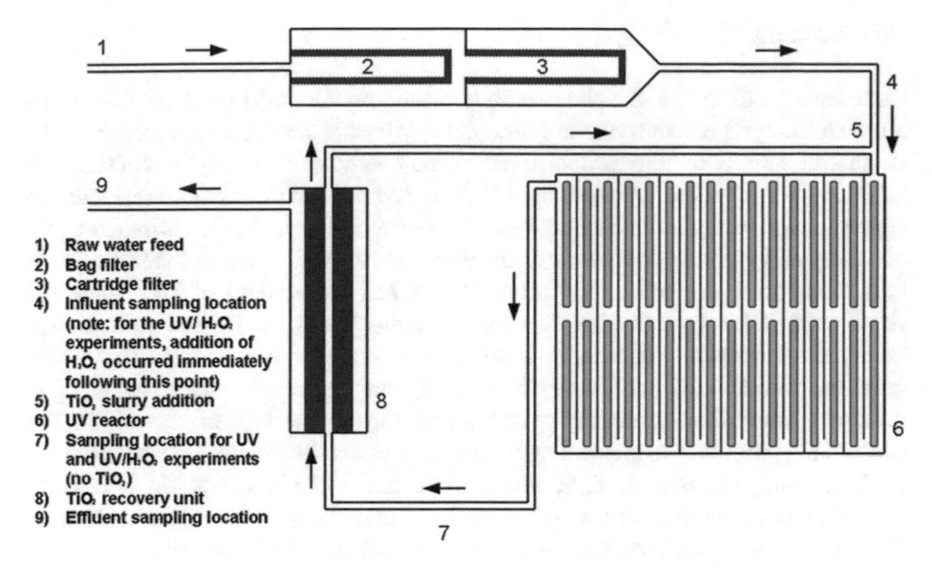

Fig. 13.10 General schematic of photocatalytic reactor membrane pilot system [25]

slurry stream. The mixed stream then passes through the reactor within the 3 mm annulus of the 32 UV lamps aligned in series, which can be individually controlled for the varying water quality. The overall hydraulic residence time for the 32 UV lamps passes between 1 and 32 s, depending on the number of UV lights being turned on. A cross-flow ceramic membrane TiO_2 recovery unit is hybridized downstream of the reactor to remove the TiO_2 from the flow stream, while allowing the treated water to exit. The retentate TiO_2 stream is recycled and remixed with the fresh TiO_2 slurry stream that enters the reactor stream.

Operational Parameters of the Photocatalytic Reactor

After the integration of the semiconductor catalyst with a photoreactor, the oxidation rates and efficiency of the photocatalytic system are highly dependent on a number of operation parameters that govern the kinetics of photomineralization and photo-disinfection. This section discusses the significance of each operation parameter on the corresponding kinetics and some recent methods to optimize the photocatalytic system via response surface analysis. The following section outlines a range of photoreactor operating parameters that affect the photocatalytic performance of TiO_2 photocatalysts in water treatment.

TiO_2 Loading

Concentration of TiO_2 in the photocatalytic water treatment system affects the overall photocatalysis reaction rate in a true heterogeneous catalytic regime, where the amount of TiO_2 is directly proportional to the overall photocatalytic reaction rate [103]. A linear dependency holds until certain extent when the reaction rate starts to aggravate and becomes independent of TiO_2 concentration. This is attributed to the geometry and working conditions of the photoreactor where the surface reaction is initiated upon light photon absorption [20]. When the amount of TiO_2 increases above a saturation level (leading to a high turbidity state), the light photon absorption coefficient usually decreases radially. However, such a light attenuation over the radial distance could not be well correlated with the Beer–Lambert law owing to the strong absorption and scattering of light photons by the TiO_2 particles [44]. The excess TiO_2 particles can create a light screening effect that reduces the surface area of TiO_2 being exposed to light illumination and the photocatalytic efficiency. Therefore, any chosen photoreactor should be operated below the saturation level of TiO_2 photocatalyst used to avoid excess catalyst and ensure efficient photon absorption. In this sense, both catalyst loading and light scattering effect can be considered as a function of optical path length in the reactor.

A large number of studies have reported the effect of TiO_2 loadings on the process efficiency [48, 61, 62, 103, 121, 247]. These results are mostly independent and a direct comparison cannot be made, as the working geometry, radiation fluxes, intensity, and wavelengths used were different. It was reported that the optimum catalyst loadings for photomineralization and photo-disinfection are varied, and mainly depend on the dimension of the photoreactor. In addition, the determination of photoreactor diameter is crucial in not only the effective photon absorption but also the water flow hydrodynamics [205]. Uniform flow region can ensure that a steady-state residence time is obtained, while turbulence flow removes catalyst deposition or reaction dead zone [204]. Reactor diameters smaller than 20–25 mm were not feasible for turbulent flow while diameter larger than 50–60 mm is impractical. This is because large diameters usually have lower saturated catalyst loading and efficiency. In this instance, the amount of catalyst should be considered. Usually the TiO_2 catalysts can be mixed uniformly with the targeted water prior to the introduction into the reactor system. During the dark homogenization period of the catalyst, a lower initial concentration of organic pollutants is observed owing to the strict adsorption of organics onto the catalyst surface [336]. Similarly, catalyst–bacterium interaction was reported in the photo-disinfection treatment of microorganisms [112].

pH

In heterogeneous photocatalytic water system, pH is one of the most important operating parameters that affect the charge on the catalyst particles, size of catalyst aggregates, and positions of conductance and valence bands. Due to the nature of

TiO_2 catalyst used, any variation in the operating pH is known to affect the isoelectric point or the surface charge of the photocatalyst used. Many reports have used the point-of-zero charge (PZC) of TiO_2 to study the pH impact on the photocatalytic oxidation performance [48, 61, 62, 247, 306]. The PZC is a condition where the surface charge of TiO_2 is zero or neutral that lies in the pH range of 4.5–7.0, depending on the catalysts used. At PZC of TiO_2, the interaction between the photocatalyst particles and water contaminants is minimal due to the absence of any electrostatic force. When operating pH < PZC(TiO_2), the surface charge for the catalyst becomes positively charged and gradually exerts an electrostatic attraction force toward the negatively charged compounds. Such polar attractions between TiO_2 and charged anionic organic compounds can intensify the adsorption onto the photon-activated TiO_2 surface for subsequent photocatalytic reactions [112, 275, 336]. This is particularly significant when the anionic organic compounds are present at a low concentration level. At pH > PZC(TiO_2), the catalyst surface will be negatively charged and will repulse the anionic compounds in water. Different pH will affect the surface charge density of the TiO_2 catalyst, according to the following water equilibrium equations (Eqs. 13.19 and 13.20):

$$\text{At pH} < \text{PZC}: \text{TiOH} + \text{H}^+ \leftrightarrow \text{TiOH}_+^2 \quad (13.19)$$

$$\text{At pH} > \text{PZC}: \text{TiOH} + \text{OH}^- \leftrightarrow \text{TiO}^- + \text{H}_2\text{O} \quad (13.20)$$

The surface charge density distribution for these TiO_2 catalyst clusters is highly dependent on the operating pH. It was reported that the distribution of TiOH is ≥80% at 3 < pH < 10; $TiO^- \geq 20\%$ at pH > 10; and $TiOH_2^+ \geq 20\%$ at pH < 3. The equilibrium constants for these reactions at different pH are $pK_{TiOH_2^+} = 2.4$ and $pK_{TiOH} = 8.0$ [162]. During photocatalytic reaction, the initial operating pH usually drops slightly from the formation of multitude intermediate by-products that may pose different chemical functional groups and affect the water pH indifferently [295].

A similar electrostatic interaction enhancement for photo-disinfection of microorganisms was observed during the photocatalytic process [112]. During the photodisinfection, the initial photoinduced damage to the microorganisms takes place on the lipopolysaccharide layer of the external cell wall and on the peptidoglycan layer. This is followed by lipid membrane peroxidation and the subsequent oxidation of the membrane protein and polysaccharides. An increased density of $TiOH_2^+$ (at low pH) can form electrostatic link with the bacteria of negatively charged surfaces, resulting in increasing rate of photo-disinfection. Herrera Melián et al. [119] observed that the bacterial inactivation rate was enhanced at pH 5.0. It should be noted that the enhanced bactericidal activity of TiO_2 at a low pH is due solely to the TiO_2-mediated photo-killing and not acidification of the cell. Heyde and Portalier [123] explained that the negligible *E. coli* reaction to acid conditions was from the presence of an acid tolerance response to the bacterium itself, which secreted acid-induced proteins for acid-shock protection. However, Rincón and Pulgarin [277] did not find any differences in *E. coli* inactivation rates when the initial pH varied between 4.0 and 9.0. To date, various types of microorganisms have been successfully inactivated using

TiO_2 photocatalysis, which include *Lactobacillus acidophilus*, *E. coli*, *Saccharomyces cerevisiae*, *Chlorella vulgaris*, *Streptococcus faecalis* and *aureus*, *Enterobacter cloacae*, total coliforms, *Candida albicans*, *Fusarium solani*, *Aspergillus niger*, and others [119, 135, 196, 212, 336].

Although the electrostatic link between the catalyst particles and microorganisms was reported to exist, a subsequent microbial cell adsorption and eventual penetration of the catalysts through the cell wall are highly dependent on the mean particle size. Since the relative sizes between the bacteria and catalysts are significantly different, the charged $TiOH_2^+$ clusters might not come into full contact with the bacteria. Sichel et al. [291] found that the TiO_2 catalyst is actually adsorbed onto fungal spores rather than the reverse setting. This is owing to the fact that the fungal spores are larger (a few order of magnitudes) than the catalyst particles. A small decrease in microbial loadings was observed during the initial dark homogenization period with catalyst particles. This is owing to the catalyst agglomeration and subsequent sedimentation of the agglomerates. The rate of bacterial adsorption in this instance was found to directly relate to the bactericidal activity of the catalyst used. It was also reported that the adsorption phenomena during the dark and irradiated phase will act indifferently and thus a strong conclusion cannot be made.

Besides, the interaction between the catalyst particles itself also exists and is dependent on the operating pH. The particle size of the same catalysts can vary from 300 nm to 4 mm depending on the distance from the PZC of TiO_2 [205]. At pH = PZC, the neutral surface charge of the catalyst particles is unable to produce the interactive rejection for solid–liquid separation. Thus, this induces catalyst aggregation where the catalyst becomes larger, leading to catalyst sedimentation [28]. This physical property is usually manipulated in the hybridized PMR system, where the pH of the treated wastewater is neutralized to pH 7 for the subsequent recovery of catalyst particles. The larger TiO_2 clusters can settle faster than the smaller one. It was reported that with such neutralization strategy, almost 97% of the catalysts can be recovered in the settling tank. The remaining TiO_2 catalysts can be recovered via the downstream MF system. Similarly, the water pH will also affect the effective separation in the PMR system where inappropriate control of water pH promotes the electrostatic repulsion, the Donnan exclusion phenomena, and thus the rejection tendency for the membrane used [225, 283]. It must be stressed that appropriate pH control strategies must be implemented at every different location of a photocatalytic water treatment process for efficient photocatalytic reaction to proceed.

Temperature

Numerous studies have been conducted on the dependence of photocatalytic reaction on the reaction temperature [45, 87, 96, 236, 274]. Although heat energy is inadequate to activate the TiO_2 surface, the understanding on such dependency could be extrapolated when operating the process under natural sunlight illumination. Most of the previous investigations stated that an increase in photocatalytic reaction temperature (>80 °C) promotes the recombination of charge carriers and disfavors the adsorption of organic compounds onto the TiO_2 surface [103]. At a reaction temperature greater than 80 °C, the photocatalytic reaction is interpreted with Langmuir–Hinshelwood (L–H) mechanism where the adsorption of reactants is disfavored resulting in KC becoming "1." This will reduce the L–H expression (Eq. 13.23) into the apparent rate equation $r = k_{\text{apparent}}C$. All these drastically reduce the photocatalytic activity of TiO_2 when the reaction temperature rises. The desorption of degraded products from the TiO_2 surface is the rate-limiting step when temperatures rise. On the contrary, a low temperature below 80 °C actually favors adsorption which is a spontaneous exothermic phenomenon, resulting in getting KC of L–H model "1," enhancing the adsorption of final reaction products. A further reaction in temperature down to 0 °C will cause an increase in the apparent activation energy. As a consequence, the optimum reaction temperature for photomineralization is reported to be in the range of 20–80 °C [205].

For photo-disinfection using TiO_2 photocatalysis, the increase in the reaction temperature increased the inactivation rate of microorganisms [274]. This is consistent with the van't Hoff–Arrhenius equation (Eq. 13.21), where the rate constant k is linearly proportional to the exponential ($-1/T$):

$$\ln\left(\frac{k_1}{k_2}\right) = \frac{E_a}{R}\left(\frac{1}{T_2} - \frac{1}{T_1}\right) \quad (13.21)$$

in which k_1 and k_2 are the constants for temperatures T_1 and T_2, E_a is the energy of activation, and R is the universal gas constant. The viability of a microbe to the catalyst activity depends on its incubation temperature, type, and resistance to temperature change. The order of resistance of microorganisms to conventional disinfection treatment is as follows: non-spore-forming bacteria < viruses < spore-forming bacteria < helminths < protozoa (oocysts). To date, there is no comprehensive study conducted to compare the effect of TiO_2 photo-disinfection on each microorganism type under different operating temperatures. Thus, the photo-disinfection using TiO_2 catalyst is usually conducted below ambient temperature of 80 °C to prevent high water heating costs (high heat capacity) [120].

Dissolved Oxygen

Dissolved oxygen (DO) plays an important role in TiO_2 photocatalysis reaction to assure that sufficient electron scavengers are present to trap the excited conduction-band electron from recombination [62]. The oxygen does not affect the adsorption on the TiO_2 catalyst surface as the reduction reaction takes place at a different location from where oxidation occurs [205]. Other roles for DO may involve the formation of other ROS, stabilization of radical intermediates, mineralization, and direct photocatalytic reactions. The total amount of DO in a reactor depends on a few technical considerations. For a photoreactor, the total delivered DO not only acts as an electron sink but also provides sufficient buoyant force for complete suspension of TiO_2 particles. Photoreactor sparging with pure oxygen in TiO_2 slurry reactor is usually a cost-ineffective solution, as the amount of DO being held up is a function of the photoreactor geometry. The difference between the two sparging media of air and oxygen is usually not very drastic as the mass transfer of oxygen to the close vicinity of the surface is the rate-dependent step [118]. Generally Henry's law can be assumed to give a good approximation of the amount of oxygen dissolved under the experimental conditions, provided that the oxygen sparging rate and the photoreactor gas holdup are known [61]. In this equilibrium law, it is also necessary to account for the decrease in oxygen solubility with increasing reaction temperature. As discussed, it is preferential to operate the photoreactor under ambient conditions to prevent elevation of cost of air or oxygen sparging for enhanced electron sink.

The presence of dissolved oxygen is also suggested to induce the cleavage mechanism for aromatic rings in organic pollutants that are present in water matrices. Wang and Hong [322] proposed that during aromatic ring cleavage in the degradation of a dioxyl compound, the molecular oxygen will follow the cleaving attack of a second hydroxyl radical. The analogous partial pressure of oxygen applied during the reaction for a closed reactor system is also important. A significantly higher partial pressure of oxygen will result in a higher initial organics photomineralization rate than its structural transformation [45, 322]. It was reported that at low oxygen pressure of 0.5 kPa, 75% of the original 2-CB compounds were transformed, but only 1% of them was mineralized to CO_2 after 5 h of UV irradiation [322]. Though elevated partial pressure of oxygen is imperative, it is difficult to quantify their specific influence on the surface activity of TiO_2, owing to the polyphasic nature of the photocatalytic liquid-phase reaction [121].

The detrimental effect of DO on the photocatalysis reaction was reported by Shirayama et al. [290]. They observed an elevated photodegradation rate of chlorinated hydrocarbons in the absence of DO. This could be explained by the strong absorption characteristics for UV photons at intensities of 185 nm and 254 nm, respectively. The DO molecules act as an inner filter in this case and cause a sharp attenuation in UV light intensity mainly at UV-C germicidal region. There is no report found to determine the DO impact on a photocatalysis process using UV-A or UV-B light sources. To date, the effect of DO on the efficiency of photo-disinfection rate has been paid little attention. The formation of various ROS under the series of redox reactions on the catalyst surface was assumed to be similar in both photomineralization and photo-disinfection reactions. If sufficient nutrients are available, the constant sparging of DO will generally promote microbial growth and offset the photo-disinfection rate. This indirectly prolongs the irradiation time necessary to achieve the desired inactivation level. It is thus recommended that the effect of DO on the microbial inactivation should be investigated thoroughly in a particular photoreactor system before improvising the DO sparging strategy.

Contaminants and Their Loading

Previous investigations have reported the dependency of the TiO_2 photocatalytic reaction rate on the concentration of water contaminants [61, 62, 247, 306]. Under similar operating conditions, a variation in the initial concentration of water contaminants will result in different irradiation times necessary to achieve complete mineralization or disinfection. Owing to the photonic nature of the photocatalysis reaction, excessively high concentration of organic substrates is known to simultaneously saturate the TiO_2 surface and reduce the photonic efficiency leading to photocatalyst deactivation [281].

Not all organic substrates will have such a profound effect on the irradiation time, and this also depends on the corresponding chemical nature of the targeted compounds for TiO_2 photocatalysis reaction. For instance, 4-chlorophenol will undergo a degradation pathway with constant evolution of intermediate(s) product (i.e., hydroquinone and benzoquinone) while oxalic acid will undergo direct transformation to carbon dioxide and water [19]. In the case of 4-chlorophenol, such evolution of intermediate(s) will further prolong the irradiation time necessary for total mineralization owing to the direct competition over unselective TiO_2 surfaces. In the development of mathematical model that represents the kinetics of mineralization while relating to the TiO_2 loading required, commonly used water quality parameters such as chemical oxygen demand (COD), total organic carbon (TOC), or dissolved organic carbon (DOC) could be more appropriate to account for such competitiveness of intermediate(s) with its predecessor compounds. Also organic substrates with electron-withdrawing nature such as benzoic acid and nitrobenzene were found to strongly adhere and to be more susceptible to direct oxidation than those with electron-donating groups [27]. Most of the TiO_2 studies conducted to date utilize a range of model organic substrates with different substituent groups but these rarely convey any useful information and merely test the photo-efficiency of a new photocatalyst or an integrated reactor column. Some field kinetics of the photomineralization of real wastewater has also been reported [206, 209, 268, 316]. It was observed that owing to the persistency of the dissolved organic in the real wastewater, a slow photomineralization kinetics is attained with prolonged irradiation times to achieve complete mineralization. Slow kinetics turnover of the photocatalytic water treatment as a stand-alone process means that a higher initial cost on the reactor volume and site area is required. Recently, this heterogeneous photocatalytic technology has been coupled with biological treatment to increase its industrial feasibility [256]. Such coupling allows the retention time in biological treatment stages to be reduced, where the nonbiodegradable compounds of the wastewater can be turned into biodegradable compounds with the aid of photocatalytic treatment.

Similarly, the photo-disinfection efficiency of various microorganisms has been assessed for the possible application of photocatalytic technology to replace the chemical disinfectant methods. In general, the mechanism involved in the microbial disinfection includes the destruction of the microbial protein structures and inhibition of their enzymatic activities [207]. Compared to the persistency during organic photomineralization, a general classification in the bacterial resistance to the disinfectant used has also been proposed. Among all, the most resistance infectious type of microorganisms is prions, followed by coccidia (*Cryptosporidium*), bacterial endospores (*Bacillus*), mycobacteria (*M. tuberculosis*), viruses (poliovirus), fungi (Aspergillus), and Gram-negative (*Pseudomonas*) and eventually Gram-positive bacteria (*Enterococcus*) [205]. Their differences in resistance are explained by the cell wall permeability, size, and complexity of the specific microorganisms. Each microorganism might also be of infectious nature, which causes epidemic diseases when they multiply in water. Most bacteria can be killed easily with TiO_2 photocatalysis, but a complete inactivation might have to be ensured as they are highly infectious. Similarly, this infectious nature can also be found in viruses (adenoviruses, enteroviruses, hepatitis A and E viruses, noroviruses and sapoviruses, rotaviruses) and the most in protozoa (*Acanthamoeba* spp., *Cryptosporidium parvum*, *Cyclospora cayetanensis*, *Entamoeba histolytica*, *Giardia intestinalis*, *Naegleria fowleri*, *Toxoplasma gondii*) [326]. All these protozoa are highly infectious in low concentration and the photocatalytic treatment should be targeted on these microorganisms as the surrogate indicators. This is to ensure adequate photocatalytic treatment to prevent the outbreak of the epidemic diseases in the treated water, if photocatalytic treatment is chosen.

Light Wavelength

The photochemical effects of light sources with different wavelength-emitting ranges will have a profound consequence on the photocatalytic reaction rate, depending on the types of photocatalysts used—crystalline phase, anatase-to-rutile composition, and any state of photocatalyst modifications. Using commercial Degussa P-25 TiO_2, which has a crystalline ratio of anatase 70/80:20/30, a light wavelength at $\lambda < 380$ nm is sufficient for photonic activation [19, 121]. The crystalline phase of rutile TiO_2 has a smaller bandgap energy of $E_B \sim 3.02$ eV, compared to the anatase TiO_2 of 3.2 eV [103, 127, 144]. This dictates that rutile TiO_2 can be activated with light wavelength of up to 400 nm, depending on the bandgap threshold for the type of rutile TiO_2 used.

For UV irradiation, its corresponding electromagnetic spectrum can be classified as UV-A, UV-B, and UV-C, according to its emitting wavelength. The UV-A range has its light wavelength spanning from 315 to 400 nm (3.10–3.94 eV), while UV-B has a wavelength range of 280–315 nm (3.94–4.43 eV) and the germicidal UV-C ranges from 100 to 280 nm (4.43–12.4 eV) [51, 84]. In most of the previous studies, the UV-A light provides light photons sufficient for photonic activation of the catalyst [27, 48, 247]. As with outdoor solar irradiation, the UV-C is usually absorbed by the atmosphere and does not reach the earth surface. Only the lamp-driven photoreactor system can utilize UV-C irradiation artificially for photonic activation of catalyst and reduction of viable microorganisms. The mechanism of UV-C cell destruction involves the direct induction on pyrimidine and purine and pyrimidine adducts on the cell DNA. However, not all microorganisms are susceptible to the UV-C radiation and some highly resistant microorganisms can survive through disinfection process. These include *Legionella pneumophila* and *Cryptosporidium parvum* oocysts [205].

The natural UV radiation that reaches the surface of the earth consists of both UV-A and UV-B spectrums. The photolysis mechanism for both UV irradiations on cell inactivation is dissimilar to the discussed UV-C mechanism. Both UV-A and UV-B irradiations can be absorbed by cellular components called intracellular chromophores. L-Tryptophan is the best known intracellular chromophore and is thought to contain unsaturated bonds such as flavins, steroids, and quinines [311]. Among these UV irradiations, the UV-A irradiation is toxic only in the presence of oxygen. The ROS or oxidative stress generated from the chromophore light absorption can damage cells and cell components, leading to lipid peroxidation, pyrimidine dimer formation, and eventually DNA lesions. The contact between the ROS and DNA results in single-strand breaks and nucleic acid modifications. Such damages on the DNA are usually lethal or mutagenic irreversible. With the presence of TiO_2 catalyst as the light sensitizers, a high degree of cell damage is seen as the amount of ROS generated increases accordingly. A few microorganisms that are resistant to UV-A photolysis have been inactivated successfully by TiO_2 photocatalysis, namely *E. cloacae*, *E. coli*, *P. aeruginosa*, and *S. typhimurium* [90].

A longer wavelength of solar irradiation ($l > 400$ nm) has also been used in solar disinfection (SODIS) study [26, 151, 196, 220, 291]. However, the photo-killing mechanism is as yet unclear as it involves a variety of microbial and a larger mixed spectrum of UV-A and solar irradiation. A similar cell destruction mechanism to the one proposed for UV-A irradiation is thought to take place in this mixed light spectrum. In the SODIS, the pathogens in the drinking water contained in PET bottles were found to be inactivated within 6 h of sunlight exposure. However, significant research and developments on the disinfection using photocatalytic mediated process need to be conducted to broaden the photoactivity of current TiO_2 catalysts used.

Light Intensity

The photonic nature of the photocatalysis reaction has outlined the dependency of the overall photocatalytic rate on the light source used. Light intensity is one of the few parameters that affect the degree of photocatalytic reaction on organic substrates. Fujishima et al. [98] indicated that the initiation of TiO_2 photocatalysis reaction rates is not highly dependent on light intensity, where a few photons of energy (i.e., as low as 1 mW/cm) can sufficiently induce the surface reaction. To achieve a high photocatalytic reaction rate, particularly in water treatment, a relatively high light intensity is required to adequately provide each TiO_2 surface-active site with sufficient photon energy required. However, when using the nominal TiO_2 particles without modifications, the surface reaction is restricted to photons with wavelengths shorter than the absorption edge of approximately 400 nm. The organic conversion in the presence of UV wavelength ($l < 400$ nm) in many studies obeyed the linear proportionality correlation to the incident radiant flux. This was evidenced by Glatzmaier et al. [110] and Glatzmaier [109], where they observed that the destruction of dioxin and polychlorinated biphenyls was significantly enhanced in the presence of high-intensity photons. A similar finding was reported in Magrini and Webb [203] where the organic decomposition rate was reported to increase with the radiation intensity.

Later, it was discovered that the dependency of the reaction rate on radiant intensity behaves indifferently under different lighting conditions [71, 149, 266]. The linear dependency of the photocatalytic reaction rate on radiant flux (∅) changed to a square root dependency ($\varnothing^{0.5}$) above certain threshold value. Such a shift in dependency form was postulated owing to the amount of photo-generated holes available during the electron/hole pair formation. In the TiO_2 catalyst used, the photoinduced generation of valence-band holes is much less than the conduction-band electrons available. In this instance, the photo-generated holes are the rate-limiting step and the detailed derivation of the square root dependency can be obtained from Malato et al. [205]. At high intensities, the dependency of the photocatalytic reaction rate on radiant flux reduced to zero ($\varnothing^0$). This was explained by the saturated surface coverage of the catalyst, resulting in a mass transfer limitation in the adsorption and desorption, thus preventing the effect of light intensity to set in. An increase in the fluid turbulency in this case might help to alleviate the mass transfer problem on the surface of the catalyst. The desorbed final products might also affect the dependency of reaction rate on radiant flux, as they might scavenge the electron acceptors and further promote the electron/hole pair recombination.

Rincón and Pulgarin [275] reported that the residual disinfecting ability of the photocatalyst largely depends on the duration of light intensity without any temporal interruptions. They investigated the effect of light intensities at 400 and 1000 W/m^2 on bacterial lethality and regrowth, and found that the higher intensity without any temporal interruptions can cause irreversible damage to the *E. coli*. In the intermittent light irradiations with constant interruptions, the bacteria were seen to regrow during the subsequent 24 or 48 h. Some studies suggested that this regrowth is due to the dark-repair mechanism where the partially damaged cells recover in the presence of nutrients [288]. Others have suggested that the damaged but not totally inactivated cells could recover its viability through photo-repairing under radiation of 300–500 nm or through resynthesis and post-replication of cells [274, 291]. It must be noted in this case that for photo-disinfection using different light intensities, a final conclusive point cannot be made directly. The disinfection results of 400 W/m^2 at 2.5-h irradiation might not be the same as the result that arose from 1000 W/m^2 for 1 h. Thus, in order to predict the minimum irradiation required at constant irradiance, preliminary studies into both the photoreactor performance and microbial consortia (different resistance) present are important.

Response Surface Analysis

From the earlier discussions on the effect of the operation parameters on the photocatalytic reaction rate, it can be seen that these parameters would affect the system indifferently. Overall, it can be interpreted that a multivariable (MV) optimization approach is actually required to optimize a photoreactor system as parameter interaction might exist. Parameter interactions refer to the relationship between operating parameters such as TiO_2 loading on pH or pH on radiant flux. For the optimization of a photoreactor system, the conventional one-parameter-at-a-time approach is mostly used to unveil the effects of one parameter after another. Although this conventional optimization approach is widely acceptable, the reported outcomes could be of insignificance and have less predictive power if the condition for one operating parameter changes.

This has led to the application of effective design of experiments (DOE), statistical analysis, and response surface analysis for photocatalytic studies [40, 60, 161, 190, 194]. Using this approach, different permutations of experimental design are involved and the operational parameters and spans are defined. As compared to the conventional one-parameter-at-a-time approach, the MV optimization approach has predetermined experimental points that are dispersed uniformly throughout the study domain; that is, only a small region is covered in the domain of conventional study. This allows the optimization process to be more time effective and enhances the identification of parameter interactions, where they can be interpreted using commercial statistical software such as Design Expert® software.

Chong et al. [60] proposed the use of Taguchi-DOE approach, together with the analysis of variance, statistical regression, and response surface analysis, to study the combined effects of four key operation parameters that affect the photocatalytic reaction rate in an annular photoreactor. They utilized 9 experimental permutations to analyze the 81 possible parameter combinations. It was reported that the interaction between the TiO_2 loading and aeration rate had a positive synergistic effect on the overall reaction rate. A response surface model was developed to correlate the reaction rate dependency on the four different parameters according to the statistical regression as shown in Eq. (13.22):

$$R_o = b_0 + \sum_{i=1}^{k} b_i X_i + \sum_{i=1}^{k} b_{ij} X_i^2 + \sum_{i_i<j}^{k} \sum_{j}^{k} b_{ij} X_i X_j \quad (13.22)$$

where R_o is the predicted response output of the photomineralization rate, and I and j are linear and quadratic coefficients, respectively. The parameters of b and k are regression coefficient and number of parameters studied in the experiment, respectively, and X_i and X_j ($i = 1, 4; j = 1, 4, i \neq j$) represent the number of independent variables in the study. This model (Eq. 13.22) is empirical and independent for each photoreactor system. Subsequent verification works are required to determine the accuracy and applicability of such a model for the prediction of photoreaction rate under the variation of its parameters. Other DOE approaches to photocatalytic reactor optimization have also been applied over the years, such as central composite design, Bayesian design, and Plackett–Burman designs [50, 140, 319].

Kinetics and Modeling

Kinetics and mechanistic studies on the photomineralization or photo-disinfection rate of the water contaminants are useful for process scaleup. The appropriate utilization of kinetic models for the interpretation of experimental data enables the design and optimization of photoreactor systems with sufficient capacity and minimal non-illuminated reactor volume. In this section, the different kinetics and rate model for both photomineralization and photo-disinfection are discussed along with the common misconceptions in kinetic modeling.

Photomineralization Kinetics

In most of the photocatalytic studies, kinetic or mechanistic studies over the irradiated TiO_2 surfaces usually only involve a single constituent model organic compound. The kinetics of different organic compounds ranging from dye molecules, pesticides, herbicides, and phenolic compounds to simple alkanes, haloalkanes, aliphatic alcohols, and carboxylic acids have been investigated [19, 103, 121, 205]. The nonselective nature of the $OH^{\bullet}$ radicals means that the disappearance rate of the studied compound with irradiation time should not be referenced as a standard for reactor design purpose. This is because numerous intermediates are formed en route to complete mineralization and neglecting this aspect is a common mistake in portraying the photomineralization kinetics. In this instance, the organic concentrations can be expressed collectively in COD or TOC to yield an in-depth understanding on the photomineralization kinetics.

As the L–H model is surface area dependent, the reaction rate is expected to increase with irradiation times since less organic substrate will remain after increased irradiation times with higher surface availability. A zero rate of degradation is associated with the total decomposition achieved. Numerous assumptions for the L–H saturation kinetics type exist and for the applicability in the rate of photomineralization, any of the four possible situations is valid: (1) reactions take place between two adsorbed components of radicals and organics; (2) the reactions are between the radicals in water and adsorbed organics; (3) reactions take place between the radical on the surface and organics in water; and (4) reaction occurs with both radical and organics in water.

Some researchers found that simpler zero- or first-order kinetics is sufficient to model the photomineralization of organic compounds. This was, however, only applicable for limited conditions where the solute concentration was inadequately low. In most kinetic studies, a plateau type of kinetic profile is usually seen where the oxidation rate increases with irradiation time until the rate becomes zero [70, 223]. The appearance of such kinetics regime usually fits the L–H scheme. According to the L–H model (Eq. 13.23), the photocatalytic reaction rate (r) is proportional to the fraction of surface coverage by the organic substrate (θ_x), k_r is the reaction rate constant, C is the concentration of organic species, and K is the Langmuir adsorption constant:

$$r = -\frac{dC}{dt} = k_r = \frac{k_r KC}{1 + KC} \tag{13.23}$$

The applicability of Eq. (13.23) depends on several assumptions, which include the following: (1) the reaction system is in dynamic equilibrium; (2) the reaction is surface mediated; and (3) the competition for the TiO_2-active surface sites by the intermediates and other reactive oxygen species is not limiting [61]. If these assumptions are valid, the reactor scheme only consists of adsorption surface sites, organic molecules and its intermediates, electron/hole pairs, and reactive oxygen species. The rate constant (k_r) for most of the photocatalytic reactions in water is usually reported to be in the order of 10^6–10^9 $(Ms)^{-1}$ [205]. This k_r-value is the proportionality constant for the intrinsic reactivity of photoactivated surface with C. Others have also showed that k_r is proportional to the power law of effective radiant flux (i.e., $\varnothing_e^n$) during the photomineralization reaction [70, 230]. The K-parameter is the dynamic Langmuir adsorption constant (M^{-1}) that represents the catalyst adsorption capacity. Equation 13.23 can be solved explicitly for t using discrete change in C from initial concentration to a reference point:

$$\ln\left(\frac{C}{C_0}\right) + K\left(C - C_0\right) = k_r K_t \tag{13.24}$$

The K-value can be obtained using a linearized form of Eq. (13.23), where $1/r$ is plotted against $1/C$:

$$\frac{1}{r_0} = \frac{1}{k_r} + \frac{1}{k_r K C_0} \tag{13.25}$$

It was reported that the real K-value obtained from the linearized plot of $1/r$ against $1/C$ is significantly smaller [205]. This was explained by the differences in adsorption-desorption phenomena during dark and illuminated period. When the organics concentration is low (in mM), an "apparent" first-order rate constant (Eq. 13.26) could be expressed where k' (min^{-1}) = $k_r K$:

$$r = -\frac{dC}{dt} = k_r KC = k'C \tag{13.26}$$

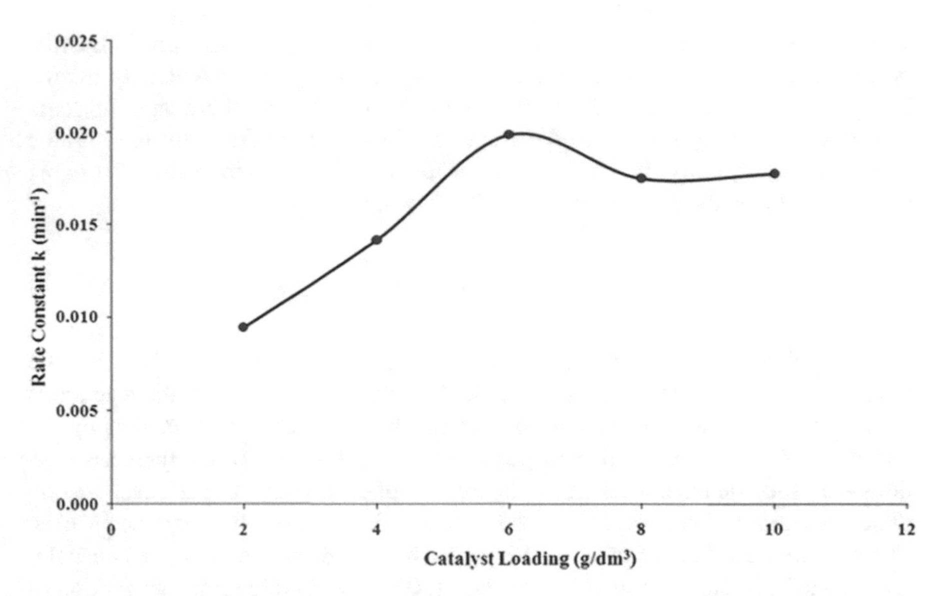

Fig. 13.11 Typical saturation kinetics plot for the degradation of organic dye molecules in an annular photoreactor system [62]

Rearranging and integration of Eq. (13.26) yield the typical pseudo-first-order model as in Eqs. (13.27) and (13.28):

$$C = C_0 e^{-k't} \tag{13.27}$$

$$\ln\left(\frac{C_0}{C}\right) = -k_r K t = -k' t \tag{13.28}$$

The apparent rate constant, however, is only served as a comparison and description for the photocatalytic reaction rate in the reactor system. Figure 13.11 shows a typical saturation kinetic plot for the degradation of organic dye molecules in an annular photoreactor system where the reaction rate increases to a point where the rate plateaus off. To interpret the maximal photomineralization rate, the tailing regime of the L–H saturation profile should be neglected. Only the slope of the tangent to the inflexion point should be used to obtain the maximal photomineralization rate. In this instance, the unit for the slope has the same chemical reaction order as the zero-order rate constant.

A lump-sum L–H saturation kinetics profile has also been used to simplify the approximation for a specific photocatalytic reactor system [224]. In such an empirical lump-sum L–H approach, the degree of organics mineralization is actually expressed in terms of TOC (Eq. 13.29):

$$r_{\text{TOC,O}} = \frac{\beta_1[\text{TOC}]}{\beta_2 + \beta_3[\text{TOC}]} \tag{13.29}$$

This (Eq. 13.29) allows the prediction of TOC degradation as a function of irradiation time. Similar reciprocal plots of $1/r$ against 1/[TOC] can be used to determine the empirical parameters, β_1, β_2, and β_3 as in Eq. (13.25). The irradiation time taken to achieve the fractional degradation of TOC can also be estimated when Eq. (13.29) is expressed as in Eq. (13.24). Such an empirical lump-sum L–H model has greatly summarized the needs for precise kinetics measurement and made a great approximation for any particular photoreactor system, provided that sufficient data are collected for the determination of rate parameters.

Photo-Disinfection Kinetics

Since the first application of semiconductor catalysts for disinfection by Matsunaga et al. [212], few studies have been found in the literature which comprehensively focus on the kinetic modeling of photo-disinfection of microorganisms in a water treatment process. Empirical kinetic models have been the mostly applied for interpretation of photo-disinfection data, because of process complexity and variability. The general expression for the empirical photo-disinfection models for demand-free condition is expressed in Eq. (13.30). In this instance, demand-free conditions assume that the catalyst concentration is constant with irradiation time:

$$\frac{dN}{dT} = -kmN^xC^nT^{m-1} \tag{13.30}$$

where dN/dt = rate of inactivation; N = number of bacterial survivors at irradiation time t; k = experimental reaction rate; C = concentration of photocatalyst used; and m, n, and x are the empirical constants. However, the most commonly employed disinfection model in photo-disinfection studies to date is the simple mechanistic Chick–Watson (C–W) model (Eq. 13.31) [51, 53]:

$$\log\frac{N}{N_o} = -k'T \tag{13.31}$$

In this C–W model, the photo-disinfection rate is expressed as a linear function of the enumerated bacteria and catalyst loading. The combined kinetic parameter of *CT* between the catalyst concentration and irradiation time required to achieve complete inactivation is widely used as a reference for process design. Other than this, this *CT*-value concept is usually used to compare the efficacy of different disinfectants used in water treatment [92]. This C–W model, however, may not always be applicable as many studies may have experienced a curvilinear or nonlinear photo-disinfection profile. Hom [126] reproduced a useful empirical modification on C–W

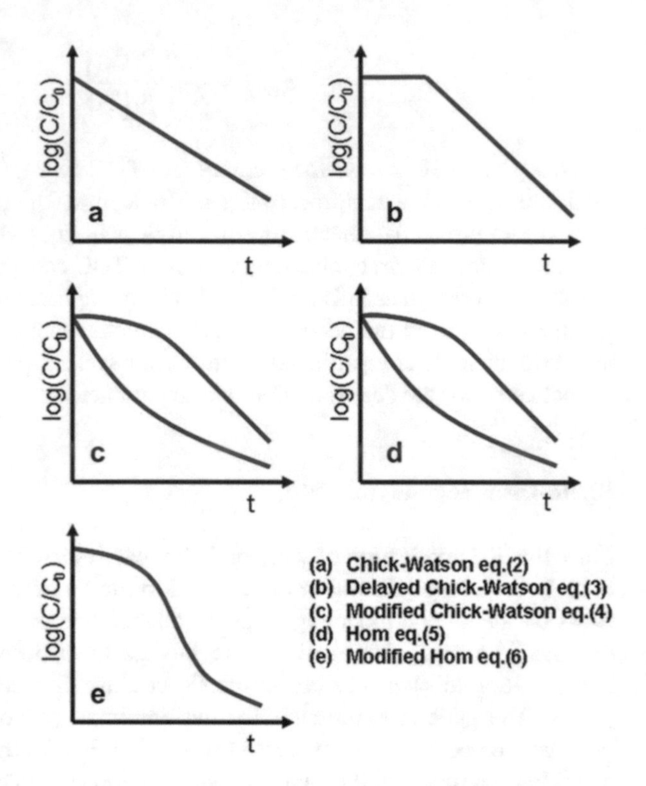

Fig. 13.12 Common nonlinearity in the photo-disinfection kinetics profile using TiO_2 catalyst, along with the appropriate disinfection models used [211]

model after having observed that the disinfection plots of natural algal–bacterial systems were curvilinear, rather than typical log-linear type (Eq. 13.32). In the Hom model, the bacterial inactivation level is predicted in a nonlinear function of C and T, depending on the empirical parameters of n and m, respectively. Since this Hom model is a two-parameter model, it should be stressed that the model is only applicable to a photo-disinfection profile with a maximum of two different nonlinear regions. For m-value greater than unity, the inactivation curve displays an initial "shoulder" while "tailing" is seen when the m-value is less than unity:

$$\log \frac{N}{N_o} = -k'C^nT^m \tag{13.32}$$

Marugán et al. [211] have outlined the common nonlinearity found in the photo-disinfection kinetics profile using Degussa P-25 TiO_2 catalyst fitted with common disinfection models (Fig. 13.12). In their study, they found three different inactivation regions in the photo-disinfection profile, namely (1) a lag or initial smooth decay, known as the "shoulder", followed by (2) a typical log-linear inactivation region and ending with (3) a long deceleration process at the end of the disinfection, which is known as the "tailing." The presence of "shoulder" was justified by the cumulative damage nature of photo-disinfection treatment on the cytoplasmic membrane rather than being instantly lethal [92]. However, the "tailing" during photo-disinfection is not well understood. Benabbou et al. [24] proposed that the "tailing" region was related to the competition for photocatalysis between organic products released from constant cell lyses and the remaining intact cells. Others have proposed that the "tailing" deviations from the log-linear reduction were due to the presence of variations in the bacterial population resistant to the disinfectant used [166]. Nevertheless, the use of CT concept or Hom model can lead to an overdesign for a photo-disinfection system [92]. A further modification to the Hom model was made to account for the simultaneous presence of shoulder, log-linear reduction, and tailing (Eq. 13.33) [52, 211]:

$$\log\frac{N}{N_o} = -k_1\left[1-\exp\left(-k_2 t\right)\right]^{k_3} \quad (13.33)$$

Equation (13.33) is known as the modified Hom model and it expands the applicability of the Hom model for the fitting of the initial shoulder, log-linear reduction, and prolonged tailing behaviors. Another detailed empirical model that can detail the different bacterial inactivation regions is a power law expression (Eq. 13.34), for the generalized differential equation (Eq. 13.30) [115]:

$$\frac{dN}{dt} = -kN^x C^n \quad (13.34)$$

Integration of Eq. (13.34) yields the rational model (Eq. 13.35). If $x = 1$, this rational model can be reverted back to the C–W model. In this instance, the rational model assumes that $x \neq 1$:

$$\log\frac{N}{N_o} = -\frac{\log\left[1+N_o^{x-1}\left(x-1\right)kC^n T\right]}{\left(x-1\right)} \quad (13.35)$$

The rational model (Eq. 13.35) can describe both "shoulder" and "tailing" characteristics for x less than or greater than unity, respectively. Similarly, the Hom model (Eq. 13.32) can also be integrated according to the rational model with the introduction of both x and $m \neq 1$ to yield the Hom-power model (Eq. 13.36):

$$\log\frac{N}{N_o} = -\frac{\log\left[1+N_o^{x-1}\left(x-1\right)kC^n T^m\right]}{\left(x-1\right)} \quad (13.36)$$

Anotai [7] reported that this Hom-power model may provide a better fit than both the Hom and rational models. However, the existence of four empirical parameters in the model may result in an over-parameterization with null physical meanings for each parameter within the model. To reduce the number of null parameters, the Selleck model (Eq. 13.37) was proposed. This model assumes that the catalyst concentration remains constant during the irradiation period:

$$\frac{dS}{dt} = \frac{kCS}{1 + KCT} \quad (13.37)$$

where S is the survival ratio = N/N_o at irradiation time t; k and K are the rate constants. In this model, the "shoulder" was assumed to be a result of cumulative effects of the chemical disinfectant on the microbial target during the contact time. A gradual decrease in the permeability of the outer cell membrane of *E. coli* as an action of the catalysts can be idealized by a series of first-order reaction steps on a single *E. coli* cell that leads to lethality. This model can be modified to account for the disinfection of coliform bacteria in water/wastewater effluent (Eq. 13.38) [115]:

$$\log \frac{N}{N_o} = -n \log \left[1 + \frac{CT}{k} \right] \quad (13.38)$$

Although empirical models provide a simple correlation of the photo-disinfection data, the mathematical nature of their relevant terms yields null or misinterpreted physical meaning. For instance, the m-value > 1 in the Hom model actually suggests that photocatalytic reactivity increases with irradiation time in the ASP. If the vitalistic assumption of bacterial population resistance distribution is valid, this actually indicates that the most resistant bacteria are killed first prior to the least resistant one [166]. This shows that the rationale for the applicability of Hom model is contradictory.

It was proposed that the mechanistic models could convey a better physical meaning in their kinetic terms. It was hypothesized that the photo-disinfection mechanism can be viewed as a pure physicochemical phenomenon and precedes in a similar way to a chemical reaction [166]. The classical C–W model is a typical example of the mechanistic photo-disinfection kinetics model. In the model, the apparent exponential decay curves for bacterial survival ratios with irradiation times follow a similar decay mechanism of a chemical reaction, and are thus applicable under both phenomena. This can be visualized when the C–W model is integrated mathematically, with N assuming to be the number of moles of reactant to yield the log-linear curve of a first-order reaction.

To formulate a mechanistic model for photo-disinfection kinetics, the occurrence of both "shoulder" and "tailing" region should be rationalized idealistically. The presence of "shoulder" can be justified mechanistically by the single-hit multiple targets or a serial phenomenon event [115]. Under mechanistic assumption, the damage to the microbial cell is viewed as cumulative rather than instantly lethal. This dictates that a large number of critical molecules need to be denatured prior to cell inactivation [115]. Severin et al. [286] proposed that the cumulative inactivation of a single bacterium can be collectively represented by a series of integer steps. These disinfection steps were thought to pass on a bacterium from one level to another in a first-order reaction with respect to the catalyst used until a finite number of lethal (l) events was reached. The microorganisms which accumulate less than the postulated number of lethal steps are considered to survive the photo-disinfection process. As for the "tailing" region, many have regarded that it was owing to the presence of a microbial subpopulation resistant to thermal sterilization. Najm [240] also suggested that the "tailing" is due to the intrinsic distribution of bacterial resistance to the sterilization method, making the bacteria more resistant, adapted, and inaccessible to heat treatment. Marugán et al. [211] reviewed that such "tailing" was not a common phenomenon in water disinfection treatment. So far, no justifiable explanation has been proposed to account for the occurrence of "tailing" in the water disinfection treatment with well-mixed conditions and equally resistant cloned bacterial population. It was suggested that the "tailing" was a result of gradual deterioration in the rate of disinfection, and total inactivation was achievable only after a sufficient retention time. The random collisions between the catalyst and bacteria can be expressed by the Poisson probability, as the number of collisions supersedes prior to the number of microbial deaths [115]. If the destruction rate of the microorganisms is assumed to be the same as for the first and lth target at the kth bacterium site, the rate of destruction for the microorganisms is given by

$$\frac{dN_k}{dt} = kCN_{k-1} - kCN_k \tag{13.39}$$

Solving for $K = 0$ to $K = l - 1$ gives the log fraction of microbial survival, not exceeding $l - 1$ at the end of the contact time:

$$\ln\frac{N}{N_o} = -kCT + \ln\left[\sum_{k=0}^{l-1}\frac{(kCT)^K}{K!}\right] \tag{13.40}$$

Lambert and Johnston [166] stated that in the formulation of all mechanistic models that take into account the "shoulder" and "tailing", an intermediate population stage is usually suggested with different rates of disinfection for each microbial state. Typically, two different microbial states for the number of damaged bacterial and undamaged bacteria (i.e., $C_{\text{dam}} + C_{\text{undam}}$) were introduced to account for such nonlinearity in bacterial survivor curves. A similar Langmuir–Hinshelwood (L–H) type of mechanistic model (Eqs. 13.41 and 13.42) can also be applied to represent the photo-disinfection kinetics to yield more meaningful kinetic terms [73, 300]. Johnston et al. [143] showed that the inoculum size of the bacteria has a large impact on resistance distribution and thus the errors associated with this term should be given attention during the mechanistic modeling:

$$\frac{dC_{\text{undam}}}{dt} = \frac{-k_1 K_{\text{undam}} C_{\text{undam}}^{\eta_{\text{undam}}}}{1 + K_{\text{undam}} C_{\text{undam}}^{\eta_{\text{undam}}} + K_{\text{undam}} C_{\text{undam}}^{\eta_{\text{undam}}}} \tag{13.41}$$

$$\frac{dC_{\text{dam}}}{dt} = \frac{k_1 K_{\text{undam}} C_{\text{undam}}^{\eta_{\text{undam}}} - k_2 K_{\text{dam}} C_{\text{dam}}^{\eta_{\text{dam}}}}{1 + K_{\text{undam}} C_{\text{undam}}^{\eta_{\text{undam}}} + K_{\text{dam}} C_{\text{dam}}^{\eta_{\text{dam}}}} \tag{13.42}$$

where k_i is the true log-linear deactivation rate constant for the reaction between generated ROS with bacteria. The pseudo-adsorption constant, K_i, represents the surface interaction between the catalyst and bacteria. This constant is similar to the adsorption equilibrium constant in the conventional L–H model. Due to the significant size differences between *E. coli* cells and catalyst agglomerates, these constants do not represent strict adsorption phenomena but a more general surface interaction during photo-disinfection. These adsorption constants allow the "shoulder" representation in the photo-disinfection kinetics data. The inhibition coefficient η_i is a power coefficient that accounts for the "tailing" in the bacterium inactivation curve. Marugán et al. [211] proposed that η_i is needed to account for the inhibition produced by the increasing concentrations in the medium of cell lysis and oxidation products competing for the ROS. This is particularly important toward the end of photo-disinfection, as high concentrations of these compounds and small numbers of viable bacteria are present in the suspension. In this instance, the η_i in the proposed L–H model essentially means that the reaction order with respect to the microbial population is higher than 1.

Both Eqs. (13.41) and (13.42) have six independent parameters that describe photo-disinfection rates. This constitutes a high risk of over-fitting the experimental data, where the statistical significance of parameters and the plausibility of the model are low. By taking into account the intrinsic kinetics of ROS attack, the catalyst–bacterium interaction and the inhibition by-products are similar for both undamaged and damaged bacteria, and the following are assumed:

$$k_1 = k_2 = k \tag{13.43}$$

$$k_{\text{undam}} = k_{\text{dam}} = K \tag{13.44}$$

$$\eta_{\text{undam}} = \eta_{\text{dam}} = \eta \tag{13.44}$$

With these, Eqs. (13.41) and (13.42) are reduced to Eqs. (13.45) and (13.46). The three independent parameters allow a simultaneous fitting of three different inactivation regimes of "shoulder," "log-linear," and "tailing." Fitting of Eqs. (13.45) and (13.46) to the experimental measurements of $(C_{\text{undam}} + C_{\text{dam}})/C_o$ can be achieved using a nonlinear regression algorithm coupled with a fifth-order Runge–Kutta numerical approach. Marugán et al. [211] showed a good fitting of such mechanistic L–H model to the photo-disinfection of *E. coli* under different loadings of Degussa P25 TiO_2 catalyst:

$$\frac{dC_{\text{undam}}}{dt} = \frac{-kKC_{\text{undam}}^{\eta}}{1 + KC_{\text{undam}}^{\eta} + KC_{\text{dam}}^{\eta}} \tag{13.45}$$

$$\frac{dC_{\text{dam}}}{dt} = k\frac{KC_{\text{undam}}^{\eta} - KC_{\text{dam}}^{\eta}}{1 + KC_{\text{undam}}^{\eta} + KC_{\text{dam}}^{\eta}} \tag{13.46}$$

Water Quality

In a full-scale plant for water treatment, a number of technical issues have to be sorted out. Water quality of the influents in the treatment plant can vary with time. In this section, the implication of TiO_2 photocatalytic processes for treatment of water sources of different qualities is discussed. This enables a fuller understanding of the effects of various key water quality parameters on the suitability of applying such advanced TiO_2 photocatalytic processes.

Turbidity

Turbidity often refers to the insoluble particulates that are present in the targeted water [314]. The presence of such insoluble particulate matters is highly detrimental to the TiO_2 photocatalysis-based process, as they can affect the optical properties and further impede the penetration of UV light by strong scattering and absorption of the rays [47, 301]. This will cause a variation in the predicted use of TiO_2 loading, UV penetration path, and light intensity. Also, excessive levels of turbidity can reduce both the photomineralization and photo-disinfection efficiency of the pollutants that are present in water owing to the shielding effects that attenuate the light penetration causing those pollutants to flee from the treatment [47, 276, 301]. Suspended solids will also shield target pollutants from oxidation reactions. In this instance, the suspended solids refer to the nonfilterable residue that is retained on the filter medium after filtration. All these factors ultimately decrease the overall photocatalytic efficiency for water treatment. To ensure rapid photocatalytic reaction rate, the turbidity of the targeted water should be kept below 5 nephelometric turbidity units (NTU) for optimal UV light utilization and photocatalytic reaction [91, 105]. Rincón and Pulgarin [274] observed that the water turbidity higher than 30 NTU will negatively affect the rate of photocatalytic disinfection. The limit of 5 NTU is arbitrary and depends on the receiving water bodies and the treatment levels required. Since TiO_2 photocatalytic processes are retrofitted to advanced water treatment stage, prior reduction in turbidity could be achieved via conventional treatment processes such as screening, filtration, sedimentation, coagulation, and flocculation. The standard for 1 NTU is a 1.0 mg/L of specified size of silica suspension and can be prepared for laboratory investigation. The turbidity standard can be calibrated with a photoelectric detector or nephelometry for the intensity of scattered light.

Inorganic Ions

The presence of inorganic ions in the targeted water for TiO_2 photocatalytic treatment is expected. To functionalize a TiO_2 water treatment process, the basic understanding of the role of these inorganic ions on the photocatalytic performance is essential. Crittenden et al. [68] reported that when a photocatalyst is utilized in either slurry or fixed-bed configuration to treat real waters with different inorganic ions, photocatalyst deactivation was usually observed. This resulted from a strong inhibition from the inorganic ions on the surface of the TiO_2 semiconductor used. Thus, the presence of these inorganic ions together with their permissible levels on the photocatalytic performance of TiO_2 in water treatment has to be determined to ensure minimal disturbances on the efficient operation of the TiO_2-based treatment process. With such data cost-effective fouling prevention with inorganic ions and photocatalyst regeneration strategies can be customized.

A number of studies have been conducted on the effects of different inorganic anions or cations on both TiO_2 photomineralization and photo-disinfection reactions [46, 114, 116, 175, 251, 273, 275, 282, 320, 329]. It must be emphasized, however, that most of these studies have concentrated on how different inorganic ions affect the rates of photocatalytic reactions with a model surrogate organic compound. The surrogate model organic compound can be biased toward its photocatalytic performance owing to its underlying chemical properties and the main constituent groups that form the compound. For instance, the model compounds with either electron-withdrawing or -donating groups will contribute to a different degree of interaction to degradation pathways [121]. Due to the zwitterionic nature of the TiO_2 particles used, it is also possible that the operating pH might have a profound effect on the selective inhibition of inorganic ions on the surface of the TiO_2 particles [114]. However, few discussions have been centered on considering how the chemical nature of the model organic used and the operating pH in the photocatalytic reactor impact the inhibition of different inorganic ions in photocatalytic water treatment.

To date, the effects of both inorganic cations (i.e., Na^+, K^+, Ca^{2+}, Cu^{2+}, Mn^{2+}, Mg^{2+}, Ni^{2+}, Fe^{2+}, Zn^{2+}, Al^{3+}) and inorganic anions (i.e., Cl^-, NO^{3-}, HCO^{3-}, ClO^{4-}, SO_4^{2-}, HPO_4^{2-}, PO_4^{3-}) on the photocatalytic water treatment have been investigated [46, 114, 116, 175, 251, 273, 275, 282, 320, 328]. A general consensus from these studies concludes that Cu^{2+}, Fe^{2+}, Al^{3+}, Cl^-, and PO_4^{3-} at certain levels may decrease photomineralization reaction rates while Ca^{2+}, Mg^{2+}, and Zn^{2+} may have negligible effects. This is because Ca^{2+}, Mg^{2+}, and Zn^{2+} are at their maximum oxidation states resulting in their inability to inhibit the photocatalysis reaction. The presence of Fe^{2+} can catalyze both the Fenton and photo-Fenton reactions. However Choi et al. [58] observed that Fe^{2+} fouled the photocatalyst surface by introducing a rusty orange color change via the formation of $Fe(OH)_3$, while PO_4^{3-} in the nominal pH range remains strongly adsorbed onto the TiO_2 surface and further inhibits its photoactivity [1, 153]. Some research groups observed that NO_3^-, SO_4^{2-}, ClO_4^-, and HCO_3^- inhibit the surface activity of the photocatalysts, while others suggest no such impact. Both NO_3^- and SO_4^{2-} have detrimental effect on the photo-disinfection rate [116]. The Cu^{2+} can enhance the photocatalytic activity at its concentration up to 0.1 mM, while further increases in its concentration reduce the reaction rate [248]. Nitrogen-containing molecules are mineralized into NH_4^+ and mostly NO_3^- ammonium ions are relatively stable and their proportion depends mainly on the initial oxidation degree of nitrogen on the irradiation time. Pollutants containing sulfur atoms are mineralized into sulfate ions. Cl^- has no inhibition on the photocatalytic degradation of trichloroethylene at a concentration up to 3.0 mM. On the contrary, the addition of oxyanion oxidants such as ClO_2^-, ClO_3^-, IO_4^-, $S_2O_8^-$, and BrO_3^- increased photoreactivity by scavenging conduction-band electrons and reducing the charge-carrier recombination [208]. The presence of salts is also known to diminish the colloidal stability as screening effects become more profound. This was followed by the double-layer compression and surface charge neutralization, which increases the mass transfer limitations and reduces surface contacts between the pollutants and catalysts. Other inorganic ions also affect photodegradation rates, where the presence of SO_4^{2-} in a TiO_2-coated glass spiral reactor could double the disappearance rate of the pesticide monocrotophos [359]. The Mn^{2+} improved the photoactivity by simultaneously increasing the electron/hole pairs and preventing their recombination [233].

Several mechanisms for fouling effects of inorganic ions on TiO_2 photoactivity have been proposed [36, 279]. These include UV screening, competitive adsorption to surface-active sites, competition for photons, surface deposition of precipitates and elemental metals, radical and hole scavenging, and direct reaction with the photocatalyst. The NO_3^- ion was reported to UV screen the photocatalyst than inhibit the TiO_2 surface [36]. The competition for surface-active sites involves the constant displacement of hydroxide ions from TiO_2 surface and thus further reduces the generation of radicals. Quantum yield is reduced as a result of the direct competition of inorganic ions for light photons (lowered number of photons entering the reactor). A similar decrease in photonic efficiency was observed when precipitates were formed and deposited onto the TiO_2 surface, blocking the accessibility of both photons and organic compounds. The ultimate inorganic anions that were determined to scavenge both the hole and radicals include Cl^-, HCO_3^-, SO_4^{2-}, and PO_4^{3-} [79]. A mechanism for Cl^- and HCO_3^- in inhibiting photocatalysis via hydroxyl radical and hole scavenging was proposed by Matthews and McEnvoy [213] and Lindner et al. [187], respectively:

$$Cl^- + OH \rightarrow Cl + OH^- \tag{13.47}$$

$$Cl^- + h^+ \rightarrow Cl \tag{13.48}$$

The Cl^- accounted for its inhibitory effect on TiO_2 photocatalysis through a preferential adsorption displacement mechanism over the surface-bound OH^- ions. This reduces the number of OH^- ions available on the TiO_2 surface, and the substituted Cl^- further increases the recombination of electron/hole pairs. Among other chlorinated molecules, Cl^- ions are readily released in the solution. This effect could be of benefit in a process where photocatalysis is associated with a biological depuration system, which is generally not efficient for chlorinated compounds [121]. Other ions such as PO_4^{3-} are known to avert adsorption of amino acids over the TiO_2 catalysts, while CO_3^{2-} and other ionic species react with OH^- radicals to compete with microorganisms and further reduce the efficiency. Thus, the presence of inorganic ions in water subjected to TiO_2 photocatalytic treatment is an important factor in determining its successful implementation.

To resolve TiO_2 photocatalyst fouling issue, preventative or regenerative strategies can be adopted, depending on the nature of the photocatalyst deactivation in the water matrix. Fouling preventative strategies by means of water pretreatment, complexation, and photocatalyst surface modifications can be addressed, while rinsing the TiO_2 surface with different chemical solutions constitutes the regenerative strategies. Water pretreatment with ion-exchange resins is one of a few methods that can be employed to ensure minimal disturbances by inorganic ions. Burns et al. [36] discussed that the operational cost of water pretreatment with ion-exchange resins can be minimized if the fouling ions were identified and selectively removed. Other preventative strategies such as complexation of fouling agents after "escorting" the ions through the reactor can be utilized, provided that the strong fouling ions are hard to remove from the feed water stream. Modification of the TiO_2 surface to increase the hydrophobicity and adsorption capacity is an upstream preventative way to enhance the rate of photodegradation in the presence of fouling ions. However, this method is unstable and impractical compared to the others, where the modifications are reported to be displaced away with time. As for the regenerative strategies, different types of chemical rinsing were reported to potentially resolubilize surface deposits, precipitates, and reduced metals. The fouled ions usually do not form strong surface complexes and can be easily displaced by an ion-exchange rinse. Abdullah et al. [1] reported that TiO_2 fouling with SO_4^{2-} and PO_4^{3-} can be displaced by NaOH, KOH, and $NaHCO_3$, while Cl^- can be easily regenerated with water. More complex mixtures of inorganic ions need to be investigated to better mimic real water matrices in photocatalytic water treatment or after investigation of water matrices with one known inorganic ion composition at a time.

Heavy and Noble Metals

Heavy metals that might be present in trace amount in the wastewater stream are highly toxic in some of their valence states [146]. Due to the pliable nature of biological treatment, these intoxicant metals can remain and permeate through the treatment process. To treat such metals, TiO_2 photocatalytic process has been reported to simultaneously convert these metals into nontoxic ionic states and further reduce them into their corresponding elemental form on the TiO_2 surface for metal recovery. Prairie et al. [265] reported that metals of Ag(I), Cr(IV), Hg(II), and Pt(II) were easily treated with TiO_2 of 0.1 wt% whereas Cd(II), Cu(II), and Ni(II) could not be removed. The extent of such metal conversion and recovery process is highly dependent on the standard reduction potential of the metals for the reduction reactions. It was reported that for an efficient removal of the metals, a positive potential of greater than 0.4 V or the flat band potential of TiO_2 was required [121, 146]. Since the rates of both oxidation (organics) and reduction (metals) on the TiO_2 surface are intrinsically interrelated, the presence of sufficient organics in the water matrices was found to facilitate metal recovery. The redox process for the metal reduction on the TiO_2 surface is given below:

$$Mn^{+} + H_2O \overset{TiO_2(h\nu)}{\rightarrow} M^{o} + nH^{+} + \frac{n}{4}O_2 \quad (13.49)$$

Herrmann et al. [122] reported that small crystallites of silver (3 and 8 nm) were initially deposited and began to agglomerate (few hundreds of nm) when the conversion increased. Since the photosensitive surface was not masked, a relatively large amount of silver was recovered, leaving behind a concentration lower than the detection limits of atomic absorption spectroscopy (≤0.01 ppm) in the solution. The effect of various factors on the photoreduction of silver on TiO_2 surface was further investigated [131]. Angelidis et al. [6] also reported that even at low concentration of metal ions, the photodeposition of metals on the TiO_2 surface from solution was still effectively performed. However, they noted that the rate of photodeposition was enhanced when Pt-loaded TiO_2 particles were used instead of unloaded TiO_2 particles. This metal photodeposition property on TiO_2 surface is particularly useful when the water legislation limit on the metal contents becomes more stringent.

Life Cycle Assessment of Photocatalytic Water Treatment Processes

In the current development of photocatalytic water/wastewater treatment processes, their possible application for the industry is still being investigated at pilot plant scale. A few pilot plants have been established to obtain feasibility data, such as the treatment efficiency, site area requirements for targeted volume, electrical energy consumption, process emissions, and chemical costs. The heterogeneous photocatalysis and photo-Fenton plants located at the INETI (Instituto Nacional de Engenharia, Tecnologia Industrial e Inovacao, Portugal) and PSA (Plataforma Solar de Almeria, Spain) are two renowned pilot plants that have delivered most of these data for technical analysis. Both plants consist of compound parabolic collectors (4.16 m^2 aperture area) exposed to sunlight, a reservoir tank, a recirculation pump, and connecting tubing and are operated in batch mode [234]. Further technical details of these plants can be found in the literature [106, 167].

In order to assess these photocatalytic processes as emerging technologies for large-scale water/wastewater treatment, a life cycle assessment (LCA) should be evaluated based on the currently available data. LCA is one of the most widely accepted tools that consider not only the environmental impact of the emergent photocatalytic water treatment, but also its technical feasibility and costs. Andreozzi et al. [5] also pointed out that the potential application of ROS-based oxidation processes entails high costs for energy and reactant consumption. Thus, to consider the feasibility of photocatalytic water treatment on the whole, a comprehensive LCA based on viable technical data should be carried out. Muñoz et al. [235] carried out a simplified LCA based on small-scale laboratory data of heterogeneous photocatalysis and other AOPs. However, they found that small-scale laboratory data interpretation in their LCA study can lead to inconclusive results.

Using LCA, the environmental burdens from a product, process, or activity via materials and energy balances can be defined and reduced as well as the waste discharges, its impacts on the environment, and the environmental improvement opportunities over the whole life cycle [67]. This holistic LCA approach in decision-making over other environmental assessment approaches includes all the burdens and impacts, and focuses on the emissions and wastes generated [15].

Muñoz et al. [234] carried out a LCA based on the two pilot plants of INETI and PSA, which treated 1 m^3 of methyphenylglycerine (MPG) to destroy nonbiodegradable and toxic compounds to a level that met the quality of aquatic ecosystems with both homogeneous photo-Fenton and heterogeneous photocatalysis. Other technically rational assumptions were also made to facilitate the estimation of energy and materials consumed and produced. Nine impact categories of the possible large-scale photocatalytic water treatment process were included in the analysis, namely global warming potential, ozone depletion potential, human toxicity potential, freshwater aquatic toxicity potential, photochemical oxidant formation potential, acidification potential, eutrophication potential, nonrenewable energy consumption, and land use.

Figure 13.13 shows the LCA results for possible large-scale water application using photocatalytic technology. The LCA results showed that the retrofitting of heterogeneous photocatalysis process to the existing biological wastewater treatment can lower eutrophication potential, but requires higher site area requirement and electricity consumption. These technical constraints are a direct result from the requirement for a large land area and the raw materials to build the parabolic collector

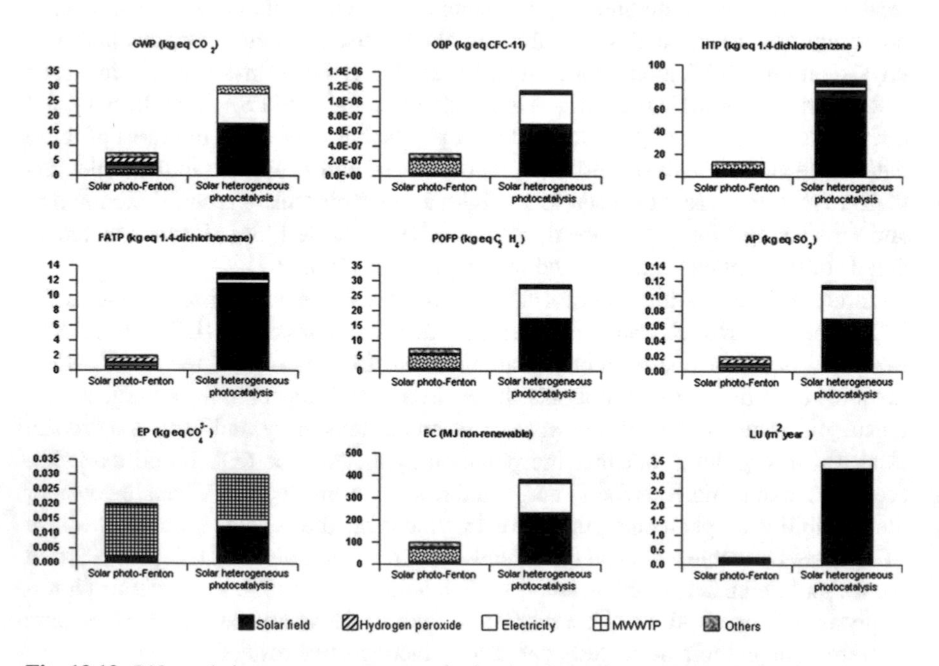

Fig. 13.13 Life cycle impact assessment results for the alternatives under study [234]

infrastructure and high power needed to pump the wastewater through the system. However, the results from the impact categories cannot be compared directly to each other as they were expressed in different measurement units. From the engineering point of view, these constraints mainly arise from the low photoactivity of the catalyst used under solar irradiation. Further materials engineering solutions and studies should be carried out to resolve such technical issues to permit the scaleup of the technology to a commercially viable process.

Future Challenges and Prospects

Semiconductor photocatalytic technology using either UV light or solar has become more prominent owing to its advantages of the use of vast additive chemicals or disinfectants and its mineralization aspects. These are particularly important, as recalcitrant organics are mineralized rather than being transformed to another phase. Coupled with the ambient operation of the process, all these make photocatalytic water treatment technology a viable alternative for commercialization in the near future. Different water contaminants, ranging from hazardous contaminants of pesticides, herbicides, and detergents to pathogens, viruses, coliforms, and spores, are effectively removed by this photocatalytic process.

The applicability of the heterogeneous photocatalytic technology for water treatment is constrained by several key technical issues that need to be further investigated. The first consideration would be whether the photocatalytic process is a pretreatment step or a stand-alone system. The nonselective reactivity on the nonbiodegradable water-soluble pollutants means that the photocatalytic process can be used effectively as a pretreatment step to enhance biodegradation of recalcitrant organic pollutants prior to biological water treatment. In such a way, the residence time and reaction volume for the biological treatment could be significantly reduced. If the photocatalytic process is used as a stand-alone treatment system, the residence time required might be prolonged for total bacterial inactivation or mineralization. As discussed, this is hindered by the slow kinetics, low photo-efficiency, and a need for continuous (without interruption) illumination to achieve the required total organic carbon removal or microbial inactivation. For the stand-alone system, the site area requirement might be proportionally from any increased reaction volume required.

In order to promote the feasibility of photocatalytic water treatment technology in the near future, several key technical constraints ranging from catalyst development to reactor design and process optimization have to be addressed. These include (1) catalyst improvement for a high photo-efficiency that can utilize wider solar spectra; (2) catalyst immobilization strategy to provide a cost-effective solid–liquid separation; (3) improvement in the photocatalytic operation for wider pH range and to minimize the addition of oxidant additives; (4) new integrated or coupling system for enhanced photomineralization or photo-disinfection kinetics; and (5) effective design of photocatalytic reactor system or parabolic solar collector for higher utilization of solar energy to reduce the electricity costs. Currently, the utilization of solar energy is limited by the photo-efficiency of the TiO_2 catalyst bandgap to only 5% of the solar spectrum. The need for continuous illumination for efficient inactivation of pathogens has diverted solar utilization to artificial UV lamp-driven process. In addition, the low-efficacy design of current solar collecting technology (0.04% capture of original solar photons) has encouraged the developmental progress of photocatalytic technology in the water treatment industry. Further pilot plant investigations with different reactor configurations are needed to ensure that the photocatalytic water technology is well established and presents vast techno-economic data for any LCA study. Finally, a large-scale photocatalytic treatment process with high-efficacy, solar-driven, and low-site-area requirements can be realized in the short future with rapid evaluation of different possible pilot plant configurations.

Acknowledgement This topic "Wastewater" is written by Xiaolei Qu, Pedro J.J. Alvarez, and Qilin Li at the Department of Civil and Environmental Engineering, Rice University, Houston, TX 77005, USA, and it was published as "Applications of nanotechnology in water and wastewater treatment" in Water Research. 47 (2013) 3931–3946.

This topic is written by Meng Nan Chong[a,b], Bo Jin[a,b,c], Christopher W.K. Chow[c], and Chris Saint[c] at [a]School of Chemical Engineering, the University of Adelaide, 5005 Adelaide, Australia; [b]School of Earth and Environmental Sciences, the University of Adelaide, Adelaide, South Australia 5005, Australia; and [c]Australian Water Quality Centre, SA Water Corporation, 5000 Adelaide, South Australia, Australia. This chapter was published as "Recent developments in photocatalytic water treatment technology: A review" in journal "Water Research". 44 (2010); 2997–3027.

References

1. Cassano, A.E., Alfano, O.M., 2000. Reaction engineering of suspended solid heterogeneous photocatalytic reactors. Catal. Today 58, 167-197.
2. WHO, 2012. Progress on Drinking Water and Sanitation. 2012 Update.
3. Cho, M., Yoon, J., 2008. Measurement of OH radical CT for inactivating Cryptosporidium parvum using photo/ferrioxalate and photo/TiO_2 systems. J. Appl. Microbiol. 104, 759-766.
4. Qu, X.L., Brame, J., Li, Q., Alvarez, J.J.P., 2013. Nanotechnology for a safe and sustainable water supply: enabling integrated water treatment and reuse. Accounts of Chemical Research 46 (3), 834-843.
5. Gernjak, W., Maldonado, M.I., Malato, S., Cáceres, J., Krutzler, T.,Glaser, A., Bauer, R., 2004. Pilot-plant treatment of olive mill wastewater (OMW) by solar TiO2 photocatalysis and solar photo-Fenton. Sol. Energy 77, 567-572.
6. Pan, B., Xing, B.S., 2008. Adsorption mechanisms of organic chemicals on carbon nanotubes. Environmental Science and Technology 42 (24), 9005-9013.
7. Ryu, J., Choi, W., Choo, K.H., 2005. A pilot-scale photocatalyst-membrane hybrid reactor: performance and characterization. Water Sci. Technol. 51, 491-497.
8. Yang, K., Xing, B.S., 2010. Adsorption of organic compounds by carbon nanomaterials in aqueous phase: Polanyi theory and its application. Chemical Reviews 110 (10), 5989-6008.
9. Curcó, D., Giménez, J., Addarak, A., Cervera-March, S., Esplugas, S., 2002. Effects of radiation absorption and catalyst concentration on the photocatalytic degradation of pollutants. Catal. Today 76, 177-188.

10. Pan, B., Lin, D.H., Mashayekhi, H., Xing, B.S., 2008. Adsorption and hysteresis of bisphenol A and 17 alpha-ethinyl estradiol on carbon nanomaterials. Environmental Science and Technology 42 (15), 5480-5485.
11. Bellobono, I.R., Morazzoni, F., Tozzi, P.M., 2005b. Photocatalytic membrane modules for drinking water purification in domestic and community appliances. Int. J. Photoenergy 7, 109-113.
12. Ji, L.L., Chen, W., Duan, L., Zhu, D.Q., 2009. Mechanisms for strong adsorption of tetracycline to carbon nanotubes: a comparative study using activated carbon and graphite as adsorbents. Environmental Science and Technology 43 (7), 2322-2327.
13. Joo, J., Kwon, S.G., Yu, T., Cho, M., Lee, J., Yoon, J., Hyeon, T., 2005. Large-scale synthesis of TiO_2 nanorods via nonhydrolytic sol-gel ester elimination reaction and their application to photocatalytic inactivation of E. coli. J. Phys. Chem. B 109, 15297-15302.
14. Chen, C.C., Lu, C.S., Chung, Y.C., Jan, J.L., 2007. UV light induced photodegradation of malachite green on TiO2 nanoparticles. J. Hazard. Mater. 141, 520-528.
15. Lin, D.H., Xing, B.S., 2008. Adsorption of phenolic compounds by carbon nanotubes: role of aromaticity and substitution of hydroxyl groups. Environmental Science and Technology 42 (19), 7254-7259.
16. Lu, C.S., Chiu, H., Liu, C.T., 2006. Removal of zinc(II) from aqueous solution by purified carbon nanotubes: kinetics and equilibrium studies. Industrial & Engineering Chemistry Research 45 (8), 2850-2855.
17. Yang, K., Wu, W.H., Jing, Q.F., Zhu, L.Z., 2008. Aqueous adsorption of aniline, phenol, and their substitutes by multi-walled carbon nanotubes. Environmental Science and Technology 42 (21), 7931-7936.
18. Fornasiero, F., Park, H.G., Holt, J.K., Stadermann, M., Grigoropoulos, C.P., Noy, A., Bakajin, O., 2008. Ion exclusion by sub-2-nm carbon nanotube pores. Proceedings of the National Academy of Sciences of the United States of America 105 (45), 17250-17255.
19. Habibi, M.H., Hassanzadeh, A., Mahdavi, S., 2005. The effect of operational parameters on the photocatalytic degradation of three textile azo dyes in aqueous TiO_2 suspensions. J. Photochem. Photobiol. A: Chem. 172, 89-96.
20. Rao, G.P., Lu, C., Su, F., 2007. Sorption of divalent metal ions from aqueous solution by carbon nanotubes: a review. Separation and Purification Technology 58 (1), 224-231.
21. Li, Y.H., Ding, J., Luan, Z.K., Di, Z.C., Zhu, Y.F., Xu, C.L., Wu, D.H., Wei, B.Q., 2003. Competitive adsorption of Pb^{2+}, Cu^{2+} and Cd^{2+} ions from aqueous solutions by multiwalled carbon nanotubes. Carbon 41 (14), 2787-2792.
22. Feng, Q.L., Wu, J., Chen, G.Q., Cui, F.Z., Kim, T.N., Kim, J.O., 2000. A mechanistic study of the antibacterial effect of silver ions on Escherichia coli and Staphylococcus aureus. Journal of Biomedical Materials Research 52 (4), 662-668.
23. Gao, W., Majumder, M., Alemany, L.B., Narayanan, T.N., Ibarra, M.A., Pradhan, B.K., Ajayan, P.M., 2011. Engineered graphite oxide materials for application in water purification. ACS Applied Materials & Interfaces 3 (6), 1821-1826.
24. Li, Y.H., Di, Z.C., Ding, J., Wu, D.H., Luan, Z.K., Zhu, Y.Q., 2005. Adsorption thermodynamic, kinetic and desorption studies of Pb^{2+} on carbon nanotubes. Water Research 39 (4), 605-609.
25. Liu, Z.Y., Bai, H.W., Lee, J., Sun, D.D., 2011c. A low-energy forward osmosis process to produce drinking water. Energy & Environmental Science 4 (7), 2582-2585.
26. Lin, Y.H., Tseng, W.L., 2010. Ultrasensitive sensing of Hg(2fl) and CH(3)Hg(fl) based on the fluorescence quenching of lysozyme type VI-stabilized gold nanoclusters. Analytical Chemistry 82 (22), 9194-9200.
27. Lu, C., Chiu, H., Bai, H., 2007. Comparisons of adsorbent cost for the removal of zinc (II) from aqueous solution by carbon nanotubes and activated carbon. Journal of Nanoscience and Nanotechnology 7 (4-5), 1647-1652.
28. Douglas, J. S., H. Habibian, C.-L. Hung, A. V. Gorshkov, H. J. Kimble, and D. E. Chang. "Quantum many-body models with cold atoms coupled to photonic crystals", Nature Photonics, 2015.

29. Koeppenkastrop, D., Decarlo, E.H., 1993. Uptake of rare-earth elements from solution by metal-oxides. Environmental Science and Technology 27 (9), 1796-1802.
30. Sukhanova, A., Devy, M., Venteo, L., Kaplan, H., Artemyev, M., Oleinikov, V., Klinov, D., Pluot, M., Cohen, J.H.M., Nabiev, I., 2004. Biocompatible fluorescent nanocrystals for immunolabeling of membrane proteins and cells. Analytical Biochemistry 324 (1), 60-67.
31. Trivedi, P., Axe, L., 2000. Modeling Cd and Zn sorption to hydrous metal oxides. Environmental Science and Technology 34 (11), 2215-2223.
32. Rivero, M.J., Parsons, S.A., Jeffrey, P., Pidou, M., Jefferson, B., 2006. Membrane chemical reactor (MCR) combining photocatalysis and microfiltration for grey water treatment. Water Sci. Technol. 53, 173-180.
33. Yean, S., Cong, L., Yavuz, C.T., Mayo, J.T., Yu, W.W., Kan, A.T., Colvin, V.L., Tomson, M.B., 2005. Effect of magnetite particle size on adsorption and desorption of arsenite and arsenate. Journal of Materials Research 20 (12), 3255-3264.
34. Auffan, M., Rose, J., Bottero, J.Y., Lowry, G.V., Jolivet, J.P., Wiesner, M.R., 2009. Towards a definition of inorganic nanoparticles from an environmental, health and safety perspective. Nature Nanotechnology 4 (10), 634-641.
35. Auffan, M., Rose, J., Proux, O., Borschneck, D., Masion, A., Chaurand, P., Hazemann, J.L., Chaneac, C., Jolivet, J.P., Wiesner, M.R., Van Geen, A., Bottero, J.Y., 2008. Enhanced adsorption of arsenic onto maghemites nanoparticles: As(III) as a probe of the surface structure and heterogeneity. Langmuir 24 (7), 3215-3222.
36. Yavuz, C.T., Mayo, J.T., Yu, W.W., Prakash, A., Falkner, J.C., Yean, S., Cong, L.L., Shipley, H.J., Kan, A., Tomson, M., Natelson, D., Colvin, V.L., 2006. Low-field magnetic separation of monodisperse Fe_3O_4 nanocrystals. Science 314 (5801), 964-967.
37. Lucas, E., Decker, S., Khaleel, A., Seitz, A., Fultz, S., Ponce, A., Li, W.F., Carnes, C., Klabunde, K.J., 2001. Nanocrystalline metal oxides as unique chemical reagents/sorbents. Chemistry-A European Journal 7 (12), 2505-2510.
38. Seven, O., Dindar, B., Aydenir, S., Metin, D., Ozinel, M.A., Icli, S., 2004. Solar photocatalytic disinfection of a group of bacteria and fungi aqueous suspensions with TiO_2, ZnO and Sahara desert dust. J. Photochem. Photobiol. A: Chem. 165, 103-107.
39. Sharma, Y.C., Srivastava, V., Singh, V.K., Kaul, S.N., Weng, C.H., 2009. Nano-adsorbents for the removal of metallic pollutants from water and wastewater. Environmental Technology 30 (6), 583-609.
40. Daus, B., Wennrich, R., Weiss, H., 2004. Sorption materials for arsenic removal from water: a comparative study. Water Research 38 (12), 2948-2954.
41. Deliyanni, E.A., Bakoyannakis, D.N., Zouboulis, A.I., Matis, K.A., 2003. Sorption of As(V) ions by akaganeite-type nanocrystals. Chemosphere 50 (1), 155-163.
42. Mayo, J.T., Yavuz, C., Yean, S., Cong, L., Shipley, H., Yu, W., Falkner, J., Kan, A., Tomson, M., Colvin, V.L., 2007. The effect of nanocrystalline magnetite size on arsenic removal. Science and Technology of Advanced Materials 8 (1-2), 71-75.
43. Hossain, Md. Faruque (2018). Photonic Thermal Energy Control to Naturally Cool and Heat the Building. *Advanced Thermal Engineering*. 131, 576–586. (Elsevier).
44. Hristovski, K.D., Westerhoff, P.K., Moller, T., Sylvester, P., 2009b. Effect of synthesis conditions on nano-iron (hydr)oxide impregnated granulated activated carbon. Chemical Engineering Journal 146 (2), 237-243.
45. Zhang, H., Quan, X., Chen, S., Zhao, H., Zhao, Y., 2006a. Fabrication of photocatalytic membrane and evaluation its efficiency in removal of organic pollutants from water. Sep. Purif. Technol. 50, 147-155.
46. Hu, J., Chen, G.H., Lo, I.M.C., 2006. Selective removal of heavy metals from industrial wastewater using maghemite nanoparticle: performance and mechanisms. Journal of Environmental Engineering-Asce 132 (7), 709-715.
47. Kim, J., Lee, C.W., Choi, W., 2010. Platinized WO(3) as an environmental photocatalyst that generates OH radicals under visible light. Environmental Science and Technology 44 (17), 6849-6854.

48. de Villoria, R.G., Hart, A.J., Wardle, B.L., 2011. Continuous high-yield production of vertically aligned carbon nanotubes on 2D and 3D substrates. ACS Nano 5 (6), 4850-4857.
49. Evgenidou, E., Fytianos, K., Poulios, I., 2005. Semiconductor sensitized photodegradation of dichlorvos in water using TiO_2 and ZnO as catalysts. Appl. Catal. B: Environ. 59, 81-89.
50. Collins, P.G., Bradley, K., Ishigami, M., Zettl, A., 2000. Extreme oxygen sensitivity of electronic properties of carbon nanotubes. Science 287 (5459), 1801-1804.
51. Crooks, R.M., Zhao, M.Q., Sun, L., Chechik, V., Yeung, L.K., 2001. Dendrimer-encapsulated metal nanoparticles: synthesis, characterization, and applications to catalysis. Accounts of Chemical Research 34 (3), 181-190.
52. Diallo, M.S., Christie, S., Swaminathan, P., Johnson, J.H., Goddard, W.A., 2005. Dendrimer enhanced ultrafiltration. 1. Recovery of Cu(II) from aqueous solutions using PAMAM dendrimers with ethylenediamine core and terminal NH_2 groups. Environmental Science and Technology 39 (5), 1366-1377.
53. Fogler, H.S., 1999. Elements of Chemical Reaction Engineering: Chapter 10: Catalysis and Catalytic Reactors. Prentice-Hall PTR Inc, p. 581-685.
54. Aragon, M., Kottenstette, R., Dwyer, B., Aragon, A., Everett, R., Holub, W., Siegel, M., Wright, J., 2007. Arsenic pilot plant operation and results. Sandia National Laboratories, Anthony, New Mexico.
55. Sylvester, P., Westerhoff, P., Mooller, T., Badruzzaman, M., Boyd, O., 2007. A hybrid sorbent utilizing nanoparticles of hydrous iron oxide for arsenic removal from drinking water. Environmental Engineering Science 24 (1), 104-112.
56. Westerhoff, P., De Haan, M., Martindale, A., Badruzzaman, M., 2006. Arsenic adsorptive media technology selection strategies. Water Quality Research Journal of Canada 41 (2), 171-184.
57. Cloete, T.E., Kwaadsteniet, M.d., Botes, M., Lopez-Romero, J.M., 2010. Nanotechnology in Water Treatment Applications. Caister Academic Press.
58. Li, D., Xia, Y.N., 2004. Electrospinning of nanofibers: reinventing the wheel? Advanced Materials 16 (14), 1151-1170.
59. Ramakrishna, S., Fujihara, K., Teo, W.E., Yong, T., Ma, Z.W., Ramaseshan, R., 2006. Electrospun nanofibers: solving global issues. Materials Today 9 (3), 40-50.
60. Bae, T.H., Tak, T.M., 2005. Effect of TiO_2 nanoparticles on fouling mitigation of ultrafiltration membranes for activated sludge filtration. J. Memb. Sci. 249, 1-8.
61. Bottino, A., Capannelli, G., D'Asti, V., Piaggio, P., 2001. Preparation and properties of novel organic-inorganic porous membranes. Separation and Purification Technology 22-23 (1-3), 269-275.
62. Maximous, N., Nakhla, G., Wong, K., Wan, W., 2010. Optimization of Al(2)O(3)/PES membranes for wastewater filtration. Separation and Purification Technology 73 (2), 294-301.
63. Pendergast, M.T.M., Nygaard, J.M., Ghosh, A.K., Hoek, E.M.V., 2010. Using nanocomposite materials technology to understand and control reverse osmosis membrane compaction. Desalination 261 (3), 255-263.
64. Ebert, K., Fritsch, D., Koll, J., Tjahjawiguna, C., 2004. Influence of inorganic fillers on the compaction behaviour of porous polymer based membranes. Journal of Membrane Science 233 (1-2), 71-78.
65. De Gusseme, B., Hennebel, T., Christiaens, E., Saveyn, H., Verbeken, K., Fitts, J.P., Boon, N., Verstraete, W., 2011. Virus disinfection in water by biogenic silver immobilized in polyvinylidene fluoride membranes. Water Research 45 (4), 1856-1864.
66. Mauter, M.S., Wang, Y., Okemgbo, K.C., Osuji, C.O., Giannelis, E.P., Elimelech, M., 2011. Antifouling ultrafiltration membranes via post-fabrication grafting of biocidal nanomaterials. Acs Applied Materials & Interfaces 3 (8), 2861-2868.
67. Zodrow, K., Brunet, L., Mahendra, S., Li, D., Zhang, A., Li, Q.L., Alvarez, P.J.J., 2009. Polysulfone ultrafiltration membranes impregnated with silver nanoparticles show improved biofouling resistance and virus removal. Water Research 43 (3), 715-723.

68. Brady-Estevez, A.S., Kang, S., Elimelech, M., 2008. A single-walled carbon-nanotube filter for removal of viral and bacterial pathogens. Small 4 (4), 481-484.
69. Ahmed, F., Santos, C.M., Vergara, R., Tria, M.C.R., Advincula, R., Rodrigues, D.F., 2012. Antimicrobial applications of electroactive PVK-SWNT nanocomposites. Environmental Science and Technology 46 (3), 1804-1810.
70. Choi, J.H., Jegal, J., Kim, W.N., 2006b. Fabrication and characterization of multi-walled carbon nanotubes/polymer blend membranes. Journal of Membrane Science 284 (1-2), 406-415.
71. Choi, H., Stathatos, E., Dionysiou, D.D., 2006a. Sol-gel preparation of mesoporous photocatalytic TiO_2 films and TiO_2/Al_2O_3 composite membranes for environmental applications. Applied Catalysis B-Environmental 63 (1-2), 60-67.
72. Wu, L., Shamsuzzoha, M., Ritchie, S.M.C., 2005. Preparation of cellulose acetate supported zero-valent iron nanoparticles for the dechlorination of trichloroethylene in water. Journal of Nanoparticle Research 7 (4-5), 469-476.
73. Wu, L.F., Ritchie, S.M.C., 2008. Enhanced dechlorination of trichloroethylene by membrane-supported Pd-coated iron nanoparticles. Environmental Progress 27 (2), 218-224.
74. Lind, M.L., Ghosh, A.K., Jawor, A., Huang, X.F., Hou, W., Yang, Y., Hoek, E.M.V., 2009a. Influence of zeolite crystal size on zeolite-polyamide thin film nanocomposite membranes. Langmuir 25 (17), 10139-10145.
75. Jeong, B.H., Hoek, E.M.V., Yan, Y.S., Subramani, A., Huang, X.F., Hurwitz, G., Ghosh, A.K., Jawor, A., 2007. Interfacial polymerization of thin film nanocomposites: a new concept for reverse osmosis membranes. Journal of Membrane Science 294 (1-2), 1-7.
76. Lind, M.L., Suk, D.E., Nguyen, T.V., Hoek, E.M.V., 2010. Tailoring the structure of thin film nanocomposite membranes to achieve seawater RD membrane performance. Environmental Science and Technology 44 (21), 8230-8235.
77. Lind, M.L., Jeong, B.H., Subramani, A., Huang, X.F., Hoek, E.M.V., 2009b. Effect of mobile cation on zeolite-polyamide thin film nanocomposite membranes. Journal of Materials Research 24 (5), 1624-1631.
78. Fernándezs-Ináñez, P., Sichel, C., Poloo López, M.I., de Cara-Garcia, M., Tello, J.C., 2009. Photocatalytic disinfection of natural well water contaminated by Fusarium solani using TiO_2 slurry in solar CPC photo-reactors. Catal. Today 144, 62-68.
79. Chin, S.S., Chiang, K., Fane, A.G., 2006. The stability of polymeric membranes in TiO_2 photocatalysis process. J. Memb. Sci. 275, 202-211.
80. Tiraferri, A., Vecitis, C.D., Elimelech, M., 2011. Covalent binding of single-walled carbon nanotubes to polyamide membranes for antimicrobial surface properties. Acs Applied Materials & Interfaces 3 (8), 2869-2877.
81. Kumar, M., Grzelakowski, M., Zilles, J., Clark, M., Meier, W., 2007. Highly permeable polymeric membranes based on the incorporation of the functional water channel protein Aquaporin Z. Proceedings of the National Academy of Sciences of the United States of America 104 (52), 20719-20724.
82. Kaufman, Y., Berman, A., Freger, V., 2010. Supported lipid bilayer membranes for water purification by reverse osmosis. Langmuir 26 (10), 7388-7395.
83. Holt, J.K., Park, H.G., Wang, Y.M., Stadermann, M., Artyukhin, A.B., Grigoropoulos, C.P., Noy, A., Bakajin, O., 2006. Fast mass transport through sub-2-nanometer carbon nanotubes. Science 312 (5776), 1034-1037.
84. Hummer, G., Rasaiah, J.C., Noworyta, J.P., 2001. Water conduction through the hydrophobic channel of a carbon nanotube. Nature 414 (6860), 188-190.
85. Pendergast, M.M., Hoek, E.M.V., 2011. A review of water treatment membrane nanotechnologies. Energy & Environmental Science 4 (6), 1946-1971.
86. Mauter, M.S., Elimelech, M., 2008. Environmental applications of carbon-based nanomaterials. Environmental Science and Technology 42 (16), 5843-5859.
87. Nednoor, P., Chopra, N., Gavalas, V., Bachas, L.G., Hinds, B.J., 2005. Reversible biochemical switching of ionic transport through aligned carbon nanotube membranes. Chemistry of Materials 17 (14), 3595-3599.

88. Hinds, B., 2012. Dramatic transport properties of carbon nanotube membranes for a robust protein channel mimetic platform. Current Opinion in Solid State & Materials Science 16 (1), 1-9.
89. Mauter, M.S., Elimelech, M., Osuji, C.O., 2010. Nanocomposites of vertically aligned single-walled carbon nanotubes by magnetic alignment and polymerization of a lyotropic precursor. Acs Nano 4 (11), 6651-6658.
90. Elimelech, M., Phillip, W.A., 2011. The future of seawater desalination: energy, technology, and the environment. Science 333 (6043), 712-717.
91. Ge, Q.C., Su, J.C., Chung, T.S., Amy, G., 2011. Hydrophilic superparamagnetic nanoparticles: synthesis, characterization, and performance in forward osmosis processes. Industrial & Engineering Chemistry Research 50 (1), 382-388.
92. Fujishima, A., Zhang, X., Tryk, D.A., 2008. TiO_2 photocatalysis and related surface phenomena. Surf. Sci. Rep. 63, 515-582.
93. Zhang, H.Z., Banfield, J.F., 2000. Understanding polymorphic phase transformation behavior during growth of nanocrystalline aggregates: insights from TiO_2. Journal of Physical Chemistry B 104 (15), 3481-3487.
94. Zhang, Z.B., Wang, C.C., Zakaria, R., Ying, J.Y., 1998. Role of particle size in nanocrystalline TiO2-based photocatalysts. Journal of Physical Chemistry B 102 (52), 10871-10878.
95. Macak, J.M., Zlamal, M., Krysa, J., Schmuki, P., 2007. Self-organized TiO_2 nanotube layers as highly efficient photocatalysts. Small 3 (2), 300-304.
96. Ni, M., Leung, M.K.H., Leung, D.Y.C., Sumathy, K., 2007. A review and recent developments in photocatalytic water-splitting using TiO_2 for hydrogen production. Renew. Sust. Energy Rev. 11, 401-425.
97. Han, X.G., Kuang, Q., Jin, M.S., Xie, Z.X., Zheng, L.S., 2009. Synthesis of titania nanosheets with a high percentage of exposed (001) facets and related photocatalytic properties. Journal of the American Chemical Society 131 (9), 3152.
98. Murakami, N., Kurihara, Y., Tsubota, T., Ohno, T., 2009. Shape controlled anatase titanium(IV) oxide particles prepared by hydrothermal treatment of peroxo titanic acid in the presence of polyvinyl alcohol. Journal of Physical Chemistry C 113 (8), 3062-3069.
99. Liu, S.W., Yu, J.G., Jaroniec, M., 2011b. Anatase TiO(2) with dominant high-energy {001} facets: synthesis, properties, and applications. Chemistry of Materials 23 (18), 4085-4093.
100. Ni, M., Leung, M.K.H., Leung, D.Y.C., Sumathy, K., 2007. A review and recent developments in photocatalytic water-splitting using TiO_2 for hydrogen production. Renewable & Sustainable Energy Reviews 11 (3), 401-425.
101. Kitano, M., Funatsu, K., Matsuoka, M., Ueshima, M., Anpo, M., 2006. Preparation of nitrogen-substituted TiO(2) thin film photocatalysts by the radio frequency magnetron sputtering deposition method and their photocatalytic reactivity under visible light irradiation. Journal of Physical Chemistry B 110 (50), 25266-25272.
102. Kominami, H., Yabutani, K., Yamamoto, T., Kara, Y., Ohtani, B., 2001. Synthesis of highly active tungsten(VI) oxide photocatalysts for oxygen evolution by hydrothermal treatment of aqueous tungstic acid solutions. Journal of Materials Chemistry 11 (12), 3222-3227.
103. Lee, J., Mackeyev, Y., Cho, M., Wilson, L.J., Kim, J.H., Alvarez, P.J.J., 2010. C(60) aminofullerene immobilized on silica as a visible light-activated photocatalyst. Environmental Science and Technology 44 (24), 9488-9495.
104. Lof, R., Van Veenendaal, M., Jonkman, H., Sawatzky, G., 1995. Band gap, excitons and Coulomb interactions of solid C 60. Journal of Electron Spectroscopy and Related Phenomena 72, 83-87.
105. Brunet, L., Lyon, D.Y., Hotze, E.M., Alvarez, P.J.J., Wiesner, M.R., 2009. Comparative photoactivity and antibacterial properties of C-60 fullerenes and titanium dioxide nanoparticles. Environmental Science and Technology 43 (12), 4355-4360.
106. Chong, M.N., Jin, B., Chow, C.W.K., Saint, C., 2010. Recent developments in photocatalytic water treatment technology: a review. Water Research 44 (10), 2997-3027.

107. Al-Bastaki, N.M., 2004. Performance of advanced methods for treatment of wastewater: UV/TiO2, RO and UF. Chemical Engineering and Processing 43 (7), 935-940.
108. Benotti, M.J., Stanford, B.D., Wert, E.C., Snyder, S.A., 2009. Evaluation of a photocatalytic reactor membrane pilot system of pharmaceuticals and endocrine disrupting compounds from water. Water Res. 43, 1513-1522.
109. Westerhoff, P., Moon, H., Minakata, D., Crittenden, J., 2009. Oxidation of organics in retentates from reverse osmosis wastewater reuse facilities. Water Research 43 (16), 3992-3998.
110. Nawrocki, J., Kasprzyk-Hordern, B., 2010. The efficiency and mechanisms of catalytic ozonation. Applied Catalysis B Environmental 99 (1-2), 27-42.
111. Orge, C.A., Orfao, J.J.M., Pereira, M.F.R., de Farias, A.M.D., Neto, R.C.R., Fraga, M.A., 2011. Ozonation of model organic compounds catalysed by nanostructured cerium oxides. Applied Catalysis B-Environmental 103 (1-2), 190-199.
112. Li, Q.L., Mahendra, S., Lyon, D.Y., Brunet, L., Liga, M.V., Li, D., Alvarez, P.J.J., 2008. Antimicrobial nanomaterials for water disinfection and microbial control: potential applications and implications. Water Research 42 (18), 4591-4602.
113. Xiu, Z.M., Ma, J., Alvarez, P.J.J., 2011. Differential effect of common ligands and molecular oxygen on antimicrobial activity of silver nanoparticles versus silver ions. Environmental Science and Technology 45 (20), 9003-9008.
114. Xiu, Z.M., Zhang, Q.B., Puppala, H.L., Colvin, V.L., Alvarez, J.J.P., 2012. Negligible particle-specific antibacterial activity of silver nanoparticles. Nano Letters 12 (8), 4271-4275.
115. Liau, S.Y., Read, D.C., Pugh, W.J., Furr, J.R., Russell, A.D., 1997. Interaction of silver nitrate with readily identifiable groups: relationship to the antibacterial action of silver ions. Letters in Applied Microbiology 25 (4), 279-283.
116. Liu, S.B., Zeng, T.H., Hofmann, M., Burcombe, E., Wei, J., Jiang, R.R., Kong, J., Chen, Y., 2011a. Antibacterial activity of graphite, graphite oxide, graphene oxide, and reduced graphene oxide: membrane and oxidative stress. Acs Nano 5 (9), 6971-6980.
117. Vecitis, C.D., Zodrow, K.R., Kang, S., Elimelech, M., 2010. Electronic-structure-dependent bacterial cytotoxicity of single-walled carbon nanotubes. Acs Nano 4 (9), 5471-5479.
118. Kang, S., Herzberg, M., Rodrigues, D.F., Elimelech, M., 2008a. Antibacterial effects of carbon nanotubes: size does matter. Langmuir 24 (13), 6409-6413.
119. Kang, S., Mauter, M.S., Elimelech, M., 2008b. Physicochemical determinants of multiwalled carbon nanotube bacterial cytotoxicity. Environmental Science and Technology 42 (19), 7528-7534.
120. Peter-Varbanets, M., Zurbrugg, C., Swartz, C., Pronk, W., 2009. Decentralized systems for potable water and the potential of membrane technology. Water Research 43 (2), 245-265.
121. Brady-Estevez, A.S., Schnoor, M.H., Kang, S., Elimelech, M., 2010. SWNT-MWNT hybrid filter attains high viral removal and bacterial inactivation. Langmuir 26 (24), 19153-19158.
122. Rahaman, M.S., Vecitis, C.D., Elimelech, M., 2012. Electrochemical carbon-nanotube filter performance toward virus removal and inactivation in the presence of natural organic matter. Environmental Science and Technology 46 (3), 1556-1564.
123. Vecitis, C.D., Schnoor, M.H., Rahaman, M.S., Schiffman, J.D., Elimelech, M., 2011. Electrochemical multiwalled carbon nanotube filter for viral and bacterial removal and inactivation. Environmental Science and Technology 45 (8), 3672-3679.
124. Theron, J., Cloete, T.E., de Kwaadsteniet, M., 2010. Current molecular and emerging nanobiotechnology approaches for the detection of microbial pathogens. Critical Reviews in Microbiology 36 (4), 318-339.
125. Vikesland, P.J., Wigginton, K.R., 2010. Nanomaterial enabled biosensors for pathogen monitoring – a review. Environmental Science and Technology 44 (10), 3656-3669.
126. Yan, J.L., Estevez, M.C., Smith, J.E., Wang, K.M., He, X.X., Wang, L., Tan, W.H., 2007. Dye-doped nanoparticles for bioanalysis. Nano Today 2 (3), 44-50.
127. Petryayeva, E., Krull, U.J., 2011. Localized surface plasmon resonance: nanostructures, bioassays and biosensing e a review. Analytica Chimica Acta 706 (1), 8-24.

128. Lei, J.P., Ju, H.X., 2012. Signal amplification using functional nanomaterials for biosensing. Chemical Society Reviews 41 (6), 2122-2134.
129. Jain, P.K., Lee, K.S., El-Sayed, I.H., El-Sayed, M.A., 2006. Calculated absorption and scattering properties of gold nanoparticles of different size, shape, and composition: applications in biological imaging and biomedicine. Journal of Physical Chemistry B 110 (14), 7238-7248.
130. Kelly, K.L., Coronado, E., Zhao, L.L., Schatz, G.C., 2003. The optical properties of metal nanoparticles: the influence of size, shape, and dielectric environment. Journal of Physical Chemistry B 107 (3), 668-677.
131. Moskovits, M., 2005. Surface-enhanced Raman spectroscopy: a brief retrospective. Journal of Raman Spectroscopy 36 (6-7), 485-496.
132. Nie, S.M., Emery, S.R., 1997. Probing single molecules and single nanoparticles by surface-enhanced Raman scattering. Science 275 (5303), 1102-1106.
133. McCreery, R.L., 2008. Advanced carbon electrode materials for molecular electrochemistry. Chemical Reviews 108 (7), 2646-2687.
134. Yang, W.R., Ratinac, K.R., Ringer, S.P., Thordarson, P., Gooding, J.J., Braet, F., 2010b. Carbon nanomaterials in biosensors: should you use nanotubes or graphene? Angewandte Chemie-International Edition 49 (12), 2114-2138.
135. Heller, I., Janssens, A.M., Mannik, J., Minot, E.D., Lemay, S.G., Dekker, C., 2008. Identifying the mechanism of biosensing with carbon nanotube transistors. Nano Letters 8 (2), 591-595.
136. Duran, A., Tuzen, M., Soylak, M., 2009. Preconcentration of some trace elements via using multiwalled carbon nanotubes as solid phase extraction adsorbent. Journal of Hazardous Materials 169 (1-3), 466-471.
137. Cai, Y.Q., Jiang, G.B., Liu, J.F., Zhou, Q.X., 2003. Multiwalled carbon nanotubes as a solid-phase extraction adsorbent for the determination of bisphenol a, 4-n-nonylphenol, and 4-tert-octylphenol. Analytical Chemistry 75 (10), 2517-2521.
138. Lisha, K.P., Anshup, Pradeep, T., 2009. Enhanced visual detection of pesticides using gold nanoparticles. Journal of Environmental Science and Health Part B-Pesticides Food Contaminants and Agricultural Wastes 44 (7), 697-705.
139. Yang, L.X., Chen, B.B., Luo, S.L., Li, J.X., Liu, R.H., Cai, Q.Y., 2010a. Sensitive detection of polycyclic aromatic hydrocarbons using CdTe quantum dot-modified TiO(2) nanotube array through fluorescence resonance energy transfer. Environmental Science and Technology 44 (20), 7884-7889.
140. Yin, H.S., Zhou, Y.L., Ai, S.Y., Chen, Q.P., Zhu, X.B., Liu, X.G., Zhu, L.S., 2010. Sensitivity and selectivity determination of BPA in real water samples using PAMAM dendrimer and CoTe quantum dots modified glassy carbon electrode. Journal of Hazardous Materials 174 (1-3), 236-243.
141. da Silva, B.F., Perez, S., Gardinalli, P., Singhal, R.K., Mozeto, A.A., Barcelo, D., 2011. Analytical chemistry of metallic nanoparticles in natural environments. TrAc Trends in Analytical Chemistry 30 (3), 528-540.
142. Tiede, K., Boxall, A.B.A., Tear, S.P., Lewis, J., David, H., Hassellov, M., 2008. Detection and characterization of engineered nanoparticles in food and the environment. Food Additives and Contaminants 25 (7), 795-821.
143. Malato, S., Fernández-Ibáñez, P., Maldonado, M.I., Blanco, J., Gernjak, W., 2009. Decontamination and disinfection of water by solar photocatalysis: recent overview and trends. Catal. Today 147, 1-59.
144. Richardson, S.D., 2008. Environmental mass spectrometry: emerging contaminants and current issues. Anal. Chem. 80, 4373-4402.
145. Suárez, S., Carballa, M., Omil, F., Lema, J.M., 2008. How are pharmaceutical and personal care products (PPCPs) removed from urban wastewaters? Rev. Environ. Sci. Biotechnol. 7, 125-138.
146. Wintgens, T., Salehi, F., Hochstrat, R., Melin, T., 2008. Emerging contaminants and treatment options in water recycling for indirect potable use. Water Sci. Technol. 57, 99-107.

147. Bradley, B.R., Daigger, G.T., Rubin, R., Tchobanoglous, G., 2002. Evaluation of onsite wastewater treatment technologies using sustainable development criteria. Clean Technol. Environ. Policy 4, 87-99.
148. Lapña, L., Cerezo, M., Garía-Augustin, P., 1995. Possible reuse of treated municipal wastewater for Citrus spp. plant irrigation. Bull. Environ. Contam. Toxicol. 55, 697-703.
149. Viessman Jr., W., Hammer, M.J., 1998. Water Supply and Pollution Control, sixth ed. Addison Wesley Longman Inc, California USA.
150. Padmanabhan, P.V.A., Sreekumar, K.P., Thiyagarajan, T.K., Satpute, R.U., Bhanumurthy, K., Sengupta, P., Dey, G.K., Warrier, K.G.K., 2006. Nano-crystalline titanium dioxide formed by reactive plasma synthesis. Vacuum 80, 11-12.
151. Gaya, U.I., Abdullah, A.H., 2008. Heterogeneous photocatalytic degradation of organic contaminants over titanium dioxide: a review of fundamentals, progress and problems. J. Photochem. Photobiol. C: Photochem. Rev. 9, 1-12.
152. Coleman, H.M., Marquis, C.P., Scott, J.A., Chin, S.S., Amal, R., 2005. Bactericidal effects of titanium dioxide-based photocatalysts. Chem. Eng. J. 113, 55-63.
153. Lu, J., Zhang, T., Ma, J., Chen, Z., 2009. Evaluation of disinfection by-products formation during chlorination and chloramination of dissolved natural organic matter fractions isolated from a filtered river water. J. Hazard. Mater. 162, 140-145.
154. Yang, H., Cheng, H., 2007. Controlling nitrite level in drinking water by chlorination and chloramination. Sep. Purif. Technol. 56, 392-396.
155. Esplugas, S., Giménez, J., Conteras, S., Pascual, E., Rodríguez, M., 2002. Comparison of different advanced oxidation processes for phenol degradation. Water Res. 36, 1034-1042.
156. Pera-Titus, M., García-Molina, V., Baños, M.A., Giménez, J., Esplugas, S., 2004. Degradation of chlorophenols by means of advanced oxidation processes: a general review. Appl. Catal. B: Environ. 47, 219-256.
157. Fujishima, A., Rao, T.N., Tryk, D.A., 2000. Titanium dioxide photocatalysis. J. Photochem. Photobiol. C: Photochem. Rev. 1, 1-21.
158. Furube, A., Asahi, T., Masuhara, H., Yamashita, H., Anpo, M., 2001. Direct observation of a picosecond charge separation process in photoexcited platinum-loaded TiO_2 particles by femtosecond diffuse reflectance spectroscopy. Chem. Phys. Lett. 336, 424-430.
159. Byrne, J.A., Eggins, B.R., 1998. Photoelectrochemistry of oxalate on particulate TiO_2 electrodes. J. Electroanal. Chem. 457, 61-72.
160. Herrmann, J.M., 1999. Heterogeneous photocatalysis: fundamentals and applications to the removal of various types of aqueous pollutants. Catal. Today 53, 115-129.
161. Vinodgopal, K., Kamat, P.V., 1992. Photochemistry on surfaces: photodegradation of 1,3-diphenylisobenzofuran over metal oxide particles. J. Phys. Chem. 96, 5053-5059.
162. Maness, P.C., Smolinski, S., Blake, D.M., Huang, Z., Wolfrum, E.J., Jacoby, W.A., 1999. Bactericidal activity of photocatalytic TiO_2 reaction: toward an understanding of its killing mechanism. Appl. Environ. Microbiol. 65, 4094-4098.
163. Bacardit, J., Stötzner, J., Chamarro, E., 2007. Effect of salinity on the photo-Fenton process. Ind. Eng. Chem. Res. 46, 7615-7619.
164. Machulek Jr., A., Moraes, J.E.F., Vautier-Giongo, C., Silverio, C.A., Friedrich, L.C., Nascimento, C.A.O., Gonzalez, M.C., Quina, F.H., 2007. Abatement of the inhibitory effect of chloride anions on the photo-Fenton process. Environ. Sci. Technol. 41, 8459-8463.
165. Neyens, E., Baeyens, J., 2003. A review of classic Fenton's peroxidation as an advanced oxidation technique. Water Res. 98, 33-50.
166. Fallmann, H., Krutzler, T., Bauer, R., Malato, S., Blanco, J., 1999. Applicability of the photo-Fenton method for treating water containing pesticides. Catal. Today 54, 309-319.
167. Gernjak, W., Krutzler, T., Malato, S., 2007. Photo-Fenton treatment of olive mill wastewater applying a combined Fenton/flocculation pretreatment. J. Sol. Energy Eng. 129, 53-59.
168. Huston, P.L., Pignatello, J.J., 1999. Degradation of selected pesticide active ingredients and commercial formulations in water by the photo-assisted Fenton reaction. Water Res. 33, 1238-1246.

169. Gogate, P.R., Pandit, A.B., 2004. A review of imperative technologies for wastewater treatment II: hybrid methods. Adv. Environ. Res. 8, 553-597.
170. Pignatello, J.J., Oliveros, E., MacKay, A., 2006. Advanced oxidation processes for organic contaminant destruction based on the Fenton reaction and related chemistry. Crit. Rev. Environ. Sci. Technol. 36, 1-84.
171. De Laat, J., Le, G.T., Legube, B., 2004. A comparative study of the effects of chloride, sulphate and nitrate ions on the rates of decomposition of H_2O_2 and organic compounds by Fe(II)/H_2O_2 and Fe(III)/H_2O_2. Chemosphere 55, 715-723.
172. Oliveros, E., Legrini, O., Hohl, M., Müller, T., Braun, A.M., 1997. Industrial wastewater treatment: large scale development of a light-enhanced Fenton reaction. Chem. Eng. Proc. 36, 397-405.
173. Torrades, F., Pérez, M., Mansilla, H.D., Peral, J., 2003. Experimental design of Fenton and photo-Fenton reactions for the treatment of cellulose bleaching effluents. Chemosphere 53, 1211-1220.
174. Dominguez, C., García, J., Pedraz, M.A., Torres, A., Galán, M.A., 1998. Photocatalytic oxidation of organic pollutants in water. Catal. Today 40, 85-101.
175. Marugán, J., Lopez-Muñoz, M.J., Gernjak, W., Malato, S., 2006. Fe/TiO_2/pH interactions in solar degradation of imidacloprid with TiO_2/SiO_2 photocatalysts at pilot-plant scale. Ind. Eng. Chem. Res. 45, 8900-8908.
176. Marugán, J., van Grieken, R., Sordo, C., Cruz, C., 2008. Kinetics of the photocatalytic disinfection of Escherichia coli suspensions. Appl. Catal. B: Environ. 82, 27-36.
177. Rincón, A.G., Pulgarin, C., 2006. Comparative evaluation of Fe^{3+} and TiO_2 photoassisted processes in solar photocatalytic disinfection of water. Appl. Catal. B: Environ. 63, 222-231.
178. Beltran-Heredia, J., Torregrosa, J., Dominguez, J.R., Peres, J.A., 2001. Comparison of the degradation of p-hydroxybenzoic acid in aqueous solution by several oxidation processes. Chemosphere 42, 351-359.
179. Torres, R.A., Nieto, J.I., Combet, E., Pétrier, C., Pulgarin, C., 2008. Influence of TiO_2 concentration on the synergistic effect between photocatalysis and high-frequency ultrasound for organic pollutant mineralization in water. Appl. Catal. B: Environ. 80, 168-175.
180. Fujishima, A., Honda, K., 1972. Electrochemical photolysis of water at a semiconductor electrode. Nature 238, 37-38.
181. Hosono, E., Fujihara, S., Kakiuchi, K., Imai, H., 2004. Growth of submicrometer-scale rectangular parallelepiped rutile TiO2 films in aqueous $TiCl_3$ solutions under hydrothermal conditions. J. Am. Chem. Soc. 126, 7790-7791.
182. Kondo, Y., Yoshikawa, H., Awaga, K., Murayama, M., Mori, T., Sunada, K., Bandow, S., Iijima, S., 2008. Preparation, photocatalytic activities, and dye-sensitized solar-cell performance of submicron-scale TiO_2 hollow spheres. Langmuir 24, 547-550.
183. Wang, R., Hashimoto, K., Fujishima, A., Chikuni, M., Kojima, E., Kitamura, A., Shimohigoshi, M., Watanabe, T., 1999. Photogeneration of highly amphiphilic TiO_2 surfaces. Adv. Mater. 10, 135-138.
184. Nagaveni, K., Sivalingam, G., Hegde, M.S., Madras, G., 2004a. Solar photocatalytic degradation of dyes. High activity of combustion synthesized nano TiO_2. Appl. Catal. B: Environ. 48, 83-93.
185. Nagaveni, K., Sivalingam, G., Hegde, M.S., Madras, G., 2004b. Photocatalytic degradation of organic compounds over combustion synthesized nano-TiO_2. Environ. Sci. Technol. 38, 1600-1604.
186. Siddiquey, I.A., Furusawa, T., Sato, M., Honda, K., Suzuki, N., 2008. Control of the photocatalytic activity of TiO_2 nanoparticles by silica coating with polydiethoxysiloxane. Dyes Pigm. 76, 754-759.
187. Byrne, J.A., Eggins, B.R., Brown, N.M.D., McKinley, B., Rouse, M., 1998b. Immobilisation of TiO2 powder for the treatment of polluted water. Appl. Catal. B: Environ. 17, 25-36.
188. Yu, J.C., Yu, J., Zhao, J., 2002. Enhanced photocatalytic activity of mesoporous and ordinary TiO2 thin films by sulphuric acid treatment. Appl. Catal. B: Environ. 36, 31-43.

189. Serpone, N., Sauvé, G., Koch, R., Tahiri, H., Pichat, P., Piccinini, P., Pelizetti, E., Hidaka, H., 1996. Standardization protocol of process efficiencies and activation parameters in heterogeneous photocatalysis: relative photonic efficiencies zr. J. Photochem. Photobiol. A: Chem. 94, 191-203.
190. Pozzo, R.L., Baltanás, M.A., Cassano, A.E., 1997. Supported titanium dioxide as photocatalyst in water decontamination: state of the art. Catal. Today 39, 219-231.
191. Yang, G.C.C., Li, C.J., 2007. Electrofiltration of silica nanoparticle containing wastewater using tubular ceramic membranes. Sep. Purif. Technol. 58, 159-165.
192. Doll, T.E., Frimmel, F.H., 2005. Cross-flow microfiltration with periodical back-washing for photocatalytic degradation of pharmaceutical and diagnostic residues-evaluation of the long-term stability of the photocatalytic activity of TiO_2. Water Res. 39, 847-854.
193. Fernández-Ibáñez, P., Blanco, J., Malato, S., de las Nieves, F.J., 2003. Application of the colloidal stability of TiO_2 particles for recovery and reuse in solar photocatalysis. Water Res. 37, 3180-3188.
194. Zhang, X., Du, A.J., Lee, P., Sun, D.D., Leckie, J.O., 2008a. TiO_2 nanowire membrane for concurrent filtration and photocatalytic oxidation of humic acid in water. J. Memb. Sci. 313, 44-51.
195. Zhao, Y., Zhong, J., Li, H., Xu, N., Shi, J., 2002. Fouling and regeneration of ceramic microfiltration membranes in processing acid wastewater containing fine TiO_2 particles. J. Memb. Sci. 208, 331-341.
196. Lee, S.A., Choo, K.H., Lee, C.H., Lee, H.I., Hyeon, T., Choi, W., Kwon, H.H., 2001. Use of ultrafiltration membranes for the separation of TiO_2 photocatalysts in drinking water treatment. Ind. Eng. Chem. Res. 40, 1712-1719.
197. Molinari, R., Palmisano, L., Drioli, E., Schiavello, M., 2002. Studies on various reactor configurations for coupling photocatalysis and membrane processes in water purification. J. Memb. Sci. 206, 399-415.
198. Xi, W., Geissen, S., 2001. Separation of titanium dioxide from photocatalytically treated water by cross-flow microfiltration. Water Res. 35, 1256-1262.
199. Zhang, X., Pan, J.H., Du, A.J., Fu, W., Sun, D.D., Leckie, J.O., 2009. Combination of one-dimensional TiO2 nanowire photocatalytic oxidation with microfiltration for water treatment. Water Res. 43, 1179-1186.
200. Chong, M.N., Vimonses, V., Lei, S., Jin, B., Chow, C., Saint, C., 2009a. Synthesis and characterisation of novel titania impregnated kaolinite nano-photocatalyst. Microporus Mesoporus Mater. 117, 233-242.
201. Kwak, S.Y., Kim, S.H., 2001. Hybrid organic/inorganic reverse osmosis (RO) membrane for bactericidal anti-fouling. 1. Preparation and characterization of TiO2 nanoparticle self-assembled aromatic polyamide thin-film-composite (TFC) membrane. Environ. Sci. Technol. 35, 2388-2394.
202. Lee, D.K., Kim, S.C., Cho, I.C., Kim, S.J., Kim, S.W., 2004. Photocatalytic oxidation of microcystin-LR in a fluidized bed reactor having TiO_2-coated activated carbon. Sep. Purif. Technol. 34, 59-66.
203. Zhu, H., Gao, X., Lan, Y., Song, D., Xi, Y., Zhao, J., 2004. Hydrogen titanate nanofibers covered with anatase nanocrystals: a delicate structure achieved by the wet chemistry reaction of the titanate nanofibers. J. Am. Chem. Soc. 126, 8380-8381.
204. Fukahori, S., Ichiura, H., Kitaoka, T., Tanaka, H., 2003. Capturing of bisphenol A photodecomposition intermediates by composite TiO2–zeolite sheets. Appl. Catal. B: Environ. 46, 453-462.
205. Kun, R., Mogyorósi, K., Dekány, I., 2006. Synthesis and structural and photocatalytic properties of TiO_2/montmorillonite nanocomposites. Appl. Clay Sci. 32, 99-110.
206. Sun, Z., Chen, Y., Ke, Q., Yang, Y., Yuan, J., 2002. Photocatalytic degradation of a cationic azo dye by TiO_2/bentonite nanocomposite. J. Photochem. Photobiol. A: Chem. 149, 169-174.
207. Xie, Z.M., Chen, Z., Dai, Y.Z., 2009. Preparation of TiO_2/sepiolite photocatalyst and its application to printing and dyeing wastewater treatment. Environ. Sci. Technol. 32, 123-127.

208. Chong, M.N., Lei, S., Jin, B., Saint, C., Chow, C.W.K., 2009b. Optimisation of an annular photoreactor process for degradation of Congo red using a newly synthesized titania impregnated kaolinite nano-photocatalyst. Sep. Purif. Technol. 67, 355-363.
209. Sun, D., Meng, T.T., Loong, T.H., Hwa, T.J., 2004. Removal of natural organic matter from water using a nano-structured photocatalyst coupled with filtration membrane. Water Sci. Technol. 49, 103-110.
210. Zhang, X., Du, A.J., Lee, P., Sun, D.D., Leckie, J.O., 2008b. Grafted multifunctional titanium dioxide nanotube membrane: separation and photodegradation of aquatic pollutant. Appl. Catal. B: Environ. 84, 262-267.
211. Bosc, F., Ayral, A., Guizard, C., 2005. Mesoporous anatase coatings for coupling membrane separation and photocatalyzed reactions. J. Memb. Sci. 265, 13-19.
212. Choi, H., Stathatos, E., Dionysiou, D.D., 2005. Sol–gel preparation of mesoporous photocatalytic TiO_2 films and TiO_2/Al_2O_3 composite membranes for environmental applications. Appl. Catal. B: Environ. 63, 60-67.
213. Choi, H., Stathatos, E., Dionysiou, D.D., 2007. Photocatalytic TiO_2 films and membranes for the development of efficient wastewater treatment and reuse systems. Desalination 202, 199-206.
214. Zhang, H., Quan, X., Chen, S., Zhao, H., Zhao, Y., 2006b. The removal of sodium dodecylbenzene sulfonate surfactant from water using silica/titania nanorods/nanotubes composite membrane with photocatalytic capability. Appl. Surf. Sci. 252, 8598-8604.
215. Artale, M.A., Augugliaro, V., Drioli, E., Golemme, G., Grande, C., Loddo, V., Molinari, R., Palmisano, L., Schiavello, M., 2001. Preparation and characterisation of membranes with entrapped TiO_2 and preliminary photocatalytic tests. Ann. Chim. 91, 127-136.
216. Bellobono, I.R.,Morazzoni, F., Bianchi, R.,Mangone, E.S.,Stancscu, R., Costache, C., Tozzi, P.M., 2005a. Solar energy driven photocatalytic membrane modules for water reuse in agricultural and food industries. Pre-industrial experience using s-triazines as model molecules. Int. J. Photoenergy 7, 87-94.
217. Kim, S.H., Kwak, S.Y., Sohn, B.H., Park, T.H., 2003. Design of TiO_2 nanoparticle self-assembled aromatic polyamide thin-film composite (TFC) membrane as an approach to solve biofouling problem. J. Memb. Sci. 211, 157-165.
218. Kleine, J., Peinemann, K.V., Schuster, C., Warnecke, H.J., 2002. Multifunctional system for treatment of wastewaters from adhesive-producing industries: separation of solids and oxidation of dissolved pollutants using doted microfiltration membranes. Chem. Eng. Sci. 57, 1661-1664.
219. Molinari, R., Pirillo, F., Falco, M., Loddo, V., Palmisano, L., 2004. Photocatalytic degradation of dyes by using a membrane reactor. Chem. Eng. Proc. 43, 1103-1114.
220. Yang, Y., Wang, P., 2006. Preparation and characterizations of a new PS/TiO_2 hybrid membrane by sol–gel process. Polymer 47, 2683-2688.
221. Albu, S.P., Ghicov, A., Macak, J.M., Hahn, R., Schmuki, P., 2007. Self-organized, free-standing TiO_2 nanotube membrane for flow through photocatalytic applications. Nano Lett. 7, 1286-1289.
222. Litter, M.I., 1999. Heterogeneous photocatalysis: transition metal ions in photocatalytic systems. Appl. Catal. B: Environ. 23, 89-114.
223. Vinodgopal, K., Wynkoop, D.E., Kamat, P.V., 1996. Environmental photochemistry on semiconductor surfaces: photosensitized degradation of a textile azo dye, Acid Orange 7, on TiO_2 particles using visible light. Environ. Sci. Technol. 30, 1660-1666.
224. Yu, Y., Yu, J.C., Yu, J.G., Kwok, Y.C., Che, Y.K., Zhao, J.C., Ding, L., Ge, W.K., Wong, P.K., 2005. Enhancement of photocatalytic activity of mesoporous TiO_2 by using carbon nanotubes. Appl. Catal. A: Gen. 289, 186-196.
225. Ishibai, Y., Sato, J., Nishikawa, T., Miyagishi, S., 2008. Synthesis of visible-light active TiO_2 photocatalyst with Pt-modification: role of TiO_2 substrate for high photocatalytic activity. Appl. Catal. B: Environ. 79, 117-121.

226. Li, H., Li, J., Huo, Y., 2006. Highly active TiO_2N photocatalysts prepared by treating TiO_2 precursors in NH_3/ethanol fluid under supercritical conditions. J. Phys. Chem. B 110, 1559-1565.
227. Shaban, Y.A., Khan, S.U.M., 2008. Visible light active carbon modified n-TiO_2 for efficient hydrogen production by photoelectrochemical splitting of water. Int. J. Hydrogen Energy 33, 1118-1126.
228. Asahi, R., Morikawa, T., Ohwaki, T., Aoki, K., Taga, Y., 2001. Visible-light photocatalysis in nitrogen-doped titanium dioxides. Science 293, 269-271.
229. Ihara, T., Miyoshi, M., Iriyama, Y., Matsumoto, O., Sugihara, S., 2003. Visible-light-active titanium dioxide photocatalyst realized by an oxygen-deficient structure and by nitrogen doping. Appl. Catal. B: Environ. 42, 403-409.
230. Irie, H., Watanabe, Y., Hashimoto, K., 2003. Nitrogen concentration dependence on photocatalytic activity of $TiO_{2-x}N_x$ powders. J. Phys. Chem. B 107, 5483-5486.
231. Chen, D., Ray, A.K., 1998. Photodegradation kinetics of 4-nitrophenol in TiO_2 suspension. Water Res. 32, 3223-3234.
232. Tsarenko, S.A., Kochkodan, V.M., Samsoni-Todorov, A.O., Goncharuk, V.V., 2006. Removal of humic substances from aqueous solutions with a photocatalytic membrane reactor. Colloid J. 68, 341-344.
233. Pozzo, R.L., Giombi, J.L., Baltanas, M.A., Cassano, A.E., 2000. The performance in a fluidized bed reactor of photocatalysts immobilized onto inert supports. Catal. Today 62, 175-187.
234. Chan, A.H.C., Chan, C.K., Barford, J.P., Porter, J.F., 2003. Solar photocatalytic thin film cascade reactor for treatment of benzoic acid containing wastewater. Water Res. 37, 1125-1135.
235. Ochuma, I.J., Fishwick, R.P., Wood, J., Winterbottom, J.M., 2007. Optimisation of degradation conditions of 1,8-diazabicyclo [5. 4.0] undec-7-ene in water and reaction kinetics analysis using a cocurrent downflow contactor photocatalytic reactor. Appl. Catal. B: Environ. 73, 259-268.
236. Pareek, V., Chong, S., Tadé, M., Adesina, A.A., 2008. Light intensity distribution in heterogeneous photocatalytic reactors. Asia-Pacific J. Chem. Eng. 3, 171-201.
237. Fu, J., Ji, M., Wang, Z., Jin, L., An, D., 2006. A new submerged membrane photocatalysis reactor (SMPR) for fulvic acid removal using a nano-structured photocatalyst. J. Hazard. Mater. 131, 238-242.
238. Molinari, R., Grande, C., Drioli, E., Palmisano, L., Schiavello, M., 2001. Photocatalytic membrane reactors for degradation of organic pollutants in water. Catal. Today 67, 273-279.
239. Chin, S.S., Lim, T.M., Chiang, K., Fane, A.G., 2007. Hybrid low pressure submerged membrane photoreactor for the removal of bisphenol A. Desalination 2002, 253-261.
240. Huang, X., Meng, Y., Liang, P., Qian, Y., 2007. Operational conditions of a membrane filtration reactor coupled with photocatalytic oxidation. Sep. Purif. Technol. 55, 165-172.
241. Meng, Y., Huang, X., Yang, Q., Qian, Y., Kubota, N., Fukunaga, S., 2005. Treatment of polluted river water with a photocatalytic slurry reactor using low-pressure mercury lamps coupled with a membrane. Desalination 181, 121-133.
242. Sopajaree, K., Qasim, S.A., Basak, S., Rajeshwar, K., 1999a. An integrated flow reactor-membrane filtration system for heterogeneous photocatalysis. Part I. Experiments and modelling of a batch-recirculated photoreactor. J. Appl. Electrochem. 29, 533-539.
243. Sopajaree, K., Qasim, S.A., Basak, S., Rajeshwar, K., 1999b. An integrated flow-reactor membrane filtration system for heterogeneous photocatalysis. Part II. Experiments on the ultrafiltration unit and combined operation. J. Appl. Electrochem. 29, 1111-1118.
244. Augugliaro, V., García-López, E., Loddo, V., Malato-Rodíguez, S., Maldonado, I., Marcí, G., Molinari, R., Palmisano, L., 2005. Degradation of lincomycin in aqueous medium: coupling of solar photocatalysis and membrane separation. Sol. Energy 79, 402-408.
245. Molinari, R., Pirilla, F., Loddo, V., Palmisano, L., 2006. Heterogeneous photocatalytic degradation of pharmaceuticals in water by using polycrystalline TiO_2 and a nanofiltration membrane reactor. Catal. Today 118, 205-213.

246. Jung, J.T., Kim, J.O., Choi, W.Y., 2007. Performance of photocatalytic microfiltration with hollow fiber membrane. Mater. Sci. Forum 544, 95-98.
247. Mozia, S., Morawski, A.W., Toyoda, M., Tsumura, T., 2009. Effect of process parameters on photodegradation of Acid Yellow 36 in a hybrid photocatalysis–membrane distillation system. Chem. Eng. J. 150, 152-159.
248. Azrague, K., Aimar, P., Benoit-Marqué, F., Maurette, M.T., 2006. A new combination of a membrane and photocatalytic reactor for the depollution of turbid water. Appl. Catal. B: Environ. 72, 197-205.
249. Camera-Roda, G., Santarelli, F., 2007. Intensification of water detoxification by integrating photocatalysis and pervaporation. J. Sol. Energy Eng. 129, 68-73.
250. Augugliaro, V., Litter, M., Palmisano, L., Soria, J., 2006. The combination of heterogeneous photocatalysis with chemical and physical operations: a tool for improving the photoprocess performance. J. Photochem. Photobiol. C: Photochem. Rev. 7, 127-144.
251. Lee, H.S., Im, S.J., Kim, J.H., Kim, H.J., Kim, J.P., Min, B.R., 2008. Polyamide thin-film nanofiltration membranes containing TiO_2 nanoparticles. Desalination 219, 48-56.
252. Meares, P., 1986. Synthetic Membranes: Science, Engineering and Applications. Springer Publisher, first ed. Peidel, Dordrecht, Netherlands.
253. Gryta, M., Tomaszewska, M., Grzechulska, J., Morawski, A.W., 2001. Membrane distillation of NaCl solution containing natural organic matter. J. Memb. Sci. 181, 279-287.
254. Tomaszewska, M., Gryta, M., Morawski, A.W., 1998. The influence of salt in solution on hydrochloric acid recovery by membrane distillation. Sep. Purif. Technol. 14, 183-188.
255. Bouguecha, S., Hamrouni, B., Dhahbi, M., 2005. Small scale desalination pilots powered by renewable energy sources: case studies. Desalination 183, 151-165.
256. Lawson, K.W., Lloyd, D.R., 1997. Membrane distillation. J. Memb. Sci. 124, 1-25.
257. Bamba, D., Athcba, P., Robert, D., Trokourey, A., Dongui, B., 2008. Photocatalytic degradation of the diuron pesticide. Environ. Chem. Lett. 6, 163-167.
258. Chong, M.N., Jin, B., Zhu, H.Y., Chow, C.W.K., Saint, C., 2009c. Application of H-titanate nanofibers for degradation of Congo red in an annular slurry photoreactor. Chem. Eng. J. 150, 49-54.
259. Malato, S., Blanco, J., Campos, A., Cáceres, J., Guillard, C., Herrmann, J.M., Fernández-Alba, A.R., 2003. Effect of operating parameters on the testing of new industrial titania catalysts at solar pilot plant scale. Appl. Catal. B: Environ. 42, 349-357.
260. Xu, Y., Langford, C.H., 2000. Variation of Langmuir adsorption constant determined for TiO_2-photocatalyzed degradation if acetophenone under different light intensity. J. Photochem. Photobiol. A: Chem. 133, 67-71.
261. Gogniat, G., Thyssen, M., Denis, M., Pulgarin, C., Dukan, S., 2006. The bactericidal effect of TiO_2 photocatalysis involves adsorption onto catalyst and the loss of membrane integrity. FEMS Microbiol. Lett. 258, 18-24.
262. Toor, A.P., Verma, A., Jotshi, C.K., Bajpai, P.K., Singh, V., 2006. Photocatalytic degradation of Direct Yellow 12 dye using UV/TiO_2 in a shallow pond slurry reactor. Dyes Pigm. 68, 53-60.
263. Rincón, A.G., Pulgarin, C., 2004. Effect of pH, inorganic ions, organic matter and H_2O_2 on E. coli K12 photocatalytic inactivation by TiO_2-implications in solar water disinfection. Appl. Catal. B: Environ. 51, 283-302.
264. Körmann, C., Bahnemann, D.W., Hoffman, M.R., 1991. Photolysis of chloroform and other organic molecules in aqueous titanium dioxide suspensions. Environ. Sci. Technol. 25, 494-500.
265. Stylidi, M., Kondarides, D.I., Verykios, X.E., 2003. Pathways of solar light-induced photocatalytic degradation of azo dyes in aqueous TiO_2 suspensions. Appl. Catal. B: Environ. 40, 271-286.
266. Herrera Melián, J.A., Doña Rodíguez, J.M., Viera Suárez, A., Valés do Campo, C., Arana, J., Pérez Peña, J., 2000. The photocatalytic disinfection of urban waste waters. Chemosphere 41, 323-327.

267. Heyde, M., Portalier, R., 1990. Acid shock proteins of Escherichia coli. FEMS Microbiol. Lett. 69, 19-26.
268. Ibáñez, J.A., Litter, M.I., Pizarro, R.A., 2003. Photocatalytic bactericidal effect of TiO_2 on Enterobacter cloacae: comparative study with other Gram (–) bacteria. J. Photochem. Photobiol. A: Chem. 157, 81-85.
269. Lonnen, J., Kilvington, S., Kehoe, S.C., Al-Touati, F., McGuigan, K.G., 2005. Solar and photocatalytic disinfection of protozoan, fungal and bacterial microbes in drinking water. Water Res. 39, 877-883.
270. Matsunaga, T., Tomoda, R., Nakajima, T., Wake, H., 1985. Photoelectrochemical sterilization of microbial cells by semiconductor powders. FEMS Microbiol. Lett. 29, 211-214.
271. Sichel, C., Tello, J., de Cara, M., Fernández-Ibáñez, P., 2007. Effect of UV solar intensity and dose on the photocatalytic disinfection of bacteria and fungi. Catal. Today 129, 152-160.
272. Blanco, J., Malato, S., de las Nieves, J., Fernández, P., 2001. Method of sedimentation of colloidal semiconductor particles, European patent application EP-1-101-737-A1, European Patent Office Bulletin 21.
273. Molinari, R., Caruso, A., Argurio, P., Poerio, T., 2008. Degradation of the drugs Gemfibrozil and Tamoxifen in pressurized and de-pressurized membrane photoreactors using suspended polycrystalline TiO_2 as catalyst. J. Memb. Sci. 319, 54-63.
274. Seffaj, N., Persin, M., Alami Younssi, S., Albizane, A., Bouhria, M., Loukili, H., Larbot, A., 2005. Removal of salts and dyes by low $ZnAl_2O_4$-TiO_2 ultrafiltration membrane deposited on support made from raw clay. Sep. Purif. Technol. 47, 36-42.
275. Fu, X., Clark, L.A., Zeltner, W.A., Anderson, M.A., 1996. Effects of reaction temperature and water vapour content on the heterogeneous photocatalytic oxidation of ethylene. J. Photochem. Photobiol. A: Chem. 97, 181-186.
276. Muradov, N.Z., Raissi, A.T., Muzzey, D., Painter, C.R., Kemme, M.R., 1996. Selective photocatalytic degradation of airborne VOCs. Sol. Energy 56, 445-453.
277. Rincón, A.G., Pulgarin, C., 2003. Photocatalytical inactivation of E. coli: effect of (continuous-intermittent) light intensity and of (suspended-fixed) TiO_2 concentration. Appl. Catal. B: Environ. 44, 263-284.
278. Herrmann, J.G., 2005. Research to protect water infrastructure: EPA's water security research program. Proc. SPIE 5781, 48.
279. Wang, Y., Hong, C.S., 2000. TiO_2-mediated photomineralization of 2-chlorobiphenyl: the role of O_2. Water Res. 34, 2791-2797.
280. Shirayama, H., Tohezo, Y., Taguchi, S., 2001. Photodegradation of chlorinated hydrocarbons in the presence and absence of dissolved oxygen in water. Water Res. 35, 1941-1950.
281. Saquib, M., Muneer, M., 2003. TiO_2-mediated photocatalytic degradation of a triphenylmethane dye (gentian violet), in aqueous suspensions. Dyes Pigm. 56, 37-49.
282. Bahnemann, D., 2004. Photocatalytic water treatment: solar energy applications. Sol. Energy 77, 445-459.
283. Bhatkhnade, D.S., Kamble, S.P., Sawant, S.B., Pangarkar, V.G., 2004. Photocatalytic and photochemical degradation of nitrobenzene using artificial ultraviolet light. Chem. Eng. J. 102, 283-290.
284. Malato-Rodíguez, S., Richter, C., Gálvez, J.B., Vincent, M., 1996. Photocatalytic degradation of industrial residual waters. Sol. Energy 56, 401-410.
285. Maruán, J., Aguado, J., Gernjak, W., Malato, S., 2007. Solar photocatalytic degradation of dichloroacetic acid with silica-supported titania at pilot-plant scale. Catal. Today 129, 59-68.
286. Radjenovíc, J., Sirtori, C., Petrovíc, M., Barceló, D., Malato, S., 2009. Solar photocatalytic degradation of persistent pharmaceuticals at pilot-scale: kinetics and characterization of major intermediate products. Appl. Catal. B: Environ. 89, 255-264.
287. Vilar, V.J.P., Maldonado, M.I., Oller, I., Malato, S., Boaventura, R.A.R., 2009. Solar treatment of cork boiling and bleaching wastewaters in a pilot plant. Water Res. 43 (16), 4050-4062.
288. Parra, S., Malato, S., Pulgarin, C., 2002. New integrated photocatalytic–biological flow system using supported TiO_2 and fixed bacteria for the mineralization of isoproturon. Appl. Catal. B: Environ. 36, 131-144.

289. WHO, Guidelines for Drinking-Water Quality First Addendum to third edition 1 Recommendations, WHO Library Cataloguing-in-Publication Data, 2006.
290. Cho, M., Chung, H., Choi, W., Yoon, J., 2004. Linear correlation between inactivation of E. coli and OH radical concentration in TiO_2 photocatalytic disinfection. Water Res. 38, 1069-1077.
291. Tyrrell, R.M., Keyse, S.M., 1990. New trends in photobiology the interaction of UVA radiation with cultured cells. J. Photochem. Photobiol. B: Biol. 4, 349-361.
292. Berney, M., Weilenmann, H.U., Siminetti, A., Egli, T., 2006. Efficacy of solar disinfection of Escherichia coli, Shigella flexneri, Salmonella typhimurium and Vibrio cholera. J. Appl. Microbiol. 101, 828-836.
293. Kehoe, S.C., Barer, M.R., Devlin, L.O., McGuigan, K.G., 2004. Batch process solar disinfection is an efficient means of disinfecting drinking water contaminated with Shigella dysenteriae type I. Lett. Appl. Microbiol. 38, 410-414.
294. McGuigan, K.G., Méndez-Hermida, F., Castro-Hermida, J.A., Ares-Mazás, E., Kehoe, S.C., Boyle, M., Sichel, C., Fernández-Ibáñez, P., Meyer, B.P., Ramalingham, S., Meyer, E.A., 2006. Batch solar disinfection inactivates oocysts of Cryptosporidium parvum and cysts of Giardia muris in drinking water. J. Appl. Microbiol. 101, 453-463.
295. Glatzmaier, G.C., Nix, R.G., Mehos, M.S., 1990. Solar destruction of hazardous chemicals. J. Environ. Sci. Health A 25, 571-581.
296. Glatzmaier, G.C., 1991. Innovative solar technologies for treatment of concentrated organic wastes. Sol. Energy Mater. 24, 672.
297. Magrini, K.A., Webb, J.D., 1990. Decomposition of aqueous organic compounds as a function of solar irradiation intensity. In: Beard, Ebadian, M.A. (Eds.), 12th ASME Int. Sol. Energy Conference. ASME, New York, pp. 159-162.
298. Karunakaran, C., Senthilvelan, S., 2005. Photooxidation of aniline on alumina with sunlight and artificial UV light. Catal. Comm. 6, 159-165.
299. Qamar, M., Muneer, M., Bahnemann, D., 2006. Heterogeneous photocatalysed degradation of two selected pesticide derivatives, triclopyr and daminozid in aqueous suspensions of titanium dioxide. J. Environ. Manage. 80, 99-106.
300. Shang, C., Cheung, L.M., Ho, C.M., Zeng, M., 2009. Repression of photoreactivation and dark repair of coliform bacteria by TiO_2-modified UV-C disinfection. Appl. Catal. B: Environ. 89, 536-542.
301. Calza, P., Sakkas, V.A., Medana, C., Baiocchi, C., Dimou, A., Pelizetti, E., Albanis, T., 2006. Photocatalytic degradation study of diclofenac over aqueous TiO_2 suspensions. Appl. Catal. B: Environ. 67, 197-205.
302. Chong, M.N., Jin, B., Chow, C.W.K., Saint, C.P., 2009. A new approach to optimise an annular slurry photoreactor system for the degradation of Congo red: statistical analysis and modelling. Chem. Eng. J. 152, 158-166.
303. Korbahti, B.K., Rauf, M.A., 2008. Application of response surface analysis to the photolytic degradation of Basic Red 2 dye. Chem. Eng. J. 138, 166-171.
304. Liu, H.L., Chiou, Y.R., 2005. Optimal decolorization efficiency of Reactive Red 239 by UV/ TiO_2 photocatalytic process coupled with response surface methodology. Chem. Eng. J. 112, 173-179.
305. Lizama, C., Freer, J., Baeza, J., Mansilla, H.D., 2002. Optimized photodegradation of reactive blue 19 on TiO_2 and ZnO suspensions. Catal. Today 76, 235-246.
306. Cho, I.H., Zoh, K.D., 2007. Photocatalytic degradation of azo dye (Reactive Red 120) in TiO_2/UV system: optimization and modelling using a response surface methodology (RSM) based on the central composite design. Dyes Pigm. 75, 533-543.
307. Jaworski, R., Pawlowski, L., Roudet, F., Kozerski, S., Petit, F., 2008. Characterization of mechanical properties of suspension plasma sprayed TiO_2 coatings using scratch test. Surf. Coat. Technol. 202, 2644-2653.
308. W. De Soto, S.A. Klein, and W. A. Beckman, "Improvement and validation of a model for photovoltaic array performance," Solar Energy, vol. 80, no. 1, pp. 78–88, Jan. 2006.

309. Cunningham, J., Sedlak, P., 1996. Kinetic studies of depollution process in TiO_2 slurries: interdependences of adsorption and UV-intensity. Catal. Today 29, 209-315.
310. Minero, C., 1999. Kinetic analysis of photoinduced reactions at the water semiconductor interface. Catal. Today 54, 205-216.
311. Monllor-Satoca, D., ómez, R., González-Hidalgo, M., Salvador, P., 2007. The "direct-indirect" model: an alternative kinetic approach in heterogeneous photocatalysis based on the degree of interaction of dissolved pollutant species with the semiconductor surface. Catal. Today 129, 247-255.
312. Minero, C., Pelizetti, E., Malato, S., Blanco, J., 1996. Large solar plant photocatalytic water decontamination: effect of operational parameters. Sol. Energy 56, 421-428.
313. Finch, G.R., Black, E.K., Labatiuk, C.W., Gyürék, L., Belosevic, M., 1993. Comparison of Giardia lamblia and Giardia muris cyst inactivation by ozone. Appl. Environ. Microbiol. 59, 3674-3680.
314. Hom, L.W., 1972. Kinetics of chlorine disinfection in an ecosystem. J. Sanit. Eng. Div. 98, 183-194.
315. Benabbou, A.K., Derriche, Z., Felix, C., Lejeune, P., Guillard, C., 2007. Photocatalytic inactivation of Escherichia coli: effect of concentration of TiO_2 and microorganism, nature, and intensity of UV irradiation. Appl. Catal. B: Environ. 76, 257-263.
316. Lambert, R.J.W., Johnston, M.D., 2000. The effect of interfering substances on the disinfection process: a mathematical model. J. Appl. Microbiol. 91, 548-555.
317. Cho, M., Chung, H., Yoon, J., 2003. Disinfection of water containing natural organic matter by using ozone initiated radical reactions. Appl. Environ. Microbiol. 69, 2284-2291.
318. Gürék, L.L., Finch, G.R., 1998. Modelling water treatment chemical disinfection kinetics. J. Environ. Eng. 124, 783-792.
319. Anotai, J., 1996. Effect of calcium ion and chemistry and disinfection efficiency of free chlorine at pH 10. Ph.D dissertation, Drexel University, Philadelphia.
320. Severin, B.F., Suidan, M.T., Engelbrecht, R.S., 1984. Series-event kinetic model for chemical disinfection. J. Environ. Eng. ASCE 110, 430-439.
321. Najm, I., 2006. An alternative interpretation of disinfection kinetics. J. Am. Water Works Assoc. 98, 93-101.
322. Johnston, M.D., Simons, E.A., Lambert, R.J.W., 2000. One explanation for the variability of the bacterial suspension test. J. Appl. Microbiol. 88, 237-242.
323. Chin, M.L., Mohamed, A.R., Bhatia, S., 2004. Performance of photocatalytic reactors using immobilized TiO_2 film for the degradation of phenol and methylene blue dye present in water stream. Chemosphere 57, 547-554.
324. Tang, C., Chen, V., 2004. The photocatalytic degradation of reactive black 5 using TiO_2/UV in an annular photoreactor. Water Res. 38, 2775-2781.
325. Rincón, A.G., Pulgarin, C., 2005. Use of coaxial photocatalytic reactor (CAPHORE) in the TiO_2 photo-assisted treatment of mixed E. coli and Bacillus sp. and bacterial community present in wastewater. Catal. Today 101, 331-344.
326. Gelover, S., Gómez, L.A., Reyes, K., Teresa Leal, M., 2006. A practical demonstration of water disinfection using TiO_2 films and sunlight. Water Res. 40, 3274-3280.
327. Crittenden, J.C.,Zhang, Y., Hand, D.W., Perram, D.L.,Marchand, E.G., 1996. Solar detoxification of fuel-contaminated groundwater using fixed-bed photocatalysts. Water Environ. Res. 68, 270-278.
328. Chen, H.Y., Zahraa, O., Bouchy, M., 1997. Inhibition of the adsorption and photocatalytic degradation of an organic contaminant in an aqueous suspension of TiO_2 by inorganic ions. J. Photochem. Photobiol. A: Chem. 108, 37-44.
329. Guillard, C., Lachheb, H., Houas, A., Ksibi, M., Elaloui, E., Hermann, J.M., 2003. Influence of chemical structure of dyes, of pH and of inorganic salts on their photocatalytic degradation by TiO_2 comparison of the efficiency of powder and supported TiO_2. J. Photochem. Photobiol. A: Chem. 158, 27-36.
330. Leng, W., Liu, H., Cheng, S., Zhang, J., Cao, C., 2000. Kinetics of photocatalytic degradation of aniline in water over TiO_2 supported on porous nickel. J. Photochem. Photobiol. A: Chem. 131, 125-132.

331. Özkan, A., Özkan, M.H., Gürkan, R., Akçay, M., Sökmen, M., 2004. Photocatalytic degradation of a textile azo dye, Sirius Gelb GC on TiO_2 or Ag-TiO_2 particles in the absence and presence of UV irradiation: the effects of some inorganic anions on the photocatalysis. J. Photochem. Photobiol. A: Chem. 163, 29-35.
332. Riga, A., Soutsas, K., Ntampegliotis, K., Karayannis, V., Papapolymerou, G., 2007. Effect of system parameters and of inorganic salts on the decolorization and degradation of Procion H-exl dyes. Comparison of H_2O_2/UV, Fenton, UV/Fenton, TiO_2/UV and TiO_2/UV/H_2O_2 processes. Desalination 211, 72-86.
333. Schmelling, D.C., Gray, K.A., Vamat, P.V., 1997. The influence of solution matrix on the photocatalytic degradation of TNT in TiO_2 slurries. Water Res. 31, 1439-1447.
334. Wang, K., Zhang, J., Lou, L., Yang, S., Chen, Y., 2004. UV or visible light induced photodegradation of AO7 on TiO_2 particles: the influence of inorganic anions. J. Photochem. Photobiol. A: Chem. 165, 201-207.
335. Wu, C.H., Huang, K.S., Chern, J.M., 2006. Decomposition of acid dye by TiO_2 thin films prepared by the solegel method. Ind. Eng. Chem. Res. 45, 2040-2045.
336. Wong, C.C., Chu, W., 2003. The direct photolysis and photocatalytic degradation of alachlor at different TiO_2 and UV sources. Chemosphere 50, 981-987.
337. Choi, W., Termin, A., Hoffman, M.R., 1994. The role of metal ion dopants in quantum-sized TiO_2: correlation between photoreactivity and charge carrier recombination dynamics. J. Phys. Chem. 98, 13669-13679.
338. Abdullah, M., Low, G.K.C., Matthews, R.W., 1990. Effects of common inorganic anions on rates of photocatalytic oxidation of organic carbon over illuminated titanium dioxide. J. Phys. Chem. 94, 6820-6825.
339. Kerzhentsev, M., Guillard, C., Herrmann, J.M., Pichat, P., 1996. Photocatalytic pollutant removal in water at room temperature: case study of the total degradation of the insecticide fenitrothion (phosphorothioic acid O, O-dimethyl-O-(3-methyl-4-nitro-phenyl) ester). Catal. Today 27, 215-220.
340. Okonomoto, K., Yamamoto, Y., Tanaka, H., Tanaka, M., Itaya, A., 1985. Heterogeneous photocatalytic decomposition of phenol over TiO_2 powder. Bull. Chem. Soc. Jpn. 58, 2015-2022.
341. Martin, S.T., Lee, A.T., Hoffmann, M.R., 1995. Chemical mechanism of inorganic oxidants in the TiO_2/UV process: increased rates of degradation of chlorinated hydrocarbons. Environ. Sci. Technol. 29, 2567-2573.
342. Zhu, H., Zhang, M., Xia, Z., Low, G.K.C., 1995. Titanium dioxide mediated photocatalytic degradation of monocrotophos. Water Res. 29, 2681-2688.
343. Mu, Y., Yu, H.Q., Zheng, J.C., Zhang, S.J., 2004. TiO_2-mediated photocatalytic degradation of Orange II with the presence of Mn2. in solution. J. Photochem. Photobiol. A: Chem. 163, 311-316.
344. Burns, R., Crittenden, J.C., Hand, D.W., Sutter, L.L., Salman, S.R., 1999. Effect of inorganic ions in heterogeneous photocatalysis. J. Environ. Eng. 125, 77-85.
345. Rizzo, L., Koch, J., Belgiorno, V., Anderson, M.A., 2007. Removal of methylene blue in a photocatalytic reactor using polymethylmethacrylate supported TiO_2 nanofilm. Desalination 211, 1-9.
346. Diebold, U., 2003. The surface science of titanium dioxide. Surf. Sci. Rep. 48, 53-229.
347. Matthews, R.W., McEnvoy, S.R., 1992. Photocatalytic degradation of phenol in the presence of near-UV illuminated titanium dioxide. J. Photochem. Photobiol. A: Chem. 64, 231.
348. Lindner, M., Bahnemann, D.W., Hirthe, B., Griebler, W.D., 1995. Novel TiO_2 powders as highly active photocatalysts. In: Stine, W.R., Tanaka, T., Claridge, D.E. (Eds.), Solar Water Detoxification; Solar Eng. ASME, New York, p. 339.
349. Kabra, K., Chaudhary, R., Sawhney, R.L., 2004. Treatment of hazardous organic and inorganic compounds through aqueous-phase photocatalysis: a review. Ind. Eng. Chem. Res. 43, 7683-7696.
350. Prairie, M.R., Evans, L.R.,Martinez, S.L., 1994. Destruction of organics and removal of heavy metals in water via TiO2 photocatalysis. In: Eckenfelder, W.W., Roth, J.A., Bowers, A.R. (Eds.), Chemical Oxidation: Technologies for the Nineties, vol. 2. Technomic Publishing Company Inc, Pennsylvania US, pp. 428-441.

351. Herrmann, J.M., Disdier, J., Pichat, P., 1988. Photocatalytic deposition of silver on powder titania: consequences for the recovery of silver. J. Catal. 113, 72-81.
352. Huang, M., Tso, E., Datye, A.K., 1996. Removal of silver in photographic processing waste by TiO_2-based photocatalysis. Environ. Sci. Technol. 30, 3084-3088.
353. Angelidis, T.N., Koutlemani, M., Poulios, I., 1998. Kinetic study of the photocatalytic recovery of Pt from aqueous solution by TiO_2, in a closed-loop reactor. Appl. Catal. B: Environ. 16, 347-357.
354. Muñoz, I., Peral, J., Ayllón, J.A., Malato, S., Passarinho, P., Domènech, X., 2006. Life cycle assessment of a coupled solar photocatalytic–biological process for wastewater treatment. Water Res. 40, 3533-3540.
355. Gernjak, W., Fuerhacker, M., Fernandez-Ibññez, P., Blanco, J., Malato, S., 2006. Solar photo-Fenton treatment process parameters and process control. Appl. Catal. B: Environ. 64, 121-130.
356. Lapertot, M., Pulgaín, C., Fernandez-Ibáñez, P., Maldonado, M.I., Pérez-Estrada, L., Oller, I., Gernjak, W., Malato, S., 2006. Enhancing biodegradability of priority substances (pesticides) by solar photo-Fenton. Water Res. 40, 1086-1094.
357. Andreozzi, R., Caprio, V., Insola, A., Marotta, R., 1999. Advanced oxidation processes (AOP) for water purification and recovery. Catal. Today 53, 51-59.
358. Muñoz, I., Rieradevall, J., Torrades, F., Peral, J., Domenech, X., 2005. Environmental assessment of different solar driven advanced oxidation processes. Sol. Energy 79, 369-375.
359. Consoli, F., Allen, D., Boustead, I., de Oude, N., Fava, J., Franklin, W., Quay, B., Parrish, R., Perriman, R., Postlethwaite, D., Seguin, J., Vigon, B., 1993. Guidelines for Life-Cycle Assessment: a Code of Practice. Society of Environmental Toxicology and Chemistry, Brussels, Belgium.
360. Azapagic, A., Clift, R., 1999. Life cycle assessment and multiobjective optimisation. J. Cleaner Prod. 7, 135-143.

Chapter 14
Water Reduction Engineering

Abstract The transpiration mechanism has been proposed for rerouting as it is the main cause of groundwater loss which is also causing significant global warming by releasing water vapor into the air. Since electrostatic force has the tendency to tug down the water, a static electricity force creating plastic tank has been proposed to install at the bottom of plants to capture the transpiration water vapor and treat it in situ by applying UV technology to meet the daily water demand throughout the world.

Keywords Groundwater loss · Transpiration mechanism · Climate change · Static electricity force · UV technology · Clean water naturally

Introduction

Plants give O_2 and take CO_2 by the process of photosynthesis to keep the global environment in balance. Plants are simply the hero for the environment; unfortunately, hero plants are also the villain for the environment which plays the significant role in causing global warming. The body of plants needs water for the reaction of biochemical metabolism for its growth [1]. This water is taken up by the cohesion-tension mechanism of the soil (groundwater) through the roots, transported by osmosis through the xylem to the leaves of the plants [2]. Interestingly only a mere 0.5–3% of water is used by plants for their metabolism and the rest of water is released into the air through stomatal cells by transpiration process [3, 4]. This process of transpiration is causing the largest loss of groundwater that is also causing global warming, since this water vapor is a notable cause for global warming. Recent studies on transpiration and groundwater relationship have been discussed along with terrestrial water fluxes where their water models reveal that streamflow is getting lower due to the plant transpiration [1, 4]. These are very interesting findings, but no mechanism has been studied yet to trap this transpiration water for meeting global water demand. In this research, therefore, a technology has been

M. F. Hossain, *Global Sustainability in Energy, Building, Infrastructure, Transportation, and Water Technology*,
https://doi.org/10.1007/978-3-030-62376-0_14

proposed to eliminate this water loss by diverting this transpiration mechanism by collecting this water vapor instead of allowing it to enter the air and transforming it into potable water and clean energy. Simply static electricity creator plastic tank near the plants has been proposed to install to trap all the water vapor as the water vapor is attracted by the force of static electricity. Just because water vapor has positive and negative charges and the electrons that end up on static electrical force have a positive charge, while water molecules have a negative charge on one side, the positive charge of static electric force and negative charge of water vapor pull each other closer together, and the positive side tugs the direction and forces the water to come down so that it can be collected in a tank and treated in situ to meet the daily water demand. Calculation revealed that only four standard oak trees can meet the total water demand for a small family throughout the year. Since the groundwater strata are getting lower fast to a finite level, and global water demand and global warming are getting dangerous putting earth on a vulnerable condition, these two vital needs must be resolved immediately. Interestingly this new finding has the total solution to solve the global water and environmental crisis for the survival of this planet which will indeed open new door in science.

Methods and Simulation

Static Electric Force Generation

To capture the water vapor from air which is released by stomatal cells of the plants during the daytime, a model has been proposed to create *Hossain static electric force* (*HsrF* = $\mathfrak{h}$) by implementing the friction of insulator into the plastic tank to pull down the water vapor into the plastic tank [5, 6]. To create *HsrF* into the plastic tank, I have implemented albanian local symmetry calculation by using MATLAB software considering gauge field symmetry and Goldstone scalar with respect to longitudinal mode of the vector [7, 8]. Thus, for each spontaneously broken particle T^{α} of the local symmetry there will be a corresponding gauge field of $A_{\mu}^{\alpha}(x)$ where *HsrF* will start to work at a local U(1) phase symmetry [9, 10]. Therefore, the model will be comprised as a complex scalar field Φ (x) of static electric charge q coupled to the EM field $A^{\mu}(x)$ which is expressed by $\mathfrak{h}$:

$$\mathfrak{h} = \frac{1}{2}\left(\partial_{\mu}\sigma\right)^{2} - v(\sigma) - \frac{1}{4}F_{\mu\nu}F^{\mu\nu} + \frac{(v+\sigma)^{2}}{2} * \left(\partial_{\mu}\Theta + qA_{\mu}\right)^{2} \quad (14.1)$$

where

$$D_{\mu}\Phi(x) = \partial_{\mu}\Phi(x) + iqA_{\mu}(x)\Phi(x)$$

$$D_{\mu}\Phi^{*}(x) = \partial_{\mu}\Phi^{*}(x) - iqA_{\mu}(x)\Phi^{*}(x) \quad (14.2)$$

And

$$V\left(\Phi^*\Phi\right)=\frac{\lambda}{2}\left(\Phi^*\Phi\right)^2+m^2\left(\Phi^*\Phi\right) \tag{14.3}$$

Suppose $\lambda > 0$ but $m^2 < 0$, so that $\Phi = 0$ is a local maximum of the scalar potential, while the minima form a degenerate circle $\Phi = \frac{v}{\sqrt{2}} * e^{i\theta}$; then

$$v=\sqrt{\frac{-2m^2}{\lambda}}, \text{any real } \theta \tag{14.4}$$

Consequently, the scalar field Φ develops a nonzero vacuum expectation value $\Phi \neq 0$, which spontaneously creates the U(1) symmetry of the static electric field. The breakdown would lead to a massless Goldstone scalar stemming from the phase of the complex field $\Phi(x)$. But for the local U(1) symmetry, the phase of $\Phi(x)$ is not just the phase of the expectation value Φ but the x-dependent phase of the dynamical $\Phi(x)$ field. To analyze this static electricity force mechanism, I have used polar coordinates in the scalar field space; thus

$$\Phi(x)=\frac{1}{\sqrt{2}}\Phi_{\mathrm{r}}(x)*e^{i\Theta(x)}, \ \text{real}\,\Phi_{\mathrm{r}}(x)>0, \ \text{real}\,\Phi(x) \tag{14.5}$$

This field redefinition is singular when $\Phi(x) = 0$, so I never used it for theories with $\langle\Phi\rangle \neq 0$, but it is alright for spontaneously broken theories where I can expect $\Phi\langle x\rangle \neq 0$ almost everywhere. In terms of the real fields $\phi_{\mathrm{r}}(x)$ and $\Theta(x)$, the scalar potential depends only on the radial field ϕ_{r},

$$V(\phi)=\frac{\lambda}{8}\left(\phi_{\mathrm{r}}^2-v^2\right)^2+\text{const}, \tag{14.6}$$

or the radial field shifted by its VEV, $\Phi_{\mathrm{r}}(x) = v + \sigma(x)$,

$$\phi_{\mathrm{r}}^2-v^2=\left(v+\sigma\right)^2-v^2=2v\sigma+\sigma^2 \tag{14.7}$$

$$V=\frac{\lambda}{8}\left(2v\sigma-\sigma^2\right)^2=\frac{\lambda v^2}{2}*\sigma^2+\frac{\lambda v}{2}*\sigma^3+\frac{\lambda}{8}*\sigma^4 \tag{14.8}$$

At the same time, the covariant derivative $D_\mu\phi$ becomes

$$D_\mu\phi=\frac{1}{\sqrt{2}}\left(\partial_\mu\left(\phi_{\mathrm{r}}e^{i\Theta}\right)+iqA_\mu*\phi_{\mathrm{r}}e^{i\Theta}\right)=\frac{e^{i\Theta}}{\sqrt{2}}\left(\partial_\mu\phi_{\mathrm{r}}+\phi_{\mathrm{r}}*i\partial_\mu\Theta+\phi_{\mathrm{r}}*iqA_\mu\right) \tag{14.9}$$

$$\begin{aligned}|D_\mu\phi|^2 &= \frac{1}{2}\left|\partial_\mu\phi_r + \phi_r * i\partial_\mu\Theta + \phi_r * iqA_\mu\right|^2\\ &= \frac{1}{2}\left(\partial_\mu\phi_r\right) + \frac{\phi_r^2}{2} * \left(\partial_\mu\Theta qA_\mu\right)^2\\ &= \frac{1}{2}\left(\partial_\mu\sigma\right)^2 + \frac{(v+\sigma)^2}{2} * \left(\partial_\mu\Theta + qA_\mu\right)^2\end{aligned} \tag{14.10}$$

Altogether,

$$\mathfrak{h} = \frac{1}{2}\left(\partial_\mu\sigma\right)^2 - v(\sigma) - \frac{1}{4}F_{\mu\nu}F^{\mu\nu} + \frac{(v+\sigma)^2}{2} * \left(\partial_\mu\Theta + qA_\mu\right)^2 \tag{14.11}$$

To confirm the creating of this static electric force ($\mathfrak{h}_{sef}$) into the static electric field properties of this *HsrF*, it has been expanded in powers of the fields (and their derivatives) and focused on the quadratic part describing the free particles:

$$\mathfrak{h}_{sef} = \frac{1}{2}\left(\partial_\mu\sigma\right)^2 - \frac{\lambda v^2}{2} * \sigma^2 - \frac{1}{4}F_{\mu\nu}F^{\mu\nu} + \frac{v^2}{2} * \left(qA_\mu + \partial_\mu\Theta\right)^2 \tag{14.12}$$

Here this *HsrF* ($\mathfrak{h}_{free}$) function obviously will suggest a real scalar particle of positive mass^2 = λv^2 involving the $A_\mu(x)$ and the $\Theta(x)$ fields to initiate to create tremendous static electricity force within the electric field of the plastic tank (Fig. 14.1).

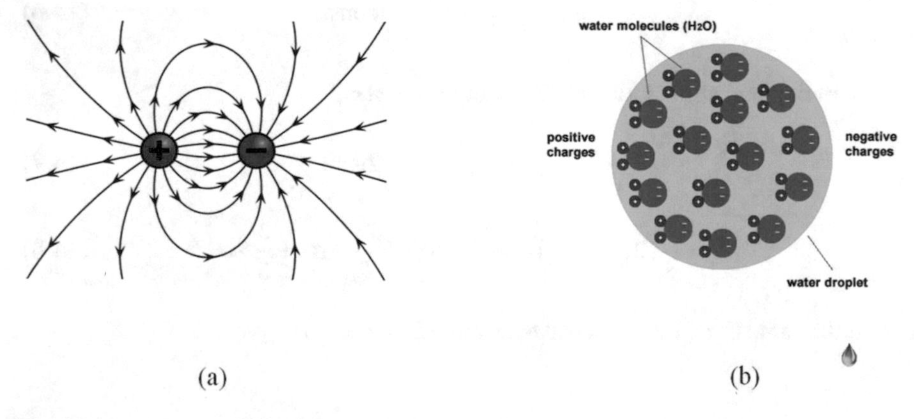

Fig. 14.1 (**a**) The creating of static electricity force and (**b**) its mechanism of conversion of static energy into an electromotive force of positive and negative charges that mobilizes the "static" electricity to tug down the water molecules

In Situ Water Treatment

Since the collected water into the plastic tank is just nothing but the liquid form of vapor, it will not require any sedimentation, coagulation, and chlorination to clean the water. Only mixing physics (UV application) and filtration will be required to treat the water to meet the US National Primary Drinking Water Standard code [cc]. It is the simplest way to treat water by using *SODIS* system (*SOlar DISinfection*), where a transparent container is filled with water and exposed to full sunlight for several hours. As soon as the water temperature reaches 50 °C with a UV radiation of 320 nm, the inactivation process will be accelerated in order to lead to complete microbiological disinfection immediately and the treated water shall be used to meet the total domestic water demand (Fig. 14.2).

Results and Discussion

To mathematically determine the electric static force proliferation around the plastic tank to confirm the tug down of the water, I have initially solved the dynamic photon proliferation by integrating *HSEF* electric field; thus, the local U(1) gauge invariant allows to add a mass term for the gauge particle under $\varnothing' \rightarrow e^{i\alpha(x)}\varnothing$. In detail it can be explained by a covariant derivative with a special transformation rule for the scalar field expressed by [11, 12]

$$\begin{aligned} \partial_\mu \rightarrow D_\mu = \partial_\mu = ieA_\mu \; [\text{covariant derivatives}] \\ A'_\mu = A_\mu + \frac{1}{e}\partial_\mu \alpha \; [A_\mu \text{ derivatives}] \end{aligned} \quad (14.13)$$

where the local U(1) gauge invariant *HSEF* for a complex scalar field is given by

$$\mathfrak{H} = (D^\mu)^\dagger (D_\mu \varnothing) - \frac{1}{4} F_{\mu v} F^{\mu v} - V(\varnothing) \quad (14.14)$$

The term $\frac{1}{4}F_{\alpha v}F^{\alpha v}$ is the kinetic term for the gauge field (heating photon) and $V(\varnothing)$ is the extra term in the *HSEF* that will be $V(\varnothing^*\varnothing) = \mu^2(\varnothing^*\varnothing) + \lambda\,(\varnothing^*\varnothing)^2$.

Therefore, the *HSEF* ($\mathfrak{H}$) under perturbations into the quantum field is initiated with the massive scalar particles ϕ_1 and ϕ_2 along with a mass μ. In this situation $\mu^2 < 0$ has an infinite number of quantum; each has been satisfied by $\phi_1^2 + \phi_2^2 = -\mu^2 / \lambda = v^2$; and the $\mathfrak{H}$ through the covariant derivatives using again the shifted fields η and ξ defined the quantum field as $\phi_0 = \frac{1}{\sqrt{2}}\left[(\upsilon + \eta) + i\xi\right]$.

$$\begin{aligned} \text{Kinetic term: } \mathfrak{H}_{(\eta,\xi)} &= (D^\mu \phi)^\dagger (D^\mu \phi) \\ &= (\partial^\mu + ieA^\mu)\phi^* \,(\partial_\mu - ieA_\mu)\,\phi \end{aligned} \quad (14.15)$$

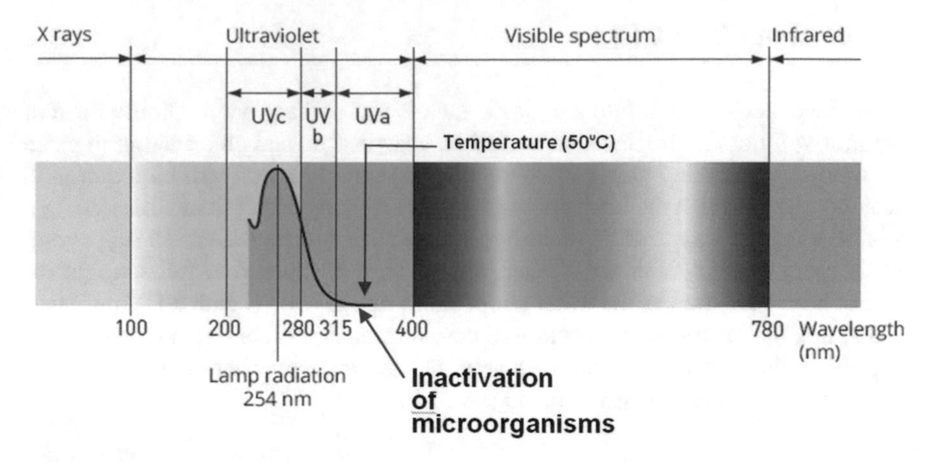

Fig. 14.2 The photo-physics radiation application for the purification of water which shows that once UV radiation of 320 nm is applied into the water, it starts to disinfect all microorganisms immediately once temperature reaches 50 °C

Thus, this expanding term in the $\mathfrak{h}$ associated to the scalar field suggests that *HSEF* electric field is prepared to initiate the proliferation of static electricity force into its quantum field to tug down the water [13, 14].

To confirm this tug down of water by static electricity force, hereby, I have readily implemented the calculation of $\bar{\varphi}\,[s_0]$ for the confirmation of the expected value of s_0 for capturing water vapor [15, 16]. Thus, the corrective functional asymptotic formulas are being used as follows:

$$\bar{\varphi}[s_0] = 2s_0(\ln 4s_0 - 2) + \ln 4s_0(\ln 4s_0 - 2) - \frac{\pi^2 - 9}{3} + s_0^{-1}\left(\ln 4s_0 + \frac{9}{8}\right) + \ldots (s_0 \gg 1); \tag{14.16}$$

$$\bar{\varphi}[s_0] = \left(\frac{2}{3}\right)(s_0 - 1)^{\frac{2}{3}} + \left(\frac{5}{3}\right)(s_0 - 1)^{\frac{5}{3}}\left(\frac{1507}{420}\right)(s_0 - 1)^{\frac{7}{3}} + \ldots (s_0 \gg 1). \tag{14.17}$$

The function $\dfrac{\bar{\varphi}[s_0]}{(s_0 - 1)}$ is thus described as $1 < s_0 < 10$; for larger s_0, it contains natural logarithmic which is s_0 to confirm the tug down of 100% water vapor by the *HSEF* into the plastic tank.

Then the application of mixing physics (UV application) and filtration for treating this water into the plastic tank as per US National Primary Drinking Water Standard code confirms that this water is potable which was analyzed by *SODIS* system (*SOlar DISinfection*).

In average 100 gallons of water is required per day per person in a standard daily life [9, 17]. Thus, it will require a total of (100_{gallons}/day/person × 4_{persons} × 365_{days}) 146,000 gallons of water per year for a small family of four persons. Since a standard oak tree can transpire 40,000 gallons (151,000 L) per year, tug down of 100% water vapor by HSEF described above will require only four standard oak trees to satisfy the total water demand for a small family.

Conclusions

Water and environmental vulnerability are the top two problems on earth where trees play a significant role in creating these problems by the process of transpiration. To mitigate these problems, transpiration mechanism has been proposed to transform and convert it into clean water to meet the global water demand and reduce the global warming by the utilization of electrostatic force to capture this transpiration water vapor and treat in situ by UV application which would indeed be a novel, integrated, and innovative field in science to console the global water demand and global warming crisis.

Acknowledgements This research was supported by Green Globe Technology under the grant RD-02018-06 for building a better environment. Any findings, predictions, and conclusions described in this chapter are solely performed by the authors and we confirm that there is no conflict of interest for publishing in a suitable journal.

References

1. Josette, M., Scott, R. The ERECTA gene regulates plant transpiration efficiency in Arabidopsis. *Nat.* 436, 866–870 (2005).
2. Lang, C. et al. Observation of resonant photon blockade at microwave frequencies using correlation function measurements. Phys. Rev. Lett. 106, 243601 (2011).
3. Reed, M., Maxwell, L. Connections between groundwater flow and transpiration partitioning. *Sci.* 353, 377-380 (2015).
4. Scott, J., Zachary, D. Terrestrial water fluxes dominated by transpiration. *Nat.* 496, 347–350 (2013).
5. Andreas, Reinhard. Strongly correlated photons on a chip. *Nat. Phot.* 6, 93-96 (2012).
6. Tame, M., McEnery, S., et al. *Quant. Plas.* 9, 329–340 (2013).
7. Douglas, S., Habibian, H., et al. Quantum many-body models with cold atoms coupled to photonic crystals. *Nat. Phot*, 9, 326-331 (2015).
8. Leijing, Y., Sheng, W., Qingsheng, Z., Zhiyong, Z., Tian, P., Yan, L. Efficient photovoltage multiplication in carbon nanotubes. *Nat. Phot.* 8, 672 – 676 (2011).
9. Langer, L., Poltavtsev, S., Bayer, M. Access to long-term optical memories using photon echoes retrieved from semiconductor spins. *Nat. Phot.* 8, 851–857 (2014).
10. Pregnolato, T., Lee, E., Song, J., Stobbe, D., Lodahl, P. Single-photon non-linear optics with a quantum dot in a waveguide. *Nat. Commun.* 6, 8655 (2015).
11. Yuwen, W., Yongyou, Z., Qingyun, Z., Bingsuo, Z., Udo, S. "Dynamics of single photon transport in a one-dimensional waveguide two-point coupled with a Jaynes-Cummings system". *Sci. Rep.* 6, 33867 (2016).
12. Li, Q., Xu, D. Recoil effects of a motional scatterer on single-photon scattering in one dimension. *Sci. Rep.* 8, 3144 (2013).

13. Yan, W., Heng, F. Single-photon quantum router with multiple output ports. *Sci. Rep.* 4, 4820 (2014).
14. Hossain, M. Solar energy integration into advanced building design for meeting energy demand and environment problem. *Inter. J. Ener. Res.* 40, 1293–1300 (2016).
15. Soto, W., Klein, S. et al. Improvement and validation of a model for photovoltaic array performance. *Sol. Ener.* 80, 78–88 (2006).
16. Zhu, Y., Xiaoyong, H., Hong, Y., Qihuang, G. On-chip plasmon-induced transparency based on plasmonic coupled nanocavities. *Sci. Rep.* 4, 3752 (2014).
17. Leijing Yang, Sheng Wang, Qingsheng Zeng, Zhiyong Zhang, Tian Pei, Yan Li & Lian-Mao Peng (2011). Efficient photovoltage multiplication in carbon nanotubes – Nature Photonics pp 672 – 676.

Part VI
Sustainable Urban and Rural Development

Chapter 15
Deurbanization and Rural Development

Abstract The concept of deurbanization and rural development is the practice of transforming urbanization into the rural area to mitigate the urban congestion to secure an ecologically balanced community. Therefore, deurbanization and rural development are being thus proposed to exercise through the implementation of sustainability tools in all sectors in our daily lives which are environmentally friendly and resource efficient throughout the life cycle to maximize the achievement of economic value to build a resilience community. Subsequently, deurbanization and rural development must be secured by implementing cutting-edge metrics of sustainability considering the following five major sectors: (1) environment, (2) energy, (3) building, (4) infrastructure and transportation, and (5) water supply. Simply deurbanization and rural development can be defined as the combined method of transformation and implementation of green planning, designing, and developing of all sectors of environment, energy, building, infrastructure, and water by conducting advanced research and environmentally friendly technology to reduce the stress on urban area in order to build a balanced environment on earth.

Keywords Urban congestion · Rural development · Sustainable environment · Clean energy · Advanced building design technology · Sustainable infrastructure and transportation · Environmental sustainability

Introduction

There has been massive development of urbanization for the past 100 years globally due to the flocking of rural habitants into the urban area for better job opportunity and lifestyle which in fact is causing severe urban congestion and global environmental crisis [1, 2]. Simply we need deurbanization and rural development in which people can migrate from urban to rural communities to perform their job from home and conduct simpler clean and green lifestyles. In recent days the communications technology has been developed tremendously and thus people from rural

M. F. Hossain, *Global Sustainability in Energy, Building, Infrastructure, Transportation, and Water Technology*,
https://doi.org/10.1007/978-3-030-62376-0_15

communities can work from home because they are connected to the urban area via internet, which certainly will confirm that most of the employment opportunities will no longer require moving to an urban area [3, 4]. Since massive urbanization is causing a severe impact on urban mobility and social lives, deurbanization and rural development is an urgent demand to restore the shrinking cities and environment. Simply, in order to secure the better ecological balanced earth, we certainly need to confirm deurbanization and sustainable rural development to mitigate the urban congestion and environmental perplexity in order to relieve severe mental stress, heat diseases, and noise pollution [5, 6]. Therefore, in this chapter a deurbanization and rural development system has been proposed that needs to be implemented in every part of the world by applying all possible advanced technological applications for rural development to facilitate modern lifestyle and entertainment to confirm an environmentally friendly rural area for the betterment of life and mental peace globally.

Methods and Materials

Environment

To secure sustainable deurbanization and rural development, the prime task is the environment of rural area that must be balanced by confirming biological systems productive and diverse for an indefinite period of time. Simply, the total biological systems need to be sustainable in order to secure for long-lived and healthy lives and cleaner ecosystem. Simply, the deurbanization and rural development considering the environmental sustainability can be defined as the durability of processes and systems including the interrelated domains of culture and politics, economics, and ecology to acquire healthy environments that will support human survival in a fresh air and survival of other creatures of earth [7, 8]. Consequently, preserving the natural resources and sustainability encounters social challenge that involves ethical consumerism, individual and local lifestyle, rural transportation and planning, as well as national and international laws that need to be authoritative precisely. While sustainable rural development is to be adopted as the holistic method of acquiring a greener and cleaner environment, it is thus essential to view sustainability as the target goal to confirm an equilibrium ecosystem for a better rural system [9, 10]. As a result, it is essential for rural sustainability to be concerned with the commitment of policymakers, investors, engineers, architects, and scientists to administer and promote necessary environmental resource conservation in securing a resilience and sustainable rural area [11, 12].

Energy

The global atmosphere is now getting to a seriously dangerous level due to increasement of all aspects of the carbon cycle in the atmosphere which has become a major problem due to the conventional energy usages in the urban area and thus urban air toxicity due to volatile organic compounds, sulfur oxide, nitrogen oxide, as well airborne pollutant substances generating photochemical smog, air pollution, and deadly chlorofluorocarbon which are causing severe effects on urban habitants and the earth's atmosphere and environment [13, 14]. Therefore, deurbanization and rural development is a must where renewable energy system is available to serve the needs of the present habitant without compromising the ability of future generations to meet their needs and thus rural area is the best place to confirm this option. Simply utilization of this renewable energy for the deurbanization and rural development will naturally refill on a scale time, and thus this sustainable energy supply will not jeopardize the system within which it is implemented to an extent of being unfit to offer needs in the coming days [15, 16]. Thus, three principles of deurbanization and rural development would be very much related to the sustainability that encompasses the four major interrelated domains: (1) culture, (2) politics, (3) economics, and (4) ecology. Here, the technologies will promote the utilization of sustainable energy including renewable energy sources for rural development, such as hydroelectricity, solar energy, wind energy, wave power, geothermal energy, and bioenergy, to confirm abundant clean energy source to secure an environmentally friendly global ecosystem.

Housing and Building

Subsequently, the focus on rural housing and building all over the world must be identified to reduce the consumption of very large portions of natural resources including water and energy. In the present day, the housing and building from the urban community are responsible for 40% of worldwide CO_2 emissions, which is equal to nine billion carbon dioxide tons yearly and by 2050 these emissions are likely to double and thus the urban area will get severely polluted [17–19]. It is essential for one to think through the clean housing and building to rural development to attain ecologically friendly building and energy efficiency which eventually will combine a massive collection of skills, approaches, and practices to cut and finally eliminate the adverse environment impacts which needs plenty of space [20, 21]. Consequently, the rural development will have the sustainable housings and buildings that will **conform to the** environmentally friendly resource-efficient lifestyle. Simply, the sustainable housing and building development for rural development will have several drives including social, economic, and environmental benefits. Thus, considering energy efficiency, toxic and waste reduction, maintenance and operation optimization of rural development, interior environmental quality improvement, material efficiency, design efficiency, and water efficiency, the technologies or practices applied in sustainable building essentially require to be focused, so as to generate a larger amassed impact.

Infrastructure and Transportation

Traditional urban infrastructure is not only causing trillions of dollars every year, but also playing a viral role in creating adverse environmental and climate perplexity [22, 23]. Thus, deurbanization and rural sustainable infrastructure shall indeed be a network to provide the "ingredients" for solving the current climatic challenges [24–26]. The main components of this approach are the roads, highways, bridges, and tunnels; management of rural development is to be achieved for climate adaptation, less heat stress, more biodiversity, food production, better air quality, sustainable energy production, clean water, and healthy soils, as well as for the more anthropocentric functions such as increased quality of life through re-creating and providing shade and shelter in and around rural area which is connected to the cities [27, 28]. Subsequently, rural infrastructure will serve to provide an ecological framework for social, economic, and environmental health of the surrounding community that would achieve more holistic road, highway, bridge, and tunnel construction and management to run the daily life smoothly in connecting rural areas to urban areas.

Water

Global environment has been impacted by the urban water cycle by accelerating evaporation which creates severe water pollution and cooling effect on earth and climate change [29, 30]. Water covers 75% of the earth's surface and out of this 97% is salty water of the oceans and only 3% is freshwater, most of which is locked up in the Antarctic ice sheet [31, 32]. Most of it is in icecaps and glaciers (69%) and groundwater (30%), while all lakes, rivers, and swamps combined only account for a small fraction (0.3%) of the earth's total freshwater reserves [33, 34]. The main source of our daily usage of urban water, either freshwater or groundwater, which is causing severe water pollution and groundwater strata has been getting lower nearly 10 m within the past few decades, which will take the groundwater to a finite level in the near future [12, 35]. Simply, the rising massive urbanization globally is creating more demand of freshwater and causing severe contamination of freshwater, thereby triggering adverse environmental impact which eventually alarms the survival of all living beings on earth. Thus, deurbanization and rural development is an urgent demand to confirm the global freshwater to be free from pollution in order to secure clean global water environment. Just because water has distinctive features that are important for life proliferation to respond in ways that eventually permit replication, it is very important to all living beings for their own existence. Simply it is an essential element to all living beings for their survival since it has many distinct properties that are critical for the proliferation of life to react in ways that ultimately allow replication where freshwater is the only source for human lives to survive. It is, therefore, vital both as a solvent in which many of the body's solutes dissolve and as an essential part of many metabolic (catabolism and anabolism) processes within the living body [30, 36]. Subsequently, in catabolism water is utilized to break the bonds within large molecules to create smaller molecules, and in anabolism, water is detached from molecules to create larger molecules. Both these processes of anabolism and catabolism cannot exist without water [25, 26]. Considering the plant kingdom, water is the fundamental element for photosynthesis and respiration where photosynthetic cells use the sun's energy to split off water's hydrogen from oxygen. Afterwards, hydrogen is mixed with carbon dioxide (CO_2) (absorbed from air or water) to form glucose to be utilized as their food and release oxygen to balance the ecosystem. Simply we need to protect our freshwater from pollution for our survival where deurbanization and rural development can help to protect this vital element, freshwater, since it can be easily recycled in rural area compared to urban area freshwater treatment which is causing several water pollution.

Results and Discussion

Environment

Rural environmental sustainability is usually measured by occurrences or junctures where the naturally befalling regenerative forces such as biomass, vegetation, atmosphere, soil, water, and solar energy intermingle with their underlying forces in the environment which is completely absent in urban area. Human activities are the major drivers of destruction of the urban environmental systems of the earth as well as its biophysical mechanism [13, 15, 26]. Therefore, the impact of a rural community on the environment is instigated by a single person or the available population, which in turn relies on complex ways like exactly what natural resources are being utilized, whether those natural resources are renewable, as well as the human activity scale in comparison to the ecosystems' carrying capacity. Accordingly, the resource consumption pattern in urban area compared to the rural area by population individually within all sectors is generating adverse effect on biodiversity, conservation biology, environmental science, and earth science. Consequently, the biodiversity loss within the environment mainly from the urban habitat fragmentation and loss generated by massive urban development as natural capital are rapidly changing all over the globe and thus all of these have a significant impact on global nitrogen, carbon, and water biogeochemical cycles [20, 29]. Subsequently, the product consumption at all scales via the consumption chain, beginning with the economic sector impact via national economies to the international economies, and the impacts of personal spending patterns and lifestyle choices, via demands of resources of particular services and goods, is seriously impacting the urban environment [1, 14]. To maintain the resource consumption, resource productivity, as well as resource intensity, it is necessary to investigate the pattern of consumption that is associated with resources to the economic, social, and environmental effects at the context or scale where rural area is the best option in comparison to the urban area. Hence, the initial world scientific evaluation on the effects of production and consumption was published in 2019 by the International Resource Panel of United Nations Environmental Programme (UNEP), which recognized the priority actions for both urban and rural areas and the findings of the study indicated that consumption by household associated with energy using, food, shelter, and mobility products is the major cause of life cycle effects of consumption and generates limited level of the existing natural resources in rural area [24, 33]. As a result, to safeguard these natural resources through the implementation of basic principles of complex ecological issues, it is necessary to undertake deurbanization and rural development to encounter challenges generated internationally through increasing urbanization and ecological degradation.

Energy

Significant advancement is being carried out in the transition of energy from fossil fuels to environmentally friendly sustainable energy systems where rural areas are having the most advantageous role in comparison to urban area. As a result, changes that need to be made on the present-day conventional energy consumption will be not only on how energy is supplied, but also on how it is used, and it is important to reduce the volume of energy needed to deliver different goods and/or services. Thus, stabilizing and decreasing the emissions of CO_2 simply requires energy efficiency and renewable energy to remain as the “twin pillars” of environmental sustainability where rural area can be the best option to support decreasing of CO_2 emission. Based on the current historical examination, the growth rate in the demand of energy has generally overtaken the rate of enhancements in energy efficiency [2, 31]. This is because of the ongoing population in urban area. Consequently, aggregate use of energy as well as correlated emissions of carbon have constantly increased, which ultimately causes deadly climate changes. As a consequence, supplies of renewable and sustainable energy which is more easily abundant in rural area would be an exigent demand to alleviate global energy demand and mitigate climate change crisis. Therefore, clean and renewable energy (and energy efficiency) are no longer niche sectors that are promoted only by governments and environmentalists; it must be promoted by the private sector by increasing the levels of investment for confirming a clean and green earth and here rural development can play a vital role.

Simply, to achieve a clean world, it is essential to look after the sustainable and renewable energy sector through application technology, carrying out advanced research, as well as via commercial application where urban development can really play no role in it. Thus, essentially much focus must be directed toward rural renewable power system such as solar, wind, biogas, and geothermal energy application to confirm a cleaner greener earth.

Housing and Building

Making the most of the renewable resources is frequently stressed in sustainable housing and building, utilizing sunlight via photovoltaic equipment, active solar, and passive solar, and utilizing trees and planets via rainwater runoff reduction, rain gardens, and green roofs can play a major role in reducing climate change [21, 23]. Simply, having appropriate synergistic design in place enables individual green building technologies to join forces to generate increasing impact where the application of this design technology would be much more easy in rural area in comparison to the urban area. Clean design or green architecture on the artistic side is the philosophy of planning a construction that is in line with the resources nearby the site and natural feature. Thus, designing clean housing and building will involve many key steps: identifying "green" materials of building from indigenous sources, reducing loads, optimizing systems, and finally producing onsite renewable energy which is abundant in rural area.

Simply, it is necessary for the housing and building sector to confirm that a dynamic clean design and development in all forms of housing and building has been enforced in order to achieve a balanced environment globally. Essentially, emphasis should be directed to the assessments of project life cycle, solicitation, pre-construction, design development, project planning, project ecology, site selection, methods, and application of construction materials to ultimately confirm a sustainable housing and building for the rural community.

Infrastructure and Transportation

Naturally, traditional transport systems which are mainly from the urban infrastructure have significant impacts on the environment, accounting for nearly 28% of world energy conventional consumption and it is causing proportional climate change and adverse environmental impact [35, 36]. While sustainable transportation refers to the broad subject of transport that should be environmentally benign in the senses of social, environmental, and climate impacts and the ability to mitigate the environmental pollution indefinitely, the urban area has limited space to adopt this technology compared to the abundant space in rural area. Components for evaluating sustainable transport include advanced vehicle technology to be used for road, water, or air transport by using renewable and clean energy with the infrastructure that should be able to accommodate the clean fuel-operated transport for roads, railways, airways, waterways, canals, and terminal pathways to mitigate energy and traffic jam crisis where rural area is the best place to have all these opportunities. Simply sustainable transport systems will make a positive contribution to the environmental, social, and economic sustainability of the communities by binding a social and economic connection where people can quickly benefit by this sustainable mobility such as zero-emission vehicle technology which is nearly impossible in the urban area whereas the rural area is the best option.

Therefore, sustainable infrastructure system and advanced transportation vehicles are needed urgently to have better, safer, and faster mobility and less environmental impact compared to traditional infrastructure and conventional vehicles where rural area will have the most advantage in comparison to the urban area.

Water

Within the world's economy, water naturally plays a significant role, with almost 70% of the freshwater being utilized by humans going to the agricultural sector, which has a larger contribution to global economy where urban development has nothing to contribute [31, 32]. For several parts of the globe, fishing in freshwater and saltwater physiques is a key food source and an important part of the global economy referred to as Blue and Brown Economy which is part of rural development. It is a fact that natural resource is becoming scarcer globally where certain places are in a vulnerable condition, as in developing world, 90% of all wastewater still goes untreated into local rivers and streams which causes a dangerous environment in the water world which is mainly caused by urban development [19, 22]. Some 150 major megacities, with roughly a third of the world's population, also suffer from medium or high water pollution [24, 26] and this is not only affecting surface freshwater bodies like rivers and lakes, but also degrading groundwater resources. Currently, about a half billion people around the world routinely drink unhealthy water resulting in some five million deaths each year caused by polluted drinking water in the urban area.

Thus, advance research and development of water is needed for sustainability of this natural resource where primary focus needs to be given to conducting emerging distributed systems for water supply and water and wastewater treatment to confirm a sustainable earth which is only possible by rural development since the rural development will have less water pollution. Consequently, much more scientifically and technologically advanced research and development for rural development must be applied considering environment friendliness: (a) physical and chemical treatment processes for water and wastewater treatment process; (b) environmental biotechnology for use in water resource management and bioremediation, and utilizing wastewater into useful product; (c) watershed and wetland management to reduce water loss; (d) advanced environmental engineering design to mitigate groundwater; and (e) sustainable water resource development as a new source of water supply.

Conclusion

The United Nations Environment Programme (UNEP) estimates that each year 2.4 million people die from air pollution and nearly 5 million people die because of water pollution, and nearly 10 million people die due to human-caused other environmental problems in urban area. The most dangerous, particularly hazardous for health, are emissions of black carbon, a component of particulate matter, which is a known cause of respiratory and carcinogenic diseases and the main contributor to global climate change due to overcongestion in the urban area. Currently the urban atmospheric CO_2 is 400 ppm and it is increasing at a rate of 2.11% yearly which is running to reach the toxic level of CO_2 concentration in the air of 60,000 ppm when all living beings will die in 30 s. If the current level of CO_2 emission is not stopped, all human race in urban area will be extinct [$\int_{400^{(2.11\%)}}^{60,000}(2017)$] in 121,017,712 years and thus it would be the end of story of human civilization in urban area. Simply deurbanization and rural development are an urgent need in order to have sustainable environment, energy, building, infrastructure, transportation, and water system for the survival of all living beings on earth.

Acknowledgements This research was supported by Green Globe Technology under the grant RD-02020-03. Any findings, conclusions, and recommendations expressed in this chapter are solely those of the author and do not necessarily reflect those of Green Globe Technology.

References

1. Hossain, Md. Faruque (2017). Green Science: Independent Building Technology to Mitigate Energy, Environment, and Climate Change. Renewable and Sustainable Energy Reviews. 73; 695-705. (Elsevier).
2. Hossain, Md. Faruque and Fara, Nowshin (2017). Integration of Wind into Running Vehicles to Meet Its Total Energy Demand. Energy, Ecology, and Environment. 2(1), 35-48. (Springer).
3. Hongyan Bao, Jutta Niggemann, Li Luo, Thorsten Dittmar, Shuh-Ji Kao. "Aerosols as a source of dissolved black carbon to the ocean", Nature Communications, 2017.
4. Schwietzke, S. et al. Upward revision of global fossil fuel methane emissions based on isotope database. Nature 538, 88–91 (2016).
5. McAvoy, D.C., Schatowitz, B., Jacob, M., Hauk, A., Eckhoff, W.S., 2002. Measurement of triclosan in wastewater treatment systems. Environ. Toxicol. Chem. 21, 1323-1329.
6. Stephens, B. B. et al. Weak northern and strong tropical land carbon uptake from vertical profiles of atmospheric CO2 science. Science 316, 1732 (2007).
7. Erb, K.-H. et al. Bias in the attribution of forest carbon sinks. Nat. Clim. Change 3, 854–856 (2013).
8. Hossain, Md. Faruque (2016). Production of Clean Energy from Cyanobacterial Biochemical Products. Strategic Planning for Energy and the Environment. 3; 6-23 (Taylor and Francis).
9. Ballantyne, A. P., Alden, C. B., Miller, J. B., Tans, P. P. & White, J. W. C. Increase in observed net carbon dioxide uptake by land and oceans during the past 50 years. Nature 488, 70–72 (2012).
10. Denman, K. L. et al. Couplings Between Changes in the Climate System and Biogeochemistry (Cambridge University Press, 2007).

11. Pasquale Borrelli, David A. Robinson, Larissa R. Fleischer, Emanuele Lugato et al. "An assessment of the global impact of 21st century land use change on soil erosion", Nature Communications, 2017.
12. Scherr, Sara J. 1995. Meeting household needs: Farmer tree-growing strategies in Western Kenya. In Tree Management in Farmer Strategies, edited by J. E. M. Arnold and P. A. Dewees. Oxford UK: Oxford University Press.
13. Emilio Garcia-Robledo, Cory C. Padilla, Montserrat Aldunate, Frank J. Stewart, Osvaldo Ulloa, Aurélien Paulmier, Gerald Gregori and Niels Peter Revsbech. Cryptic oxygen cycling in anoxic marine zones. PNAS (2017) 114 (31) 8319-8324.
14. Hossain, Md. Faruque (2018). Green Science: Advanced Building Design Technology to Mitigate Energy and Environment. Renewable and Sustainable Energy Reviews. 81 (2), 3051-3060. (Elsevier).
15. Bauer, J. E. et al. The changing carbon cycle of the coastal ocean. Nature 504, 61–70 (2013).
16. Ciais, P. & Sabine, C. Chapter 6: Carbon and other biogeochemical cycles in Climate Change 2013 The Physical Science Basis (eds T. Stocker, D. Qin. & G.K. Platner) (Cambridge University Press, 2013).
17. Hossain, Md. Faruque (2018). Green Science: Decoding Dark Photon Structure to Produce Clean Energy. Energy Report. https://doi.org/10.1016/j.egyr.2018.01.001. (Elsevier).
18. Hossain, Md. Faruque (2018). Transforming Dark Photon into Sustainable Energy. International Journal of Energy and Environmental Engineering. https://doi.org/10.1007/s40095-017-0257-1. (Springer).
19. Hossain, Md. Faruque (2016). Theoretical Modeling for Hybrid Renewable Energy: An Initiative to Meet the Global Power. Journal Sustainable Energy Engineering. 4; 5-36. (Wiley).
20. Betts, R. A., Jones, C. D., Knight, J. R., Keeling, R. F. & Kennedy, J. J. El Nino and a record CO2 rise. Nat. Clim. Change 6, 806–810 (2016).
21. Hossain, Md. Faruque (2017). Design and Construction of Ultra-Relativistic Collision PV Panel and Its Application into Building Sector to Mitigate Total Energy Demand. Journal of Building Engineering. 9, 147-154. (Elsevier).
22. Hossain, Md. Faruque (2016). Solar Energy Integration into Advanced Building Design for Meeting Energy Demand. International Journal of Energy Research. 40, 1293-1300. (Wiley).
23. Parazoo et al. "Contrasting carbon cycle responses of the tropical continents to the 2015–2016 El Niño", Science, 2017.
24. Canadell, J. G. et al. Contributions to accelerating atmospheric CO2 growth from economic activity, carbon intensity, and efficiency of natural sinks. Proc. Natl. Acad. Sci. U.S.A. 104, 18866–18870 (2007).
25. Davis, S. J. & Caldeira, K. Consumption-based accounting of CO2 emissions. Proc. Natl. Acad. Sci. U.S.A. 107, 5687–5692 (2010).
26. Duce, R. A. et al. Impacts of atmospheric anthropogenic nitrogen on the open ocean. Science 320, 893–897 (2008).
27. Hossain, Md. Faruque (2016). Theory of Global Cooling. Energy, Sustainability, and Society. 6:24. (Springer).
28. Li, W. et al. Reducing uncertainties in decadal variability of the global carbon budget with multiple datasets. Proc. Natl. Acad. Sci. U.S.A. 113, 13104–13108 (2016).
29. Fairhead, James, and Melissa Leach. 1996. Misreading the African Landscape: Society and Ecology in a Forest-Savanna Mosaic. Cambridge: Cambridge University Press.
30. Pickup, G., G.N. Bastin, and V.H. Chewings. 1998. Identifying Trends in Land Degradation in Non-Equilibrium Rangelands. Journal of Applied Ecology 35:365-377.
31. Gonzalez-Gaya, B. et al. High atmosphere-ocean exchange of semivolatile aromatic hydrocarbons. Nat. Geosci. 9, 438–442 (2016).
32. Postel, S. L., Daily, G. C. & Ehrlich, P. R. Human appropriation of renewable fresh water. Science 271, 785 (1996).

33. Hossain, Md. Faruque (2017). Application of Advanced Technology to Build a Vibrant Environment on Planet Mars. International Journal of Environmental Science and Technology. 14 (12), 2709–2720. (Springer).
34. Hossain, Md. Faruque (2017). Invisible Transportation Infrastructure Technology to Mitigate Energy and Environment. Energy, Sustainability, and Society. 7:27. (Springer).
35. Liu, Z. et al. Reduced carbon emission estimates from fossil fuel combustion and cement production in China. Nature 524, 335–338 (2015).
36. Mason Earles, J., Yeh, S. & Skog, K. E. Timing of carbon emissions from global forest clearance. Nat. Clim. Change 2, 682–685 (2012).

Chapter 16
Sustainable Cities and Communities

Abstract Photon energy has been implemented to design all buildings and houses where at least 25% of its exterior curtain skin wall could be used as the acting photovoltaics (PV) panel to trap the solar energy to transform into electricity to satisfy its net energy demand without any outsource connection in order to develop sustainable cities and communities. Given the current rate of conventional fuel consumption, atmospheric greenhouse gas emissions (GHGs) are increasing rapidly where the building and housing sector of traditionally built cities and communities alone is responsible for 40% of GHG emission. These GHGs ultimately cause environmental vulnerability such as climate change, stratospheric ozone depletion, acid rain, flooding, and air toxicity which threaten the survival of all living beings in urban communities. Therefore, the mechanism of photo-physical transformation by the acting PV panel of the buildings and houses of exterior skin in all cities and communities in response to solar radiation shall indeed would be a cutting-edge technology to confirm to develop sustainable cities and communities.

Keywords Global environmental vulnerability · Solar radiation · PV panel acting as building skin · Clean energy production · Climate change mitigation · Sustainable cities and community

Introduction

Massive development of conventional cities and communities throughout the world is consuming fossil energy tremendously. Consequently, it is causing severe environmental perplexity such as acid deposition, stratospheric ozone depletion, and climate change severely where traditional cities and community development alone are responsible for 40% of this environmental and climate change disaster [1–3]. Besides, conventional energy deposition is getting to a finite level. At present the total amount of fossil fuel reserved in the whole world is 36,600 EJ (crude oil 1.65×10^{11} t or unit energy 4.2×10^{10} J/t is equivalent to 6930 EJ; natural gas

M. F. Hossain, *Global Sustainability in Energy, Building, Infrastructure, Transportation, and Water Technology*,
https://doi.org/10.1007/978-3-030-62376-0_16

1.81×10^{14} t or unit energy 3.6×10^{7} J/m^{3} is equivalent to 6500 EJ; high-quality coal 4.90×10^{11} t or unit energy 3.1×10^{10} J/t is equivalent to 15,000 EJ; low-quality coal 4.3×10^{11} t or 1.9×10^{10} J/t unit energy is equivalent to 8200 EJ) [4–6]. The annual energy consumption worldwide was 283 EJ in 1980, 347 EJ in 1990, 400 EJ in 2000, and 511 EJ in 2010, and it will be 590 EJ in 2025. This rate is expected to increase to 607 EJ in the year 2020, 702 EJ in 2030, 855 EJ in 2040, and 988 EJ in 2050 [7–9]. In the year 2017 the total fossil energy consumption was 560 EJ where cities and community development sector alone consumed 224 EJ. This means if the current level of fossil fuel consumption continues, the total fossil fuel energy source will be run out in 65 years. Since the fossil fuel causes severe environmental disaster and is also getting to a finite level rapidly, clean and renewable energy source is an urgent demand for the development of sustainable cities and communities. Here, the solar radiation has been defined as a sustainable energy to be implemented in the building and house exterior skin for all cities and communities by the process of photo-physical reaction to act as PV panel to produce the clean energy to satisfy the total energy demand without any outsource connection. The average solar radiation on the earth's surface is 1366 W/m^{2}, commonly known as solar constant (Fig. 16.1). The radius of earth is $(2/\pi) \times 10^{7}$ m and thus the total solar radiation reaching the earth is $1366 \times (4/\pi) \times 10^{14} \cong 1.73 \times 10^{17}$ W [6, 10]. There are 86,400 s a day, with an average of 365.2422 days a year. Total annual solar radiation energy per year = $1.73 \times 10^{17} \times 86{,}400 \times 365.2422 = 5.46 \times 10^{24}$ J, equivalent to 5,460,000 EJ/year (Fig. 16.1). The world energy consumption in 2017 was 5.60×10^{20} J = 560 EJ, of which nearly 40% was spent by the building sector. Only 0.01% of the solar energy reaching the earth can meet the global energy demand where the building and housing sector of cities and community can play a vital role in harvesting solar energy by its exterior skin to fulfill the mission of clean energy technology for the development of cities and communities.

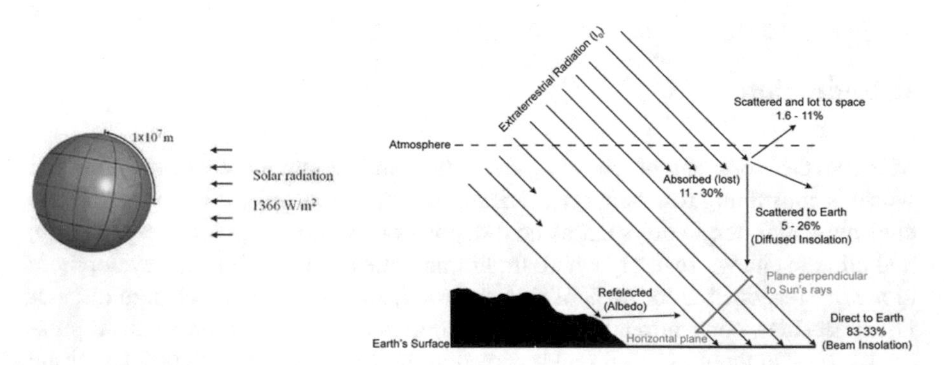

Fig. 16.1 (**a**) Shows the yearly solar irradiance arriving on the surface of earth. The mean solar energy on the earth is 1366 W/m^{2}. The length of the meridian of earth is 10,000,000 m. The total solar irradiance that arrives at the surface of earth per year is 5,460,000 EJ calculatively. (**b**) The effect of the atmosphere on the solar radiation reaching the earth's surface

In the past several decades, huge research has been performed on solar energy and its supply technology to the national grid for commercial application as the source of alternative energy technology [11–14]. Shi et al. showed that solar energy can be harvested by installing massive solar panels in particular places in the tropical or subtropical area and then supplied to the national grid as a source of sustainable energy technology [15]. Gleyzes et al. suggested an advanced mechanism of solar panel by the application of graphene silicon surface that can have maximum 30% efficiency to capture the solar energy to convert into clean energy [16]. Reinhard et al. showed that the breakdown of photon energy and its application by quantum machines into a PV panel can produce tremendous amount of clean energy [17]. All these research findings are indeed interesting, but these technologies required additional place and technology and supply mechanism to utilize the solar energy, and none of these investigations revealed that this solar energy can be utilized directly by the buildings and houses of the cities and communities itself by using its exterior curtain wall to act as the PV panel to produce energy. In this research, therefore, an innovative technology has been proposed to design all buildings and houses of the cities and communities to have at least 25% of the exterior curtain walls to be used as the photovoltaic (PV) panel to capture the solar radiation and then convert it into clean energy to meet the total energy demand for a building and house.

Methods and Materials

The buildings and houses of all cities and communities are proposed to be designed in such a way wherein 25% of the exterior curtain walls are to be built with solar panels. Prior to that this solar panel acting as curtain wall skin must be determined the factors involved, angle, latitude, longitude, and coordinate transformation in a Cartesian coordinate system, to ensure that the maximum solar thermal radiation can be captured by the panel (Fig. 16.2). Considering angle, the acting solar panel needs to be designed for the effect of latitude and module tilt on the solar radiation received throughout the year and the module should be facing south in the northern hemisphere and north in the southern hemisphere [18–20]. Cartesian coordinates for the horizon system need to be used accordingly where south should be x, west should be y, and zenith should be z [16, 21, 22]. These positions of the celestial body are to be determined by two angles, height h and azimuth angle A; Cartesian coordinates for the equatorial system of z'-axis point to the North Pole; east-west y'-axis and x'-axis have to be perpendicular to both directions. Then the position of the celestial body is to be determined by declination δ and hour angle ω; refer to figure in the preceding texts.

Since light is an electromagnetic wave which is produced when an electric charge vibrates due to a hot object, the acting PV panel installation must follow the longitude and latitude, polar coordinates, and three-dimension axis (x, y, z) to get maximum sunlight obtainable on the building the whole year considering the Stefan–Boltzmann laws. The clarification of Stefan–Boltzmann laws is that the electromagnetic waves follow the equal-partition radiation intensity once it is emitted on the plane of the PV panel [23, 24].

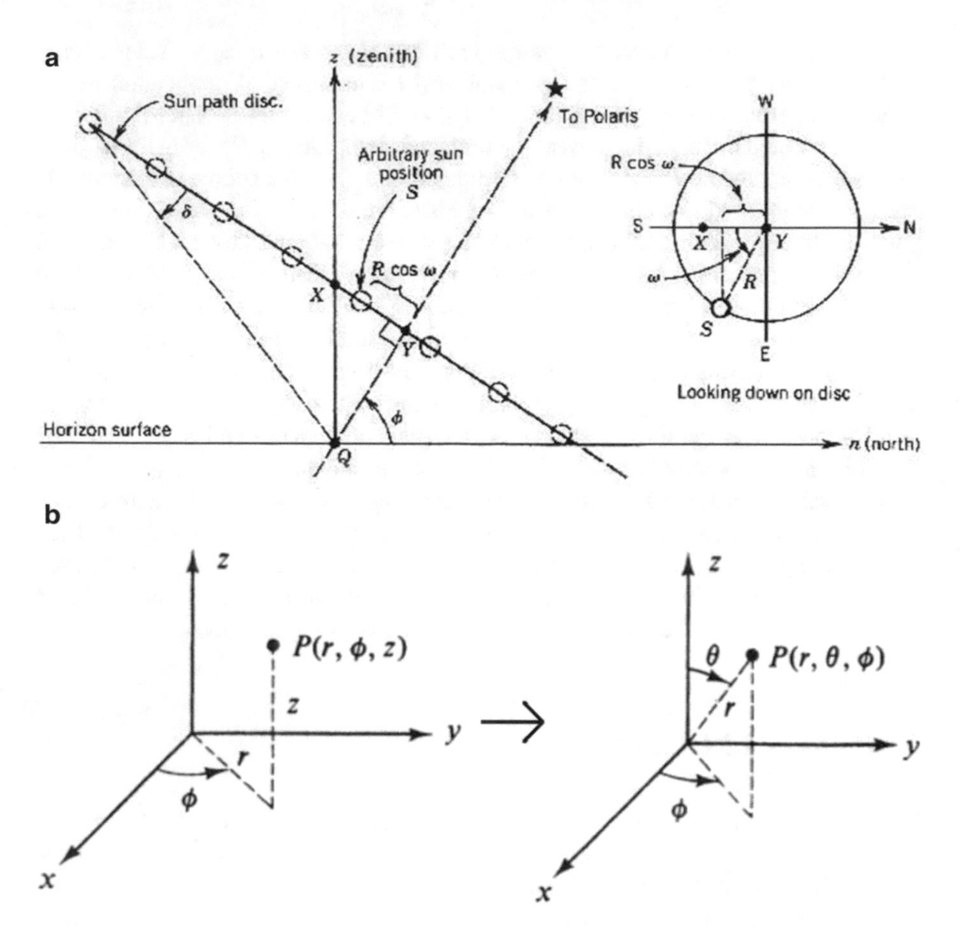

Fig. 16.2 (**a**) Photovoltaic (PV) panel angle has been calculated by considering the impact of latitude and module tilt on the solar energy received within the year. (**b**) The Cartesian coordinate has been utilized for the horizon clarification considering the use of conventions where south is *x*, west is *y*, and *z* is the position of a celestial body and it is determined by two angles of the equatorial systems

Once the angles and Cartesian coordinates of the solar panel are determined, the panel wiring diagram is focused on ensuring that the photovoltaic DC current can transmit into AC current at high efficiency [25, 26]. Thus, two of the Kyocera plates are mounted on aluminum bars to carry light from photovoltaic modules DC; then it is multiplied into the short-circuit current by 125%; this value is used for all 80% efficiency; and then 125% AC current ofcontinuous flow is obtained by using three parallel circuits.

Thereafter, parameters of the current–voltage (*I*–*V*) characteristic are necessarily explained considering two/single-diode equivalent circuit model of the PV cell [27–29]. Subsequently, the photovoltaic array at various parameters (*voltage proliferation, transformation rate, and PVVI curves*) and least control strategy are being used to confirm the active electricity production (I_{v+}) from solar energy to the utilization of it by the building itself [5, 12].

The next step is to determine the photovoltaic current production by I_{pv} calculation from the model of one diode (Fig. 16.3a), considering *I–V–R* relationship (Fig. 16.3b), and use the illumination received by the photovoltaic array to convert from DC to AC and then use it for domestic energy and low voltage current demand (Fig. 16.3c).

Since the solar thermal energy is taken by the photo-physical reaction of photovoltaic panel (PV) which is a continuum flow of photons into the PV panel, it is necessary to determine the excellent photo-physical reaction in the curtain wall-acting PV panel. For this, it is essential to clarify solar thermal conductivity and solar cell antireflective coatings by analyzing quantum electrodynamics, the most effective fields in modern physics to capture much more solar energy [30–32]. Subsequently, classical statistical physics has been used to determine the radiation energy density of inner surface on its (curtain wall) considering the electromagnetic wave and Maxwell-Boltzmann constant statistics (Fig. 16.4). Therefore, a mathematical model of photovoltaic dynamic, in this research, has been developed to capture maximum solar energy by the acting PV panel of building exterior skin

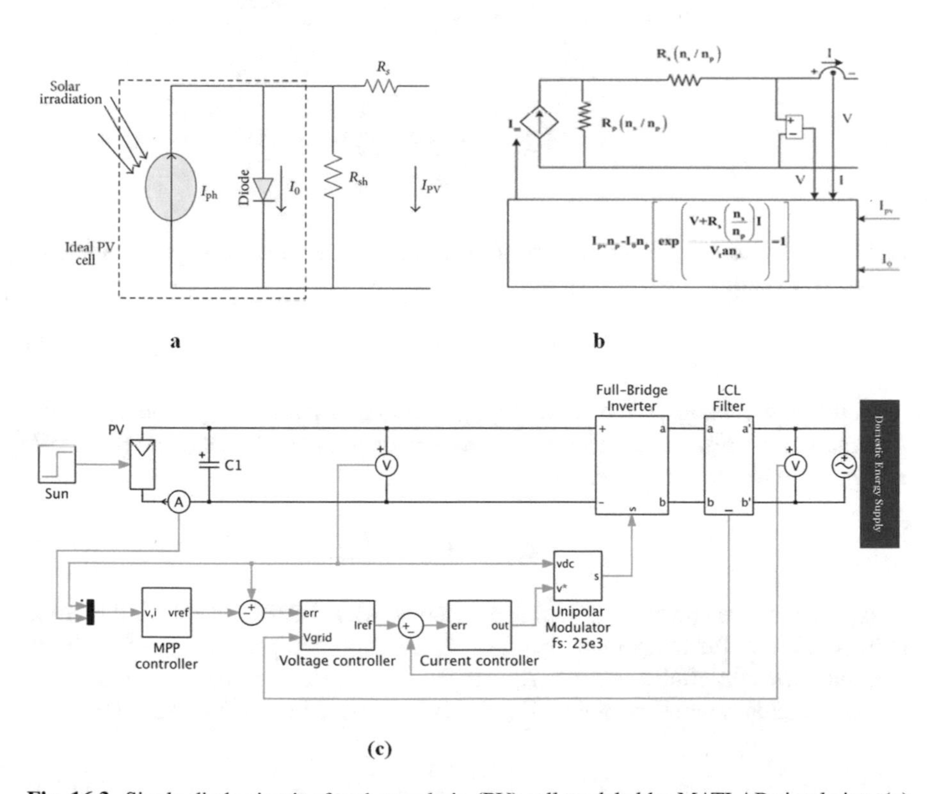

Fig. 16.3 Single-diode circuit of a photovoltaic (PV) cell modeled by MATLAB simulation, (**a**) the photovoltaic current production, (**b**) the model with a diode considering *I–V–R* relationship (**c**), the conversion process of DC to AC for the use of domestic energy and low voltage current demand for the building

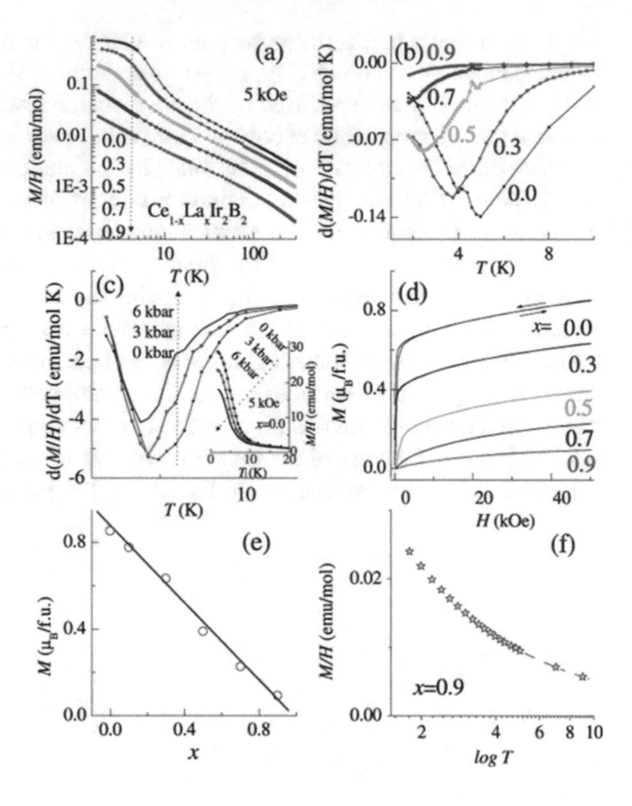

Fig. 16.4 Photonic thermal activation mechanism in various parameters of PV cells shows the charge generation into the PV cells [6, 33, 34]

curtain wall where the following equation calculates the energy output of a photovoltaic (PV) cell:

$$P_{\text{PV}} = \eta_{\text{pvg}}\, A_{\text{pvg}}\, G_{\text{t}} \tag{16.1}$$

In this equation, η_{pvg} refers to the PV generation efficiency, A_{pvg} refers to the PV generator area (m^2), and G_{t} refers to the solar radiation in a tilted module plane (W/m^2). η_{pvg} can be further defined as

$$\eta_{\text{pvg}} = \eta_{\text{r}}\eta_{\text{pc}}\left[1-\beta\left(T_{\text{c}}-T_{\text{cref}}\right)\right] \tag{16.2}$$

η_{pc} refers to the power conditioning efficiency; when MPPT is applied, it is equal to 1; β refers to the temperature coefficient (0.004–0.006 per °C); η_{r} refers to the reference module efficiency; and T_{cref} refers to the reference cell temperature in °C. The reference cell temperature (T_{cref}) can be obtained from the relation below:

$$T_{\text{c}} = T_{\text{a}} + \left(\frac{\text{NOCT}-20}{800}\right)G_{\text{t}} \tag{16.3}$$

T_a refers to the ambient temperature in °C, G_t refers to the solar irradiance in a tilted module plane (W/m²), and NOCT refers to the standard operating cell temperature in Celsius (°C) degree. The total irradiance in the solar cell, considering both standard and diffuse solar irradiance, can be estimated by the following equation:

$$I_t = I_b R_b + I_d R_d + (I_b + I_d) R_r \tag{16.4}$$

The solar cell is essentially a P-N junction semiconductor able to produce electricity via the PV effect, which is interconnected in a series-parallel configuration to form a photovoltaic (PV) cell [22, 35, 36]. Besides, to improve the efficiency of the resulting photovoltaic (PV) cell, graphene is integrated into the PV module [15, 37, 38].

Using a standard single diode, as depicted in Fig. 16.2, for a cell with N_s series-connected arrays and N_p parallel-connected arrays, the cell current must be related to the cell voltage as

$$I = N_p \left[I_{ph} - I_{rs} \left[\exp\left(\frac{q(V + IR_s)}{AKTN_s} - 1 \right) \right] \right] \tag{16.5}$$

where

$$I_{rs} = I_{rr} \left(\frac{T}{T_r} \right)^3 \exp\left[\frac{E_G}{AK} \left(\frac{1}{T_r} - \frac{1}{T} \right) \right] \tag{16.6}$$

In Eqs. (16.5) and (16.6), q refers to the electron charge (1.6×10^{-9} C), K refers to Boltzmann's constant, A refers to the diode idealist factor, and T refers to the cell temperature (K). IR_s refers to the cell reverse saturation current at T, T_r refers to the cell referred temperature, I_{rr} refers to the reverse saturation current at T_r, and E_G refers to the bandgap energy of the semiconductor used in the cell. The photocurrent I_{ph} varies with the cell's temperature and radiation as follows:

$$I_{ph} = \left[I_{SCR} + k_i (T - T_r) \frac{S}{100} \right] \tag{16.7}$$

I_{SCR} refers to the cell short-circuit current at the reference temperature and irradiance, k_i refers to the short-circuit current temperature coefficient, and S refers to the solar irradiance (mW/cm²). The I–V characteristics of the photovoltaic (PV) cell can be derived using a single-diode model which includes an additional shunt resistance concurrent with the optimal shunt diode model as follows:

$$I = I_{ph} - I_D \tag{16.8}$$

$$I = I_{ph} - I_0 \left[\exp\left(\frac{q(V + R_s I)}{AKT} - 1 \right) - \frac{V + R_s I}{R_{sh}} \right] \tag{16.9}$$

I_{ph} refers to the photocurrent (A), I_D refers to the diode current (A), I_0 refers to the inverse saturation current (A), A refers to the diode constant, q refers to the charge of the electron (1.6×10^{-9} C), K refers to Boltzmann's constant, T refers to the cell temperature (°C), R_s refers to the series resistance (ohm), R_{sh} refers to the shunt resistance (Ohm), I refers to the cell current (A), and V refers to the cell voltage (V). The output current of the PV cell using the diode model can be described as follows:

$$I = I_{pv} - I_{D1} - I_{D2} - \left(\frac{V + IR_s}{R_{sh}} \right) \tag{16.10}$$

where

$$I_{D1} = I_{01} \left[\exp\left(\frac{V + IR_s}{a_1 V_{T1}} \right) - 1 \right] \tag{16.11}$$

$$I_{D2} = I_{02} \left[\exp\left(\frac{V + IR_s}{a_2 V_{T2}} \right) - 1 \right] \tag{16.12}$$

I_{01} and I_{02} are the reverse saturation currents of diode 1 and diode 2, respectively, and V_{T1} and V_{T2} are the thermal voltages of the respective diodes. The diode idealist constants are represented by a_1 and a_2. The simplified model of the photovoltaic (PV) system model is presented below:

$$v_{oc} = \frac{V_{oc}}{cKT / q} \tag{16.13}$$

$$P_{max} = \frac{\frac{V_{oc}}{cKT / q} - \ln\left(\frac{V_{oc}}{cKT / q} + 0.72 \right)}{\left(1 + \frac{V_{oc}}{nKT / q} \right)} \left(1 - \frac{V_{oc}}{\frac{V_{oc}}{I_{SC}}} \right) \left(\frac{V_{oc0}}{1 + \beta \ln \frac{G_0}{G}} \right) \left(\frac{T_0}{T} \right)^{y} I_{sc0} \left(\frac{G}{G_0} \right)^{a} \tag{16.14}$$

where ν_{oc} refers to the normalized value of the open-circuit voltage V_{oc} related to the thermal voltage $V_t = nkT/q$, K refers to Boltzmann's constant, n refers to the idealist factor ($1 < n < 2$), T refers to the temperature of the photovoltaic (PV) module in Kelvin, α refers to the factor responsible for all the nonlinear effects on which the photocurrent depends, q refers to the electron charge, γ refers to the factor representing all the nonlinear temperature-voltage effects, while β refers to a photovoltaic (PV) module technology-specific dimensionless coefficient. Equation (16.14) only represents the maximum energy output of a single photovoltaic (PV) module while a real system consists of several photovoltaic (PV) modules connected in series and in parallel. Therefore, the equation of total power output for an array with N_s cells connected in series and N_p cells connected in parallel with power P_M for each module would be for PV panel

$$P_{\text{array}} = N_s N_p P_M \tag{16.15}$$

Naturally the movement of photon flux applied to the solar panel will be activated by the photo-physical reactions to deliver energy-level charges [39, 40]. Since the energy density of the solar radiation considering the photon wave frequency has been modeled by using the classical statistical physics in Fig. 16.5a, the maximum solar energy formation considering a single photon excitation at a rate of 1.4 eV with an energy value of 27.77 MW/m^2 eV has been determined in Fig. 16.5b.

Results and Discussion

The result of the PV model is determined by the I–V equation of PV cells in the single-diode mode. The I–V relationship equation in the PV panel can be expressed as

$$I = I_L - I_O \left\{ \exp^{\left[\frac{q(V + I_{Rs})}{AkT_c}\right]} - 1 \right\} - \frac{(V + I_{Rs})}{R_{sh}} \tag{16.16}$$

I_L represents the photon-generating current, I_o represents the saturated current in the diode, R_s represents resistance in a series, A represents the diode passive function, k (= 1.38×10^{-23} W/m^{2}K) represents Boltzmann's constant, q (= 1.6×10^{-19} C) represents the charge amplitude of an electron, and T_c represents the functional cell

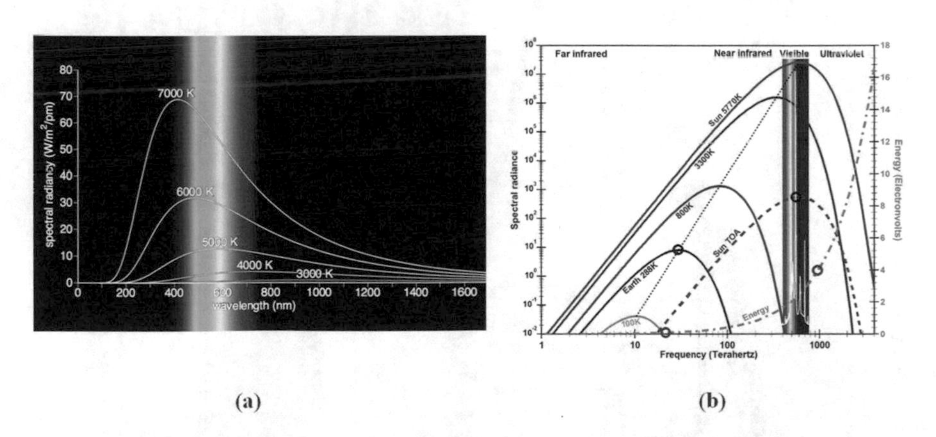

(a) (b)

Fig. 16.5 The thermal energy density of the solar radiation frequencies shown by the classical statistical physics and the figure depicts the solar radiation at various temperatures. (**a**) The spectral irradiance of the light in difference wavelength, (**b**) the radiation in difference frequencies at different temperatures where the maximum irradiance by sun nearly at 5800 K (actual 5770 K) power is equivalent to 6.31×10^7(W/m^2); peak E is 1.410 (eV); peak λ is 0.88 (μm); peak μ is 2.81×10^7 (W/m^2 eV)

temperature. Subsequently, the *I–q* relationship in the PV cells varies owing to the diode current and/or saturation current, which can be expressed as [41, 42]

$$I_{\rm O} = I_{\rm Rs}\left(\frac{T_{\rm C}}{T_{\rm ref}}\right)^3 \exp\left[\frac{qE_{\rm G}\left(\frac{1}{T_{\rm ref}}-\frac{1}{T_{\rm c}}\right)}{KA}\right] \tag{16.17}$$

where IR_s represents the saturated current considering the functional temperature and solar irradiance and qE_G represents the bandgap energy in the silicon and graphene PV cell considering the normal, normalized, and perfect modes at various reflectance band structures of photonic energy emission (Fig. 16.6).

Considering a PV module, the *I–V* equation, apart from the *I–V* curve, is a conjunction of *I–V* curves among all cells of the PV panel. Therefore, the equation can be rewritten as follows to determine the *V–R* relationship:

$$V = -IR_s + K\log\left[\frac{I_{\rm L} - I + I_{\rm O}}{I_{\rm O}}\right] \tag{16.18}$$

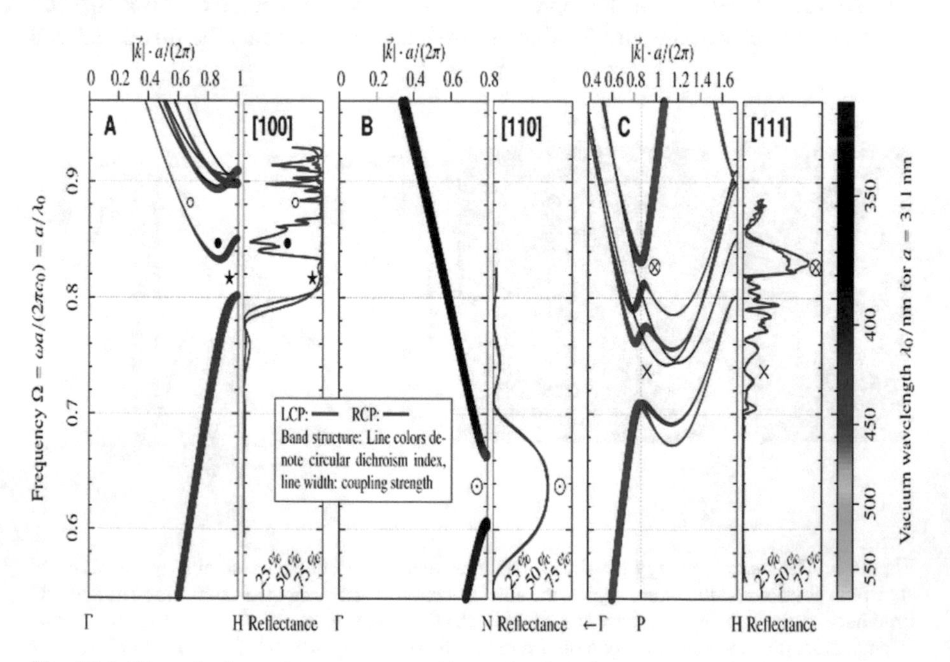

Fig. 16.6 Photonic thermal energy conversion modes in unit area versus frequency at various reflectance band structures of photonic energy emission by glazing wall skins [43, 44]

Here, K is as a constant $\left(=\frac{AkT}{q}\right)$ and I_{mo} and V_{mo} are the current and voltage in the PV panel. Therefore, the relationship between I_{mo} and V_{mo} shall be the same as the PV cell I–V relationship:

$$V_{mo} = -I_{mo}R_{Smo} + K_{mo}\log\left(\frac{I_{Lmo} - I_{mo} + I_{omo}}{I_{omo}}\right) \tag{16.19}$$

where I_{Lmo} represents the photon-generated current, I_{omo} represents the saturated current into the diode, R_{smo} represents the resistancc in series, and K_{mo} represents the factorial constant. Once all non-series (Ns) cells are interconnected in series, then the series resistance shall be counted as the summation of each cell series resistance $R_{smo} = N_s \times R_s$, and the constant factor can be expressed as $K_{mo} = N_s \times K$. There is a certain amount of current flow into the series-connected cells; thus, the current flow in Eq. (16.5) remains the same in each component, i.e., $I_{omo} = I_o$ and $I_{Lmo} = I_L$. Thus, the module $I_{mo} - V_{mo}$ equation for the N_s series of connected cells will be written as

$$V_{mo} = -I_{mo}N_S R_S + N_S K\log\left(\frac{I_L - I_{mo} + I_o}{I_o}\right) \tag{16.20}$$

Similarly, the current–voltage calculation can be rewritten for the parallel connection once all N_p cells are connected in parallel mode and can be expressed as follows [45–47]:

$$V_{mo} = -I_{mo}\frac{R_s}{N_p} + K\log\left(\frac{N_{sh}I_L - I_{mo} + N_p I_o}{N_p I_o}\right) \tag{16.21}$$

Because the photon-generated current primarily will depend on the solar irradiance and relativistic temperature conditions of the PV panel, the current can be calculated using the following equation:

$$I_L = G\left[I_{SC} + K_I\left(T_C - T_{ref}\right)\right] \times V_{mo} \tag{16.22}$$

where I_{sc} represents PV current at 25 °C and KW/m^2, K_I represents the relativistic PV panel coefficient factor, T_{ref} represents the PV panel's functional temperature, and G represents the solar energy in kW/m^2.

Conversion of Electricity

To convert solar thermal energy into electricity, a single-diode circuit has been used consisting of a small disk of semiconductor attached by wire to a circuit consisting of a positive and a negative film of silicon placed under a thin slice of glass and attached to graphene, using an exterior curtain wall skin of all cities and community buildings and houses. Necessarily, the PV panel shall have an *open-circuit point* where the current and voltage are at the maximum, open-circuit voltage V_{oc}, and *maximum power point* that can be clarified using the maximum current and voltage calculation immediately upon capturing the nonequilibrium photons [48–50]. The power delivered by a PV panel will thus have the capability to attain a maximum value at the points (I_{mp}, V_{mp}) [51–53]. It is confirmed by the PV panel's capability of such current–voltage flow to get net energy production by the PV panel by analyzing circuit models of the PV module: (a) normal, (b) normalized, and (c) perfect modes where perfect modes are the best option (Fig. 16.7).

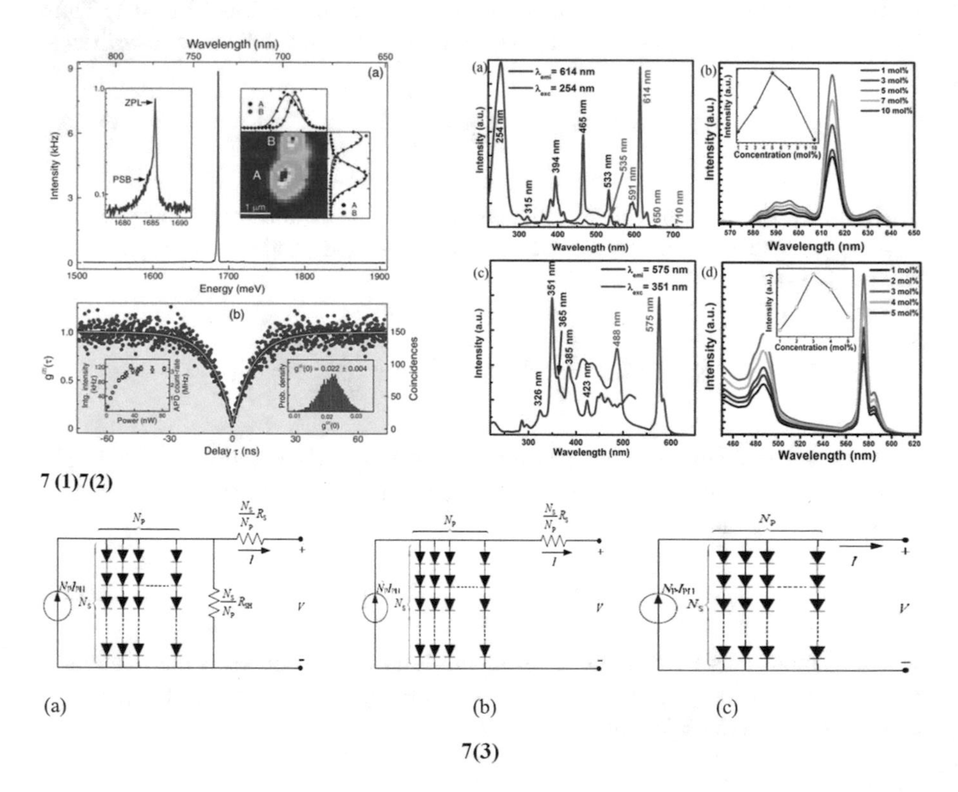

Fig. 16.7 Solar cell current–voltage characteristic features for conceptual function of (1) net optimal solar energy production intensity with respect to the delay of coincidental factor and (2) intensity of solar energy production in various wavelengths; (3) the current–voltage modules of the current source near the short-circuit point and as a voltage source in the vicinity of the open-circuit point at the condition of the PV module are (a) normal, (b) normalized, and (c) perfect modes

To establish a connection between the number of light-quanta solar energy by steady state, the intensity of solar irradiance is considered to convert it into electricity energy by PV panel [54–56]. The number of stationary states of light quanta is a certain type of polarization which frequency is in the range of ν_r to $\nu_r + d\nu_r$ [17, 57, 58]. From this the maximum solar radiation can be achieved at 1.4 eV with an energy value of 27.77 mW/m^2 eV based on an average of 5-h solar irradiance harvesting in a day's peak levels, which is the equivalent of 27,770 kW/year or 7.6 kW/day energy [59, 60]. Due to physical principles, there are losses in the conversion of solar energy into DC power and convertion of direct current into alternating current (AC). This ratio of AC to DC is called "derating factor," which is typically 0.8 [61–63]. Thus, the surface texture of selective solar metal is excellent in energy conversation [26, 27, 46], since the current net conversion by solar panels is 125% higher level with an efficiency of 80% [11, 13, 14] of solar panels which means that (27,770 × 1.25 × 0.8) = 27,770 kW/year or 7.6 kW/day. Energy remains equal to the initial solar energy before the introduction to the solar panel. Necessarily, the maximum solar irradiance is depicted as 1.4 eV with an energy value of 27.77 mW/m^2 eV in a year an average of 5 h a day for 365 days [47, 64, 65]. A standard residential house requires an average 6 kW/day [35, 45, 66]. Since the produced energy is equivalent to 27,770 kW/year or 7.6 kW/day, in fact it will meet the energy demand for a residential house required, 6 kW/day, by using only one solar panel of 1 m^2. The average monthly energy consumption rate of commercial offices or buildings is about 10,000 kW/day for a footprinting 32 m × 31 m with 30 m (10 floors) [67–69]. In the calculation of a building with an average of 32 m × 31 m footprint and with a height of 30 m, the total installed 1 m^2 PV panels should be 1195 units (945 + 250) with a capacity of 7.6 kW/unit energy production that can provide a total energy × 1195 = 9082 kW/day to meet the daily energy demand of about 10,000 kW/day for a commercial office or building.

Savings on Energy Cost

On the other hand, the net cost for 30 years' energy purchase from a traditional utility source for a standard industry (100 people capacity) at 0.12/kWh of 4000 kWh per month is (30 × 12 × 4000 × 0.12) $172,800. This difference between traditional energy use and curtain wall-assisted PV panel energy production clearly indicates the cost saving of $68,400 once curtain wall-assisted PV panel is used as the energy source.

Conclusions

In recent decades, concern about the deadly risks of greenhouse gases (GHGs) has been growing due to their level of accumulation in the atmosphere and the adverse impact on earth due to the conventional development of the cities and communities throughout the world. Given that conventional energy consumption by these traditional cities and communities is the final factor that leads to environmental perplexity and climate change, the intelligent deployment of solar energy, in effect, for a better methodological application to mitigate global environmental perplexity and climate changes is an urgent demand. In addition, the limited nature of the fossil fuel reserve poses a serious challenge for future energy supply as the consumption of fossil fuels will end in the next 65 years. In this study, it is therefore proposed to ensure an economical, reliable, and sustainable energy technology to meet the future energy demand which is also climate friendly in order to secure sustainable cities and communities throughout the world. Simply capturing solar thermal energy using exterior curtain wall of buildings and houses demonstrated in this research shall indeed be the innovative technology to meet the total energy demand to secure the development of sustainable cities and communities throughout the world.

Acknowledgements This research was supported by Green Globe Technology, Inc. under the grant RD-02018-03 for building a better environment. Any findings, predictions, and conclusions described in this chapter are solely those of the authors, who confirm that the review has no conflicts of interest for publication in a suitable journal.

References

1. Artemyev, N., Jentschura, U. D., Serbo, V. G., Surzhykov, A. Strong Electromagnetic Field EFFECTS in Ultra-Relativistic Heavy-Ion Collisions. Eur. Phys. J. C 72, 1935 (2012).
2. Birnbaum, K. M. et al. Photon blockade in an optical cavity with one trapped atom. Nature 436, 87–90 (2005).
3. Xiao, Y. F. et al. Asymmetric Fano resonance analysis in indirectly coupled microresonators. Phys. Rev. A 82, 065804 (2010).
4. Busch, K., von Freymann, G., Linden, S., Mingaleev, S. F., Tkeshelashvili, L. and Wegener, M. Periodic nanostructures for photonics. Phys. Rep. 444, 101 (2007).
5. Chang, D. E., Sørensen, A. S., Demler, E. A. and Lukin, M. D. A single-photon transistor using nanoscale surface plasmons. Nature Physics. 3, 807–812 (2007).
6. Cheng, M. and Song, Y. Fano resonance analysis in a pair of semiconductor quantum dots coupling to a metal nanowire. Opt. Lett. 37, 978–980 (2012).
7. Armani, D. K., Kippenberg, T. J., Spillane, S. M. and Vahala, K. J. Ultra-high-Q toroid microcavity on a chip. Nature 421, 925 (2003).
8. Chen, J., Wang, C., Zhang, R. and Xiao, J. Multiple plasmon-induced transparencies in coupled-resonator systems. Opt. Lett. 37,5133–5135 (2012).
9. Gupta, N., Singh, S.P., Dubey, S.P. and Palwalia, D.K., Fuzzy logic controlled three-phase three-wired shunt active power filter for power quality improvement, (2011) *International Review of Electrical Engineering (IREE)*, 6 (3), pp. 1118-1129.

10. Liao, J. Q. and Law, C. K. Correlated two-photon transport in a one-dimensional waveguide side-coupled to a nonlinear cavity. Phys. Rev. A 82, 053836 (2010).
11. Dayan, B. et al. A photon turnstile dynamically regulated by one atom. Science 319, 1062–1065 (2008).
12. Douglas, J. S., Habibian, H., Hung, C., Gorshkov, A., Kimble, H. and Chang, D. "Quantum many-body models with cold atoms coupled to photonic crystals", Nature Photonics, 2015.
13. Gould, R. J. "Pair Production in Photon-Photon Collisions", Physical Review, 03/1967.
14. Guerlin, C. et al. Progressive field-state collapse and quantum non-demolition photon counting. Nature 448, 889 (2007).
15. Shi, T., Fan, S. and Sun, C. P. Two-photon transport in a waveguide coupled to a cavity in a two-level system. Phys. Rev. A 84, 063803 (2011).
16. Gleyzes, S. et al. Quantum jumps of light recording the birth and death of a photon in a cavity. Nature 446, 297 (2007).
17. Reinhard, A. "Strongly correlated photons on a chip", Nature Photonics, 2011.
18. Eichler, J., Stöhlker, Th. Radiative electron capture in relativistic ion-atom collisions and the photoelectric effect in hydrogen-like high-Z systems, Phys. Rep. 439, 1 (2007).
19. Hossain, Md. Faruque. (2016). Solar Energy Integration into Advanced Building Design for Meeting Energy Demand. International Journal of Energy Research. 40, 1293-1300.
20. Hossain, Md. Faruque (2017). Design and Construction of Ultra-Relativistic Collision PV Panel and Its Application into Building Sector to Mitigate Total Energy Demand. Journal of Building Engineering. https://doi.org/10.1016/j.jobe.2016.12.005.
21. Englund, D. et al. Resonant excitation of a quantum dot strongly coupled to a photonic crystal nanocavity. Phys. Rev. Lett. 104, 073904 (2010).
22. Yan, W. and Fan, H. "Single-photon quantum router with multiple output ports", Scientific Reports, 2014.
23. Beyer, H F, T Gassner, M Trassinelli, R Heß, U Spillmann, D Banaś, K-H Blumenhagen, F Bosch, C Brandau, W Chen, Chr Dimopoulou, E Förster, R E Grisenti, A Gumberidze, S Hagmann, P-M Hillenbrand, P Indelicato, P Jagodzinski, T Kämpfer, Chr Kozhuharov, M Lestinsky, D Liesen, Yu A Litvinov, R Loetzsch, B Manil, R Märtin, F Nolden, N Petridis, M S Sanjari, K S Schulze, M Schwemlein, A Simionovici, M Steck, Th Stöhlker, C I Szabo, S Trotsenko, I Uschmann, G Weber, O Wehrhan, N Winckler, D F A Winters, N Winters, and E Ziegler. "Crystal optics for precision x-ray spectroscopy on highly charged ions—conception and proof", Journal of Physics B Atomic Molecular and Optical Physics, 2015.
24. Li, Qiong, D. Z. Xu, C. Y. Cai, and C. P. Sun. "Recoil effects of a motional scatterer on single-photon scattering in one dimension", Scientific Reports, 2013.
25. Hossain, Md. Faruque. (2018). Photonic Thermal Energy Control to Naturally Cool and Heat the Building. Advanced Thermal Engineering. 131, 576–586.
26. Liao, J. Q. and Law, C. K. Correlated two-photon scattering in cavity optomechanics. Phys. Rev. A 87, 043809 (2013).
27. Li, Q., Xu, D. Z., Cai, C. Y. and Sun, C. P. "Recoil effects of a motional scatterer on single-photon scattering in one dimension", Scientific Reports, 2013.
28. Lo, P., Xiong, H., and Zhang, W. "Breakdown of Bose-Einstein Distribution in Photonic Crystals", Scientific Reports, 2015.
29. Sayrin, C. et al. Real-time quantum feedback prepares and stabilizes photon number states. Nature 477, 73 (2011).
30. Han, Z. and Bozhevolnyi, S. I. Plasmon-induced transparency with detuned ultracompact Fabry-Pérot resonators in integrated plasmonic devices. Opt. Express 19, 3251–3257 (2011).
31. Hossain, Md. Faruque (2017). Green Science: Independent Building Technology to Mitigate Energy, Environment, and Climate Change. Renewable and Sustainable Energy Reviews. https://doi.org/10.1016/j.rser.2017.01.136.
32. Hossain, Md. Faruque. (2018). Green Science: Advanced Building Design Technology to Mitigate Energy and Environment. Renewable and Sustainable Energy Reviews. 81 (2), 3051-3060.

33. Hossain, Md. Faruque. (2018). Transforming Dark Photon into Sustainable Energy. International Journal of Energy and Environmental Engineering. https://doi.org/10.1007/s40095-017-0257-1.
34. Huang, J. F., Shi, T., Sun, C. P. and Nori, F. Controlling single-photon transport in waveguides with finite cross section. Phys. Rev. A 88, 013836 (2013).
35. Yan, W., Huang, J. and Fan, H. "Tunable single-photon frequency conversion in a Sagnac interferometer", Scientific Reports, 2013.
36. Yu, G.. "A novel two-mode MPPT control algorithm based on comparative study of existing algorithms", Solar Energy, 2004.
37. Shen, J. T. and Fan, S. Strongly correlated two-photon transport in a one-dimensional waveguide coupled to a two-level system. Phys. Rev. Lett. 98, 153003 (2007).
38. Yang, L., Wang, S., Zeng, Q., Zhang, Z., Pei, T., Li, Y. and Peng, L. (2011). Efficient photovoltage multiplication in carbon nanotubes. Nature Photonics pp 672 – 676.
39. Huang, Y., Min, C. and Veronis, G. Subwavelength slow-light waveguides based on a plasmonic analogue of electromagnetically induced transparency. Appl. Phys. Lett. 99, 143117 (2011).
40. Lang, C. et al. Observation of resonant photon blockade at microwave frequencies using correlation function measurements. Phys. Rev. Lett. 106, 243601 (2011).
41. Jentschura, U., Hencken, K. and Serbo, V. Revisiting unitarity corrections for electromagnetic processes in collisions of relativistic nuclei. The European Physical Journal C, 58(2), pp. 281-289. (2008).
42. Lei, C. U. and Zhang, W. M. A quantum photonic dissipative transport theory. Ann. Phys. 327, 1408 (2012).
43. Zhang, W. M., Lo, P. Y., Xiong, H. N., Tu, M. W. Y. and Nori, F. General Non-Markovian Dynamics of Open Quantum Systems. Phys. Rev. Lett. 109, 170402 (2012).
44. H Rauh. "Optical transmittance of photonic structures with linearly graded dielectric constituents", New Journal of Physics, 07/26/2010.
45. Zhu, Yu, Xiaoyong Hu, Hong Yang, and Qihuang Gong. "On-chip plasmon-induced transparency based on plasmonic coupled nanocavities", Scientific Reports, 2014.
46. Tang, Jing, Weidong Geng, and Xiulai Xu. "Quantum Interference Induced Photon Blockade in a Coupled Single Quantum Dot-Cavity System", Scientific Reports, 2015.
47. Etienne Saloux, Alberto Teyssedou, Mikhaïl Sorin. "Explicit model of photovoltaic panels to determine voltages and currents at the maximum power point", Solar Energy, 2011.
48. Joannopoulos, J. D., Villeneuve, P. R. and Fan, S. Photonic crystals: putting a new twist on light. Nature 386, 143 (1997).
49. Sánchez Muñoz, C., Laussy, F., Valle, E., Tejedor, C. and González-Tudela, A. (2018). Filtering multiphoton emission from state-of-the-art cavity quantum electrodynamics. Optica, 5 (1), 14-26.
50. Tame, M. S., McEnery, K. R., Özdemir, Ş. K., Lee, J., Maier, S. A. and Kim, M. S. "Quantum plasmonics", Nature Physics, 2013.
51. Kai Hencken. "Transverse Momentum Distribution of Vector Mesons Produced in Ultraperipheral Relativistic Heavy Ion Collisions", Physical Review Letters, 01/2006.
52. Klein, S. A. Calculation of flat-plate collector loss coefficients. *Solar Energy*, 17:79–80, 1975.
53. Roy, D. Two-photon scattering of a tightly focused weak light beam from a small atomic ensemble: An optical probe to detect atomic level structures. Phys. Rev. A 87, 063819 (2013).
54. Marco T. Manzoni, Darrick E. Chang, James S. Douglas. "Simulating quantum light propagation through atomic ensembles using matrix product states", Nature Communications.
55. Poshakinskiy, Alexander V., Alexander, and Poddubny N. "Biexciton-mediated superradiant photon blockade", Physical Review A, 2016.
56. Matteo Mariantoni, H. Wang, Radoslaw C. Bialczak, M. Lenander et al. "Photon shell game in three-resonator circuit quantum electrodynamics", Nature Physics, 2011.
57. O'Shea, D., Junge, C., Volz, J. and Rauschenbeutel, A. Fiber-optical switch controlled by a single atom. Phys. Rev. Lett. 111, 193601. (2013).

58. Ruiz, Alberto. "Partial Recovery of a Potential from Backscattering Data", Communications in Partial Differential Equations, Springer Tracts in Modern Physics, 2014.
59. Kofman, A. G., Kurizki, G. and Sherman, B. Spontaneous and Induced Atomic Decay in Photonic Band Structures. J. Mod. Opt. 41, 353 (1994).
60. Zhou, W.. "A novel model for photovoltaic array performance prediction", Applied Energy, 2007.
61. Kolchin, P., Oulton, R. F. and Zhang, X. Nonlinear quantum optics in a waveguide: Distinct single photons strongly interacting at the single atom level. Phys. Rev. Lett. 106, 113601 (2011).
62. Okamoto, Hiroyuki, Kenzo Yamaguchi, Masanobu Haraguchi, and Toshihiro Okamoto. "Development of plasmonic racetrack resonators with a trench structure", Plasmonics Metallic Nanostructures and Their Optical Properties X, 2012.
63. Reiserer, Andreas Kalb, Norbert Rempe, Gerhard Ritter, Stephan. "A quantum gate between a flying optical photon and a single trapped atom. (RESEARCH: LETTER) (Report)", Nature, April 10 2014.
64. Longo, P., Schmitteckert, P. and Busch, K. Few-photon transport in low-dimensional systems. Phys. Rev. A 83, 063828 (2011).
65. Yuwen Wang, Yongyou Zhang, Qingyun Zhang, Bingsuo Zou, Udo Schwingenschlogl. "Dynamics of single photon transport in a one-dimensional waveguide two-point coupled with a Jaynes-Cummings system", Scientific Reports, 2016.
66. Valluri, S. R., Becker, U., Grün, N. and Scheid, W. Relativistic Collisions of Highly-Charged Ions, J. Phys. B: At. Mol. Phys. 17, 4359 (1984).
67. Lü, X., Zhang, W., Ashhab, S., Wu, Y. and Nori, F. "Quantum-criticality-induced strong Kerr nonlinearities in optomechanical systems", Scientific Reports, 2013.
68. Najjari, B., Voitkiv, A., Artemyev, A. and Surzhykov, A. Simultaneous electron capture and bound-free pair production in relativistic collisions of heavy nuclei with atoms, Phys. Rev. A 80, 012701 (2009).
69. Tu, M. W. Y. and Zhang, W. M. Non-Markovian decoherence theory for a double-dot charge qubit. Phys. Rev. B 78, 235311 (2008).

Chapter 17
Rapid Connectivity Within the Urban and Rural Area

Abstract There has been expanded huge development of mass urbanization recently throughout the world, quickening the mobilization of habitants from rural area to the urban area that will eventually cause severe environmental impact on the urban area. We certainly need the development of rapid connectivity within the urban and rural areas to mitigate urban congestion and environment perplexity. For hundreds of years, people have been getting huge interest to live in urban area to conduct their business and job in a reasonable distance which is causing severe congestion of urban area resulting in severe urban mobility condition. Simply if the rapid connectivity is developed by advanced transportation technology within the urban and rural areas, people will be interested to live in rural area while working in urban area. Therefore, in this study a ***high-speed flying train*** technology is being proposed to have a rapid and resilience communication within the urban and rural areas to get a relief from the urban congestion where people can reach the urban work area in a reasonable time and come back home in the rural area after work in a convenient time. Simply, a model of 3D *k-omega* flying train is being developed considering the aerodynamic impact during the takeoff, landing, and flying stage of the train. Considering the application of this technology to design a high-speed flying train would indeed be an advanced invention that will not only facilitate the rapid communication within the urban and rural areas but also significantly mitigate urban congestion.

Keywords Urban congestion · Urban-rural connectivity · 3D numerical simulation · High-speed flying train · Environmental sustainability

Introduction

Today, nearly 54% of the world's population lives in urban areas and it is expected to increase nearly 70% by the year 2050 [1–3]. The estimate confirms that urbanization and gradual shift in residence of the human population from rural to urban areas, combined with the overall growth of the world's population, could add another 2.5 billion people to urban areas each year [4–6]. Thus, the recent widespread

M. F. Hossain, *Global Sustainability in Energy, Building, Infrastructure, Transportation, and Water Technology*,
https://doi.org/10.1007/978-3-030-62376-0_17

concern about the massive urbanization has increased due to the limitations of urban mobility, as well as the urban's finite ability to absorb habitants due to the flocking of the citizens into the urban area. The urban area covers 3% of the world's land surface and the vast land on earth surface remains uninhabited which is still having less pollution in air and environment [7, 8]. Simply, if a rapid communication within the urban and rural area is developed in order to help people work in urban area from rural area by employing advanced connectivity technology, it will indeed keep the urban environment physically, chemically, biologically, and environmentally balanced. The recent study suggested that the urban transportation infrastructure consumes 1% of the earth surface which is nearly 1/3 of the total urban area and made the urban system terribly congested causing severe environmental perplexity [9–11]. To address this severe connectivity problem, the ***high-speed flying train*** technology would be an innovative transportation technology which will conveniently transport the people from rural to urban and urban to rural area and thus in this study a model of computational 3D numerical simulation for a flying train has been presented considering the implementation of 3D k-omega aerodynamic impact during the takeoff, landing, and flying stage of the train. Subsequently, the numerical aspect comprises its Reynolds-averaged Navier–Stokes (RANS) mechanism that has been analyzed in order to confirm that this flying train is highly safe during the takeoff, landing, and flying stage. This study, thus, aims to introduce a 3D numerical simulation of propulsive and impulsive force, and its 3D k-omega mechanism to address the aerodynamics of takeoff, landing, and flying velocity to confirm that the performance of the flying train is perfect enough to travel smoothly.

Methods and Materials

In order to design a ***high-speed flying train***, initially the guideway model system has been conducted by adopting Bernoulli-Euler equation to confirm a safe takeoff and landing of the high-speed flying train considering the levitation and lateral force control (Fig. 17.1).

Hence, a free body diagram has been shown to prepare the guideway model where equal-interval force (d) is considered at a various level of speed v, where m = train weight, c = damping coefficient, EI_y = flexural rigidity in the y direction, EI_z = flexural rigidity in the z direction, l = car length, m_w = lumped mass of magnetic wheel, m_v = distributed mass of the rigid train body, and $\theta_{i=x,y,z}$ = midpoint rotation components of the rigid train body. Considering these, the following equations of motion of the train on the guideway have been formulated:

$$m\ddot{u}_{y,j} + c_y \dot{u}_{y,j} + EI_y \ddot{u}_{y,j} = \sum_{k=1}^{K} \left[G_{y,k}\left(i_k, h_{y,k}\right) \varphi_j (x_k, t) \right] \tag{17.1}$$

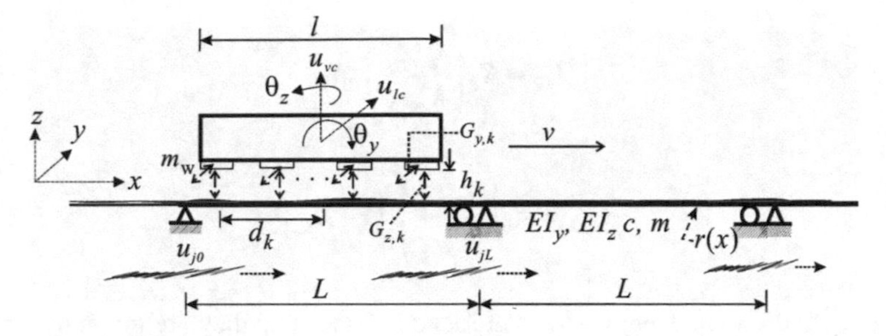

Fig. 17.1 Free body diagram shows the guideway force considering weight and motion of the train where the guideway functions by a series of equal-distant concentrated masses to levitate propulsive force of the train to the guideway which is induced by the motion force to allow takeoff and landing on longitudinal direction

$$m\ddot{u}_{z,j}+c_z\dot{u}_{z,j}+EI_z u_{z,j}=p_0-\sum_{k=1}^{K}\left[G_{z,k}\left(i_k,h_{z,k}\right)\varphi_j(x_k,t)\right] \tag{17.2}$$

and

$$\varphi_j\left(x_k,t\right)=\delta\left(x-x_k\right)\left[H\left(t-t_k-\frac{(j-1)L}{v}\right)-H\left(t-t_k-\frac{jL}{v}\right)\right] \tag{17.3}$$

together with the following boundary conditions with lateral (y direction) support movements:

$$\begin{aligned} &u_{y,j}\left(0,t\right)=u_{yj0}\left(t\right),\quad u_{y,j}\left(L,t\right)=u_{yjL}\left(t\right),\\ &EI_z\overset{¤}{u}_{z,j}\left(0,t\right)=EI_z\overset{¤}{u}_{z,j}\left(L,t\right)=0 \end{aligned} \tag{17.4}$$

$$\begin{aligned} &u_{z,j}\left(0,t\right)=u_{z,j}\left(L,t\right)=0\\ &EI_y\overset{¤}{u}_{y,j}\left(0,t\right)=EI_y\overset{¤}{u}_{y,j}\left(L,t\right)=0 \end{aligned} \tag{17.5}$$

where (y)′(y)/x, (y)(y)/t, $u_{z,j}(x, t)$ = vertical deflection of the jth span, $u_{y,j}(x, t)$ = lateral deflection of the jth span, L = span length, K = number of magnets attached to the rigid levitation frame, (y) = Dirac's delta function, $H(t)$ = unit step function, $k = 1, 2, 3, \ldots$, kth moving wheel, $t_k = (k - 1)d/v$ = arrival time of the kth wheel, x_k = position of the kth wheel on the guideway, and $(G_{y,k}, G_{z,k})$ = lateral guidance and uplift levitation forces of the kth lumped force in the vertical and lateral directions [12–14].

Since the train will fly by aerodynamics force, this guidance kinetic force is tuned by the controlled motion of the flying train which is adopted by the lateral guidance force ($G_{y,k}$) and the uplift levitation force ($G_{z,k}$) to keep and guide the kth force as flying which is expressed as

$$G_{y,k} = K_0 \left(\frac{i_k(t)}{h_{z,k(t)}} \right)^2 K_{k,z} \tag{17.6}$$

$$G_{y,k} = K_0 \left(\frac{i_k(t)}{h_{z,k(t)}} \right)^2 \left(1 - K_{y,k}\right) \tag{17.7}$$

where $\kappa_{y,k}$ and $\kappa_{z,k}$ represent induced guidance factors and they are given by

$$K_{y,k} = \frac{\chi_k \times h_{y,k}}{W(1+\chi_k)}, \quad K_{z,k} = \frac{\chi_k \times h_{y,k}}{W(1+\chi_k)} \tag{17.8}$$

In Eqs. (17.6) and (17.7), $\kappa_0 = \mu_0 N_0^2 A_0 / 4$ = coupling factor, $\chi_k = \pi h_{y,k} h_{z,k} / 4h$, W = pole width, μ = vacuum permeability, N_0 = number of turns of the windings, A_0 = pole face area, $i_n(t) = i_0 + \iota_n(t)$ = electric current, $\iota_n(t)$ = deviation of current, and (i_0, h_{y0}, h_{z0}) = desired current and air gaps around a specified nominal operating point of the wheels at *static* equilibrium. And the uplift levitation ($h_{y,k}$) and lateral guidance ($h_{z,k}$) gaps are, respectively, given by

$$h_{y,k}(t) = h_{y0} + u_{l,k}(t) - u_{y,j}(x_k), \quad u_{l,k}(t) = u_{lc}(t) + d_k \theta_z \tag{17.9}$$

$$h_{z,k}(t) = h_{z0} + u_{v,k}(t) - u_{z,j}(x_k) + r(x_k), \quad u_{v,k}(t) = u_{vc}(t) + d_k \theta_y \tag{17.10}$$

where $(u_{l,k}, u_{v,k})$ = displacements of the kth wheel in the y and z directions, (u_{lc}, u_{vc}) = midpoint displacements of the rigid train, (θ_y, θ_z) = midpoint rotations of the rigid car, $r(x)$ = irregularity of guideway, and d_k = location of the kth wheel to the midpoint of the guideway. As indicated in Eqs. (17.6)–(17.8), the motion-dependent nature and guidance factors ($\kappa_{y,k}$, $\kappa_{z,k}$) dominate the control forces of the guideway system which is expressed as

$$M_0 \ddot{u}_{lc} = g(t) + \sum_{k=1}^{K} G_{y,k}, \quad I_T \ddot{\theta}_z = g(t) \times l + \sum_{k=1}^{K} \left[G_{y,k} d_k \right] \tag{17.11}$$

$$M_0 \ddot{u}_{vc} = p_0 + \sum_{k=1}^{K} G_{z,k}, \quad I_T \ddot{\theta}_y = -\sum_{k=1}^{K} \left[G_{z,k} d_k \right] \tag{17.12}$$

in which $M_0 = m_v l + K m_w$ = lumped mass of the flying train, $g(t)$ = control force to tune the lateral response of the flying train, I_T = total mass moment of inertia of the flying train, and $p_0 = M_0 g$ = lumped weight of the flying train.

Since the flying train's performance in the air mainly depends on the aerodynamic force, the propelling forces, drag force, stability, and its proper control capabilities have also been calculated mathematically in order to control flying train run in the sky smoothly [15–17]. Thus, it has been conducted to implement the Mach number contours for a safe physical model of a flying train by implementing a 3D CFD, *k-omega* turbulence model for flying train (Fig. 17.2).

As the 3D standard *k-omega* turbulence model depends on the flying train's velocity mode, the empirical model of flying train has been presented in order to assure its accurate dissipation rates for turbulence kinetic energy to run the plane in the sky smoothly which is expressed as

$$\psi = \left[-\left(\frac{v_\infty r_o^2}{4}\right)\frac{r_o}{r} + \left(\frac{3v_\infty r_o}{4}\right)r - \left(\frac{v_\infty}{2}\right)r^2\right]\sin^2\theta \tag{17.13}$$

$$v_r = v_\infty\left[1 - \frac{3}{2}\left(\frac{r_o}{r}\right) + \frac{1}{2}\left(\frac{r_o}{r}\right)^3\right]\cos\theta \tag{17.14}$$

$$v_\theta = -v_\infty\left[1 - \frac{3}{4}\left(\frac{r_o}{r}\right) - \frac{1}{4}\left(\frac{r_o}{r}\right)^3\right]\sin\theta \tag{17.15}$$

Here, the equation confirms the second coupled hierarchy formulation of the flying train to adequately find the solution to standard *k-omega* turbulence with shear flow corrections. The numerical solution will, therefore, be considered as an entirely limited volume scheme of the Reynolds-averaged Navier–Stokes compressible force that neutralizes the flying train's total motion [18, 19]. Simply this numerical solution can be adopted for the baseline solutions to the flying train's motion control by incorporating 9618 NACA airfoils to achieve maximum aerodynamic performance [13, 20, 21].

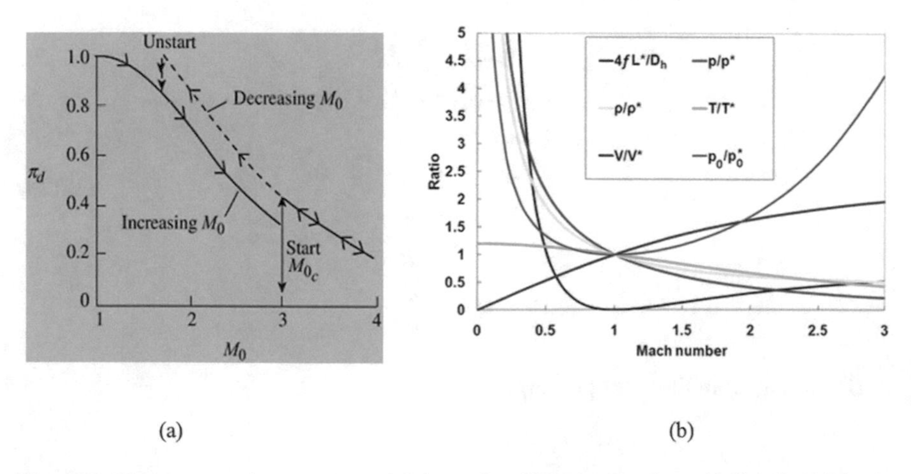

Fig. 17.2 The conceptual mechanism of flying train which simulated considering the (**a**) implementation of force control M_0 and (**b**) the Mach number contours for a safe physical model of a flying train speeding which is implemented by a 3D CFD, k-omega kinetic mode ratio

Consequently, the generator modeling of the flying train has been conducted to power the flying train consistently by using the induction synchronous model of energy transformation principles based on a *d*–*q* referencing framework as below:

$$V_q = -R_s i_q - L_q \frac{di_q}{dt} - \omega L_d i_d + \omega \lambda_m \tag{17.16}$$

$$V_d = -R_s i_d - L_d \frac{di_q}{dt} + \omega L_q i_q \tag{17.17}$$

in which electronic torque is indicated by

$$T_e = 1.5\rho\left[\lambda i_q + \left(L_d - L_q\right) i_d i_q\right] \tag{17.18}$$

In the equation, q axis inductance is represented by L_q, while L_d *indicates* d axis inductance. i_q is the current from the q axis while i_d current from d axis. V_q represents voltage from the q axis, and the energy from the d axis is indicated by V_d. ω_r represents the velocity of the angular rotor, λ the induced flux amplitude, and p the number of poles. Thus, the squirrel cage induction generator can also be used in modeling the generator using the equation below:

$$\begin{bmatrix} V_{qs} \\ V_{ds} \\ V_{qr} \\ V_{dr} \end{bmatrix} = \begin{bmatrix} R_s + pL_s & 0 & pL_m & 0 \\ 0 & R_s + pL_s & 0 & pL_m \\ pL_m & -\omega_r L_m & R_r + pL_r & -\omega_r L_r \\ \omega_r L_m & pL_m & \omega_r L_r & R_r + pL_r \end{bmatrix} \begin{bmatrix} i_{qs} \\ i_{ds} \\ i_{qr} \\ i_{dr} \end{bmatrix} \tag{17.19}$$

Beginning from the stator position

$$\begin{aligned} \lambda_{ds} &= L_s i_{ds} + L_m i_{dr} \\ \lambda_{qs} &= L_s i_{qs} + L_m i_{dr} \\ L_s &= L_{ls} + L_m \\ L_r &= L_{lr} + L_m \\ V_{ds} &= R_s i_{ds} + \frac{d}{dt}\lambda_{ds} \\ V_{qs} &= R_s i_{qs} + \frac{d}{dt}\lambda_{qs} \end{aligned} \tag{17.20}$$

Beginning from the rotor position

$$
\begin{aligned}
\lambda_{dr} &= L_{\mathrm{r}} i_{dr} + L_{\mathrm{m}} i_{ds} \\
\lambda_{qr} &= L_{\mathrm{r}} i_{qr} + L_{\mathrm{m}} i_{qs} \\
V_{dr} &= R_{\mathrm{r}} i_{dr} + \frac{d}{dt}\lambda_{dr} + \omega_{\mathrm{r}}\lambda_{qr} \\
V_{qr} &= R_{\mathrm{r}} i_{qr} + \frac{d}{dt}\lambda_{qr} - \omega_{\mathrm{r}}\lambda_{dr}
\end{aligned}
\tag{17.21}
$$

To determine the air flux gaps

$$
\begin{aligned}
\lambda_{d\mathrm{m}} &= L_{\mathrm{m}}\left(i_{ds} + i_{dr}\right) \\
\lambda_{qr} &= L_{\mathrm{m}}\left(i_{qr} + i_{qs}\right)
\end{aligned}
\tag{17.22}
$$

In the above equation, R_{s}, R_{r}, ω_{r}, i_d, i_q, V_d, V_q, λ_d, λ_q, L_m, L_{ls}, and L_{lr} indicate the resistance from wind by the stator and the fluxes. Energy conversion diagram-implemented inverter is often used in the preparation of the energy output. This is done through the calculation in Simulink-MATLAB.

Results and Discussion

To develop a guideway for safe takeoff and landing of the flying train, a cross-sectional force has been analyzed from the guideway where the propulsion force is calculated to run the flying train in the elliptical loops along the guideway [22–24]. So, the levitation and guidance force are formed and create its own motion to control the takeoff and landing. Concretely it can be explained that due to the direction of the kinetic force back-and-forth movement of the guideway, the propelling of the train is run by activated force of attraction and repulsion force [25–27]. Simply, once the levitation force passes through the train, an electric current is produced in the guideway and the vehicle is levitated by the force of attraction, which will pull up on the sky in the vehicle, as well as reversely by repulsion force, the train will be pushed down smoothly to touch the ground [28, 29]. Simply, to determine the levitation and lateral force for the vehicle's takeoff and landing, an electromagnetic force induction is analyzed to confirm the most efficient and economical way to implement this force properly and it has been expressed by the following block diagram (Fig. 17.3).

Naturally it can be explained that when an electric current flow passes through the propulsion force, a kinetic energy is produced which works for the attraction and repulsion force and the flying train's propel [19, 30, 31]. Necessarily, the flying train's speed and flight height are calculated to adjust the polarity shift in the propulsion and the kinetic force of the flying train [11, 32, 33]. As the train flies with its maximum height by the act of kinetic force, the flying train will need its control on

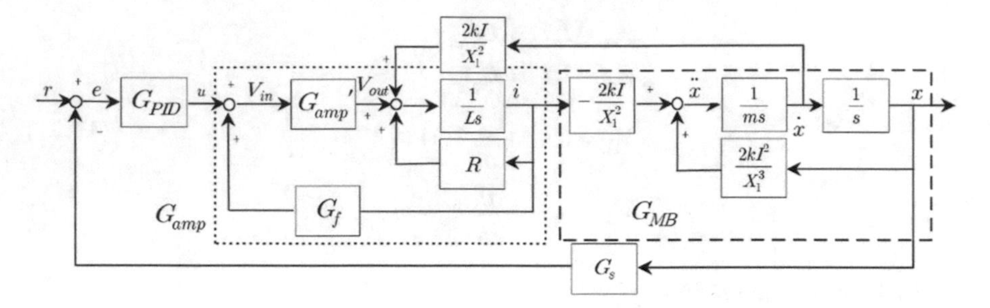

Fig. 17.3 Block diagram to mathematically determine the levitation and lateral force for the flying train's takeoff and landing control

the sky. Thus, the model is developed by kinetic force producing energy for the flying train control mechanism by governing the mechanical power control as below:

$$P_w = \frac{1}{2} C_p(\lambda,\beta)\rho A V^3 \tag{17.23}$$

where ρ is the air density (kg/m^3), C_p is the power coefficient, A is the intercepting area of the flying train (m^2), V is the average speed (m/s), and λ is the tip speed ratio [31]. The theoretical maximum value of the power coefficient C_p is 0.593; C_p is also known as Betz's coefficient. Mathematically,

$$\lambda = \frac{R\omega}{V} \tag{17.24}$$

R is the radius of the flying train (m), ω is the speed (rad/s), and V is the average speed (m/s) and thus the net energy required by the flying train is calculated as

$$Q_w = P \times (\text{Time}) \, [\text{kWh}] \tag{17.25}$$

Since wind velocity has some impact on flying train, a direct measurement of wind speed has also been calculated considering any particular motion when the train will be flying and thus the motion of the flying train has been determined considering its maximum flying height to run the flying train in the sky smoothly:

$$v(Z)\ln\left(\frac{Z_r}{Z_0}\right) = v(Z_r)\ln\left(\frac{Z}{Z_0}\right) \tag{17.26}$$

where Z_r is the flying height (m), Z is the height to be determined, Z_0 is the measure of surface roughness (0.1–0.25 for crop land), $v(Z)$ is the flying train speed at height z (m/s), and $v(Z_r)$ is the wind speed at the reference height z (m/s) (Fig. 17.4). The power output in terms of the flying train shall be estimated using the following equation:

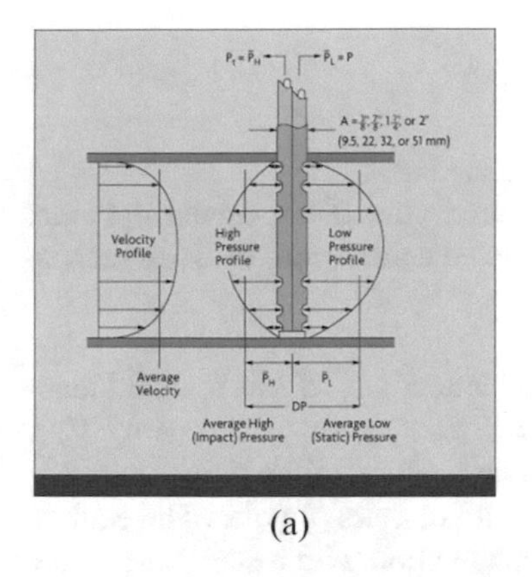

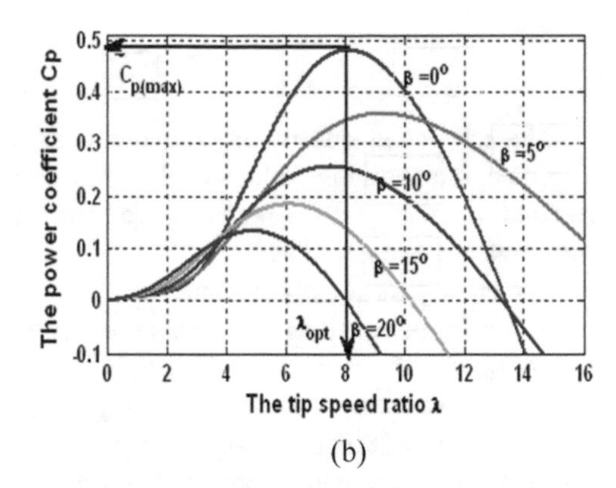

(a) (b)

Fig. 17.4 (**a**) The flying train's power control mechanism considering its maximum flying height, (**b**) relationship between mechanical power generation of the flying train speeds at different power coefficients where it is suggested that the maximum values of C_p are achieved for the curve associated with $\beta = 2°$. From this curve, the maximum value of C_p ($C_{p,max} = 0.5$) is obtained for $\lambda_{opt} = 0.91$. This value (λ_{opt}) represents the optimal speed ratio

$$P_w(v) = \begin{cases} \dfrac{v^k - v_C^k}{v_R^k - v_C^k} \cdot P_R & v_C \le v \le v_R \\ P_R & v_R \le v \le v_F \\ 0 & v \le v_C \text{ and } v \ge v_F \end{cases} \tag{17.27}$$

where P_R is the rated power, v_C is the cut-in flying train speed, v_R is the rated wind speed, v_F is the rated cutout speed, and k is the Weibull shape factor [34, 35]. When the kinetic energy of the flying train is near to zero, the power coefficient is maximized for an optimal TSR [30] which is calculated as

$$\omega_{opt} = \frac{\lambda_{opt}}{R} V_{wn} \tag{17.28}$$

which will give

$$V_{wn} = \frac{R\omega_{opt}}{\lambda_{opt}} \tag{17.29}$$

where ω_{opt} is the optimal speed of flying train in rad/s, λ_{opt} is the optimal tip speed ratio, R is the radius of the flying train in meters, and V_{wn} is the speed in m/s (Fig. 17.4).

Hence, the flying train speed is controlled by its mechanical powers; thus, the computational applicable in airstream-driven energy generation has been calculated considering the repulsive and attracting force of variable velocity of flying train and is expressed as

$$V_d = -R_s i_d - L_d \frac{di_q}{dt} + \omega L_q i_q$$

The automated energy torque is herewith calculated as

Since the variable velocity of the flying train is the direct-driven dynamo, airstream-driven power schemes have been calculated considering its *d–q* synchronous and are expressed as

The energy torque is calculated as

In the above equation, the q axis maximum induction is L_q, d axis mutual induction is L_d, q air flow of energy is i_q, i_d is the d axis flow, V_q is the q axis energy, V_d is the d axis energy, ω_r is the rate of change of angular position of the flying train, λ is the amplitude of flux prompted, and p is the quantity of pairs of poles. The equivalence that follows, which can be applicable in conjunction with a static *d–q* frame of reference for dynamic modeling, is relevant with regard to a squirrel cage induction dynamo (SCIG):

The equations that are applicable for the stator side are

The equations that are applicable for the rotor side are

The equations that apply to the air gap flux connotation are

where R_s, R_r, L_m, L_{ls}, L_{lr}, ω_r, i_d, i_q, V_d, V_q, λ_d, and λ_q are the stator resistance, generator resistance, fascinating inductance, stator leakage inductance, rotor leakage inductance, electrical rotor velocity, and flow of energy, correspondingly, of the *d–q* model. Obtaining of the torque energy (T_t) in terms of consistent speed has been calculated as

$$P_w = \frac{1}{2}\rho A C_p(\lambda,\beta)\left(\frac{R\omega_{opt}}{\lambda_{opt}}\right)^3$$
$$T_t = \frac{1}{2}\rho A C_p(\lambda,\beta)\left(\frac{R}{\lambda_{opt}}\right)^3 \omega_{opt} \tag{17.30}$$

Simply through the adoption of these principles, this study adequately suggests that flying train takeoff, landing, and flying in the sky depend on the levitation, propulsive, and aerodynamic force which can be controlled by implanting the model of 3D k-omega numerical simulation to facilitate a united power on the stator position to run the flying train smoothly on the sky.

Conclusion

The conventional transport infrastructure systems to connect within the urban and rural area all over the world are causing adverse traffic congestion and environmental impacts in urban areas. Thus, the solution of flying train for balancing rapid link within the urban and rural area and controlling traffic shall indeed be a new thought for the rapid connectivity within the cities and rural areas. Naturally, the implementation of flying train is an advanced technology by confirming the computation model of introducing 3D numerical simulation of propulsive force, impulsive force, and its 3D k-omega mechanism to control the aerodynamics that will indeed end up the dependability on the old-fashioned connectivity within the cities and rural areas. Simply, if the urban and rural connectivity thoughts are modernized by this innovative technology of flying train, it will be a new epoch of science to dispatch unembellished transportation system within the urban and rural communication.

Acknowledgements This research was supported by Green Globe Technology under the grant RD-02020-03. Any findings, conclusions, and recommendations expressed in this chapter are solely those of the author and do not necessarily reflect those of Green Globe Technology.

References

1. Adriá Junyent-Ferré, Oriol Gomis-Bellmunt, Andreas Sumper, Marc Sala, Montserrat Mata, Modeling and control of the doubly fed induction generator wind turbine, *Simulation Modelling Practice and Theory, vol. 18*, 2010, pp. 1365-1381.
2. M. F. Hossain (2020). Application of wind energy into the transportation sector. International Journal of Precision Engineering and Manufacturing-Green Technology. DOI: https://doi.org/10.1007/s40684-020-00235-1. (Springer).
3. Tamalouzt, Salah, Toufik Rekioua, and Rachid Abdessemed. "Direct torque and reactive power control of grid connected doubly fed induction generator for the wind energy conversion", 2014 International Conference on Electrical Sciences and Technologies in Maghreb (CISTEM), 2014.
4. Ali M. Eltamaly, Hassan M. Farh, Maximum power extraction from wind energy system based on fuzzy logic control, *Electric Power Systems Research, vol. 97*, 2013, pp. 144– 150.
5. Chierchie, F. and Paolini, E. (2013). Real-Time Digital PWM with Zero Baseband Distortion and Low Switching Frequency. *IEEE Transactions on Circuits and Systems I: Regular Papers*, 60(10), pp. 2752-2762.
6. De, Juan, and Janaina Goncalves de Oliveir. "Electric Machine Topologies in Energy Storage Systems", Energy Storage, 2010.
7. Belgacem, Kh.; Mezouar, A. and Massoum, A. "Fuzzy Logic Control of Double-Fed Induction Generator Wind Turbine", International Review on Modelling & Simulations, 2013.
8. Casella, I., Capovilla, C., Sguarezi Filho, A., Jacomini, R., Azcue-Puma, J. and Ruppert, E. (2013). An ANFIS Power Control for Wind Energy Generation in Smart Grid Scenario Using Wireless Coded OFDM-16-QAM. *Journal of Control, Automation and Electrical Systems*, 25(1), pp. 22-31.
9. Chikouche, T., Hadjeri, S., Mezouar, A., Terras, T., A New State-Space Nonlinear Control Approach of a Doubly Fed Induction Motor Using Variable Gain PI and Fuzzy Logic Controllers, (2013) *International Review on Modelling and Simulations (IREMOS)*, 6 (1), pp. 59-67.
10. M. F. Hossain. "Transforming dark photons into sustainable energy", International Journal of Energy and Environmental Engineering, 2018
11. Terrafugia's Transition street-legal airplane continues flight and drive testing". Terrafugia. Retrieved 30 July 2012.
12. Boroujeni, H. Z., M. F. Othman, A. H. Shirdel, R. Rahmani, P. Movahedi, and E. S. Toosi. "Improving waveform quality in direct power control of DFIG using fuzzy controller", Neural Computing and Applications, 2015.

13. M. F. Hossain (2017). Design and Construction of Ultra-Relativistic Collision PV Panel and Its Application into Building Sector to Mitigate Total Energy Demand. Journal of Building Engineering. 9, 147-154.
14. Mohammad Pichan, Hasan Rastegar, Mohammad Monfared, Two fuzzy-based direct power control strategies for doubly-fed induction generators in wind energy conversion systems, *Energy, vol. 51*, 2013, pp. 154–162.
15. M. F. Hossain and Nowshin Fara (2016). Integration of Wind into Running Vehicles to Meet Its Total Energy Demand. Energy, Ecology, and Environment. 2(1), 35-48. (Springer Nature).
16. M. F. Hossain, Solar Energy Integration into Advanced Building Design for Meeting Energy Demand and Environment Problem, Int. J. Energy Res. 17 (2016) 49–55.
17. Tien, H., Scherer, C., Scherpen, J. and Muller, V. (2016). Linear Parameter Varying Control of Doubly Fed Induction Machines. *IEEE Transactions on Industrial Electronics*, 63(1), pp. 216-224.
18. Chakib, R., Essadki, A. and Cherkaoui, M. (2014). Modeling and control of a wind system based on a DFIG by active disturbance rejection control. *International Review on Modelling and Simulations (IREMOS)*, 7(4), p. 626.
19. M. F. Hossain (2017). Green Science: Independent Building Technology to Mitigate Energy, Environment, and Climate Change. Renewable and Sustainable Energy Reviews. 73; 695-705.
20. Bhandari, B., Lee, KT., Lee, GY. et al. Int. J. of Precis. Eng. and Manuf.-Green Tech. (2015) 2: 99.
21. Chien-Hsiung Tsai, Lung-Ming Fu, Chang-Hsien Tai, Yen-Loung Huang, Jik-Chang Leong, Computational aero-acoustic analysis of a passenger car with a rear spoiler, Applied Mathematical Modelling, Volume 33, Issue 9, September 2009, Pages 3661-3673.
22. A. Gaillard, P. Poure, S. Saadate, M. Machmoum. Variable Speed DFIG Wind Energy System for Power Generation and Harmonic Current Mitigation, *Renewable Energy, vol. 34*, 2009, pp. 1545-1553.
23. Beltran, B., Benbouzid, M., Ahmed-Ali, T., Mangel, H., DFIG-based wind turbine robust control using high-order sliding modes and a high gain observer, (2011) *International Review on Modelling and Simulations (IREMOS),* 4 (3), pp. 1148-1155.
24. Chang, S. (2003). "Evaluating Disaster Mitigations: Methodology for Urban Infrastructure Systems." Nat. Hazards Rev., https://doi.org/10.1061/(ASCE)1527-6988(2003)4:4(186), 186-196.
25. Abdelmalek, Samir, Linda Barazane, Abdelkader Larabi, and Hocine Belmili. "Contributions to diagnosis and fault tolerant control based on proportional integral observer: Application to a doubly-fed induction generator", 2015 4th International Conference on Electrical Engineering (ICEE), 2015.
26. Abdullah, M. A., Yatim, A. H. M., and Chee Wei, T., "A Study of Maximum Power Point Tracking Algorithms for Wind Energy system," Proc. of IEEE First Conference on Clean Energy and Technology (CET), pp. 321-326, 2011.
27. de Sherbinin, Alex, Marc A Levy, Erica Zell, Stephanie Weber, and Malanding Jaiteh. "Using satellite data to develop environmental indicators", Environmental Research Letters, 2014.
28. Dürr, M., Cruden, A., Gair, S., and McDonald, J. R., "Dynamic Model of a Lead Acid Battery for Use in a Domestic Fuel Cell System," Journal of Power Sources, Vol. 161, No. 2, pp. 1400-1411, 2006.
29. S.J. Thompson, Congressional Research Service, "High Speed Ground Transportation (HGST): Prospects and Public Policy," Apr. 6, 1989, p. 5.
30. Abdullah Asuhaimi B. Mohd Zin, Mahmoud Pesaran H. A, Azhar B. Khairuddin, Leila Jahanshaloo, Omid Shariatu, An overview on doubly fed induction generators' controls and contributions to wind based electricity generation, *Renewable and Sustainable Energy Reviews, vol. 27*, 2013, pp. 692-708.
31. Benoît Robyns, Bruno Francois, Philippe Degobert, Jean Paul Hautier, *Vector control of induction machines* (Springer-Verlag, London, 2012).
32. D.L. Mitchell and D.U. Gubser, "Magnetohydrodynamic Ship Propulsion with Superconducting Magnets, " Journal of Superconductivity, vol. 1, No. 4, 1988, p. 349.

33. Soedibyo, S., Pradipta, A., Suyanto, S., Ridwan, M., Zulkarnain, G. and Ashari, M. (2017). Active and Reactive Power Control in 20 kV Grid Connected Distributed Generation System. *International Review of Automatic Control (IREACO)*, 10(3), p. 211.
34. Bhandari, Binayak, Shiva Raj Poudel, Kyung-Tae Lee, and Sung-Hoon Ahn. "Mathematical modeling of hybrid renewable energy system: A review on small hydro-solar-wind power generation", International Journal of Precision Engineering and Manufacturing-Green Technology, 2014. April 2014, Volume 1, Issue 2, pp. 157-173.
35. Cardoso, J., Casella, I., Filho, A., Costa, F. and Capovilla, C. (2016). SCIG wind turbine wireless controlled using morphological filtering for power quality enhancement. *Renewable Energy*, 92, pp. 303-311.

Part VII
Environment

Chapter 18
Air

Abstract Currently, global carbon emission and sequestration rates are not in equilibrium level; therefore, the global environment is currently under severe vulnerable condition, which threatens the survival of all living beings on earth. The annual global carbon emission considering all forms of CO_2 emissions, including the burning of fossil fuel and terrestrial land-use change, along with other related factors and the global carbon sequestration by all sources, such as the ocean sink, earth sink, absorption by terrestrial vegetation, and other associated factors, has been estimated in this report to determine the acceleration of atmospheric CO_2 concentration. The estimate shows that the atmospheric CO_2 concentration has been increasing at a rate of 2.11% ppm per year for the past several years, and it continues to rapidly increase. It is well established that the toxic level of atmospheric CO_2 concentration is 60,000 ppm at which all living beings will die within 30 min. If the current level of CO_2 emission is not stopped, all human race on earth will be extinct [$\int_{400^{(2.11\%)}}^{60,000}(2017)$] in 121,017,712 years since the toxic level of CO_2 will reach 60,000 ppm within this time and thus it would be the end of story of human civilization on earth.

Keywords CO_2 emission · Air quality · Environmental vulnerability · Climate change · Survival period of life on earth

Introduction

The annual average atmospheric CO_2 concentration is currently 400 ppm [1, 2]. Since the industrial revolution of the early 1990s, fossil fuels, deforestation, land-use change, and external forces have become the key sources of anthropogenic CO_2 that have caused the concentration of CO_2 in the atmosphere to rapidly increase and change the global climate (Fig. 18.1).

M. F. Hossain, *Global Sustainability in Energy, Building, Infrastructure, Transportation, and Water Technology*,
https://doi.org/10.1007/978-3-030-62376-0_18

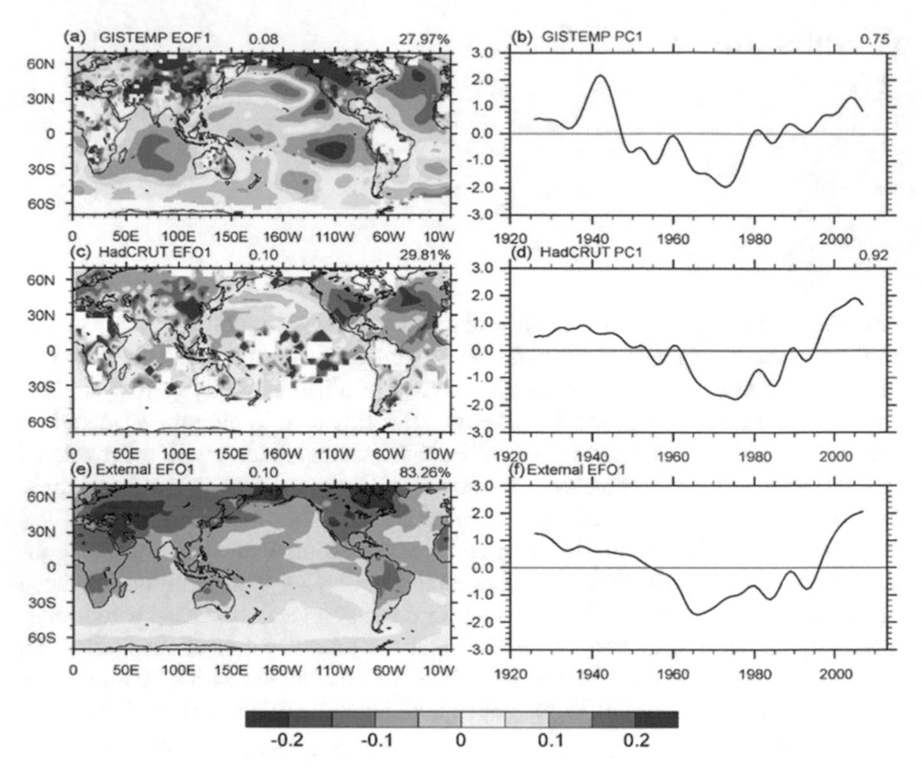

Fig. 18.1 The leading EOF pattern (in °C) and associated standardized principal component (PC) of global climate change based on 8-year low-pass filtered annual anomalies after the long-term trend for 1920–2015 are removed from (**a**, **b**) GISTEMP, (**c**, **d**) HadCRUT, and (**e**, **f**) external forcing based on 18 CMIP5 models. Values shown at the top right of the left panels represent the percentage variance explained by each mode, and those of the right panels represent the correlation coefficient with PC1 of external forcing

In this chapter, I have calculated the "global carbon budget" and the "global CO_2 concentration," and, by determining the amount of CO_2 being input to the atmosphere relative to the global CO_2 balance, the rate of acceleration of atmospheric CO_2 emissions annually. Furthermore, I have prepared detailed data sets using MATLAB software and have calculated the annual total global carbon emissions from the preindustrial period (1750) to the modern period (2015). All CO_2 emissions from the burning of fossil fuel and land-use change from 1750 to the industrial revolution and from 1870 to the modern era have been analyzed in detail by incorporating data from the Fifth Assessment Report of the Intergovernmental Panel on Climate Change (IPCC AR5) into MATLAB software to ensure the most accurate estimates of CO_2 emissions into the atmosphere since this gas is the major driver of climate change. The estimated CO_2 emissions from burning fossil fuel are referred to as E_{FF}; Gt/year, and the emissions resulting from land-use change are E_{LUC}; Gt/year. The sequestration and the absorption of CO_2 by the oceans, which is referred to as S_{OCEAN}; Gt/year, and by the terrestrial vegetation, which is referred to as S_{LAND};

Gt/year, have also been estimated to determine the growth rate of the atmospheric CO_2 concentration (G_{ATM}; Gt/year) by using the following equation describing the CO_2 balance among the atmosphere, ocean, and land:

$$E_{FF} + E_{LUC} = G_{ATM} + S_{OCEAN} + S_{LAND} \quad (18.1)$$

$$G_{ATM} = E_{FF} + E_{LUC} - S_{OCEAN} - S_{LAND} \quad (18.2)$$

The atmospheric G_{ATM} is calculated in parts per million per year (ppm/year), which can be converted into the total mass of carbon per year (GtC/year). Assessments of environmental vulnerability have been conducted globally for the past several decades, and numerous studies have been performed [2–4]. To date, little research has been done on the adverse impacts of toxic levels of CO_2 in the atmosphere on life on earth. Thus, the purpose of this report is to identify the yearly growth rate of the CO_2 concentration in the atmosphere and its impact on the future state of the global environmental to determine the survival period of humanity and other living organisms.

Methods and Simulation

Global CO_2 emissions, absorption, and sequestration were analyzed by interpreting reports from several organizations (CDIAC, IEA, UNEP, USDoE, ECE, EIA, PBL, NEAA, NEDO, NOAA, and NASA), and the data were incorporated into MATLAB software to develop the data set. To accurately calculate annual global carbon estimates, I considered all data up to the year 2015 and the projected fossil energy emissions for 2016, and from the projected total carbon estimates for 2016, the annual growth rate of the atmospheric concentration of CO_2 was determined [5–7].

CO_2 Emissions from Fossil Fuel

The yearly growth rate in CO_2 emissions was estimated from the difference between two consecutive years, which was divided by the first-year emissions per the following equation:

$$\left[\frac{E_{FF(t_{0+1})-E_{FF(t_0)}}}{E_{FF(t_0)}}\right] \times 100\% \,/\, \text{year} \quad (18.3)$$

In general, a simple calculation can characterize the yearly CO_2 emission growth rate. However, to accurately determine the growth rate over multiple decades, I applied a leap-year factor to confirm the net annual growth rate of carbon (E_{FF}) by using its logarithm equivalent in the following equation:

$$\frac{1}{E_{FF}} \frac{d(\ln E_{FF})}{dt} \tag{18.4}$$

Here, I calculated the pertinent CO_2 emission growth rates considering multi-decadal periods by implementing a nonlinear drift into $\ln(E_{FF})$ in Eq. (18.4) and by calculating the yearly growth percentage. Thus, I fitted the logarithm of E_{FF} into the equation rather than directly using E_{FF} to ensure an accurate growth rate estimate and satisfy Eq. (18.3) [8–10].

CO_2 Emissions from the Land-Use Change (E_{LUC})

The emissions reported here (E_{LUC}) include CO_2 fluxes from deforestation, forest degradation, and abandonment of agricultural land associated with modern civilization and were calculated by implementing dynamic global vegetation modeling (DGVM) bookkeeping simulations in MATLAB [11–13]. The simulations were prepared with DGVMs, in which I initially clarified the historical changes in land use followed by the atmospheric CO_2 concentrations [10, 14]. Therefore, I implemented a time series of the distribution of preindustrial land cover by allocating the estimated variance into the first simulation and the dynamic evolution of biomass soil carbon to the prescribed land-cover change [15–17]. All the DGVMs here represent complete vegetation growth and decay processes as well as decomposition of dead organic matter to determine the response to increasing atmospheric CO_2 levels [18–20].

Ocean CO_2 Sink

The CO_2 sequestered by the ocean from 1959 to 2015 was calculated by combining seven global oceans' biogeochemical cycle models and this approach can be used to comprehensively analyze the physical, chemical, and biological processes that are directly impacted by the concentration of CO_2 at the ocean surface as well as the air-sea CO_2 fluxes [21, 22]. Thus, the ocean CO_2 sequestration is normalized by accurate observational values by dividing the individual yearly values by the modeled average for 1990–1999 and then multiplying the result by an observation-based calculation of 2.2 GtC/year [23–25]. Therefore, the oceanic CO_2 sequestration per year (t) in GtC/year is calculated as follows:

$$S_{OCEAN}(t) = \frac{1}{n} \sum_{m=1}^{m=n} \frac{S^{m}_{OCEAN}(t)}{S^{m}_{OCEAN^{(1990-1999)}}} \times 2.2 \tag{18.5}$$

where n represents the number of variables; m represents the factors; and t represents the period. The normalization is considered when the ratio or the sequestration of CO_2 in the 1990s is assumed to be underestimated and is initially triggered by diffusion, which relies on the CO_2 gradient. Therefore, the ratio is considered a naturally appropriate approach that accounts for the time dependence of the CO_2 gradient in the oceans [8, 26, 27].

CO_2 Absorption by Terrestrial Vegetation and the Earth

The variations in the CO_2 emissions from fossil fuels (E_{FF}) and land-use change (E_{LUC}) as well as the growth rate of the atmospheric CO_2 concentration (G_{ATM}) and ocean CO_2 sequestration (S_{OCEAN}) can be accounted for to determine the net sequestration of CO_2 by the terrestrial vegetation (S_{LAND}) by considering Eq. (18.1). Therefore, this type of sequestration can be computed as the CO_2 remaining from the mass balanced budget, which is expressed as follows:

$$S_{LAND} = E_{FF} + E_{LUC} - \left(G_{ATM} + S_{OCEAN}\right) \tag{18.6}$$

Here, S_{LAND} is computed from the remainder of the estimates and includes all perturbed carbon from fossil fuels, land-use change, and CO_2 atmospheric growth rate. The computation of S_{LAND} in Eq. (18.6) with the budget from the DGVMs can be used to calculate E_{LUC} by subtracting the impact of land-use changes, which will provide an independent calculation of a consistent S_{LAND}. Thus, it can represent an appropriate understanding of the role of the terrestrial vegetation in determining the response to CO_2 and climate variability (Fig. 18.2).

Calculation of the Growth Rate of the Atmospheric CO_2 Concentration (G_{ATM})

The total growth rate of the concentration of atmospheric CO_2 was calculated for 1956–2016; that is, the annual increase in the concentration of atmospheric CO_2 was calculated by comparing records of CO_2 emissions and sequestration from multiple sources (DEP, NASA, NOAA/ESRL, 2015, Scripps, NEDO, and UNDP), considering that the ocean boundary layers and the air are very well mixed and very much in a functional band with time period and latitude [28–30]. The growth rate unit, ppm/year, is then converted from GtC/year to be consistent with the other components.

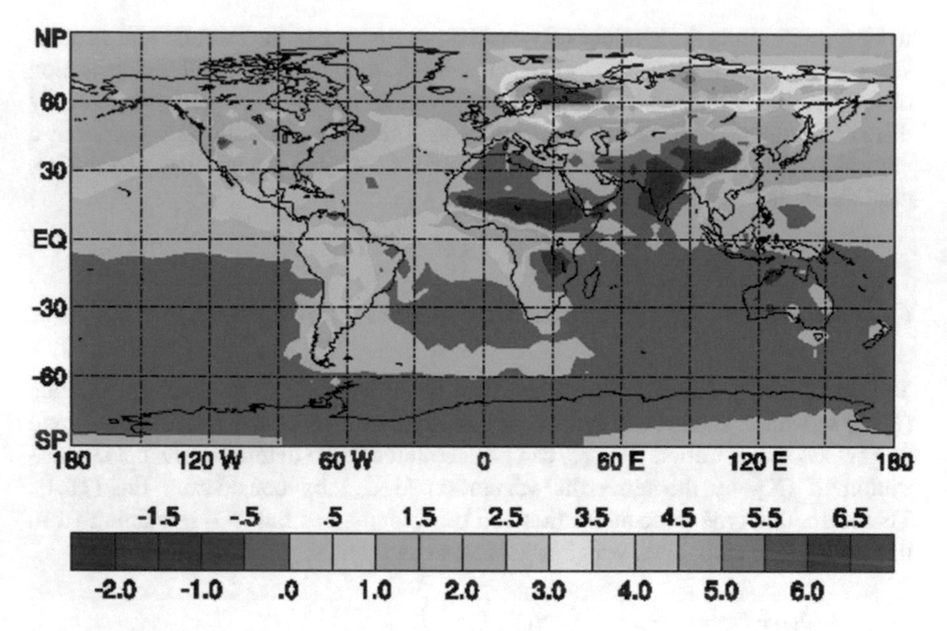

Fig. 18.2 Annual mean mixing ratio of CO_2 (ppmv), according to the PBL, as simulated in MATLAB using terrestrial biosphere surface fluxes. The concentration was initialized to be globally uniform, and the model was run on a coarse 7.2 × 9 grid with 9 levels for 10 years. The end of this "*spin up: run" was then used as the initial condition for a further 4-year integration on a 4 × 5 grid with 17 levels. All results presented here are for the final 3 years of the 4 × 5 run

Results and Discussion

The average global carbon emissions from 1875 to 2016 have been estimated by numerous scientists from NASA, NOAA, DEP, NASA, NOAA/ESRL, Scripps, NEDO, UNDP, and several other agencies (Fig. 18.2). Over this timescale, 91% of the total emissions ($E_{FF} + E_{LUC}$) was from the burning of fossil fuels and other industrial activities, and 9% resulted from land-use change [31, 32]. These emissions were captured by sinks within the atmosphere (44%), the oceans (26%), and the land (30%), which account for the annual growth of the CO_2 concentration in the atmosphere [27, 33]. Here, except for the land-use component, carbon emissions have grown since 1959 with important inter-annual differences in the growth rate of atmospheric CO_2 concentration and CO_2 sequestration by the terrestrial vegetation (Fig. 18.4) and considering decadal differences in all terms (Table 18.1).

Table 18.1 The results using bookkeeping technology and budgeted residuals from DGVMs implemented in MATLAB software, and its inverse calculations for 1960–1969, 1970–1979, 1980–1989, 1990–1999, and 2000–2009 shown in GtC/year

Mean (GtC/year)	1960–1969	1970–1979	1980–1989	1990–1999	2000–2009	2006–2015	2016
Carbon emissions from land-use change (ELUC)							
Bookkeeping DGVMs	1.5 ± 0.5 1.2 ± 0.3	1.3 ± 0.5 1.2 ± 0.3	1.4 ± 0.5 1.2 ± 0.2	1.6 ± 0.5 1.2 ± 0.2	1.0 ± 0.5 1.1 ± 0.2	1.0 ± 0.5 1.3 ± 0.3	1.3 ± 0.5 1.2 ± 0.4
Residual terrestrial sink (S_{LAND})							
Budget residual DGVMs	1.7 ± 0.7 1.2 ± 0.5	1.7 ± 0.8 2.2 ± 0.5	1.6 ± 0.8 1.7 ± 0.6	2.6 ± 0.8 2.3 ± 0.5	2.6 ± 0.8 2.8 ± 0.6	3.1 ± 0.9 2.8 ± 0.7	1.9 ± 0.9 1.0 ± 1.4
Total land fluxes ($S_{LAND} - E_{LUC}$)							
Budget ($E_{EF} - G_{ATM} - S_{OCEAN}$) DGVMs	0.2 ± 0.5 −0.2 ± 0.7	0.4 ± 0.6 1.1 ± 0.5	0.1 ± 0.6 0.4 ± 0.5	1.0 ± 0.6 1.1 ± 0.3	1.6 ± 0.6 1.8 ± 0.4	2.2 ± 0.7 1.7 ± 0.5	0.6 ± 0.7 −0.1 ± 1.4
Inversions (CTE2016-FT/Jena CarboScope/(CAM))	–/–/–	–/–/–	–/0.2/0.9	–/1.0/1.9	1.5/1.6/2.5	2.2/2.3/3.4	1.9/2.6/2.6

CO_2 Emissions from Fossil Fuels, Land-Use Change, and Other Factors

Global CO_2 emissions from burning fossil fuels and industrial activities increased at an average of 3.1 ± 0.2 GtC/year per decade in the 1960s and by an average of 9.3 ± 0.5 GtC/year from 2006 to 2015 (Table 18.1 and Fig. 18.3). The emission growth rate decreased in the 1960s and 1990s being 4.5% per year in the 1960s (1960–1969), 2.8% per year in the 1970s (1970–1979), 1.9% per year in the 1980s (1980–1989), and 1.1% per year in the 1990s (1990–1999). Growth started to increase again in the 2000s at an average rate of 3.5% per year, but it has decreased to 1.8% per year in recent decades (2006–2015). However, the CO_2 emissions from land-use change have remained nearly constant at approximately 1.3 ± 0.5 GtC/year during 1960–2015. The reduction in emissions from land conversion suggests that significant emissions occurred in the 1990s and 2000s as calculated by bookkeeping and DGVM methods in MATLAB software (Table 18.1). A change in tropical deforestation does not account for this reduction, but the flux in the 1990s was like that in the 2000s, suggesting a decrease in E_{LUC} from 2011 to 2015 relative to that from 2001 to 2010.

Ocean and Terrestrial Vegetation CO_2 Sinks

The global CO_2 sequestration by the ocean was estimated based on the mean CO_2 sequestration value for the 1990s combined with the differences in ocean CO_2 sequestration for 1959–2015 including all seven-global ocean biogeochemical cycles. For the 1990s, the mean ocean CO_2 sequestration was estimated to be approximately 2.2 ± 0.4 GtC/year based on indirect observational data from the

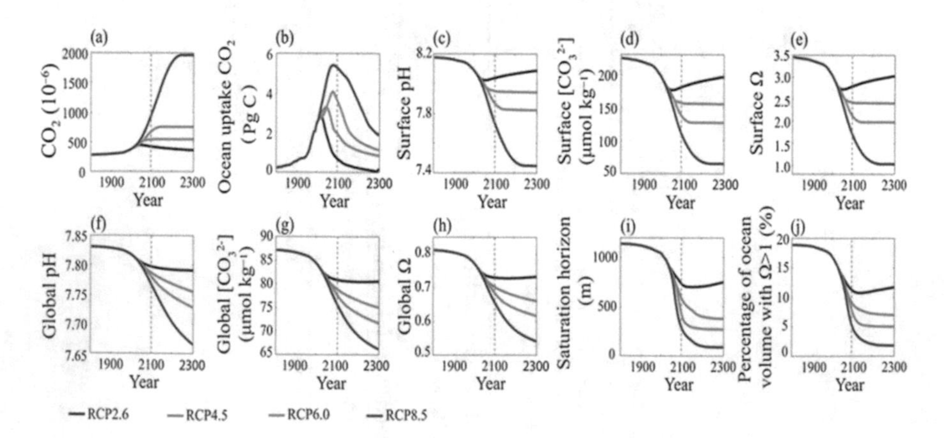

Fig. 18.3 The impact of global scale carbon emission (**a**) on earth, (**b**) ocean, (**c–e**) earth surface in various parameters, (**f–j**) ocean pH and various parameters of RCP2.6, RCP4.5, RCP6.0, and RCP8.5

IPCC and considering the timescale for CFC penetration [14, 27]. This CO_2 sequestration value is comparable to that of 2.0 ± 0.5 GtC/year calculated for the 1990s and 1.9–2.5 GtC/year for 1990–2009 [18, 21] with a ±0.3 to ±0.7 GtC/year range of uncertainty. Therefore, this method can accurately calculate observation-based ocean sequestration by implementing the sea-air pCO_2 (sea-air pCO_2 variable) × (gas transfer coefficient), and the results are completely consistent with the observed results [21, 34].

Ultimately, the calculation shows that residual terrestrial CO_2 sequestration increased from 1.7 ± 0.7 GtC/year in the 1960s to 3.1 ± 0.9 GtC/year from 2006 to 2015 with an inter-annual variability of up to 2 GtC/year. It also indicates reduced land sequestration during El Ni n̂ o and overaccumulation by ocean sequestration, which accounts for the increasing growth rate of atmospheric CO_2 (Fig. 18.4). The high CO_2 sequestration by the land from 2006 to 2015 is comparable to that of the

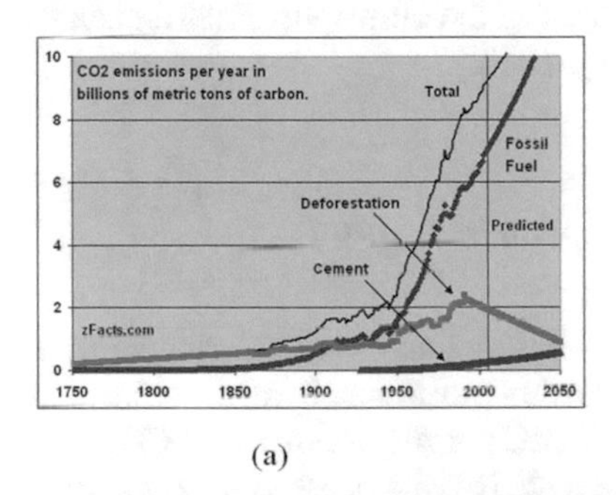

(a)

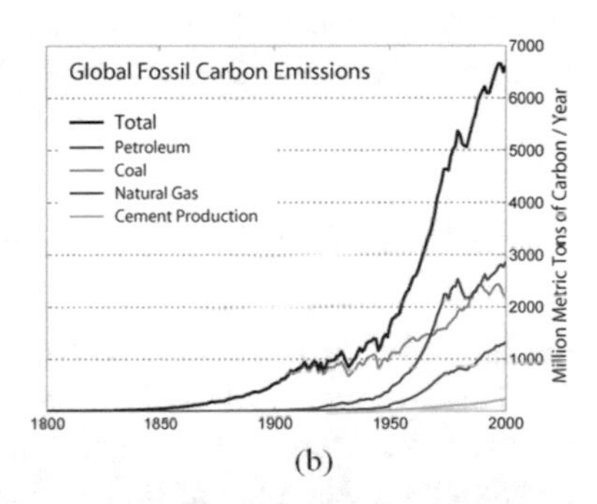

(b)

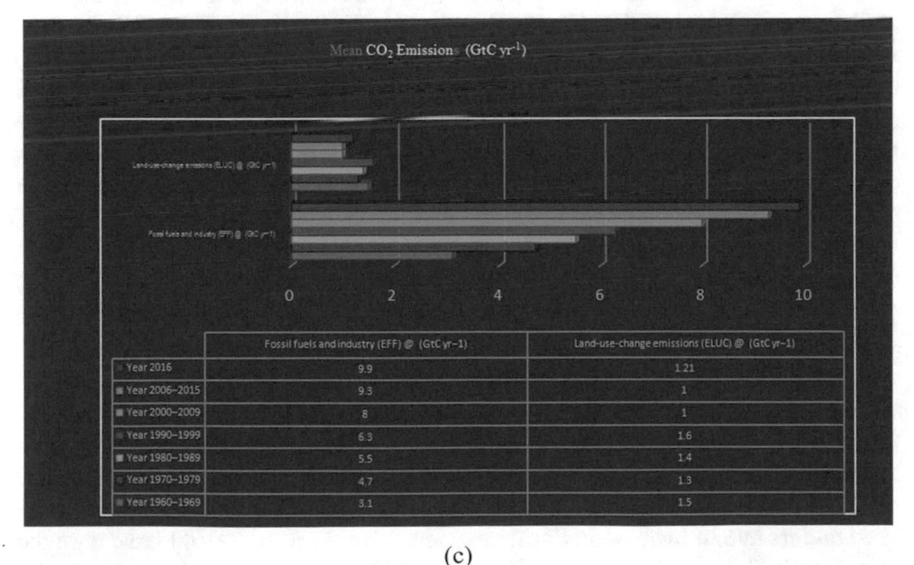

	Fossil fuels and industry (EFF) @ (GtC yr−1)	Land-use-change emissions (ELUC) @ (GtC yr−1)
Year 2016	9.9	1.21
Year 2006–2015	9.3	1
Year 2000–2009	8	1
Year 1990–1999	6.3	1.6
Year 1980–1989	5.5	1.4
Year 1970–1979	4.7	1.3
Year 1960–1969	3.1	1.5

(c)

Fig. 18.4 (**a**) Total CO_2 emission of earth; (**b**) CO_2 budget from burning fossil fuels, e.g., gas, oil, and coal; (**c**) total estimated CO_2 emissions from both fossil fuels and land-use change from 1960 to 1969 through 2016

1960s as determined by DGVMs by incorporating the estimated budget residual in response to the increase in cumulative atmospheric CO_2 (Fig. 18.4). This DGVM mean of 2.8 ± 0.7 GtC/year is technically the observational average for 2006–2015, which is estimated from the budget residual (Fig. 18.5).

Subsequently, the entire CO_2 land flux ($S_{LAND} - E_{LUC}$), which is extended by atmospheric inversion, is greatly beneficial to the calculation of the global budget, G_{ATM}, along with the assumption of relative uncertainty in S_{OCEAN} and E_{FF} due to the inversions. The total land flux has the same magnitude considering the decadal mean, which for 2006–2015 accounted for three inversions of 2.2, 2.3, and 3.4 GtC/year relative to the total flux calculation of 2.1 ± 0.7 GtC/year using Eq. (18.1). The total land sequestrations by these inversions are 1.8, 1.8, and 3.0 GtC/year including the average of all reverse flux variables, 0.45 GtC/year. Interestingly, the inter-annual variation within the inversions matched the residual-based S_{LAND} very closely (Fig. 18.5). Therefore, the total land fluxes determined by DGVM confirm that the calculation of the carbon budget and its inversion in the atmosphere has a decadal mean of 1.7 ± 0.5 GtC/year (Table 18.1 and Fig. 18.5).

Cumulative CO_2 Emissions and Atmospheric Impact

In this study, the total emission from fossil fuel burning and land-use change from 1870 to 2015 was calculated as 555 ± 55 GtC. Within the atmosphere, the emissions are partitioned among the atmosphere (235 ± 5 GtC per the atmospheric CO_2 concentrations of 288 ppm in 1870 and 399.1 ppm in 2015), the ocean (160 ± 20 GtC), and the land (160 ± 60 GtC). The total emissions during the preindustrial development period of 1750–1869 were 3 GtC for E_{FF} and approximately 45 GtC for E_{LUC} (rounded to the nearest 5), of which 10 GtC was emitted in the 1850–1870 period and 30 GtC in the 1750–1850 period for an average of 25 GtC, which revealed that the atmospheric CO_2 concentration growth rate during this period was 25 GtC; that of ocean sequestration was 20 GtC; and that of terrestrial land sequestration was 5 GtC [13, 35].

Based on the calculations in this study, the growth of the atmospheric CO_2 concentration (G_{ATM}) is projected to increase in 2016 to 6.7 ± 1.1 GtC (3.15 ± 0.53 ppm). Combining the projected E_{FF} and G_{ATM} reveals that summing the sequestration by the land and ocean and subtracting the emissions from land-use change ($S_{LAND} + S_{OCEAN} - E_{LUC}$) only yield approximately 3 GtC. S_{OCEAN} accounted for 3.0 GtC in 2015 and is expected to increase in 2016 [31, 34], and E_{LUC} accounted for 1.3 GtC in 2015 with a decadal average of 1.0 GtC/year. Here, the remaining terrestrial sequestration, S_{LAND}, for 2016 is, as expected, far below the mean balance of E_{LUC} for 2006–2015 [17, 24]. Therefore, to determine the total carbon accumulated and its rate of increase in the atmosphere since 1750 or 2016, I have used the atmospheric CO_2 concentrations of 288 ± 3 or 399 ± 3 ppm, respectively. The variation of ±3 ppm (converted to ±1σ) was obtained from IPCC data, and the rate of growth of the atmospheric CO_2 concentration was calculated using MATLAB soft-

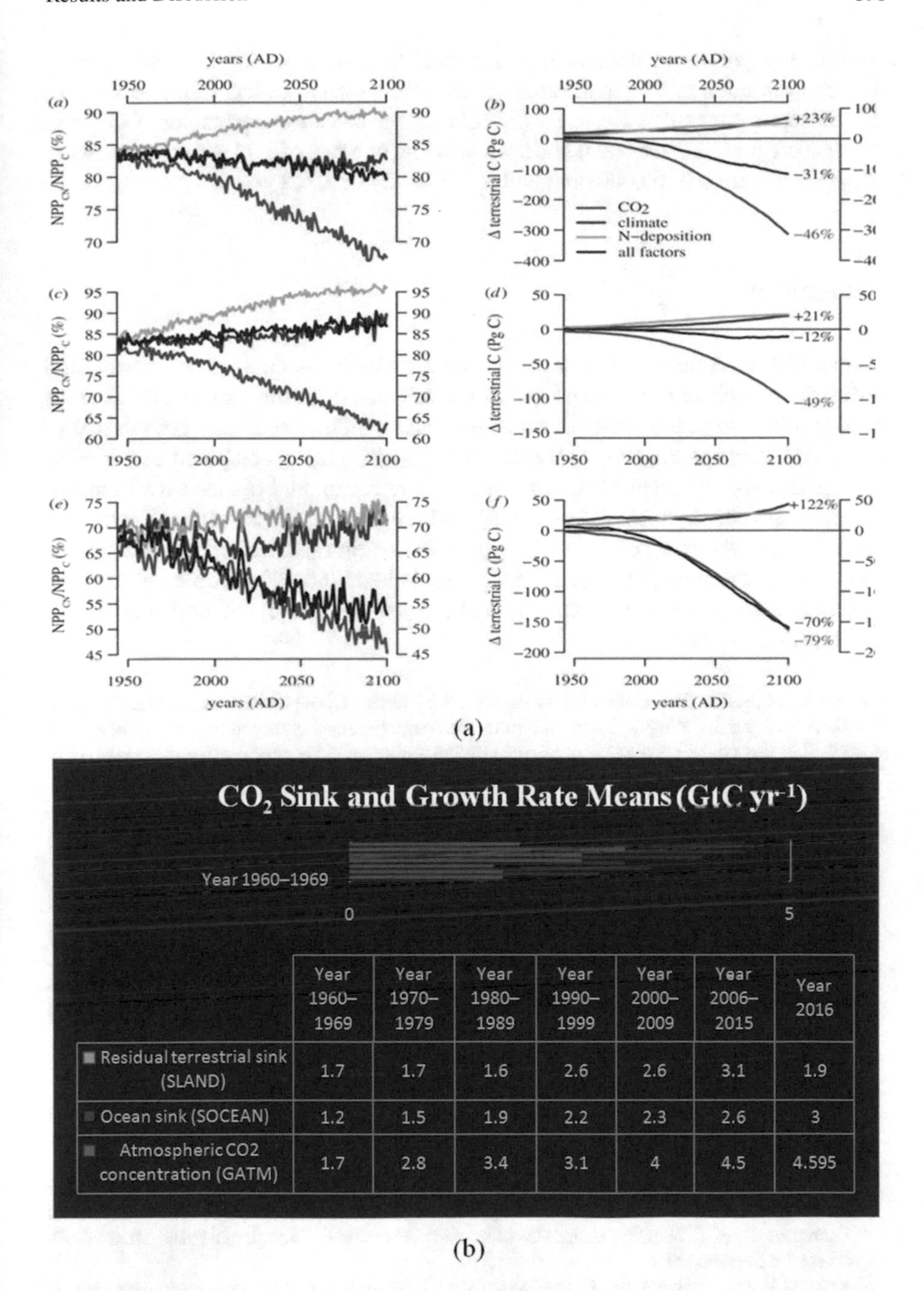

	Year 1960–1969	Year 1970–1979	Year 1980–1989	Year 1990–1999	Year 2000–2009	Year 2006–2015	Year 2016
Residual terrestrial sink (SLAND)	1.7	1.7	1.6	2.6	2.6	3.1	1.9
Ocean sink (SOCEAN)	1.2	1.5	1.9	2.2	2.3	2.6	3
Atmospheric CO2 concentration (GATM)	1.7	2.8	3.4	3.1	4	4.5	4.595

Fig. 18.5 (**a**) CO_2 sink, N-depletion, climate dynamics, and all other factors in terrestrial environment; (**b**) CO_2 sinks and annual growth rate of CO_2 emissions in GtC/year from 1960 to 1969 through 2016. The calculation $[(4.595 - 4.500/4.500)\times 100\% = 2.11\%]$ revealed atmospheric CO_2 concentration increasing at a rate of 2.11% annually

ware by comparing the decadal and individual annual values for the past 10 years. The concentration of CO_2 in the atmosphere was shown to be increasing at a rate of 2.11% ppm annually. Because the atmospheric concentration of CO_2 was 399.4 ± 0.1 ppm in 2016 and it has been growing at a rate of 2.11% per year, a toxic CO_2 concentration of 60,000 ppm will result in 121,017,712 years.

Conclusion

Global CO_2 emissions and their redistribution within the atmosphere, ocean, and terrestrial biosphere through absorption are calculated for the past several decades to determine the future atmospheric CO_2 concentration. The annual growth rate of the global atmospheric CO_2 concentration (G_{ATM}) (ppm) is computed considering the global carbon budget of the atmosphere, the ocean, and the land over the last several years and is shown to be currently 2.11% annually. If we cannot reduce this current level (400 ppm) of annual CO_2 growth in a timely manner, the atmospheric CO_2 concentration will eventually reach a toxic level (60,000 ppm), at which time the entire human race and all other living beings on the planet will perish in 30 min in 121,017,712 years.

Acknowledgements This research was supported by Green Globe Technology under the grant RD-02017-06 for building a better environment. Any findings, predictions, and conclusions described in this chapter are solely performed by the authors and we confirm that there is no conflict of interest for publishing in a suitable journal.

References

1. Andreas Reinhard. "Strongly correlated photons on a chip", Nature Photonics, 12/18/2011.
2. Douglas, J. S., H. Habibian, C.-L. Hung, A. V. Gorshkov, H. J. Kimble, and D. E. Chang. "Quantum many-body models with cold atoms coupled to photonic crystals", Nature Photonics, 2015.
3. G. Baur, K. Hencken, D. Trautmann. Revisiting unitarity corrections for electromagnetic processes in collisions of relativistic nuclei. Phys. Rep. 453, 1 (2007).
4. G. Baur, K. Hencken, D. Trautmann, S. Sadovsky, Y. Kharlov. Dense laser-driven electron sheets as relativistic mirrors for coherent production of brilliant X-ray and γ-ray beams. Phys. Rep. 364, 359 (2002).
5. Alvar R. Garrigues, Li Yuan, Lejia Wang, Eduardo R. Mucciolo, Damien Thompon, Enrique del Barco, Christian A. Nijhuis. "A Single-Level Tunnel Model to Account for Electrical Transport through Single Molecule- and Self-Assembled Monolayer-based Junctions", Scientific Reports, 2016.
6. Xiao, Y. F. et al. Asymmetric Fano resonance analysis in indirectly coupled microresonators. Phys. Rev. A 82, 065804 (2010).
7. Md. Faruque Hossain. "Solar energy integration into advanced building design for meeting energy demand and environment problem". International Journal of Energy Research, 2016.

8. J. Eichler, Th. Stöhlker. Radiative electron capture in relativistic ion-atom collisions and the photoelectric effect in hydrogen-like high-Z systems, Phys. Rep. 439, 1 (2007).
9. W. De Soto, S.A. Klein, and W. A. Beckman, "Improvement and validation of a model for photovoltaic array performance," Solar Energy, vol. 80, no. 1, pp. 78–88, Jan. 2006.
10. O. Khaselev. "A Monolithic Photovoltaic-Photoelectrochemical Device for Hydrogen Production via Water Splitting", Science, 04/17/1998.
11. Tame, M. S., K. R. McEnery, Ş. K. Özdemir, J. Lee, S. A. Maier, and M. S. Kim. "Quantum plasmonics", Nature Physics, 2013.
12. Yan, Wei-Bin, and Heng Fan. "Single-photon quantum router with multiple output ports", Scientific Reports, 2014.
13. Zhang, W. M., Lo, P. Y., Xiong, H. N., Tu, M. W. Y. & Nori, F. General Non-Markovian Dynamics of Open Quantum Systems. Phys. Rev. Lett. 109, 170402 (2012).
14. Zhu, Yu, Xiaoyong Hu, Hong Yang, and Qihuang Gong. "On-chip plasmon-induced transparency based on plasmonic coupled nanocavities", Scientific Reports, 2014.
15. N. D. Benavides and P. L. Chapman, "Modeling the effect of voltage ripple on the power output of photovoltaic modules," IEEE Trans. Ind. Electron., vol. 55, no. 7, pp. 2638– 2643, Jul. 2008.
16. S. A. Klein. Calculation of flat-plate collector loss coefficients. *Solar Energy*, 17:79–80, 1975.
17. Tu, M. W. Y. & Zhang, W. M. Non-Markovian decoherence theory for a double-dot charge qubit. Phys. Rev. B 78, 235311 (2008).
18. L. Langer, S. V. Poltavtsev I. A. Yugova, M. Salewski, D. R. Yakovlev, G. Karczewski, T. Wojtowicz, I. A. Akimov & M. Bayer. Access to long-term optical memories using photon echoes retrieved from semiconductor spins. Nature Photonics 8, 851–857 (2014).
19. Leijing Yang, Sheng Wang, Qingsheng Zeng, Zhiyong Zhang, Tian Pei, Yan Li & Lian-Mao Peng (2011). Efficient photovoltage multiplication in carbon nanotubes – Nature Photonics pp 672 – 676.
20. Md. Faruque Hossain. Theory of global cooling. 2016. Energy, Sustainability and Society. 6-24.
21. Najjari, A. B. Voitkiv, A. Artemyev, A. Surzhykov. Simultaneous electron capture and bound-free pair production in relativistic collisions of heavy nuclei with atoms, Phys. Rev. A 80, 012701 (2009).
22. T. Pregnolato, E. H. Lee, J. D. Song, S. Stobbe, and P. Lodahl. "Single-photon non-linear optics with a quantum dot in a waveguide", Nature Communications, 2015.
23. Peter Arnold. "Photon Emission from Ultrarelativistic Plasmas", Journal of High Energy Physics, 11/27/2001.
24. U. Becker, N. Grün, W. Scheid. K-shell ionisation in relativistic heavy-ion collisions, J. Phys. B: At. Mol. Phys. 20, 2075 (1987).
25. Yuwen Wang, Yongyou Zhang, Qingyun Zhang, Bingsuo Zou, Udo Schwingenschlogl. "Dynamics of single photon transport in a one-dimensional waveguide two-point coupled with a Jaynes-Cummings system", Scientific Reports, 2016.
26. Kai Hencken. "Transverse Momentum Distribution of Vector Mesons Produced in Ultraperipheral Relativistic Heavy Ion Collisions", Physical Review Letters, 01/2006.
27. Li, Qiong, D. Z. Xu, C. Y. Cai, and C. P. Sun. "Recoil effects of a motional scatterer on single-photon scattering in one dimension", Scientific Reports, 2013.
28. Lo, Ping-Yuan, Heng-Na Xiong, and Wei-Min Zhang. "Breakdown of Bose-Einstein Distribution in Photonic, Crystals", Scientific Reports, 2015.
29. Reed M. Maxwell, Laura E. Condon. Connections between groundwater flow and transpiration partitioning. Science 22 Jul 2016: Vol. 353, Issue 6297, pp. 377-380.
30. Tobias D. Wheeler & Abraham D. Stroock. The transpiration of water at negative pressures in a synthetic tree. 2008. Nature 455, 208–212.
31. Jaivime Evaristo, Scott Jasechko & Jeffrey J. McDonnell. Global separation of plant transpiration from groundwater and streamflow. 2015. Nature 525, 91–94.

32. Josette Masle, Scott R. Gilmore & Graham D. Farquhar. The ERECTA gene regulates plant transpiration efficiency in Arabidopsis. 2005. Nature 436, 866–870.
33. Masujima. "Calculus of Variations: Applications", Applied Mathematical Methods in Theoretical Physics, 08/19/2009.
34. Scott Jasechko, Zachary D. Sharp, Peter J. Fawcett. Terrestrial water fluxes dominated by transpiration. 2013. Nature 496, 347–350.
35. N. Artemyev, U. D. Jentschura, V. G. Serbo, A. Surzhykov. Strong Electromagnetic Field EFFECTS in Ultra-Relativistic Heavy-Ion Collisions. Eur. Phys. J. C 72, 1935 (2012).

Chapter 19
Water

Abstract Water is the fundamental component in our day-by-day life which should be moderated and is ought to be without contaminants to keep the worldwide condition and living beings protected. This report subsequently recognizes the emerging contaminants (ECs) in water and suggests the course for future to ensure contaminant-free water for our day-by-day needs. These ECs are omnipresent in the amphibian condition, for the most part is from the release of city wastewater effluents. Their essence is of worry because of the conceivable natural effect (e.g., endocrine interruption) to biota inside the earth. To all the more likely comprehend their destiny in water and in the earth, a normalized approach is expected to guarantee that agent information is accomplished and encourages a superior comprehension of spatial and transient patterns of EC event. This report therefore portrays the effect of conventional wastewater treatment into nature. Ordinary wastewater treatment has a lack of suspended particulate issue investigation because of further planning prerequisites and an absence of good scientific methodologies which brings about a poisonous amphibian condition. In the customary innovation, a large portion of treated slop is applied straightforwardly to rural land without examination of ECs which results in dirt tainting, therefore causing harm to the uncovered earthbound life-forms. Along these lines, this report recommended forsaking the customary wastewater treatment instrument and discovering better innovation to get a universe of contamination-free water to make sure about a superior situation for earth.

Keywords Natural water · Emerging contaminant · Conventional water treatment · Environmental impact · Clean water

Introduction

In surface waters, a broad range of inorganic and organic contaminants have traditionally originated from industrial or agricultural chemical wastes that causes environmental pollutions. Besides, a greater number of municipal wastewater treatment

M. F. Hossain, *Global Sustainability in Energy, Building, Infrastructure, Transportation, and Water Technology*,
https://doi.org/10.1007/978-3-030-62376-0_19

chemicals and pharmaceuticals described as ECs such as 17β-estradiol (E2), 17α-ethinylestradiol (EE2), and diclofenac also cause water contamination. To date, millions of different industries alone have been responsible for water contamination and toxicity globally, with concentrations up to a maximum of 6.5 mg/L for the antibiotic ciprofloxacin [1, 2]. In this report, single-compound acute toxicity testing (including crustaceans, algae, and bacteria) has been conducted under controlled laboratory conditions that has found median effective concentrations (EC_{50}s—concentration at which the toxicological response to an organism is halfway between a normal and maximum response for a preset time period) for a number of these ECs to be <1 mg/L [3, 4]. Such effect concentrations classify the chemical as potentially very toxic to aquatic organisms [5, 6]. The presence of these chemicals in the environment is more concerning considering that they do not appear individually, but as a complex mixture, which could lead to unwanted synergistic effects. The ubiquity of a high number of potentially toxic ECs in the environment underpins the need to better understand their occurrence, fate, and ecological impact. This report describes current knowledge on the occurrence of ECs in wastewaters and surface waters using the standard data set as a model to represent global water contamination. From the data set and wider literature, areas of concern considered to be understudied are discussed considering spatial and temporal variability of ECs in wastewater and river water, partitioning of ECs to solid matter during wastewater treatment, fate of ECs in environmental waters, and toxicological impact of ECs within the environment. Finally, recommendation for future water environmental monitoring approaches is proposed for contaminant- and toxic-free water world.

Materials and Methods

The presence of ECs in the water environment is mainly attributed to the discharge of treated wastewater from wastewater treatment plants (WwTPs) which is contaminated by chemicals where aquatic environment gets polluted which has the impact on dissolved oxygen in the body of water and causes rapid growth of deadly microorganisms. Conventional secondary processes (activated sludge and trickling filters) represent the most extensively used and studied processes and are therefore focused on in this report. However, this report is not designed to remove ECs which are discharged to receiving surface waters including rivers, lakes, and coastal waters but describes the impact of chemicals such as chlorine, acetaminophen, amitriptyline, EMDP, dosulepin, fluoxetine, norfluoxetine, triclosan, ofloxacin and ciprofloxacin, NSAIDs, β-blockers, antidepressants, and the antiepileptic carbamazepine on influent wastewaters [7, 8]. Some researchers suggested that removal of chemicals like chlorine and acetaminophen varies broadly from low (<50%) to high (>80%) due to their different physicochemical properties such as susceptibility to microbiological attack in the water environment by receiving surface waters in the ng to mg/L range which leads to surface water getting severely contaminated [9, 10]. Thus, in this research a large variation in influent wastewater concentrations (equivalent to more than an order of magnitude) has been observed for some ECs where chlorine and acetaminophen were found within influent wastewater at mean concentrations ranging from 6924 to 492,340 ng/L [11, 12]. This indicates spatial and/or temporal variations in their usage. However, dilution from industrial inputs, degradation within the upstream sewer, and rainfall and sampling mode will all contribute to this variability. The use of inappropriate sampling strategies is considered the greatest weakness of reported occurrence data. Ort et al. gave an excellent account of uncertainties associated with different sampling modes. Current approaches tend to use sampling of low inter-day frequency and often no intraday repetition. This approach has limitations as it only yields a snapshot of EC concentration for a specific point of time. Time- and volume/flow-proportional composite samplers are also reported, but to a lesser extent [13, 14]. The latter are preferred as it accounts for fluctuations in flow. The relatively high cost and logistical issues associated with this sampling mode have resulted in very few studies applying flow-proportional sampling report. Also, a major uncertainty of 24-h composite sampling is chemical stability. This is not often investigated but is known to be significant for some chemicals [15, 16]. Due to uncertainties of existing sampling methods, there is lack of understanding on spatial and temporal variations in EC concentrations. Consequently, on the other hand, studies have been able to tentatively assess spatial distribution of microorganisms within the water catchment [17, 18]. Such studies must rely heavily on sampling due to the large number of sites which are monitored at approximately the same time. Despite the uncertainties associated with sampling report, information attained from such studies is extremely valuable. It was used as a primary indicator of sites within the catchment that require more detailed study. These can then be subject to more robust sampling protocols to facilitate greater understanding of EC fate and occurrence during wastewater treatment and in the environment at these predetermined sites of interest.

Results and Discussion

It is anticipated that concentrations of ECs in receiving wastewater vary throughout the day. The collected hourly composite samples (i.e., one sample every 15 min to create an hourly composite) for influent wastewater over 24-h sampling periods were used to investigate within-day variability of chlorine and acetaminophen concentrations. Samples were chilled upon collection but their stability at 4 °C was not investigated or referenced to. However, significant degradation (>15%) has been observed for some ECs stored at 4 °C for only 12 h. Despite this uncertainty, due to the presence of chlorine and acetaminophen in the water the microbial contaminants showed variation in concentration throughout the day [19, 20]. A peak in receiving load was observed between 07:00 and 9:00 h [7, 17, 21]. Therefore, the response of wastewater treatment process (WwTP) to this daily spike in receiving concentration and volumetric load has been analyzed to get the influence of microorganism on water contamination in different concentrations of microorganism and time (Fig. 19.1).

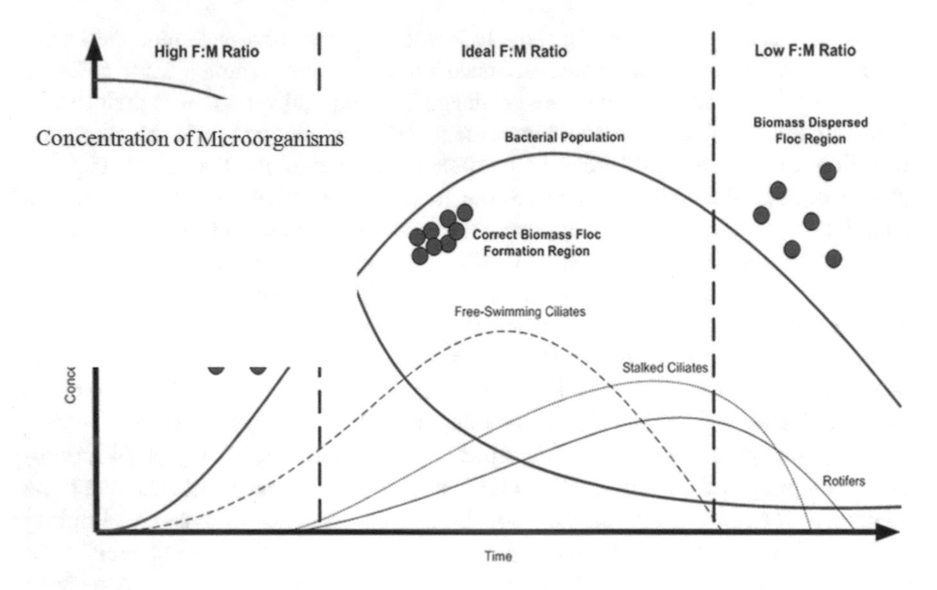

Fig. 19.1 The concentration of microorganism at various sludge ages and the concentration of microorganism production in wastewater at times in related to various types of waste

Wide reports of 19 countries using a variety of time-, flow-, and volume-proportional sampling attempted to investigate inter-day variability (i.e., over a 1-week period) of chemicals in influent wastewater [3, 22, 23]. Collated reports revealed a trend of recreational use for some compounds [24, 25]. Both organic and inorganic chemicals showed elevated levels of water contaminants. In contrast, it is anticipated that also hospital-dispensed chemicals such as chlorine, acetaminophen, amitriptyline, EMDP, dosulepin, fluoxetine, norfluoxetine, triclosan, ofloxacin, and ciprofloxacin and anticancer chemicals are observed at higher concentrations in the wastewater body. Inter-day variability of EC concentration is difficult to fully appreciate without conducting flow-proportional sampling. This is most significant when comparing weekday and weekend periods where wastewater flows can vary notably. Differences in weekday/weekend flow will be specific for the catchment in question. For example, a residential (commuter) or tourist area will see increased flow during weekends. On the other hand, a catchment containing industrial works may observe a reduction in flow and dilution during weekends. Rainfall will also result in changes to wastewater flow between days. This illustrates that reporting load (mass per day) over the traditional approach of concentration (mass per liter) to overcome temporal variations in flow is more suited to describe findings.

Emerging contaminant analysis in wastewaters tends to be undertaken on the aqueous phase only (i.e., a pre-filtered sample). Particulate-phase analysis is not routinely undertaken because up to 1 g of dry solids is often required for each analysis and additional sample extraction is needed [26, 27]. However, the report here is essential as some chemicals have a high affinity to particulate matter. Numerous chemicals including chlorine, acetaminophen, amitriptyline, EMDP, dosulepin, fluoxetine, norfluoxetine, triclosan, ofloxacin, and ciprofloxacin have been found to be within the particulate phase of influent wastewater at significant concentrations (>20% of the total concentration). Consequently, particulate-phase determination is necessary to correctly report influent concentration for some chemicals. This is also essential for back calculations of population usage by wastewater-based epidemiology [28, 29]. Partitioning of ECs to suspended matter in final effluents is even less studied due to the very low solid concentrations encountered (typically 10–30 mg/L for secondary processes). Nevertheless, some researchers found that the final effluent of secondary processes had >20% of the total triclosan, ofloxacin, and ciprofloxacin concentration to be within the particulate phase, despite very low suspended solid concentrations. Particulate-phase concentrations were equivalent to, and in the range of, 26–296 ng/L, respectively. This provides a route for their release into the environment which goes unmonitored and the environmental fate of these particulate-bound chemicals is unknown.

Emerging contaminants have also been reported in riverine sediments due to the rapid growth of bacterial contamination (Fig. 19.2). Determining toxicological impact here is essential as sediments can act as a sink for their accumulation. However, this is notoriously difficult to ascertain due to the complexity of the system. Benthic organisms can be exposed to ECs within the sediment itself, in interstitial water, and in overlying water [4, 13]. This makes experimental design and setup critical for reproducing representative conditions. A study investigating toxicity of

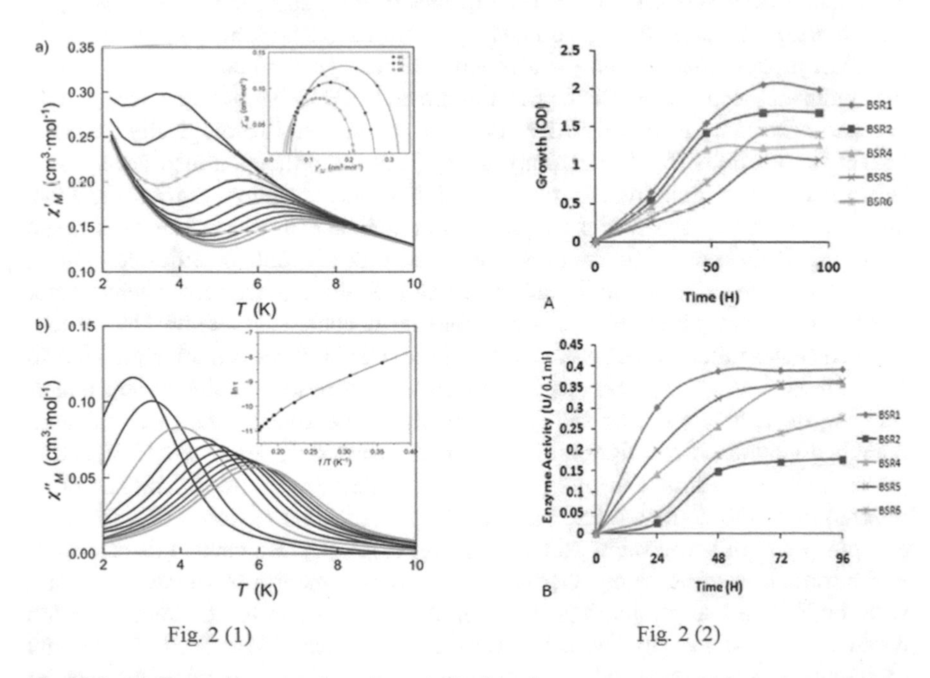

Fig. 2 (1) Fig. 2 (2)

Fig. 19.2 (1) Shows the contaminant bacterial ignition in response to contaminant water temperature per unit areas, (2) shows the contaminant bacterial growth during a certain period of time where enzyme activities were initiated by the water contamination

diclofenac to the crustacean *Hyalella azteca* prepared synthetic sediment at a 1:3 ratio with water [8, 24]. The authors failed to report or investigate aqueous/liquid partitioning of diclofenac, whether equilibrium conditions were present, or if desorption occurred throughout the test. Consequently, reporting toxicological effect concentrations for sediments can have considerable uncertainties. An improved study of biological response information with chemical analysis for both aqueous and particulate phases has been performed [2, 3]. Their findings suggested that sorption to sediments resulted in a reduction of bioavailability and toxicity. On the other hand, the accumulation (and increased concentration) of readily sorbed compounds in sediments within the environment could compensate for this reduction in toxicity. Interestingly, the activity of benthic invertebrate also resulted in increased desorption, leading to improved bioavailability [12, 18]. Numerous mechanisms take place, illustrating the complexity of ascertaining sediment toxicity. Studies over long-time periods which simulate representative steady-state riverine sediment conditions are required for a range of indicator species. These firstly need to be supported with chemical analysis to assess the behavior of the EC(s) in question. Such studies would help assess chronic and multi-generation impact of sediment contamination on benthic organisms such as bacteria which ultimately impacts global ecology and finally affects the human being (Fig. 19.2).

To better understand their pathways of removal during wastewater treatment, particulate-phase analysis is needed as well as analysis of the biomass (either suspended or attached) of the process [8, 30]. Ideally, corresponding aqueous and particulate determinations should be undertaken for each sampling point such that a complete process mass balance is attained. This can give valuable information on the dominant mechanisms which govern removal [7, 31]. Removal can vary greatly between ECs from a physically driven process (adsorption) to biologically mediated enzymatic reactions (biodegradation) [2, 8]. Their identification also needs to be supported with information of process conditions and operation, nutrient removal, and complementary analysis of the physical/biological characterization of biomass [2, 4]. With this knowledge, the operation of the process could be adjusted to favor their removal [4, 9]. Such information can also be used to identify where further research efforts may be needed. For example, those chemicals removed by biodegradation suggest that further investigation of possible biotransformation products in final effluents is needed, and those removed by adsorption require further understanding of their fate during and following sludge treatment where ECs undergo microbially mediated reactions during secondary wastewater treatment [6, 9] and in the environment which can be an effective technology. Biodegradation is often referred to as the dominant fate pathway for the removal of some ECs from the aqueous phase of wastewaters and surface waters [2, 32]. However, this can result in the formation of numerous degradation or transformation products. Therefore, with the ECs in wastewater, it is expected that a great number of transformation products (of unknown toxicity and persistence) will exist in the final effluent and receiving surface waters need to be removed to avoid chronic impact on the water environment and living beings on earth [9, 12].

Conclusions

Global water is getting contaminated due to lack of knowledge for water conservation and application of conventional wastewater treatment process which needs to be resolved urgently to protect the environment and living beings. Currently, the application of conventional technology for the treatment of wastewater and water management is backdated which is causing severe environmental pollution and loss of water. Therefore, removal performance of different WwTPs at various operational conditions needs to be re-evaluated to conserve and treat water to mitigate water perplexity throughout the world. Necessarily, further research and sustainable technology are needed to determine the fate and removal of ECs during treatment sophisticatedly considering its impact across the complete ecosystem and living beings on earth.

Acknowledgement This topic is written by Bruce Petrie[a], Ruth Barden[b], and Barbara Kasprzyk-Hordern[a], at the [a]Department of Chemistry, University of Bath, Bath BA2 7AY, UK; [b]Wessex Water, Bath BA2 7WW, UK. This chapter was originally published as "A review on emerging contaminants in wastewaters and the environment: Current knowledge, understudied areas and recommendations for future monitoring" in the journal "Water Research". 72 (2015); 3–27. The report has been modified from its original research to fit the publication in this book chapter under the topic "Water."

References

1. Bagnall, J.P., Malia, L., Lubben, A.T., Kasprzyk-Hordern, B., 2012b. Stereoselective biodegradation of amphetamine and methamphetamine in river microcosms. Water Res. 47, 5708-5718.
2. Bauer, J. E. *et al.* The changing carbon cycle of the coastal ocean. *Nature* **504**, 61–70 (2013).
3. Andreozzi, R., Caprio, V., Ciniglia, C., De Champdoré, M., Lo Giudice, R., Marotta, R., Zuccato, E., 2004. Antibiotics in the environment: occurrence in Italian STPs, fate, and preliminary assessment on algal toxicity of amoxicillin. Environ. Sci. Technol. 38, 6832-6838.
4. Bridges, E.M., and L.R. Oldeman. 1999. Global assessment of human-induced soil degradation. *Arid Soil Research and Rehabilitation* 13 (4):319-325.
5. Cleaver, Kevin, and Gotz Schreiber. 1994. Reversing the Spiral: The Population, Agriculture and Environment Nexus in Sub-Saharan Africa. Washington, D.C.: World Bank.
6. Davis, S. J. & Caldeira, K. Consumption-based accounting of CO2 emissions. *Proc. Natl. Acad. Sci. U.S.A.* **107**, 5687–5692 (2010).
7. Arnell, N. *et al.* Climate Change 1995: Impacts, Adaptations, and Mitigation of Climate Change (eds R. T. Watson, M. C. Zinyowera, & R. H. Moss) 325–363 (Cambridge University Press, 1996).
8. Berardi, Gigi. 2002. Commentary on the challenge to change: participatory research and professional realities. *Society and Natural Resources* 15:847-852.
9. Betts, R. A., Jones, C. D., Knight, J. R., Keeling, R. F. & Kennedy, J. J. El Nino and a record CO2 rise. *Nat. Clim. Change* **6**, 806–810 (2016).
10. van der Werf, G. R. *et al.* Climate regulation of fire emissions and deforestation in equatorial Asia. *Proc. Natl. Acad. Sci. U.S.A.* **105**, 20350–20355 (2008).
11. Duce, R. A. *et al.* Impacts of atmospheric anthropogenic nitrogen on the open ocean. *Science* **320**, 893–897 (2008).

12. de Sherbinin, A. 2002. Land-Use and Land-Cover Change. In *A CIESIN Thematic Guide*. Palisades, NY: Center for International Earth Science Information Network of Columbia University.
13. Canadell, J. G. *et al.* Contributions to accelerating atmospheric CO2 growth from economic activity, carbon intensity, and efficiency of natural sinks. *Proc. Natl. Acad. Sci. U.S.A.* **104**, 18866–18870 (2007).
14. Dlugokencky, E. & Tans, P. Trends in atmospheric carbon dioxide. *National Oceanic & Atmospheric Administration, Earth System Research Laboratory (NOAA/ESRL)* http://www.esrl.noaa.gov/gmd/ccgg/trends/global.
15. Maitima, J.M., Robin Reid, L.N. Gachimbi, A. Majule, H. Lyaruu, D. Pomery, S. Mugatha, S. Mathai, and S. Mugisha. 2004. A methodological guide on how to identify trends and linkages between changes in land use, biodiversity and land degradation. Nairobi: International Livestock Research Institute.
16. McAvoy, D.C., Schatowitz, B., Jacob, M., Hauk, A., Eckhoff, W.S., 2002. Measurement of triclosan in wastewater treatment systems. Environ. Toxicol. Chem. 21, 1323-1329.
17. Ciais, P. & Sabine, C. Chapter 6: Carbon and other biogeochemical cycles in *Climate Change 2013 The Physical Science Basis* (eds T. Stocker, D. Qin. & G.K. Platner) (Cambridge University Press, 2013).
18. Petrie, Bruce, Ruth Barden, and Barbara Kasprzyk-Hordern. "A review on emerging contaminants in wastewaters and the environment: Current knowledge, understudied areas and recommendations for future monitoring", Water Research, 2015.
19. Hossain, Md. Faruque (2016). Theory of Global Cooling. Energy, Sustainability, and Society. 6:24. (Springer).
20. Le Quéré, C. *et al.* Global carbon budget 2016. *Earth Syst. Sci. Data* **8**, 605–649 (2016).
21. Hongyan Bao, Jutta Niggemann, Li Luo, Thorsten Dittmar, Shuh-Ji Kao. "Aerosols as a source of dissolved black carbon to the ocean", Nature Communications, 2017.
22. Chevallier, F. On the statistical optimality of CO2 atmospheric inversions assimilating CO2 column retrievals. *Atmos. Chem. Phys.* **15**, 11133–11145 (2015).
23. Conroy, Michael J. 1996. Mapping of Species Richness for Conservation of Biological Diversity: Conceptual and Methodological Issues. *Ecological Applications* 6 (3):763-773.
24. Corinne Le Quéré, Robbie M. Andrew, Pierre Friedlingstein, Stephen Sitch et al. "Global Carbon Budget 2017", Earth System Science Data Discussions, 2017.
25. Dietzenbacher, E., Pei, J. & Yang, C. Trade, production fragmentation, and China's carbon dioxide emissions. *J. Environ. Econ. Manage.* **64**, 88–101 (2012).
26. Denman, K. L. *et al. Couplings Between Changes in the Climate System and Biogeochemistry* (Cambridge University Press, 2007).
27. Hossain, Md. Faruque (2018). Green Science: Advanced Building Design Technology to Mitigate Energy and Environment. Renewable and Sustainable Energy Reviews. 81 (2), 3051-3060. (Elsevier).
28. Hossain, Md. Faruque (2016). Production of Clean Energy from Cyanobacterial Biochemical Products. Strategic Planning for Energy and the Environment. 3; 6-23 (Taylor and Francis).
29. Hossain, Md. Faruque, and Mukai, Hiroshi (2000). Importance of Nutrients (N, P, and NO3+NO2) in Growth of the Surfgrass, Phyllospadix iwatensis Makino.
30. Ballantyne, A. P., Alden, C. B., Miller, J. B., Tans, P. P. & White, J. W. C. Increase in observed net carbon dioxide uptake by land and oceans during the past 50 years. *Nature* 488, 70–72 (2012).
31. Bebbington, A.J., and S.P.J. Batterbury. 2001. Transnational livelihoods and landscapes: political ecologies of globalization. *Ecumene* 8 (4):369-380.
32. Box, P. 2002. Spatial Units as Agents: Making the Landscape an Equal Player in Agent-Based Simulations. In *Integrating Geographic Information Systems and Agent-Based Modeling Techniques for Simulating Social and Ecological Processes*, edited by R. H. Gimblett. New York: Methuen.

Chapter 20
Land

Abstract The major contribution of human exercises to the environment is quickening the land-use and land cover (LULC) change on earth severely. Simply the land surface warms up and cools down quicker in contrast to water and air balance on earth; thus, the planet land interacts with its ecological equilibrium intensely and causes environmental vulnerability globally. The reflectivity, topography, elevation, and latitude of lands thus have adverse impacts because land latitude loses its capability to mitigate solar radiations that reach the earth surface. Thus, the planetary albedo reflexivity as well as the sort of land cover that obtains energy from the sun influences the number of energies that is mirrored to earth that causes global warming. As a result, the misuse of land management and land-use practices deliver a substantial influence on natural resources comprising animals, plants, nutrients, soil, and water.

Keywords Misuse of land · Afforestation · Surface heating and cooling · Natural resource consumption · Land-use management

Introduction

In international environment change, LULC change is a key issue. During the 1972 Stockholm Conferences on the Human Environment and the 1992 United Nations Conferences on Environment and Development (UNCED) the community of scientific research called for a practical study of land-use changes [1, 2]. Simultaneously, the International Geosphere and Biosphere Program (IGBP) as well as the International Human Dimension Program (IHDP) co-arranged a working group to promote research activity and to set up research agenda for land-use and land-cover changes. Where other factors are incorporated into the necessity basis to drive different developmental index for water and land resources, land cover/land-use mapping is a vital element. Naturally based on NRSA (1989), land cover refers to artificial covers, rock/soil, water bodies, natural vegetation, and others found on the

M. F. Hossain, *Global Sustainability in Energy, Building, Infrastructure, Transportation, and Water Technology*,
https://doi.org/10.1007/978-3-030-62376-0_20

land. On the other hand, land use refers to the activities of human beings and various uses which are conducted over land. One of the most essential earth system properties is land cover which is defined as biotic and abiotic component assemblage on the surface of the earth. While land cover refers to that which covers the earth surface, land use is defined as the way in which land cover is transformed. Land cover is comprised of bare soil, forest, grassland snow, and water. Conversely, land use includes wildlife management areas, recreation area, built-up land, and agricultural land. Land use describes the activities of man on land, which are directly associated to the land [3, 4]. Land cover mirrors the biophysical condition of the surface of the earth and the immediate subsurface, hence embracing water, vegetation, and soil material. Land cover and land use are self-motivated, and transformations may include the intensity of nature of change, although it may as well involve spatial and time aspects. In addition, land cover and land use involve the alteration either indirect or direct of natural habitations and their influence on the area ecology. Moreover, population pressure which results in the intense use of land minus proper management practices is mainly the cause of land degradations [2, 5]. Because of overpopulation, people tend to shift toward sensitive areas on earth, and use land without taking into consideration the slope as well as the erodibility, which eventually results in severe soil erosion. Road construction effect as well as other similar landscape disturbances on landslides and erosion, and other mass movements on earth, are well recognized [6, 7]. Evidently, the extent of the land underneath forest is getting reduced due to human activities. Similarly, the land being utilized for cultivation is decreasing as well. On the other hand, instantaneously, land underneath built-up region is constantly increasing. In recent times the function of the property promoters and the real estate individuals has been bringing a severe tragedy to agricultural land and forest area. Clearly, this is a harmful land management condition. Therefore, understanding the existing state is essential for this kind of study on land-use and land-cover change detection, to make a bright future. Simply the use of land cover or land-use change investigation, as well as human activity assessments and biophysical capacities, would greatly increase the method to comprehend the key causes and extent of land degradation as well as loss of variety [8, 9]. Land use, however, involves dynamic and essential data for planning of operative land measurement program, and thus requires more practicality to deal with degradation's root cause. This report therefore offers an overall framework and guide for this analysis. The report also discusses the exact information that can be acquired from land-use investigation and provides a theoretical framework for recognizing the key causes of land-use change. Thus, it offers a set of common problems as well as research questions and finally discusses the analytical framework implications for the choice of approaches and result interpretations. This is based on intuitions as well as capabilities acquired from enduring research and development activities carried out by team members of LUCID project across numerous sites within East Africa, and from other land-use transformations and research on key causes [10–12]. Through offering an overall analytical framework for research design of land use and the gathering and analysis of data, this report matches other LUCID project procedural papers.

Materials and Methods

In this research, the root-cause investigation has been conducted in a way in which trends in land management affect land biodiversity and degradation by using land spatial analysis regarding rich societal and environmental information. The following are the examples of available data from root-cause and land-use analyses: leverage points within the system to enhance land management; identification of serious conditions, groups, and locations; scenarios of the influence of likely program or policies; determination of the paste program's and policies' effect on land management as well as environment so as to acquire lessons learned from cautionary and positive experiences; and recognition of the elements of landscape that are changing, key reasons for those particular changes, as well as their influence on land biodiversity and degradation. To completely understand the trends as well as their causes, a multi-scale method is essentially conducted because the environmental and societal processes work at varied scales and different information types exist at different scales. For instance, depending on the analysis scale, the part of land that is being utilized particularly for cultivation in a semiarid area is likely to appear different. This has been demonstrated by several researchers while plot-level analysis discloses that farmers have in fact left several of their fields uncultivated because of low productivity resulting in those fields reverting to bush [13, 14]. Using at the plot and household level, fieldwork is carried out to acquire particular information considering the sample analyzation as the representation of wider areas [15, 16]. Land management practices like gender of head, ethnicity, labor availability, income activities, household assets, indicator plants, yield, fuelwood collection or fertilizer application, and soil conservation; household and individual information like wealth, ethnic group, or farm size; animal species counts; water quantity and quality; erosion estimates; soil properties; and plant species composition are the examples of information that has also been analyzed by conducting mapping and spatial process and identification. Here, in case a specific study site is not placed within its comprehensive context, and in case the higher level forces influence at international, regional, and national levels disturbing the local land use as well as management factors are left out, the answers are often incomplete. Different kinds of information regarding land biodiversity and degradation that are of global interest are being conducted by higher level analysis. Root-cause and land-use analyses at both the global and regional levels may make available information on, for instance, (1) status of wildlife migration corridors that pass through national borders and (2) land-use/land-cover change impact on the local to international climate. Conservation of land use like afforestation or deforestation affects the quantity of atmospheric dust within the atmosphere, as well as the carbon amount stocked in vegetation within the soil. In addition, the parameters of land surface like surface roughness, soil moisture, leaf area index, and albedo are extremely sensitive to the conservation of land cover that affect the regional and local climate directly. (3) Root-cause and land-use comparison amid sites makes available new information on the root causes of land-use modification, recognition of common driving forces, and common patterns and land-use change process. These kinds of analyses allow the development of upcoming land-use change scenarios as well as its effect on the environment; (4) private and public investments, governance, markets and prices, and agricultural or land policies; (5) trends in internationally important biodiversity, including the changing of habitats, specific ecosystems, or flora; and (6) vegetation removal off-site impact. For instance, watershed deforestation may cause surface water siltation downstream and erosion.

Although the framework proposes that a linear, simplex causative chain has been conducted, it does not embrace the connections amid variables within spatial distribution. Furthermore, it does not expressly embrace temporal trends or spatial scale thought. There is an additional assumption that indicators same as a political action as a response indicator trail from the past indicator even though not different elements live. In summary, it makes available a logical framework for demonstrating presumed factors, and then offers very little assistance in the initial recognition of the significant variables, their relation, or whichever way they connect to the issue as well as its achievable resolution for the land management due to the intensification.

Here, the intensification analysis refers to subevolution in the use of land and agricultural techniques in the direction of higher application of capital and labor to the land in response to increasing demand for agricultural products and population densities. Within the continent, several farming systems are intimate with rising population densities as well as modifications in their system of farming, as indicated by intensification, same as decreasing usage of fallow, as well as embracing upper yielding of stringent crops [17, 18]. Nevertheless, intensification studies do not usually adopt factors same as those of land distribution amid households or the wider socioeconomic framework in which agricultural change occurs (Fig. 20.1). The analysis of these human measurements, like differentiation amid households and financial achievement in diversification ways, discloses the forces behind dynamic management of soil and land.

Notwithstanding the fact that motivating forces of land-use modification are many-faceted as time goes by, these motivating forces may change in relative impact, which will differ in the context. The land-use modification examination demands theoretical contexts as well as analytical approaches conducted here that are both broad enough to accommodate socio-environment interaction dynamics at different scales, and should also be flexible enough to capture the temporal underlying forces of these procedures to have precise information about the land change and management globally [5, 19].

Results and Discussion

Based on the above research, it is evident that this report commended the wide selection of approaches accessible to carry out root-cause and land-use analyses regarding dynamic land variety and degradation. As a result, the analysis findings offered a detailed data and sources of data through the examination of the complicated systems same as atmosphere/society association causing land-use and management

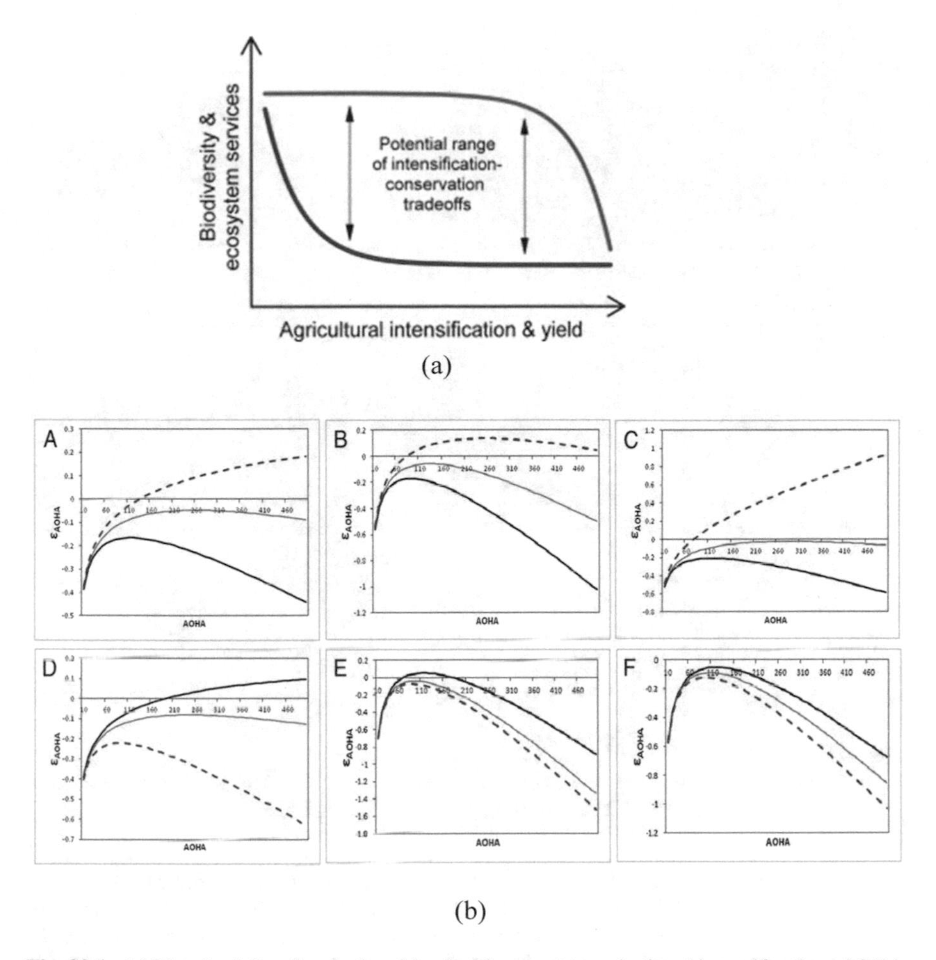

Fig. 20.1 (**a**) The elasticity of agricultural land with respect to agricultural intensification εAOHA, and (**b**) as estimated with model 1 (GOV = CORC in **A**), model 2 (GOV = ROL in **B**), model 3 (GOV = ACC in **C**), model 4 (GOV = PA in **D**), model 5 (GOV = ESI in **E**), and model 6 (GOV = EPI in **F**). In computing the elasticity, the governance scores are set at high (sample maximum, red dashed line), moderate (sample mean, blue solid line), and low (sample minimum, black solid line) values. When conventional governance scores are high, intensification leads to agricultural expansion (thus signaling a Jevons paradox), whereas when conventional governance scores are low or moderate, intensification leads to agricultural contraction (thus signaling land sparing). Conversely, when the environmental governance indicator takes high or moderate values, land sparing occurs. For low values of the environmental governance indicator, either a Jevons paradox occurs (**D**, **E**) or the intensity of the land-sparing effect is reduced (**F**)

change. Figure 20.2 shows the primary data variability as well as information that revealed the concerns in dynamic land degradation, variety, and land use. In the examination, secondary information and information from, for instance, literature review, different government statistics, and population censuses could complement the above. In addition, the information is slightly different from every other source. Several spatial sampling methods were urged by the spatial distribution data to be useful, particularly

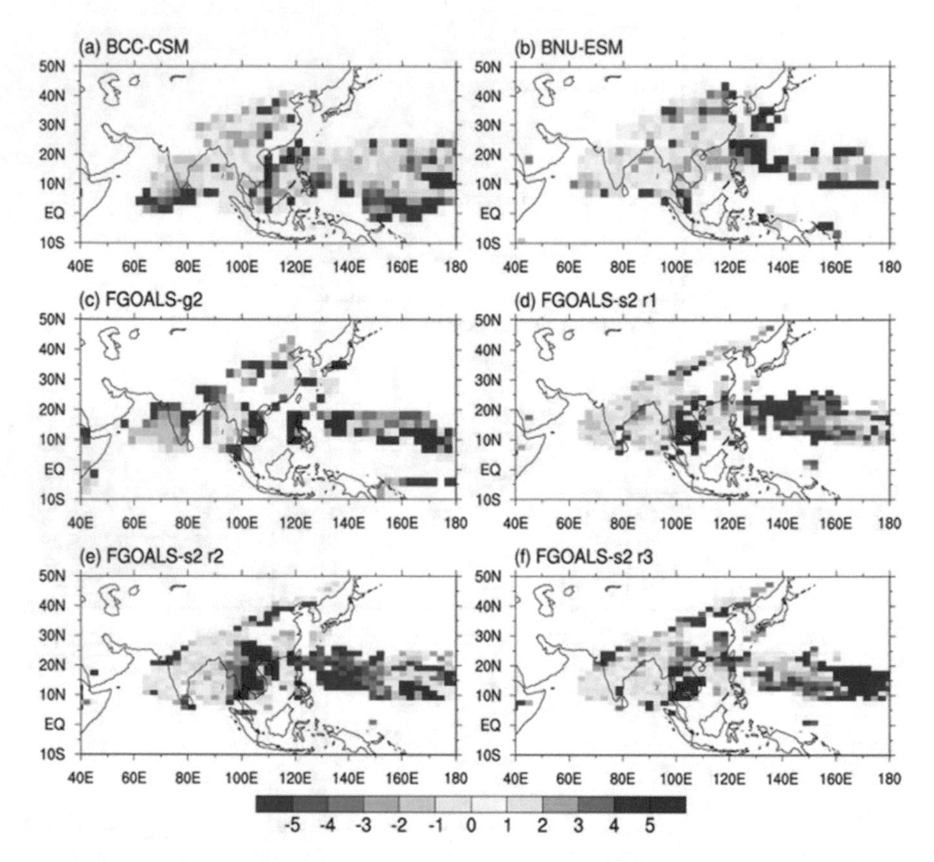

Fig. 20.2 Spatial distributions of the global land use derived from (**a**) GPCP, (**b**) CMAP, (**c**) BCC-CSM, (**d**) BNU-ESM, (**e**) FGOALS-g2, (**f**) FGOALS-s2 r1, (**g**) FGOALS-s2 r2, and (**h**) FGOALS-s2 r3. "r1," "r2," and "r3" denote the three realizations of FGOALS-s2, respectively. The data of coupled models are derived from the twentieth-century historical climate simulations during 1981–2015. The results are presented with the grid of individual model/observation. Note that pentad 30 is the end of May. The numbers shown in the upper right corner of each plot are the spatial pattern correlation coefficients with GPCP land-use change data and with CMAP data

in representing the ecological variability that differs across houses more probably than the socioeconomic characteristics [20, 21]. Line transect that must cut across different applicable variables or several ecosystems is one method that is usually used in geographical and ecological field analysis (Fig. 20.2). To acquire information on the phenomena distribution, the systematic or random-purpose samples will afterwards be analyzed on the transect line. Conversely, a random point distribution across a neighborhood offers a less prejudiced demonstration of the real; nonetheless, transport to every other point is tough as well if protection is disturbed.

A difference in this is thus done here as a clustered sampling, in which the points to look are easier to attain since they are clustered, though the clusters are chosen haphazardly. An associate method to cluster sampling, stratified sampling, captures each socioeconomic or ecological variability purpose location, same as farms or villages, and the samples are haphazardly selected within a spatial institution of interest. Same as in agroecosystem, that sample represents those within associate system, and in a systematic way, valid similarities and dissimilarities might be generated amid ecosystems [22–24]. Therefore, each contextualized website-level data within the regional, nation, and global frameworks for spatial distribution is analyzed by microlevel if data on higher level root causes of land-use change is shifting land managers to the native level. By their nature, land-use analysis and root-cause analysis are multifaceted and complicated reasons, and as such, the taken, analyzed, and conferred data may essentially remove this complexness to determine the real root causes of land-use change.

These opposing information convenience types at different scales shall, therefore, represent the detailed interpretations and processes of land-use change all over the world. Lastly, soil erosion impact represented an important force driving the LULC modification, where social action, power, social scale, and above all human activities are liable for worldwide land-use alteration (Fig. 20.3). As a result, the motivating forces to alter soil erosion in numerous scales were identified as the chief objective of land-use management. These findings disclosed a significant empirical goal of providing corrected interpretations as well as new information of soil erosion which eventually is part of land-use modification.

Therefore, it is essential that the causes of land management and land-use changes must be known in the course of a multi-scale, multidisciplinary framework applying a style of nonspatial and spatial methods. Similar to social differentiation, as well as the role of power and politics, the analytical context puts more emphasis on bound factors that need to be expressly thought of. Moreover, the past examination into the root causes as well as the environmental effect of land management and land-use changes and that of the LUCID project has revealed significant procedures that play an oversized role of shifting land connections to land degradation and variety [25, 26]. Although these procedures play a very essential part in managing selections of shifting land management and land use, in broad land-use change analyses, this is typically unnoticed. Simply, how national governments can greatly influence land use by basically changing who can access what land is one of the striking findings from the LUCID root-cause examinations conducted in this research. Inspired or discouraged migration via settlement schemes developments, or though permitting individuals from other parts to access land or possess land use rights; altered the regulations of land tenure, like privatizing past public land, changing traditional land tenure systems that lead to changed rights over land or group ranches delimitation; and gazetted or degazetted as secured areas, changing who gets access to what land has been the case elsewhere for civil strife and insecurity. The civil strife and overall insecurity are regarded as temporary phenomena in root causes and land use, despite the fact that they may have a key effect on land use and land management. This unfortunately is necessarily not the case, and long-term and regional analyses must often consider their impacts. The effect on land use as well as management in the LUCID research relied on the length and severity of the unrest, but it also included the reduced investment of households in developing land as well as delay in government

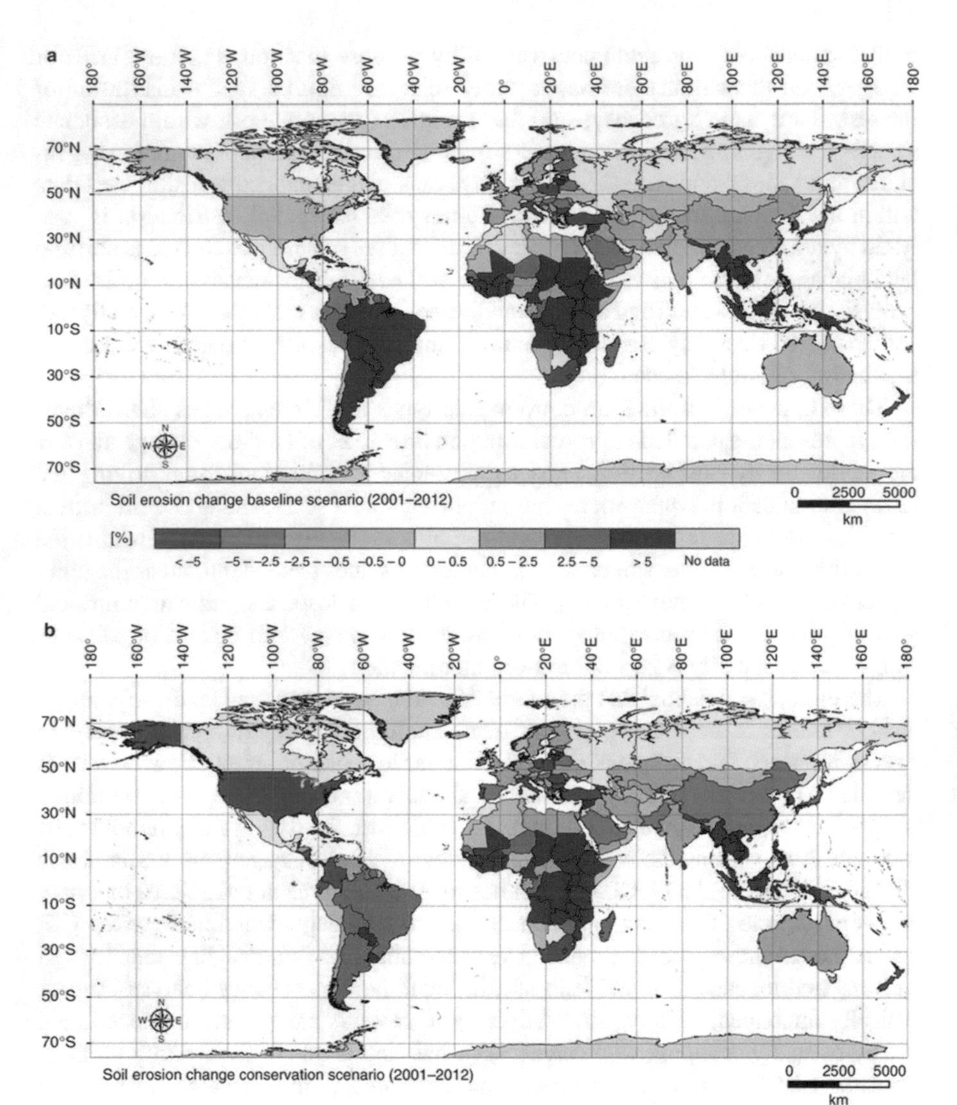

Fig. 20.3 Country-specific changes of the annual average soil erosion. Based on the above, panels (**a**) and (**b**) have a common legend, with the chromatic scale representing the decrease or increase of the yearly average rates of soil erosion in percentage. The rates are acquired through comparing the pixel-based values in each of the 202 countries that are under study. Panel (**a**) shows soil erosion change amid 2001 and 2012 based on the baseline scenario. The existing delta amid the two periods being observed relies only on LULC change outline joining the agricultural inventory information with the satellite-derived LULC data. Panel (**b**) displays soil erosion alteration amid 2001 and 2012 based on the conservation scenario. The percentage change of soil erosion, in this instance, is caused by the extenuating impact of soil preservation practices as well as the joined impact of LULC modification

investment in the construction of roads and other infrastructure causing slower economic growth and land-use modification than expected; farmers' out-migration to capital city and local urban areas; and uncertainty of trade in agricultural products as well as other commodities that impact income diversification. It is because within the rural areas, households often get involved in earning income from nonagricultural sources. Usually, this involves the adults, especially men leaving their farms to find work in other places like large farms or urban centers, or family members engaging in local off-farm work like service, commerce, or small-scale manufacturing. This can have a great impact on the local land use given that labor is pulled out from farms, and as a result cultivation is less intensive or extensive, as limited labor is invested in the farm as well as in the management of soil. Simply, the earth's landscape has considerably been changed by human activities including agriculture, urban development, and deforestation [27, 28]. This kind of land disturbance affects essential ecosystem services and processes, which eventually impact the environment severely causing calamities such as draught. Engaging in exhaustive agriculture can possibly cause severe environmental consequences. For instance, deforestation together with agriculture, urban sprawl, as well as other human activities has considerably changed and broken the vegetative cover of the earth. This kind of disturbance is altering the global atmospheric concentration of the heat-trapping gas and affecting the global climate through altering the energy equilibrium on the earth surface. These issues include wildlife habitat loss, pollution of water, and pollution of air. In addition to nutrient sediments, urban runoff often has toxic contaminants, which can not only pollute water but also cause variation in temperatures and stream flow that causes habitat destruction as well as species extinctions [29–31].

Conclusion

In conclusion, in several developing countries modification of land-use trends is very rapid. The course of modification and rates are fluctuating, and before corrective program and policies can be effective in the long term across wide areas, it is a must that the root-cause pattern of these trends is identified, and their effect on land biodiversity and degradation is understood. Research of this study is being based on one or the other aspects of social or ecological system to make analysis of the ecological or social trends across space to allow for root cause, process, and pattern generalization, and also permit the learning of general lessons. Considering the basic modification processes as well as the effect of their being effectual on the environment, much attention has indeed been directed toward corrective approaches to address land-use change concerns like land degradation. As a result, this specific objective is significantly improved through the adoption of a mutual analytical framework to carry out such cross-site comparisons and analysis. This report has therefore sketched one such framework, which requires the adoption of a system method to understand the connection of environmental and social processes over space and time, and eventually cause climate changes.

References

1. Bebbington, A.J., and S.P.J. Batterbury. 2001. Transnational livelihoods and landscapes: political ecologies of globalization. *Ecumene* 8 (4):369-380.
2. Box, P. 2002. Spatial Units as Agents: Making the Landscape an Equal Player in Agent-Based Simulations. In *Integrating Geographic Information Systems and Agent-Based Modeling Techniques for Simulating Social and Ecological Processes*, edited by R. H. Gimblett. New York: Methuen.
3. Ballantyne, A. P., Alden, C. B., Miller, J. B., Tans, P. P. & White, J. W. C. Increase in observed net carbon dioxide uptake by land and oceans during the past 50 years. *Nature* 488, 70–72 (2012).
4. Biot, Yvan, Piers M. Blaikie, Cecile Jackson, and Richard Palmer-Jones. 1995. Rethinking Research on Land Degradation in Developing Countries. Washington, D.C.: World Bank.
5. Arnell, N. *et al.* Climate Change 1995: Impacts, Adaptations, and Mitigation of Climate Change (eds R. T. Watson, M. C. Zinyowera, & R. H. Moss) 325–363 (Cambridge University Press, 1996).
6. Evans, T.P., and E.F. Moran. 2002. Spatial integration of social and biophysical factors related to landcover change. *Population and Development Review* 28 (Supplement).
7. FAO. 2003. Data Sets, Indicators and Methods to Assess Land Degradation in Drylands: Report of the LADA e-mail Conference 9 October–4 November 2002. In *World Soil Resources Reports*. Rome: Food and Agriculture Organisation.
8. Bagnall, J.P., Malia, L., Lubben, A.T., Kasprzyk-Hordern, B., 2012. Stereoselective biodegradation of amphetamine and methamphetamine in river microcosms. Water Res. 47, 5708-5718.
9. Barbier, Edward B. 1997. The economic determinants of land degradation in developing countries. *Phil. Trans. R Soc. London* 352:891-899.
10. Bridges, E.M., and L.R. Oldeman. 1999. Global assessment of human-induced soil degradation. *Arid Soil Research and Rehabilitation* 13 (4):319-325.
11. Hossain, Md. Faruque (2016). Theory of Global Cooling. Energy, Sustainability, and Society. 6:24. (Springer).
12. Vörösmarty, C., Fekete, B., Meybeck, M. & Lammers, R. Global system of rivers: Its role in organizing continental land mass and defining land to ocean linkages. *Global Biogeochem. Cycles* **14**, 599–621 (2000).
13. Bauer, J. E. *et al.* The changing carbon cycle of the coastal ocean. *Nature* **504**, 61–70 (2013).
14. Ewel, K.C. 2001. Natural resource management: the need for interdisciplinary collaboration. *Ecosystems* 4:716-722.
15. Fairhead, James, and Melissa Leach. 1996. *Misreading the African Landscape: Society and Ecology in a Forest-Savanna Mosaic*. Cambridge: Cambridge University Press.
16. Geist, H.J., and E.F. Lambin. 2001. What Drives Tropical Deforestation? A Meta-Analysis of Proximate and Underlying Causes of Deforestation Based on Subnational Case Study Evidence. Louvain-la-Neuve, Belgium: LUCC International Project Office, University of Louvain.
17. Achard, F. *et al.* Determination of tropical deforestation rates and related carbon losses from 1990 to 2010. *Glob. Change Biol.* **20**, 2540–2554 (2014).
18. Diouf, A., and E. F. Lambin. 2001. Monitoring land-cover changes in semi-arid regions: Remote sensing data and field observations in the Ferlo, Senegal. *Journal of Arid Environments* 48:129–148.
19. Campbell, David J. 1998. Towards an Analytical Framework for Land Use Change. In *Carbon and Nutrient Dynamics in Tropical Agro-Ecosystems*, edited by Bergstrom and Kirschmann. Wallingford, UK: CAB International.
20. Lambin, E.F., M.D.A. Rounsevell, and H.J. Geist. 2000. Are Agricultural Land-Use Models Able to Predict Changes in Land Use Intensity? *Agriculture, Ecosystems and Environment* 82:321-331.

21. Runnstrom, M. C. 2003. Rangeland Development of the Mu Us Sandy Land in Semiarid China: An Analysis using Landsat and NOAA Remote Sensing Data. *Land Degradation and Development* 14:189–202.
22. de Sherbinin, A. 2002. Land-Use and Land-Cover Change. In *A CIESIN Thematic Guide*. Palisades, NY: Center for International Earth Science Information Network of Columbia University.
23. Hossain, Md. Faruque, and Mukai, Hiroshi (2000). Importance of Nutrients (N, P, and NO3+NO2) in Growth of the Surfgrass, Phyllospadix iwatensis Makino.
24. van der Werf, G. R. *et al.* Climate regulation of fire emissions and deforestation in equatorial Asia. *Proc. Natl. Acad. Sci. U.S.A.* **105**, 20350–20355 (2008).
25. Irwin, Elena G., and Jacqueline Geoghegan. 2001. Theory, data, methods: Developing spatially explicit economic models of land use change. *Agriculture, Ecosystems and Environment* 85:7-23.
26. Qi, J., Weltz M., Huete A.R., Sorooshian S., Bryant R., Kerr Y.H., and Moran M.S. 2000. Leaf area index estimates using remotely sensed data and BRDF models in a semiarid region. *Remote Sensing of Environment* 73 (1):18-30.
27. Mills, G.A., Gravell, A., Vrana, B., Harman, C., Budzinski, H., Mazella, N., Occlka, T., 2014. Measurement of cnvironmental pollutants using passive sampling devices e an updated commentary on the current state of the art. Env. Sci. Process. Impact 16, 369-373.
28. Nagendra, Harini, Darla K. Munroe, and Jane Southworth. 2004. From pattern to process: Landscape fragmentation and the analysis of land use/land cover change. *Agriculture, Ecosystems and Environment* 101:111-115.
29. Moran, M.S., Kerr Y., Hymer D.C., and Qi J. 2002. Comparison of ERS-2 SAR and Landsat TM imagery for monitoring agricultural crop and soil conditions. *Remote Sensing of Environment* 79 (2-3):243-252.
30. Pasquale Borrelli, David A. Robinson, Larissa R. Fleischer, Emanuele Lugato et al. "An assessment of the global impact of 21st century land use change on soil erosion", Nature Communications, 2017
31. Pijanowski, B.C., D. G. Brown, G. Manik, and B. Shellito. 2002. Using Neural Nets and GIS to Forecast Land Use Changes: A Land Transformation Model. In *Computers, Environment and Urban Systems*.

Part VIII
Sustainable Planet

Chapter 21
Climate Control

Abstract A theory is being proposed to control the global climate change by conducting natural heat and cooling the earth surface in order to help all living beings to live in a relentless comfort temperature condition throughout the seasons. To achieve this, photon particle has been remodeled by implementing Bose-Einstein (*B-E*) dormant photonic dynamics of the earth surface plane. Simply, the proposed decoded *B-E* photons will be induced by the photonic bandgap of earth surface to convert the solar photons into cooling-state photons, here named as the *Hossain cooling photon* (HcP^-) which eventually will cool the earth surface. Interestingly, this HcP^- could also be converted into thermostated photons named as the *Hossain thermal photon* (HtP^-) by implementing Higgs boson ($H \rightarrow \gamma\gamma^-$) electromagnetic quantum fields utilized by earth's electromagnetic force. Because $H \rightarrow \gamma\gamma^-$ quantum field of earth surface plane has the extreme small length and weak force which will enforce the electrically charged HcP^- quanta to get voracious to convert it into HtP^- in order to heat the earth surface naturally. The formation of HcP^- from the photon particles and then the conversion of HcP^- to HtP^- have been proved by a set of mathematical tests which reveal the feasibility of the deformed photons (HcP^- and HtP^-) that are actively doable on the earth surface to cool and heat the earth naturally for ultimately controlling the global climate change.

Keywords Reformation of Bose · Einstein photon dynamics · Higgs boson BR ($H \rightarrow \gamma\gamma^-$) quantum fields · Hossain cooling photon (HcP^-) · Hossain thermal photon (HtP^-) · Control of earth temperature

Introduction

Global heating and cooling mechanism during the season varies tremendously due to the anthropogenic activities throughout the world which causes adverse effect on the comfortable life of all living beings on earth. Consequently, trillions of dollars are being spent each year to make our houses, offices, and premises comfortable

M. F. Hossain, *Global Sustainability in Energy, Building, Infrastructure, Transportation, and Water Technology*,
https://doi.org/10.1007/978-3-030-62376-0_21

which in fact is creating adverse environmental and atmospheric vulnerability. This is because conventional heating mechanism devours fossil energy that delivers CO_2, which is the primary player in climate change. Consequently, it causes severe environmental and atmospheric nonequilibrium, and results in tremendous fluctuation of temperature throughout the session [1–3]. At the same time, traditional cooling systems form chlorofluorocarbons (CFCs), which creates holes in the ozone layer, a protection shield that defends the maximum high-frequency ultraviolet rays of the Sun in order to reduce the cause of skin cancer and severe reproduction problem in all mammalians [4–6]. To mitigate this dangerous environmental and atmospheric vulnerability, in this study, I propose a natural cooling and heating mechanism for earth surface by implementing Bose-Einstein photon dynamics and Higgs boson ($H \rightarrow \gamma\gamma^{-}$) quantum. Simply this novel mechanism will decode the photons (solar energy) into cooling-state photons once photon flux from solar irradiance penetrates the earth surface. This is because solar photons will be absorbed by nano-point-break waveguides of photon band edges (PBEs) by utilizing the quantum electrodynamics (QED) to form cooling-state photons in order to cool the earth surface naturally [7, 8]. Mediated by earth surface, the cooling-state photons can then be converted into the heating-state photons via the photonic radiations (PR) irradiated by the quantum of Higgs bosons ($H \rightarrow \gamma\gamma^{-}$) by utilizing the electromagnetic fields of earth to heat the earth surface naturally [9–11]. This cooling and heating conversion mechanism of earth surface is indeed a noble science to control global climate change naturally and mitigate the global energy demand and environmental perplexity dramatically for securing a sustainable earth.

Methods and Simulation

Cooling Mechanism of Earth Surface

Once the solar photons penetrate the earth surface it would be deformed into the cooling-state photons by the cliques of nano-point-break waveguide of earth surface plane by creating point defects in the photon particle [12–14]. Simply, the cluster of photon bandgap (PBG) waveguides will be defected once the photon particles hit the earth plane surface [15–17]. These nano-point defect and PBG waveguide will deform the quantum dynamic mechanism of photon in order to transform it into cooling-state photons on the earth surface to cool the earth naturally. This research work provides a mechanism to form cooling-state photons using the transformation of thermal photons by using MATLAB software. The calculation revealed that the thermal waveguides that are embedded in the earth surface plane act as photons' reservoirs in order to conduct electrodynamics of the cool-state photon which can be calculated by the following *Hossain* equation [18–20]:

$$H = C + \Sigma \omega_{ci} a_i^{\dagger} a_i + \sum_{K} \omega_k b_k^{\dagger} b_k + \sum_{ik} \left(V_{ik} a^{\dagger} b_k + V_{ik}^{*} b_k^{\dagger} a_i \right), \tag{21.1}$$

where $a_i\left(a_i^{\dagger}\right)$ and $b_k\left(b_k^{\dagger}\right)$ present the operator of the nano-point-break modules and the photodynamic modules of the photons' nanostructures, and the co-efficient, V_{ik}, presents the magnitude of the photon modules within the nano-breakpoint and photon nanostructures. Figure 21.1 shows the transmission contours and the spectrum of photon plane wave and pulses to confirm the calculation of cooling state of photons to cool the earth surface.

Simply cooling-state photonic modules are proposed here that form HcP^- from solar irradiance to cool the earth surface, considering the internuclear distances, where Q_1 denotes dissociation of $H(n = 1) + H(n = 2, \ldots, \infty)$ and Q_2 into the $H(n = 2, l = 1) + H(n = 2, \ldots, \infty)$, where n and l are, respectively, the principal and angular momentum quantum numbers of the cooling-state photons. The following photon energy-level diagram considering time (min), temperature (°C), and photon intensity (a.u.) by the module which are converted from the solar photons and by the mechanism of dissociative photons confirms the photon point-break activation to cool the solar photons into cooling-state photons (Fig. 21.2).

This cooling-state photon can be clarified by their current and volt (I–V) characteristics as expressed by

$$I = I_{\mathrm{L}} - I_{\mathrm{O}}\left\{\exp^{\left[\frac{q(V+I_{\mathrm{Rs}})}{AkT_{\mathrm{C}}}\right]} - 1\right\} - \frac{\left(V + I_{\mathrm{Rs}}\right)}{R_{\mathrm{s}}}, \tag{21.2}$$

where I_{L} presents the cooling-state photons' current generation, I_{o} presents the unsaturated current, and R_{s} denotes the series resistances. Here A represents the inactive mode of the diodes, k (= 1.38×10^{-23} W/m^2 K) is the Boltzmann's constant, q (= 1.6×10^{-19} C) considers the function of active charges of the electrons, and T_{c} defines the dynamic function of cell thermal condition. Consequently, the I–q function in the photonic cells will be varied within the diode, which can be expressed by [21–23]

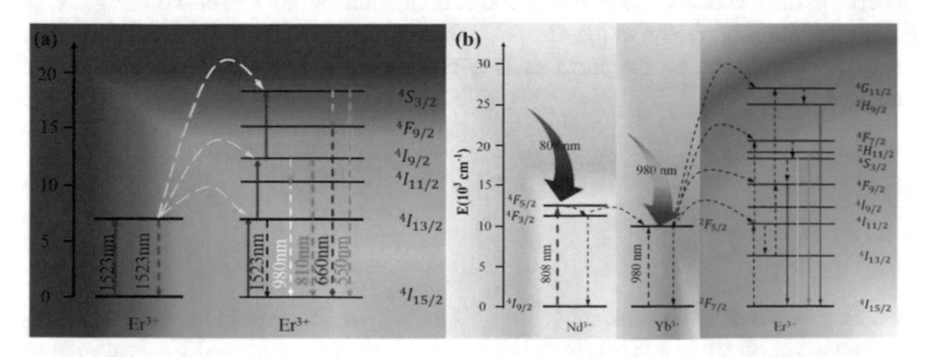

Fig. 21.1 **(a, b)** The array of the photonic probable density (standardized to its pick value) while the functions of x and t act on the occurrence of the probable distributions of the reflection and transmission pulse of heating photons

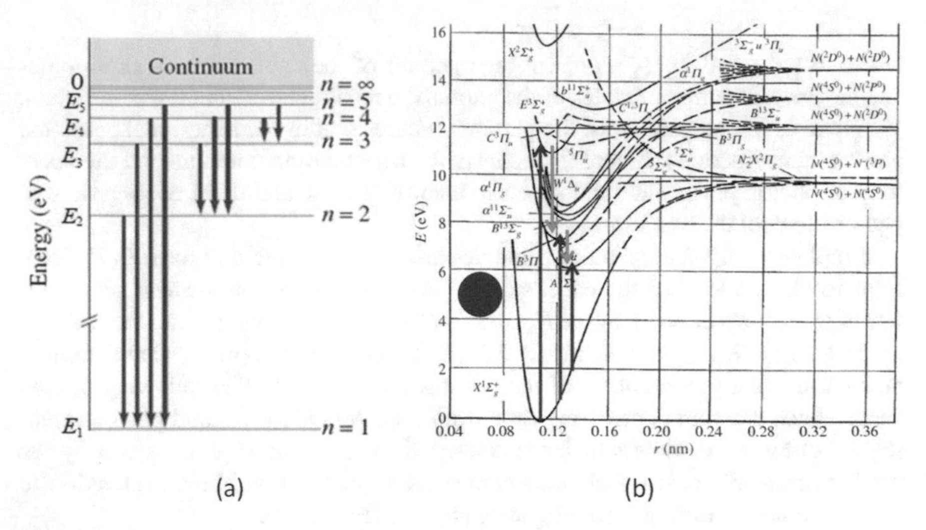

Fig. 21.2 Transformation mechanism of photon energy-level diagram and dissociative photons. (**a**) Total cooling photonic energy systems as a function of internuclear distance (a.u., atomic units). Red and blue are the cooling series of excited states of cooling photon with $^1\Pi_u$ symmetry where there are semiclassical pathways for dissociative cooling ionization by absorption of cooling photons. (**b**) Direct cooling ionization leading to cooling-state photons leading to cooling energy through the lowest cooling excited states leading to cooling of the earth surface

$$I_O = I_{Rs}\left(\frac{T_C}{T_{ref}}\right)^3 \exp\left[\frac{qEG\left(\frac{1}{T_{ref}}-\frac{1}{T_C}\right)}{KA}\right]. \tag{21.3}$$

In Eq. (21.3), IR_s defines the unsaturation current that relies on the dynamic function of temperature mode and photon speed, and qEG defines the photonic bandgap energy of the electron per unit area of the cooling-state photons in the cell (Fig. 21.3).

In this cooling photon mode, the *I–V* relations integrate the *I–V* curve of all cells in photon emission on the earth surface that are in *V–R* relationship that can be expressed by

$$V = -IR_S + K\log\left[\frac{I_L - I + I_O}{I_O}\right], \tag{21.4}$$

where K represents the constant $\left(=\frac{AkT}{q}\right)$ and I_{mo} and V_{mo} are the current and voltage on the earth surface. Therefore, the correlation between I_{mo} and V_{mo} is similar as the *I–V* relationship on the earth surface plane:

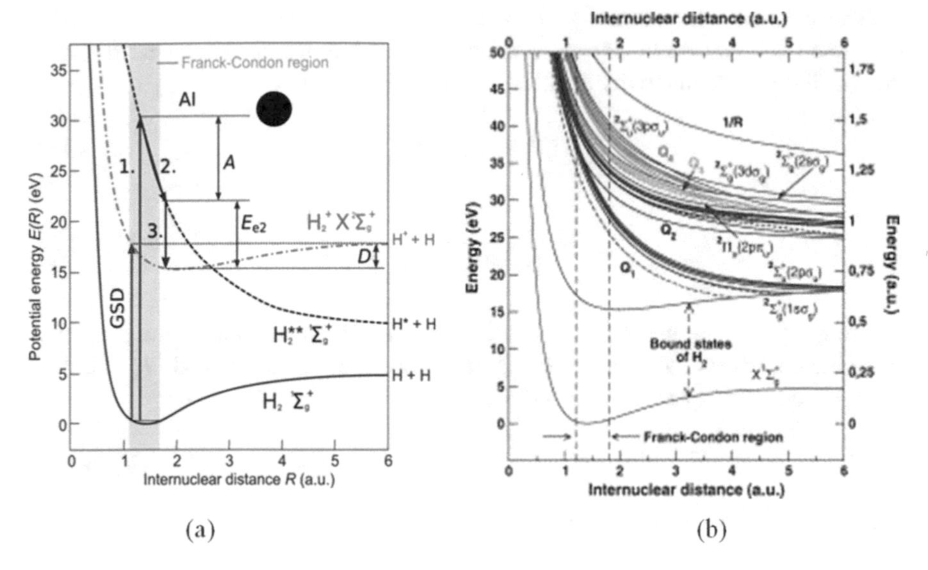

Fig. 21.3 Cooling-state photon's eV characteristic, (**a**) cooling-state continuum energy dynamic formation, (**b**) earth surface plane photon dynamics in the normalized condition

$$V_{\text{mo}} = -I_{\text{mo}} R_{\text{Smo}} + K_{\text{mo}} \log\left(\frac{I_{\text{Lmo}} - I_{\text{mo}} + I_{\text{omo}}}{I_{\text{omo}}}\right), \tag{21.5}$$

where I_{Lmo} defines the currents from the photon generation, I_{omo} is the unsaturated current into the diode, R_{Smo} represents the series resistances, and K_{mo} represents the constant. The resistance from all non-series (Ns) earth surface is connected into the series, the total resistances are expressed as $R_{\text{Smo}} = N_s \times R_s$, and the constant is $K_{\text{mo}} = N_s \times K$. The current dynamics into the series-connected earth surface is the same in each component, i.e., $I_{\text{omo}} = I_o$ and $I_{\text{Lmo}} = I_L$. Thus, the I_{mo}–V_{mo} relationship in the N_s connected cells is given by

$$V_{\text{mo}} = -I_{\text{mo}} N_S R_S + N_S K \log\left(\frac{I_L - I_{\text{mo}} + I_o}{I_o}\right). \tag{21.6}$$

Accordingly, once all N_p are interconnected in parallel to the earth surface, the I_{mo}–V_{mo} relationship is considered as [8, 24]

$$V_{\text{mo}} = -I_{\text{mo}} \frac{R_s}{N_p} + K \log\left(\frac{N_{\text{sh}} I_L - I_{\text{mo}} + N_P I_o}{N_P I_o}\right). \tag{21.7}$$

This is because the photon-generated cooling thermal conductivity relies mainly on photon flux and relativistic thermal state of the photon emission on earth surface, and the thermal conductivity could be expressed as follows:

$$I_{\mathrm{L}} = G\left[I_{\mathrm{sc}} + K_{\mathrm{I}}\left(T_{\mathrm{cool}}\right)\right]V_{\mathrm{mo}}, \tag{21.8}$$

$$T_{\mathrm{cool}} = \left(\frac{I_{\mathrm{L}}}{\left(G * V_{\mathrm{mo}}\right)\times\left(I_{\mathrm{sc}} + K_{\mathrm{I}}\right)}\right),$$

where I_{sc} is the photonic currents per unit area at 25 °C, K_{I} defines the relativistic photonic coefficient, T_{cool} is the cooling thermal of the photons, and G is the solar thermal conductivity per unit area [25, 26].

Heating Mechanism of Earth Surface

To convert cooling-state photon into heating-state photon the local Higgs quantum field on the surface of earth plane has been used. Thus, I have simulated albanian local symmetries by implementing Higgs boson electromagnetic field of earth to create comfortable heat on earth surface once it is needed [27–29]. Simply, due to the penetration of solar light into the earth surface, it will break the gauge field symmetries of earth plane, and the Goldstone scalar particles would become the longitude modes of the vector Boson [24, 30, 31] where the local symmetries of each spontaneously break down the particle T^{α} with respect to gauge field of $A_{\mu}^{\alpha}(x)$. This Higgs quantum field will begin to form into local U(1)-phase symmetries and will create heat [32, 33]. Therefore, this mechanism can be comprised of the perplex scalar fields $\Phi(x)$ of electrically charged q coupled with the electromagnetic field $A^{\mu}(x)$ that is defined by the below Lagrangian equation:

$$\mathcal{L} = -\frac{1}{4}F_{\mu\nu}F^{\mu\nu} + D_{\mu}\Phi^{*}D^{\mu}\Phi - V\left(\Phi^{*}\Phi\right), \tag{21.9}$$

where

$$D_{\mu}\Phi(x) = \partial_{\mu}\Phi(x) + iqA_{\mu}(x)\Phi(x),$$

$$D_{\mu}\Phi^{*}(x) = \partial_{\mu}\Phi^{*}(x) - iqA_{\mu}(x)\Phi^{*}(x), \tag{21.10}$$

and

$$V\left(\Phi^{*}\Phi\right) = \frac{\lambda}{2}\left(\Phi^{*}\Phi\right)^{2} + m^{2}\left(\Phi^{*}\Phi\right). \tag{21.11}$$

Here $\lambda > 0$ but $m^2 < 0$; therefore, $\Phi = 0$ is a local highest value of the scalar potentials, and the minimum form of deformed circle is $\Phi = \frac{v}{\sqrt{2}} * \mathrm{e}^{i\theta}$ with respect to the following equation:

$$v = \sqrt{\frac{-2m^2}{\lambda}} \quad \text{for any real } \theta. \tag{21.12}$$

Subsequently, the scalar fields Φ will develop a nonzero expected value $\langle \Phi \rangle \neq 0$ that instinctually will create the U(1) symmetries into the electromagnetic field of earth. The deformation of these symmetries will thus create a massless Goldstone scalar from the phase of the perplex fields $\Phi(x)$. Nevertheless, in local U(1) symmetries, the phase of $\Phi(x)$ is sx-dependent phase of the dynamic $\Phi(x)$ fields rather than the phase of the expected value of earth surface $\langle \Phi \rangle$.

To determine this heating strategy, I have expressed the scalar field spaces with respect to the polar coordinate:

$$\Phi(x) = \frac{1}{\sqrt{2}} \Phi_{\mathrm{r}}(x) * \mathrm{e}^{i\Theta(x)}, \quad \text{real}\, \Phi_{\mathrm{r}}(x) > 0, \text{real}\, \Phi(x). \tag{21.13}$$

Since the scalar field in this mechanism is singular at $\Phi(x) = 0$, it would be indeed applicable in the theory of $\langle \Phi \rangle \neq 0$ since it is considered as very much adequate for instinctually broken theory, where $\Phi\langle x \rangle \neq 0$ is the expectation for most of the earth surface plane. Considering the real field of $\phi_{\mathrm{r}}(x)$ and $\Theta(x)$, the scalar potential relies duly on the radial field ϕ_{r}:

$$V(\phi) - \frac{\lambda}{8} \left(\phi_{\mathrm{r}}^2 - v^2\right)^2 + \text{const.} \tag{21.14}$$

Once the radial fields are shifted by the variable scalars, $\Phi_{\mathrm{r}}(x) = v + \sigma(x)$, then it can be expressed as

$$\phi_{\mathrm{r}}^2 - v^2 = \left(v + \sigma\right)^2 - v^2 = 2v\sigma + \sigma^2, \tag{21.15}$$

$$V = \frac{\lambda}{8} \left(2v\sigma - \sigma^2\right)^2 = \frac{\lambda v^2}{2} * \sigma^2 + \frac{\lambda v}{2} * \sigma^3 + \frac{\lambda}{8} * \sigma^4. \tag{21.16}$$

Meanwhile, the covariant derivative $D_\mu \phi$ becomes

$$D_\mu \phi = \frac{1}{\sqrt{2}} \left(\partial_\mu \left(\phi_{\mathrm{r}} \mathrm{e}^{i\Theta}\right) + iqA_\mu * \phi_{\mathrm{r}} \mathrm{e}^{i\Theta}\right) = \frac{\mathrm{e}^{i\Theta}}{\sqrt{2}} \left(\partial_\mu \phi_{\mathrm{r}} + \phi_{\mathrm{r}} * i\partial_\mu \Theta + \phi_{\mathrm{r}} * iqA_\mu\right), \tag{21.17}$$

$$\begin{aligned}
\left|D_\mu \phi\right|^2 &= \frac{1}{2}\left|\partial_\mu \phi_r + \phi_r * i\partial_\mu \Theta + \phi_r * iqA_\mu\right|^2 \\
&= \frac{1}{2}\left(\partial_\mu \phi_r\right) + \frac{\phi_r^2}{2} * \left(\partial_\mu \Theta q A_\mu\right)^2 \\
&= \frac{1}{2}\left(\partial_\mu \sigma\right)^2 + \frac{\left(v+\sigma\right)^2}{2} * \left(\partial_\mu \Theta + qA_\mu\right)^2 .
\end{aligned} \tag{21.18}$$

The Lagrangian is then given by

$$\mathcal{L} = \frac{1}{2}\left(\partial_\mu \sigma\right)^2 - v\left(\sigma\right) - \frac{1}{4}F_{\mu\nu}F^{\mu\nu} + \frac{\left(v+\sigma\right)^2}{2} * \left(\partial_\mu \Theta + qA_\mu\right)^2 . \tag{21.19}$$

To conduct the heat $\left(\mathcal{L}_{\text{heat}}\right)$ into the magnetic field of earth surface into this Lagrangian, I have expanded $\mathcal{L}_{\text{heat}}$ as the energy series of the field into the earth and then I have extracted the quadratic parts by defying the free particles:

$$\mathcal{L}_{\text{heat}} = \frac{1}{2}\left(\partial_\mu \sigma\right)^2 - \frac{\lambda v^2}{2} * \sigma^2 - \frac{1}{4}F_{\mu\nu}F^{\mu\nu} + \frac{v^2}{2} * \left(qA_\mu + \partial_\mu \Theta\right)^2 . \tag{21.20}$$

Thus, it will obviously initiate to form comfortable heating photons into the quantum field of the earth plane, and the free particle (with Lagrangian $\mathcal{L}_{\text{free}}$) shall act as the real scalar particle to deliver heat on the earth surface considering intensity and wavelength (Fig. 21.4).

Results and Discussion

Cooling Mechanism of Earth Surface

To mathematically confirm the deliberation of cooling-state photons on the earth surface plane, I have determined the dynamics of photonic proliferations by merging Eqs. (21.15) and (21.16). Due to the cool unit area requirement $J(\omega)$ and the steady weak coupling limits, the earth surface plane absorbs the proliferated photons [8, 34]. Therefore, $J(\omega)$ of the quantum field areas is defined by the density of states (DOS) of photon formed on the earth surface plane by the cool photon magnitudes $V(\omega)$ within the photon band (PB). Besides, photon delivery follows the Weisskopf-Wigner theory [26, 35, 36]. Subsequently, all the proliferated *HcP*s shall pass the dynamic-state mode (1D, 2D, and 3D) in the earth surface, as shown in Table 21.1 [40–42].

In the 3D earth surface plane, Ω_C is the standard frequency cutoff which can avoid the bifurcation of the DOS. Naturally, the 1D and 2D earth surface plane conducts a sharp frequency cutoff at Ω_d to avoid negative DOS. Here, $Li_2(x)$ and

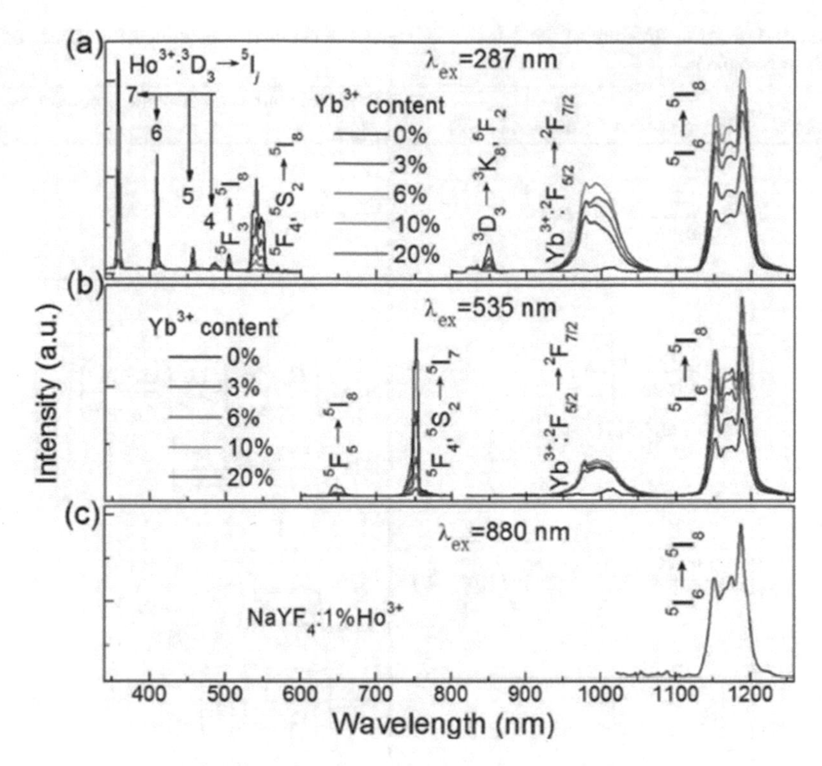

Fig. 21.4 Illustration of heating energy-level photon activation; (**a**) total energy intensity of the a.u. as a function of photon absorbance with respect to wavelengths of 287 nm, (**b**) 535 nm, and (**c**) 880 nm

$e_{rfc}(x)$ are the bi-logarithmic and additional variable functions. Consequently, the DOS of various earth surface planes, here named as $\varrho_{PC}(\omega)$, is confirmed by estimating the photonic energy frequencies and energy functions of Maxwell's theory in the nanostructure [25, 43, 44]. In a 1D earth surface plane, the DOS is thus provided by $\varrho_{PC}(\omega) \propto \frac{1}{\sqrt{\omega-\omega_e}}\Theta(\omega-\omega_e)$, where $\Theta(\omega - \omega_e)$ where ω_e denotes the frequency of PBE at the used DOS.

Then the bandgap has been used to precisely determine the qualitative states of the non-Weisskopf-Wigner modes and the cool-state photons calculated by the isotropic classification of the earth surface plane (Fig. 21.5). In the earth surface plane, the bandgap is closer to the DOS and it is determined by $\varrho_{PC}(\omega) \propto \frac{1}{\sqrt{\omega-\omega_e}}\Theta(\omega-\omega_e)$. This DOS is thereafter determined considering the electromagnetic field vectors on the earth surface [45–47]. In 2D and 1D earth surface plane, the cool-state photons' DOS shows a real logarithmic divergence close to the PBE, which is $\varrho_{PC}(\omega) \propto -[\ln|(\omega - \omega_0)/\omega_0| - 1]\Theta(\omega - \omega_e)$, where ω_e defines the primary point of the peak of DOS distribution of ground-state photon emissions [24, 30].

Table 21.1 Photon structure of the densities of states (DOS) in various dimension modes of the earth surface plane

Photons	Unit area $J(\omega)$ for different DOS	Reservoir-induced self-energy correction $\Sigma(\omega)$
1D	$\frac{C}{\pi}\frac{1}{\sqrt{\omega-\omega_e}}\Theta(\omega-\omega_e)$	$-\frac{C}{\sqrt{\omega_e-\omega}}$
2D	$-\eta\left[\ln\left\lvert\frac{\omega-\omega_0}{\omega_0}\right\rvert-1\right]$ $\Theta(\omega-\omega_e)\Theta(\Omega_d-\omega)$	$\eta\left[\mathrm{Li}_2\left(\frac{\Omega_d-\omega_0}{\omega-\omega_0}\right)-\mathrm{Li}_2\left(\frac{\omega_0-\omega_e}{\omega_0-\omega}\right)\right.$ $\left.-\ln\frac{\omega_0-\omega_e}{\Omega_d-\omega_0}\ln\frac{\omega_e-\omega}{\omega_0-\omega}\right]$
3D	$\chi\sqrt{\frac{\omega-\omega_e}{\Omega_C}}\exp\left(-\frac{\omega-\omega_e}{\Omega_C}\right)\Theta(\omega-\omega_e)$	$\chi\left[\pi\sqrt{\frac{\omega_e-\omega}{\Omega_C}}\exp\left(-\frac{\omega-\omega_e}{\Omega_C}\right)\right.$ $\left.erfc\sqrt{\frac{\omega_e-\omega}{\Omega_C}}-\sqrt{\pi}\right]$

The unit areas $J(\omega)$ and the proliferation of self-energy inductions on the earth surface reservoir $\Sigma(\omega)$, demonstrated by the photonic dynamic on the standard relativistic earth surface plane, vary among the photonic structures. The variables C, η, and χ function as coupled forces between the point break and earth surface plane in 1, 2, and 3 dimensions [37–39]

Therefore, the density of state (DOS) and the expected density of state (EDOS) of deformed photon to convert into the cool state will confirm the (1) total reproduction of DOS(T) at orbitals, (2) EDOS of the fourth level of earth surface quantum atoms, and (3) EDOS of the earth surface quantum [10, 25]. As defined above, $J(\omega)$ denotes the DOS field produced on the earth surface plane by the standard cooling photon magnitudes $V(\omega)$ in the PB and on the earth surface plane [48–50]:

$$J(\omega)=\varrho(\omega)\left|V(\omega)\right|^2. \tag{21.21}$$

Therefore, I have considered, here, the PB as the frequency of ω_c and the proliferative cooling photon dynamic is $\langle a(t)\rangle = u(t,t_0)\langle a(t_0)\rangle$, where the function $u(t,t_0)$ describes the photon structure. $u(t,t_0)$ is determined by clarifying the integral–differential equation given in Eq. (21.18):

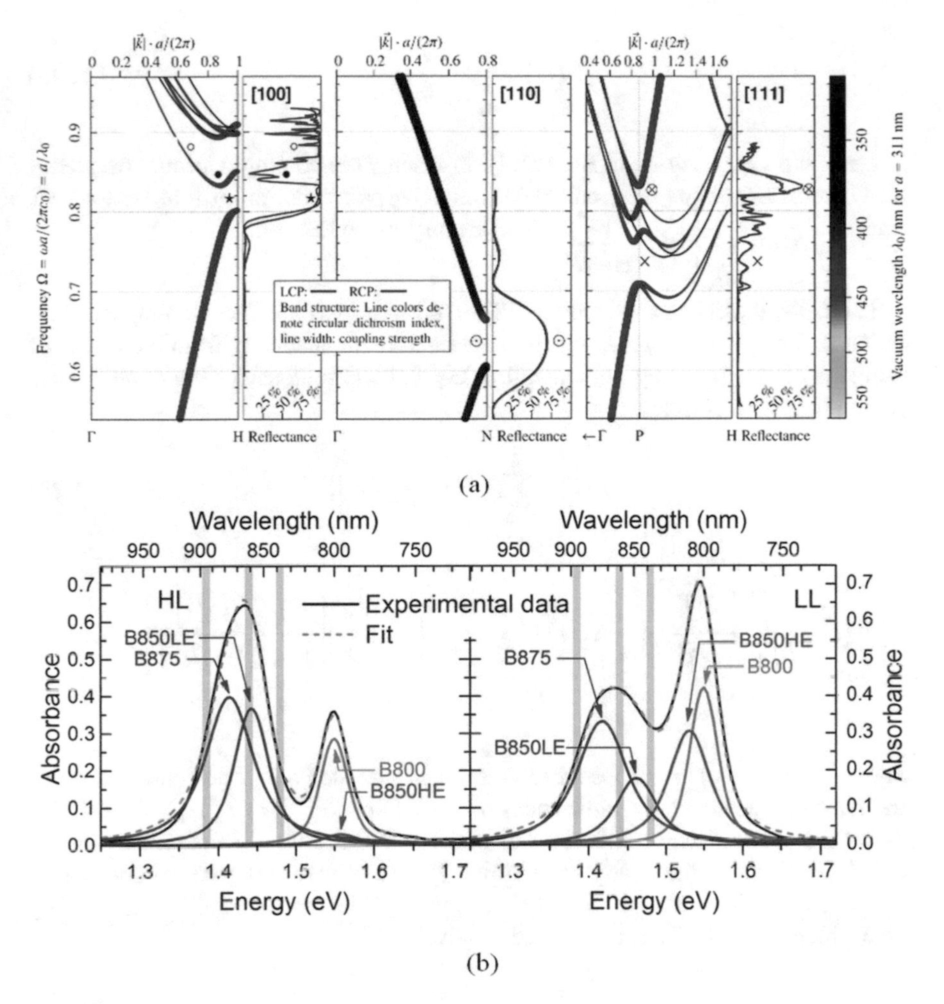

Fig. 21.5 The photon band structures and photon transformation mode. (**a**) Unit reflectance vs. frequencies at different band structures of photons on earth surface plane. (**b**) The photon modes of magnitude to deliver the thermal energy on the earth surface plane in relation to absorption and wavelength

$$u(t,t_0) = \frac{1}{1-\Sigma'(\omega_b)} e^{-i\omega(t-t_0)} + \int_{\omega_e}^{\infty} d\omega \frac{J(\omega)e^{-i\omega(t-t_0)}}{[\omega-\omega_c-\Delta(\omega)]^2+\pi^2 J^2(\omega)}, \quad (21.22)$$

where $\Sigma'(\omega_b) = [\partial\Sigma(\omega)/\partial\omega]_{\omega=\omega_b}$ and $\Sigma(\omega)$ defines the PB photonic self-energy corrections inducted into the reservoir:

$$\Sigma(\omega)=\int_{\omega_e}^{\infty}d\omega'\frac{J(\omega')}{\omega-\omega'}. \tag{21.23}$$

Here, the frequency ω_b in Eq. (21.2) represents the cooling photonic frequency mode in the PBG ($0<\omega_b<\omega_e$), calculated under the pole condition $\omega_b-\omega_c-\Delta(\omega_b)=0$, where $\lesssim\Delta(\omega)=\mathcal{P}\left[\int d\omega'\frac{J(\omega')}{\omega-\omega'}\right]$ is a primary-value integral.

In details, it can be explained that PB is referred to as the Fock cooling determination n_0, i.e., $\rho(t_0)=|n_0\rangle\langle n_0|$, which is obtained mathematically from the real-time quantum field [51–53] by implementing Eq. (21.21) to consider the cooling-state photon induction at time t:

$$\rho(t)=\sum_{n=0}^{\infty}\mathcal{P}_n^{(n_0)}(t)|n_0\rangle\langle n_0|, \tag{21.24}$$

$$\mathcal{P}_n^{(n_0)}(t)=\frac{[v(t,t)]^n}{[1+v(t,t)]^{n+1}}[1-\Omega(t)]^{n_0}\times\sum_{k=0}^{\min\{n_0,n\}}\binom{n_0}{k}\binom{n}{k}\left[\frac{1}{v(t,t)}\frac{\Omega(\mathrm{t})}{1-\Omega(\mathrm{t})}\right]^k, \tag{21.25}$$

where $\Omega(t)=\frac{|u(t,t_0)|^2}{1+v(t,t)}$. These results revealed that the Fock-state cooling photons are indeed induced into dynamic states of earth plane $\mathcal{P}_n^{(n_0)}(t)$ of $|n_0\rangle$. Figure 21.5 shows the plots of the proliferation of photon deliberation $\mathcal{P}_n^{(n_0)}(t)$ in the prime state $|n_0=5\rangle$ and in the steady-state limits $\mathcal{P}_n^{(n_0)}(t\to\infty)$. Thus, the deliberation of the generated cool-state photons shall eventually reach a nonequilibrium cooling state which will cool the earth surface finally.

Heating Mechanism of Earth Surface

In the proposed theory, I have utilized the electromagnetic field of earth surface formed by Higgs boson quantum dynamics. Here, the local U(1) gauge-variable QED shall allow another mass in terms of gauge particles $\varnothing'\to e^{i\alpha(x)}\varnothing$, which is the cool-state photons that could be converted into thermal state photons in order to heat the earth surface at a comfort level (Fig. 21.6). This mechanism can be tested by the variable derivatives with the specific transformational rules for the scalar field, written by [55–57]

$$\partial_\mu\to D_\mu=\partial_\mu=ieA_\mu \quad [\text{covariant derivatives}],$$

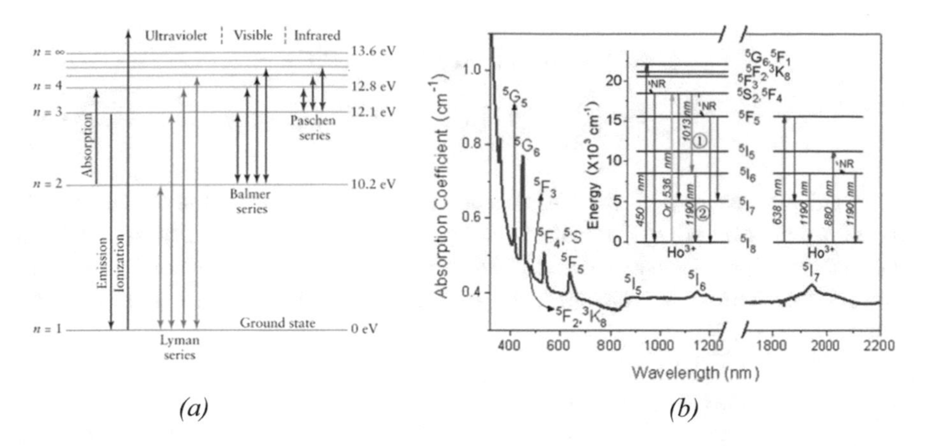

Fig. 21.6 The photon receptor where electron is heated on the surface of earth considering (**a**) transforming of the curve step to reflection symmetry step for the determination of wavelength-induced heating photon production. (**b**) Shows the functional coincidence rate of heating photon production considering the detuning parameters of wavelength and absorption coefficient into the band structure of earth plane [1, 24, 54]

$$A_{\mu}^{'} = A_{\mu} + \frac{1}{e}\partial_{\mu}\alpha \quad \left[A_{\mu} \text{ derivatives}\right]. \tag{21.26}$$

Here, the local U(1) gauge-invariant Lagrangian is considered as the perplex scalar field that is expressed by

$$\mathcal{L} = \left(D^{\mu}\right)^{\dagger}\left(D_{\mu}\varnothing\right) - \frac{1}{4}F_{\mu\nu}F^{\mu\nu} - V\left(\varnothing\right). \tag{21.27}$$

The term $\frac{1}{4}F_{\alpha\nu}F^{\alpha\nu}$ is the kinetic mechanism in the gauge field for considering thermal photons and $V(\varnothing)$ denotes the additional term written as $V(\varnothing^*\varnothing) = \mu^2(\varnothing^*\varnothing) + \lambda(\varnothing^*\varnothing)^2$.

As per the equation of the Lagrangian $\mathcal{L}$, the perturbations of the quantum field initiate the production of massive scalar particles ϕ_1 and ϕ_2 and a mass μ. Here, $\mu^2 < 0$ shall be admitted as an infinite number of quanta in order to satisfy the equation $\phi_1^2 + \phi_2^2 = -\mu^2 / \lambda = v^2$. In terms of the shifted fields η and ξ, the quantum field is clarified as $\phi_0 = \frac{1}{\sqrt{2}}\left[\left(\upsilon + \eta\right) + i\xi\right]$, and the variable derivative of the Lagrangian then confirms the following:

$$\begin{aligned} \text{Kinetic term: } \mathcal{L}_{\text{kin}}\left(\eta,\xi\right) &= \left(D^{\mu}\phi\right)^{\dagger}\left(D^{\mu}\phi\right) \\ &= \left(\partial^{\mu} + ieA^{\mu}\right)\phi^{*}\left(\partial_{\mu} - ieA_{\mu}\right)\phi \end{aligned} \tag{21.28}$$

Final term to the second order: $V(\eta, \xi) = \lambda\, \upsilon^2\eta^2$. Thus, the full Lagrangian can be written as

$$\mathcal{L}_{\text{kin}}(\eta,\xi) = \frac{1}{2}(\partial_\mu \eta)^2 - \lambda \upsilon^2 \eta^2 + \frac{1}{2}(\partial_\mu \xi)^2 - \frac{1}{4}F_{\mu\nu}F^{\mu\nu} + \frac{1}{2}e^2 \upsilon^2 A_\mu^2 - e\upsilon A_\mu (\partial^\mu \xi) + \text{int. terms.} \tag{21.29}$$

Here, η is massive, ξ is massless (as before), μ is the mass for the quanta, and A_μ is defined up to a term $\partial_\mu \alpha$, as is the evidence for Eq. (21.27). Naturally, A_μ and ϕ can be changed spontaneously, so Eq. (21.28) could be rewritten to confirm the deliberation of the thermal photon particle spectrum within the quantum field of earth surface:

$$\begin{aligned}\mathcal{L}_{\text{scalar}} &= (D^\mu \phi)^\dagger (D^\mu \phi) - V(\phi^\dagger \phi) \\ &= (\partial^\mu + ieA^\mu)\frac{1}{\sqrt{2}}(v+h)(\partial_\mu - ieA_\mu)\frac{1}{\sqrt{2}}(v+h) - V(\phi^\dagger \phi)\end{aligned} \tag{21.30}$$

$$= \frac{1}{2}(\partial_\mu h)^2 + \frac{1}{2}e^2 A_\mu^2 (v+h)^2 - \lambda v^2 h^2 - \lambda v h^3 - \frac{1}{4}\lambda h^4 + \frac{1}{4}\lambda h^4. \tag{21.31}$$

The redeveloped term of the Lagrangian of the scalar field thus revealed which Higgs boson quantum field could be initiated to produce thermal photons on the earth surface plane.

To determine this heating photon conversion on the earth surface plane, I further calculated the isotropic distributions of kinetic on the differential cones with respect to the angle θ from the vertical axis. The differential between θ and $\theta + d\theta$ is $\frac{1}{2}\sin\theta\, d\theta$. The differential photon density at energy $\in$ and angle θ is then given by

$$dn = \frac{1}{2}n(\in)\sin\theta\, d\in d\theta. \tag{21.32}$$

Naturally, the speed of function of the high-energy photons is estimated as c $(1 - \cos\theta)$ with respect to the absorption of photon per unit area considering the following equation:

$$\frac{d\tau_{\text{abs}}}{dx} = \int\int \frac{1}{2}\sigma n(\in)(1-\cos\theta)\sin\theta\, d\in d\theta. \tag{21.33}$$

Rewriting these variables as integral over s instead of θ, by (21.31) and (21.33), I have determined

$$\frac{d\tau_{\text{abs}}}{dx} = \pi r_0^2 \left(\frac{m^2c^4}{E}\right)^2 \int_{\frac{m^2c^4}{E}}^{\infty} \in^{-2} n(\in)\bar{\varphi}\left[s_0(\in)\right] de, \tag{21.34}$$

where

$$\bar{\varphi}\left[s_0(\in)\right]=\int_1^{s_0(\in)} s\bar{\sigma}(s)ds,\quad \bar{\sigma}(s)=\frac{2\sigma(s)}{\pi r_0^2}. \tag{21.35}$$

This result defines the dimensional variable $\bar{\varphi}$ and dimensionless cross section $\bar{\sigma}$. The variable $\bar{\varphi}[s_0]$ is calculated based on a detailed graphical frame for $1 < s_0 < 10$. I calculated $\bar{\varphi}$ by a functional asymptotic calculation:

$$\bar{\varphi}[s_0]=\frac{1+\beta_0^2}{1-\beta_0^2}\ln\omega_0-\beta_0^2\ln\omega_0-\ln^2\omega_0-\frac{4\beta_0}{1-\beta_0^2}+2\beta_0+4\ln\omega_0\ln(\omega_0+1)-L(\omega_0),$$

where $s_0 - 1 \ll 1$ or $s_0 \gg 1$,

$$\beta_0^2=\frac{1-1}{s_0},\quad \omega_0=\frac{(1+\beta_0)}{(1-\beta_0)},\quad \text{and } L(\omega_0)=\int_1^{\omega_0}\omega^{-1}\ln(\omega+1)d\omega. \tag{21.36}$$

The last integral can be written as

$$(\omega+1)=\omega\left(\frac{1+1}{\omega}\right),\quad L(\omega_0)=\frac{1}{2}\ln^2\omega_0+L'(\omega_0),$$

where

$$\begin{aligned} L'(\omega_0)&=\int_1^{\omega_0}\omega^{-1}\ln\left(1+\frac{1}{\omega}\right)d\omega \\ &=\frac{\pi^2}{12}-\sum_{n=1}^{\infty}(-1)n^{-1}n^{-2}\omega_0^{-n} \end{aligned} \tag{21.37}$$

This confirmed presentation of the heating photon deliberation readily shall allow the calculation of $\bar{\varphi}[s_0]$ to determine the accurate value of heating photon production s_0. Therefore, the correct functional asymptotic formula is expressed as follows to confirm the heating photon production:

$$\begin{aligned} &\bar{\varphi}[s_0]=2s_0(\ln 4s_0-2)+\ln 4s_0(\ln 4s_0-2)- \\ &\frac{(\pi^2-9)}{3}+s_0^{-1}\left(\ln 4s_0+\frac{9}{8}\right)+\cdots+(s_0\gg 1), \end{aligned} \tag{21.38}$$

$$\bar{\varphi}[s_0]=\left(\frac{2}{3}\right)(S_0-1)^{\frac{3}{2}}+\left(\frac{5}{3}\right)(S_0-1)^{\frac{5}{2}}-\left(\frac{1507}{420}\right)(S_0-1)^{\frac{7}{2}}+\cdots+(s_0-1\ll 1). \tag{21.39}$$

The function $\frac{\bar{\varphi}[s_0]}{(s_0-1)}$ is revealed in Fig. 21.5 for $1 < s_0 < 10$; at larger s_0, it defines a normal logarithm function of s_0. The energy-law spectra of the thermal photon is thus written in the form $n(\in) \propto \in^m$ for two systems in a pristine state and for a system with photonic band heating on the earth surface plane.

Therefore, the solar irradiance absorption spectrum shall confirm a peak energy cutoff with $m > 0$, which is the derived form of the thermal photon spectra at a peak energy cutoff.

Thus, the formation of the spectrum is clarified as follows:

$$n(\in) = D\in^{\beta}, \quad \in<\in_m, \quad \beta \le 0 \tag{21.40}$$

$$= 0, \quad \in>\in_m . \tag{21.41}$$

Thus, I have determined

$$\frac{d\tau_{\text{abs}}}{dx} = \pi r_0^2 D\left(\frac{m^2c^4}{E}\right)^{1+\beta} \times \begin{cases} 0, & E < E_m, \\ F_\beta(\sigma_m), & E > E_m, \end{cases} \tag{21.42}$$

where

$$\sigma_m = \frac{E}{E_m} = \frac{\in_m E}{m^2c^4}, \tag{21.43}$$

$$F_\beta(\sigma_m) = \int_1^{\sigma_m} s_0^{\beta-2}\bar{\varphi}[s_0]\,ds_0. \tag{21.44}$$

Again, by Eqs. (21.40) and (21.41), I have obtained the asymptotic forms:

$$\beta = 0: \quad F_\beta(\sigma_m) \to A_\beta + \ln^2\sigma_m - 4\ln\sigma_m + \ldots,$$

$$\beta \ne 0: \quad F_\beta(\sigma_m) \to A_\beta + 2\beta^{-1}\sigma_m^{\beta}\left(\ln 4\sigma_m - \beta^{-1} - 2\right) + \ldots, \quad \sigma_m > 10 \tag{21.45}$$

$$\text{All } \beta: \quad F_\beta(\sigma_m) \to \left(\frac{4}{15}\right)(\sigma_m - 1)^{\frac{5}{2}} + \left[\frac{2(2\beta+1)}{21}\right](\sigma_m - 1)^{\frac{7}{2}} + \ldots, \quad \sigma_m - 1 \ll 1. \tag{21.46}$$

Here, the plots $\sigma_m^{-\beta} F_\beta(\sigma_m)$ for β = 0–3.0A_β in 0.5–A_β intervals determine the integral of the region of plane [16, 19]. The value is clarified as $A_\beta = 8.111$ ($\beta = 0$), 13.53 ($\beta = 0.5$), 9.489 ($\beta = 1.0$), 15.675 ($\beta = 1.5$), 34.54 ($\beta = 2.0$), 85.29 ($\beta = 2.5$), and 222.9 ($\beta = 3.0$). Consequently, I have calculated the thermal photons in terms of photonic spectra for both negative and positive indexes:

$$n(\in)=0, \quad \in<\in_0$$
$$=C_{\in}^{-\alpha} \text{ or } D_{\in}^{\beta}, \quad \in_0<\in<\in_m \tag{21.47}$$
$$=0, \quad \in>\in_m$$

I then obtained

$$\left(\frac{d\tau_{\text{abs}}}{dx}\right)_{\alpha}=\pi r_0^2 C\left(\frac{m^2c^4}{E}\right)^{1-\alpha}$$
$$\times\begin{cases}0, & E<E_m, \\ \left[F_\alpha(1)-F_\alpha(\sigma_m)\right], & E_m<E<E_0, \\ \left[F_\alpha(\sigma_0)-F_\alpha(\sigma_m)\right], & E>E_0,\end{cases} \tag{21.48}$$

$$\left(\frac{d\tau_{\text{abs}}}{dx}\right)_{\beta}=\pi r_0^2 D\left(\frac{m^2c^4}{E}\right)^{1+\beta}$$
$$\times\begin{cases}0, & E<E_m, \\ \left[F_\beta(\sigma_{\text{m}})\right], & E_m<E<E_0, \\ \left[F_\beta(\sigma_{\text{m}})-F_\beta(\sigma_0)\right], & E>E_0.\end{cases} \tag{21.49}$$

In these functional variables, the heating photon spectra on the earth surface plane can be properly defined by asymptotic formula. The term $\Gamma_{\gamma}^{\text{LPM}}$ defines the photonic obligation to the irradiated light per unit area [53, 58, 59]:

$$\Gamma_{\gamma}\equiv\frac{dn_{\gamma}}{dVdt}. \tag{21.50}$$

Adding all the contributions $\Gamma_{\gamma}^{\text{LPM}}$, the rate of thermal photon production is confirmed as $O(\alpha_{\text{EM}}\alpha_s)$. Here, it has been calculated by the following equation for the polarized emitted rate $\Gamma_{\gamma}^{\text{LPM}}$ at the thermodynamically controlled equilibrium of the plasma surface at temperature T and photo-physical reaction μ:

$$\frac{d\Gamma_{\gamma}^{\text{LPM}}}{d^3k}=\frac{d_{\text{F}}q_{\text{s}}^2\alpha_{\text{EM}}}{4\pi^2k}\int_{-\infty}^{\infty}\frac{dp_{\parallel}}{2\pi}\int\frac{d^2p_{\perp}}{(2\pi)^2}A(p_{\parallel},k)\,\text{Re}\left\{2p_{\perp}\bullet f(p_{\perp};,p_{\parallel};,k)\right\}, \tag{21.51}$$

where d_{F} presents the variable strategy of the photon particles [N_c in $SU(N_c)$], q_s presents the albanian charge of the photonic quark $k\equiv|k|$, and the kinetic functional mode $A(p_{\parallel}, k)$ is irradiated particles which are expressed by

$$A\left(p_{\parallel},k\right)\equiv\begin{cases}\dfrac{n_{\rm b}\left(k+p_{\parallel}\right)\left[1+n_{\rm b}\left(p_{\parallel}\right)\right]}{2p_{\parallel}\left(p_{\parallel}+k\right)}, & \text{scalars,}\\[2ex] \dfrac{n_{\rm f}\left(k+p_{\parallel}\right)\left[1-n_{\rm f}\left(p_{\parallel}\right)\right]}{2\left[p_{\parallel}\left(p_{\parallel}+k\right)\right]^{2}}\left[p^{2}+\left(p_{\parallel}+k\right)^{2}\right], & \text{fermions,}\end{cases} \tag{21.52}$$

with

$$n_{\rm b}\left(p\right)\equiv\frac{1}{\exp\left[\beta\left(p-\mu\right)\right]-1},\quad n_{\rm f}\left(p\right)\equiv\frac{1}{\exp\left[\beta\left(p-\mu\right)\right]+1}. \tag{21.53}$$

The function $f(p_{\perp}; p_{\parallel}, k)$ in Eq. (21.51) is integrated to resolve the below equation that suggests that the heating photon production on the earth surface is feasible [26, 35, 58]:

$$2p_{\perp}=i\delta Ef\left(p_{\perp};,p_{\parallel};,k\right)+\frac{\pi}{2}C_{\rm F}g_{\rm s}^{2}m_{\rm D}^{2}\int\frac{d^{2}q_{\perp}}{\left(2\pi\right)^{2}}\frac{dq_{\parallel}}{2\pi}\frac{dq^{0}}{2\pi}2\pi\delta\left(q^{0}-q_{\parallel}\right)$$
$$\times\frac{T}{|q|}\left[\frac{2}{\left|q^{2}-\Pi_{\rm L}\left(Q\right)\right|^{2}}+\frac{\left[1-\left(q^{0}/\|q_{\parallel}\|\right)^{2}\right]^{2}}{\left|\left(q^{0}\right)^{2}-q^{2}-\Pi_{\rm T}\left(Q\right)\right|^{2}}\right]\left[f\left(p_{\perp};,p_{\parallel};,k\right)-f\left(q+p_{\perp};,p_{\parallel};,k\right)\right]. \tag{21.54}$$

In Eq. (21.54), $C_{\rm F}$ represents the quark [$C_{\rm F}=\left(N_{\rm c}^{2}-1\right)/2N_{\rm c}=4/3$ in QCD], $m_{\rm D}$ represents the lead-order Debye mass, and δE is considered as the energy variable among quasiparticles, which determines the photonic emissions and the state of the thermodynamic temperature equilibrium:

$$\delta E\equiv k^{0}+E_{p}sign\left(p_{\parallel}\right)-E_{p+k}sign\left(p_{\parallel}+k\right). \tag{21.55}$$

For an SU(N) gauge model with $N_{\rm s}$ complex scalars and $N_{\rm f}$ Dirac fermions, the Debye mass in the fundamental here is thus represented by the following equation [25]

$$m_{\rm D}^{2}=\frac{1}{6}\left(2N+N_{\rm s}+N_{\rm f}\right)g^{2}T^{2}+\frac{N_{\rm f}}{2\pi^{2}}g^{2}\mu^{2}. \tag{21.56}$$

In order to conduct accurate calculation further, the heating photon energy emission rate is $p_{\parallel}>0$; I therefore matriculated the distributions of $n(k+p_{\parallel})[1\pm n(p_{\parallel})]$ in the integral which contains $A(p_{\parallel}, k)$ in Eq. (21.51), which confirms the distribution of heating photon production, and it is expressed by the following equation:

$$n_{\mathrm{b}}(-p) = -\left[1+\bar{n}_{\mathrm{b}}(p)\right],\quad n_{\mathrm{f}}(-p) = \left[1-\bar{n}_{\mathrm{f}}(p)\right], \tag{21.57}$$

where $n(p) \equiv 1/[e\beta(p+\mu) \mp 1]$ is the definite antiparticle function; thus, the variable $A(p_{\|}, k)$ in this interval can be expressed by

$$A\left(p_{\|},k\right) \equiv \begin{cases} \dfrac{n_{\mathrm{b}}\left(k-\left|p_{\|}\right|\right)\bar{n}_{\mathrm{b}}\left(\left|p_{\|}\right|\right)}{2\left|p_{\|}\right|\left(k-\left|p_{\|}\right|\right)}, & \text{scalars,} \\ \dfrac{n_{\mathrm{f}}\left(k-\left|p_{\|}\right|\right)\bar{n}_{\mathrm{f}}\left(\left|p_{\|}\right|\right)}{2\left[\left|p_{\|}\right|\left(k-\left|p_{\|}\right|\right)\right]^2}\left[p_{\|}^2+\left(k-\left|p_{\|}\right|\right)^2\right], & \text{fermions,} \end{cases} \tag{21.58}$$

Thus, the energy E_p of a hard quark with momentum $|p|$ is explicitly given by

$$E_p = \sqrt{p^2+m_\infty^2} \simeq |p| + \frac{m_\infty^2}{2|p|} \simeq \left|p_{\|}\right| + \frac{p_\perp^2+m_\infty^2}{2\left|p_{\|}\right|}, \tag{21.59}$$

Here the asymptotic thermal "mass" is

$$m_\infty^2 = \frac{C_{\mathrm{f}} g^2 T^2}{4}. \tag{21.60}$$

Substituting the explicit form of E_p into definition (21.60), it is thus expressed as

$$\delta E = \left[\frac{p_\perp^2+m_\infty^2}{2}\right]\left[\frac{k}{p_{\|}\left(k+p_{\|}\right)}\right]. \tag{21.61}$$

In the above, it has been derived from the explicit forms of Eqs. (21.52) and (21.55). Figure 21.6 shows the leading-order heating photoemission rates that maximize the heating photon generation into the given electron time of flight of the heating-state photons on the earth surface plane to heat the earth naturally at a comfort level.

Conclusions

Due to the climate change, the enormous fluctuation of global temperature on the earth surface throughout the year is indeed an alarming danger since it creates uncomfortable thermal dynamics on the earth surface for all living beings against their comfortable life. Strikingly, this theory revealed a mechanism to deform the solar photons into cool-state photons (HcP^{-}) by integrating the Bose-Einstein (B-E)

photonic distribution theory to cool the earth surface naturally during the hot season. This cooling-state photon could likewise be reconverted into the thermal state photons (HtP^{-}) by the application of Higgs boson [BR ($H \rightarrow \gamma\gamma^{-}$)] quantum fields of earth surface to heat the earth surface naturally during the cold season. This mechanism of cooling and heating the earth surface naturally would indeed be the cutting-edge science to control global change totally.

Acknowledgements This research has been supported by Green Globe Technology, Inc. under the grant RD-02019-01 for building a better earth. Any findings, predictions, and conclusions described in this chapter are solely performed by the authors and it is confirmed that there is no conflict of interest for publishing this research in a suitable journal.

References

1. C. Gopal, M. Mohanraj, P. Chandramohan, P. Chandrasekar, Renewable energy source water pumping systems - A literature review, Renew. Sustainable Energy Rev. 25 (2013) 351–370 DOI: https://doi.org/10.1016/j.rser.2013.04.012.
2. G. Baur, K. Hencken, D. Trautmann, Revisiting unitarity corrections for electromagnetic processes in collisions of relativistic nuclei, Phys. Rep. 453 (2007) 1–27.
3. G. Wang, Ke Zhao, Jiangtao Shi, Wei Chen, Haiyang Zhang, Xinsheng Yang, Yong Zhao. "An iterative approach for modeling photovoltaic modules without implicit equations", Applied Energy, 2017.
4. A. Belkacem, H. Gould, B. Feinberg, R. Bossingham, W. E. Meyerhof, Semiclassical dynamics and relaxation, Phys. Rev. Lett. 71 (1993) 1514–1517.
5. A.K. Agger, A.H. Sørensen, Atomic and molecular structure and dynamics, Phys. Rev. A 55 (1997) 402–413.
6. N. Gupta, S.P. Singh, S.P. Dubey, D.K. Palwalia, Fuzzy logic controlled three-phase three-wired shunt active power filter for power quality improvement, IREE, 6 (2011) 1118–1129.
7. J. Park, H. Kim, Y. Cho, C. Shin, Simple modeling and simulation of photovoltaic panels using Matlab/Simulink, Adv. Sci. Technol. Lett. 73 (2014) 147–155.
8. J. Eichler, T. Stöhlker, Radiative electron capture in relativistic ion-atom collisions and the photoelectric effect in hydrogen-like high-Z systems, Phys. Rep. 439 (2007) 1–99.
9. E. Kamal, M. Koutb, A.A. Sobaih, B. Abozalam, An intelligent maximum power extraction algorithm for hybrid wind-diesel-storage system, Int. J. Electr. Power Energy Syst. 32 (2010) 170–177.
10. F. Martin. "Single Photon-Induced Symmetry Breaking of H2 Dissociation", Science, 02/02/2007
11. G. Sivasankar, V.S. Kumar, Improving low voltage ride through of wind generators using STATCOM under symmetric and asymmetric fault conditions, IREMOS 6 (2013) 1212–1218.
12. A.N. Celik, N. Acikgoz, Modelling and experimental verification of the operating current of mono-crystalline photovoltaic modules using four- and five-parameter models, Appl. Energy 84 (2007) 1–15.
13. A. Reinhard, T. Volz, M. Winger, A. Badolato, K.J. Hennessy, E.L. Hu, A. Imamoğlu, Strongly correlated photons on a chip, Nat. Photonics 6 (2012) 93–96.
14. A. Soedibyo, F.A. Pamuji, M. Ashari, Grid quality hybrid power system control of micro-hydro, wind turbine and fuel cell using fuzzy logic, IREMOS 6 (2013) 1271–1278.
15. B. Boukhezzar, H. Siguerdidjane, Nonlinear control with wind estimation of a DFIG variable speed wind turbine for power capture optimization, Energy Convers. Manag. 50 (2009) 885–892.

16. B. Najjari, A.B. Voitkiv, A. Artemyev, A. Surzhykov, Simultaneous electron capture and bound-free pair production in relativistic collisions of heavy nuclei with atoms, Phys. Rev. A 80 (2009) 012701.
17. B. Robyns, B. Francois, P. Degobert, J.P. Hautier, Vector control of induction machines, Springer-Verlag, London, 2012.
18. Chihhui Wu. (2012). "Metamaterial-based integrated plasmonic absorber/emitter for solar thermophotovoltaic systems", Journal of Optics.
19. G. Baur, K. Hencken, D. Trautmann, S. Sadovsky, Y. Kharlov, Dense laser-driven electron sheets as relativistic mirrors for coherent production of brilliant X-ray and γ-ray beams, Phys. Rep. 364 (2002) 359–450.
20. H. Faida, J. Saadi, Modelling, control strategy of DFIG in a wind energy system and feasibility study of a wind farm in Morocco, IREMOS 3 (2010) 1350–1362.
21. J.S. Douglas, H. Habibian, C.-L. Hung, A.V. Gorshkov, H.J. Kimble, D.E. Chang, Quantum many-body models with cold atoms coupled to photonic crystals, Nat. Photonics 9 (2015) 326–331.
22. J.J. Soon, K.S. Low, Optimizing photovoltaic model parameters for simulation, IEEE International Symposium on Industrial Electronics (2012) 1813–1818.
23. M. Laine, "Thermal 2-loop master spectral function at finite momentum", Journal of High Energy Physics, 2013.
24. K.G. Sharma, A. Bhargava, K. Gajrani, Stability analysis of DFIG based wind turbines connected to electric grid, IREMOS 6 (2013) 879–887.
25. M. F. Hossain (2018). Transformation of dark photon into sustainable energy. *International Journal of Energy and Environmental Engineering*. 9; 99-110. (Springer Nature).
26. M. F. Hossain (2017). Design and construction of ultra-relativistic collision PV panel and its application into building sector to mitigate total energy demand. Journal of Building Engineering. 9, 147-154.
27. M.S. Tame, K.R. McEnery, Ş.K. Özdemir, J. Lee, S.A. Maier, M. S. Kim, Quantum plasmonics, Nat. Phys. (2013).
28. M.W.Y. Tu, W.M. Zhang, Non-Markovian decoherence theory for a double-dot charge qubit, Phys. Rev. B 78 (2008) 235311.
29. Simon C Huot (2006). "Photon and dilepton production in supersymmetric Yang-Mills plasma", Journal of High Energy Physics.
30. M. F. Hossain (2018). Sustainable design and build. (Book; 468 Pages Monograph Published by Elsevier). ISBN: 9780128167229.
31. W.M. Zhang, P.Y. Lo, H.N. Xiong, M.W.Y. Tu, F. Nori, General Non-Markovian Dynamics W. Xiao, W.G. Dunford, A. Capal, A Novel Modeling Method for Photovoltaic Cells, 35th Annual IEEE Power Electronics Specialists Conference, Aachen, Germany (2004) 1950–1956.
32. M. F. Hossain (2018). Green Science: Advanced Building Design Technology to Mitigate Energy and Environment. *Renewable and Sustainable* Energy Reviews. 81 (2), 3051-3060. (Elsevier).
33. M. F. Hossain (2018). Photonic thermal energy control to naturally cool and heat the building. *Applied Thermal Engineering*. 131, 576–586. (Elsevier).
34. L. Langer, S.V. Poltavtsev, I.A. Yugova, M. Salewski, D.R. Yakovlev, G. Karczewski, T. Wojtowicz, I.A. Akimov, M. Bayer, Access to long-term optical memories using photon echoes retrieved from semiconductor spins. Nat. Photonics 8 (2014) 851–857.
35. M. F. Hossain (2017). Green science: Advanced building design technology to mitigate energy and environment. Renewable and Sustainable Energy Reviews. 81; 3051-3060.
36. N.D. Benavides, P.L. Chapman, Modeling the effect of voltage ripple on the power output of photovoltaic modules, IEEE Trans. Ind. Electron. 55 (2008) 2638–2643.
37. V. Cardoso. "Quasinormal modes of Schwarzschild black holes in four and higher dimensions", Physical Review D, 02/2004.
38. W.B. Yan, H. Fan, Single-photon quantum router with multiple output ports, Sci. Rep. 4 (2014) 4820.

39. Y.T. Tan, D.S. Kirschen, N. Jenkins, A model of PV generation suitable for stability analysis, IEEE Trans. Energy Convers. 19 (2004) 748–755.
40. L. Yang, S. Wang, Q. Zeng, Z. Zhang, T. Pei, Y. Li, L.M. Peng, Efficient photovoltage multiplication in carbon nanotubes, Nat. Photonics 5 (2011) 672–676.
41. M.C. Güçlü, J. Li, A.S. Umar, D.J. Ernst, M.R. Strayer, Electromagnetic lepton pair production in relativistic heavy-ion collisions, Ann. Phys. 272 (1999) 7–48.
42. Y.F. Xiao, M. Li, Y.C. Liu, Y. Li, X. Sun, Q. Gong, Asymmetric Fano resonance analysis in indirectly coupled microresonators, Phys. Rev. A 82 (2010) 065804.
43. M. F. Hossain (2018). Green science: Decoding dark photon structure to produce clean energy. *Energy Reports*. 4; 41-48. (Elsevier).
44. M. F. Hossain (2018). Photon application in the design of sustainable buildings to console global energy and environment. *Applied Thermal Engineering*. 141; 579-588. (Elsevier).
45. Szafron, Robert, and Andrzej Czarnecki (2016). “High-energy electrons from the muon decay in orbit: Radiative corrections”, Physics Letters B.
46. T. Ghennam, E.M. Berkouk, B. Francois, A vector hysteresis current control applied on three-level inverter. Application to the active and reactive power control of doubly fed induction generator based wind turbine, IREE 2 (2007) 250–259.
47. T. Pregnolato, E.H. Lee, J.D. Song, S. Stobbe, P. Lodahl, Single-photon non-linear optics with a quantum dot in a waveguide, Nat. Commun. 6 (2015) 8655.
48. M. F. Hossain (2018). Global environmental vulnerability and the survival period of all living beings on earth. *International Journal of Environmental Science and Technology*. DOI : https://doi.org/10.1007/s13762-018-1722-y (Springer Nature).
49. M. F. Hossain (2018). Photon energy amplification for the design of a micro PV panel. *International Journal of Energy Research*. DOI: https://doi.org/10.1002/er.4118. (Wiley).
50. M. F. Hossain (2017). Green science: Independent building technology to mitigate energy, environment, and climate change. Renewable and Sustainable Energy Reviews. 73; 695-705.
51. M. F. Hossain (2017). Application of advanced technology to build a vibrant environment on Planet Mars. *International Journal of Environmental Science and Technology*. 14 (12), 2709–2720. (Springer Nature).
52. M. F. Hossain (2016). Solar energy integration into advanced building design for meeting energy demand. *International Journal of Energy Research*. 40, 1293-1300. (Wiley).
53. U. Becker, N. Grün, W. Scheid, K-shell ionisation in relativistic heavy-ion collisions, J. Phys. B: At. Mol. Phys. 20 (1987) 2075.
54. Q. Li, D.Z. Xu, C.Y. Cai, C.P. Sun, Recoil effects of a motional scatterer on single-photon scattering in one dimension, Sci. Rep. 3 (2013).
55. M. F. Hossain and N. Fara (2016). Integration of wind into running vehicles to meet its total energy demand. *Energy, Ecology, and Environment*. 2(1), 35-48. (Springer Nature).
56. M. F. Hossain (2016). Production of clean energy from cyanobacterial biochemical products. *Strategic Planning for Energy and the Environment*. 3; 6-23 (Taylor and Francis).
57. Y. Zhu, X. Hu, H. Yang, Q. Gong, On-chip plasmon-induced transparency based on plasmonic coupled nanocavities, Sci. Rep. 4 (2014).
58. M. F. Hossain (2016). Theory of global cooling. energy, sustainability, and society. 6, 1-5. 20166:24.
59. W. De Soto, S.A. Klein, W.A. Beckman, Improvement and validation of a model for photovoltaic array performance, Sol. Energy 80 (2006) 78–88.

Chapter 22
Global Environmental Equilibrium

Abstract The transpiration mechanism has been proposed to reroute in order to convert this transpired water vapor into clean energy and potable water to balance the global environmental equilibrium ultimately. Since electrostatic force has the tendency to tug down the water, a static electricity force creating plastic tank has been proposed to be installed at the bottom of plants to capture the transpiration water vapor and treat it in situ by applying UV technology to meet the domestic water demand. A given amount of collected water (H_2O) is also decomposed into oxygen (O_2) and energy-state hydrogen gas (H_2) by implementing electrolysis (anode oxidation: $2H_2O(l) = O_2(g) + 4H^+(aq) + 4e^-$ $e° = +1.23$ V; cathode reduction $2H^+(aq) + 2e^- = H_2(g)$ $e° = 0.00$ V) to produce clean energy. Since the ideal potential at the water electrolysis cell ($e^{\circ}_{cell} = e^{\circ}_{cathode} - e^{\circ}_{anode}$) is −1.23 V at 77 °F and pH 0 ($[H^+] = 1.0$ M) at 77 °F with pH 7 ($[H^+] = 1.0 \times 10^{-7}$ M), the thermodynamic level of cell potential is to be acquired from ideal condition of free energy calculations by employing the formula $\Delta G° = -n\text{Fe}°$ to collect the activated hydrogen energy for domestic use to meet the energy demand.

Keywords Transpiration · Water vapor · Static electricity force · Capturing water vapor · UV technology · Potable water · Clean energy · Climate change mitigation

Introduction

Plants give O_2 and take CO_2 by the process of photosynthesis to keep the global environment in balance [1, 2]. Plants are simply the hero for the environment; unfortunately, hero plants are also the villain for the environment which plays a significant role in causing global warming. The body of plants needs water for the reaction of biochemical metabolism for its growth [3, 4]. This water is taken up by the cohesion-tension mechanism of the soil (groundwater) through the roots, and transported by osmosis through the xylem to the leaves of the plants [3, 5]. Interestingly only a mere 0.5–3% of water is used by plants for their metabolism and the rest of the water is released into the air through stomatal cells by transpiration process [6–8]. This process of transpiration is causing the largest loss of groundwater that is also causing global warming, with water vapor being a significant reason.

M. F. Hossain, *Global Sustainability in Energy, Building, Infrastructure, Transportation, and Water Technology*,
https://doi.org/10.1007/978-3-030-62376-0_22

In this research, therefore, a technology has been proposed to eliminate this water loss by diverting this transpiration mechanism by collecting this water vapor instead of allowing it to enter the air and transforming it into potable water and clean energy. Simply static electricity creator plastic tank near the plants has been proposed to be installed to trap all the water vapor as the latter is attracted by the force of static electricity [9–11]. This is because water vapor has positive and negative charges and the electrons that end up on static electrical force have a positive charge, while water molecules have a negative charge on one side; the positive charge of static electric force and negative charges of water vapor pull each other closer together; the positive side tugs the direction and forces the water to get collected in a tank and be treated in situ to meet the domestic water demand [12–14]. A certain amount of water shall be allowed to pass through an electrolysis process to produce clean energy (hydrogen) and the by-product O_2 will go into the air to balance the environment.

Calculation revealed that just five standard oak trees can meet the total water and energy demand for a small family throughout the year. Since the groundwater strata are getting to a finite level faster and global energy demand and global warming are getting to a seriously dangerous level putting earth on a vulnerable condition, these three vital issues must be resolved immediately [15–17]. Interestingly this new finding has the total solution to solve the global water, energy, and environmental crisis for the survival of this planet which will indeed open a new door in science.

Materials, Methodology, and Simulations

Generating Static Electric Forces

This research intends to create a *Hossain static electric force/field* (HSEF = $\mathfrak{H}$) that would allow water to be harvested from the air after it has been released through transpiration from plants. In the suggested model friction created by the contact of the insulator and the plastic tank will draw in water vapor [6, 7]. When attempting to incorporate the HSEF within the plastic tank, we can use MATLAB software to perform calculations of albanian local symmetry that permit the consideration of data field symmetry and also the Goldstone scalar as it relates to the longitudinal mode assumed by the vector [17, 18]. Thus, every particle T^{α} in the local symmetry that undergoes spontaneous breaking has a matching gauge field of $A_{\mu}^{\alpha}(x)$ with HSEF commencing functioning at local U(1)-phase symmetry [19–21]. The model therefore comprises a complex scalar field $\Phi(x)$ having a static electric charge of q that integrates with the EM field $A^{\mu}(x)$. The model's expression integrates $\mathfrak{H}$

$$= -\frac{1}{4} F_{\mu\nu} F^{\mu\nu} + D_{\mu}\Phi^{*} D^{\mu}\Phi - V\left(\Phi^{*}\Phi\right) \tag{22.1}$$

with

$$D_\mu \Phi(x) = \partial_\mu \Phi(x) + iqA_\mu(x)\Phi(x)$$

$$D_\mu \Phi^*(x) = \partial_\mu \Phi^*(x) - iqA_\mu(x)\Phi^*(x) \tag{22.2}$$

and

$$V(\Phi^*\Phi) = \frac{\lambda}{2}(\Phi^*\Phi)^2 + m^2(\Phi^*\Phi) \tag{22.3}$$

An assumption is made that $\lambda > 0$ but $m^2 < 0$, implying that $\Phi = 0$ is a representation of the local maximum for scalar potential with the minimal creating a general circle $\Phi = \frac{v}{\sqrt{2}} * e^{i\theta}$, whereby

$$v = \sqrt{\frac{-2m^2}{\lambda}}, \text{ any real } \theta \tag{22.4}$$

Resulting from this, the scalar field Φ creates a nonzero vacuum expectation value $\langle\Phi\rangle \neq 0$ that assists in the creation of U(1) symmetry adopted by the magnetic field. The subsequent breakdown creates a Goldstone scalar arising from the complex field $\Phi(x)$ phase. Nevertheless, with reference to the local U(1) symmetry, $\Phi(x)$ is not just covering the expectation value $\langle\Phi\rangle$ as the x-dependent variable in the dynamic $\Phi(x)$ field. Through confirmation of the mechanisms used with static electricity forces, this research uses polar coordinates within the appropriate scalar field space. Thus

$$\Phi(x) = \frac{1}{\sqrt{2}}\Phi_r(x) * e^{i\Theta(x)}, \quad \text{real } \Phi_r(x) > 0, \text{real } \Phi(x) \tag{22.5}$$

Field redefinition is still singular because $\Phi(x) = 0$. Because of this, this research did not employ it for any theories where $\langle\Phi\rangle \neq 0$. Nonetheless it would be suited to any theory which undergoes spontaneous breakage as it is predicted that $\Phi\langle x\rangle \neq 0$ will be universally observed. In terms of relation to the real fields $\Theta(x)$ and $\phi_r(x)$, the only factor impacting the scalar potential is the radial field ϕ_r:

$$V(\phi) = \frac{\lambda}{8}(\phi_r^2 - v^2)^2 + \text{const}, \tag{22.6}$$

or the radial field that is caused to shift by its VEV, $\Phi_r(x) = v + \sigma(x)$,

$$\phi_r^2 - v^2 = (v + \sigma)^2 - v^2 = 2v\sigma + \sigma^2 \tag{22.7}$$

$$V = \frac{\lambda}{8}(2v\sigma - \sigma^2)^2 = \frac{\lambda v^2}{2} * \sigma^2 + \frac{\lambda v}{2} * \sigma^3 + \frac{\lambda}{8} * \sigma^4 \tag{22.8}$$

At the same time, the covariant derivative $D_\mu \phi$ consists of

$$D_\mu \phi = \frac{1}{\sqrt{2}}\left(\partial_\mu \left(\phi_r e^{i\Theta}\right) + iqA_\mu * \phi_r e^{i\Theta}\right) = \frac{e^{i\Theta}}{\sqrt{2}}\left(\partial_\mu \phi_r + \phi_r * i\partial_\mu \Theta + \phi_r * iqA_\mu\right) \tag{22.9}$$

$$\begin{aligned} \left|D_\mu \phi\right|^2 &= \frac{1}{2}\left|\partial_\mu \phi_r + \phi_r * i\partial_\mu \Theta + \phi_r * iqA_\mu\right|^2 \\ &= \frac{1}{2}\left(\partial_\mu \phi_r\right) + \frac{\phi_r^2}{2} * \left(\partial_\mu \Theta qA_\mu\right)^2 \\ &= \frac{1}{2}\left(\partial_\mu \sigma\right)^2 + \frac{\left(v+\sigma\right)^2}{2} * \left(\partial_\mu \Theta + qA_\mu\right)^2 \end{aligned} \tag{22.10}$$

The total will be

$$\mathfrak{h} = \frac{1}{2}\left(\partial_\mu \sigma\right)^2 - v\left(\sigma\right) - \frac{1}{4} F_{\mu\nu}F^{\mu\nu} + \frac{(v+\sigma)^2}{2} * \left(\partial_\mu \Theta + qA_\mu\right)^2 \tag{22.11}$$

Through confirmation that the static electric force ($\mathfrak{h}_{sef}$) is created and included within the electrical field attributes of HSEF, the force undergoes expansion in relation to the different powers and powers derived from them with the greatest influence deriving from the quadratic section illustrating free particles:

$$\mathfrak{h}_{sef} = \frac{1}{2}\left(\partial_\mu \sigma\right)^2 - \frac{\lambda v^2}{2} * \sigma^2 - \frac{1}{4} F_{\mu\nu}F^{\mu\nu} + \frac{v^2}{2} * \left(qA_\mu + \partial_\mu \Theta\right)^2 \tag{22.12}$$

The HSEF ($\mathfrak{h}_{free}$) function derived is predicted to create a scalar particle having a positive mass^2 = λv^2 involving the fields $A_\mu(x)$ and $\Theta(x)$ to initiate a very significant force of static electricity within the electric field of the plastic tank as shown in Fig. 22.1:

On-Site Water Treatment

The water collected in the plastic tank is water vapor reverted to liquid. It does not require coagulation, sedimentation, or addition of chlorine in order to be clean. To adhere to the requirements of the US National Primary Drinking Water Standard code it simply requires UV treatment, mixing physics, and filtration. This is the easiest way to treat water using a *solar disinfection* (SODIS) system whereby a transparent container is filled with water and exposed to direct sunlight for two or three hours. When the temperature of the water reaches 50 °C under the application of UV radiation around 320 nm the inactivation process is accelerated, causing robust bacteriological disinfection. Water treated in this way can then be employed domestically (Fig. 22.2).

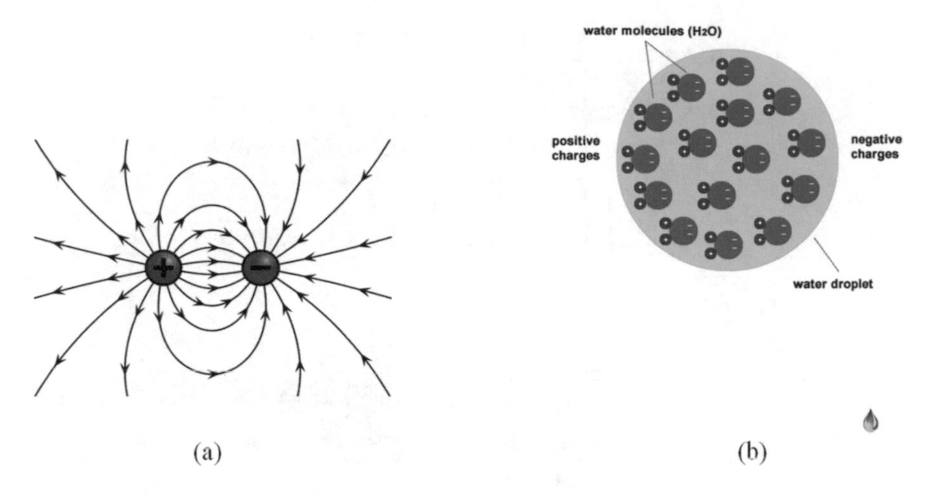

Fig. 22.1 (**a**) The static electricity force being created, and (**b**) the means whereby static energy is seen in the electromotive force combining positive and negative charges, gathering the "static" electricity together to draw in the water molecules

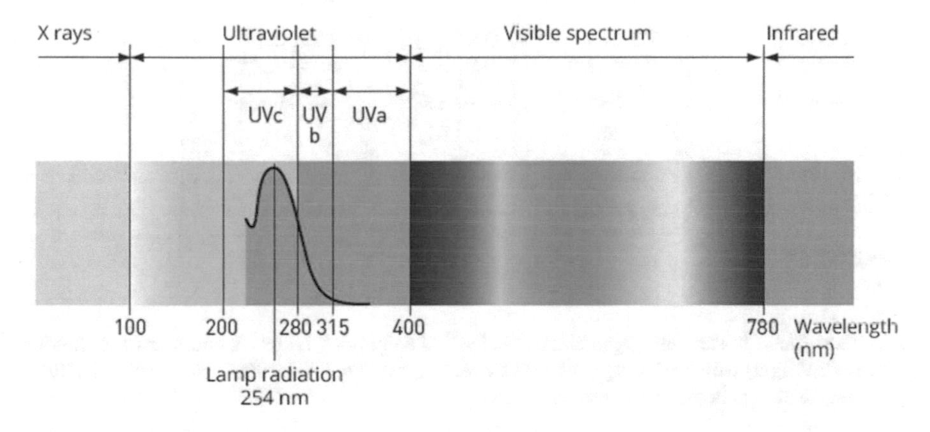

Fig. 22.2 Employing of photo-physics radiation to purify water, illustrating that when a UV radiation of 320 nm is applied on water it commences disinfection of microorganisms as soon as the temperature reaches 50 °C

Clean Energy Production

A certain proportion of water will be employed to generate clean energy using *electrolysis*, a means whereby water is converted to hydrogen energy. Instead of using standard approaches, this research proposes a direct water electrolysis system on the basis of a novel integrated monolithic photo-electrochemical (PEC)/PV device illustrated in Fig. 22.4. This device has similarities with a $GaInP_2/GaAs$ *p*/*n* tandem cell device. With this mechanism, a solid-state tandem cell comprises a GaA lower cell with a link to a $GaInP_2$ upper cell created using a tunnel diode. The upper *p*/*n*

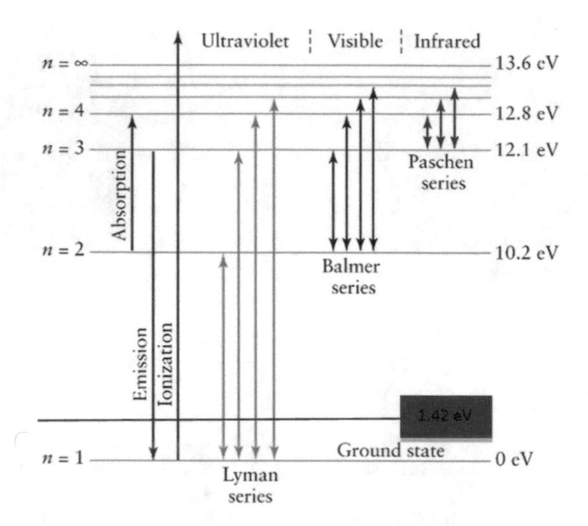

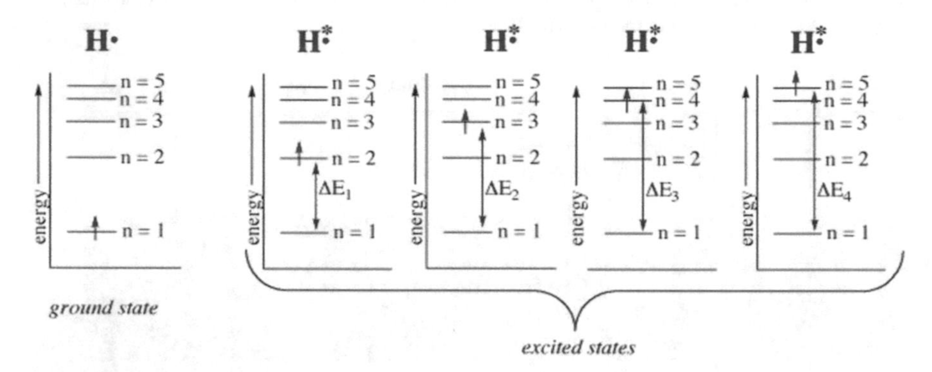

Fig. 22.3 Electron-state hydrogen energy is clarified employing radiation emissions for photon energy (UV light) with the bandgap of 1.42 eV that allows it to be clarified using photo-electrolysis that creates excited-state hydrogen

$GaInP_2$ junction with a bandgap of 1.83 eV is usually set up for absorption of the solar spectrum's visible element. With the lower *p*/*n* $GaInP_2$ junction's bandgap being around 1.42 eV, this junction is responsible for absorption of the close-to-infrared element of the spectrum passing through the junction to allow the stimulated radiant energy to undertake conduction of electrolysis as illustrated in Fig. 22.3:

Theoretically the ideal solar electrical energy efficiency for the currently combined bandgaps will be achieved if the standard solid-state tandem cell that implements the PEC Schottky-style junction is created at the highest level of the *p*/*n* junction. This would result in a voltage-biased PEC device which has an integral PV device. When illumination is applied, electrons travel toward surfaces that are illuminated and the holes travel in the direction of ohmic contact to split the water. The water splits under illumination. If this is to occur, the PEC device can only have

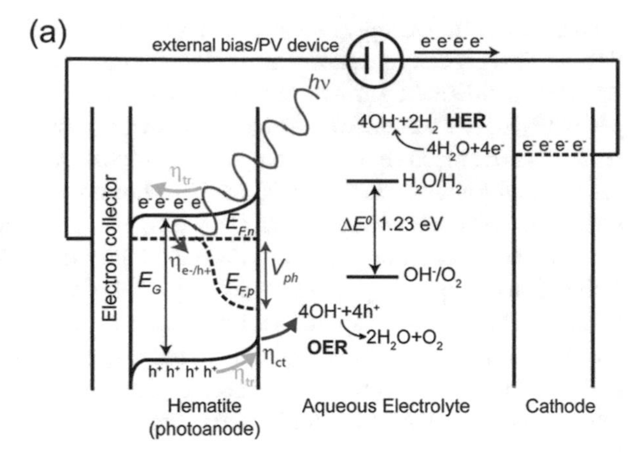

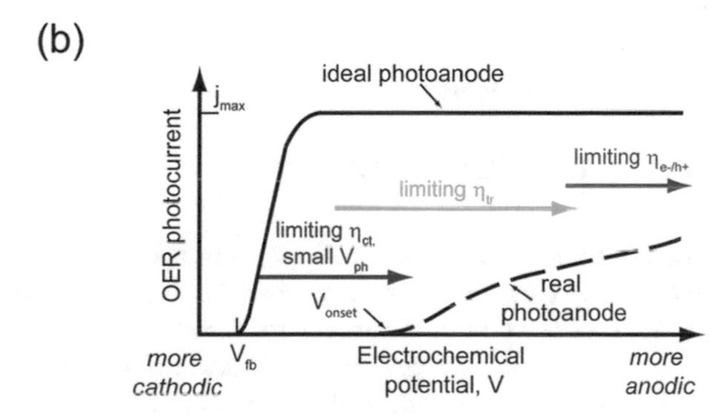

Fig. 22.4 (**a**) Monolithic bias PEC/PV device, schematic diagram; (**b**) monolithic PEC/PV photo-electrolysis device, idealized photoanode energy level

a light input. The p-$GaInP_2$/GaAs employed in this research were created employing atmospheric pressure organometallic vapor-phase epitaxy. This methodology uses an upper layer of an epitaxial cell, p-$Ga_{0.52}In_{0.48}P$ (as p-$GaInP_2$), that is 4.0 ± 0.5 μm thick. Via a low resistivity and cell-in tunnel junction (TJ) it is connected in series to a GaAs *p*/*n* lower cell based on a GaA substrate to achieve the best possible result. Standard chemical/electrochemical processes are used with the sample surface being coated with a platinum catalyst [22, 23]. A fiber-optic light source with a 150 W tungsten-halogen bulb is employed to generate the light required for photo-electrolysis. The surface light irradiance levels are measured through a calibrated PV-tandem cell mounted within the electrode holder within the cell that takes measurements when the photoelectrodes are exposed to light [7, 19].

With the PEC/PV configuration, the GaA lower cell is supplied with a high enough voltage to allow appropriate functioning of the configuration. The voltage surpasses the energetic mismatch occurring with water redox reactions at the band edges for $GaInP_2$. In addition, the need for extra voltage is factored in which is required for overcoming overvoltage loss arising from H_2/O_2 evolution reactions. The full photovoltage output comes close to the thermodynamic levels needed for natural splitting of water (1.23 V). It encompasses the polarization losses μ_a (anodic processes) and μ_c (cathodic processes). In Fig. 22.1b an idealized diagram can be seen showing the energy level at which water is split photolytically employing this device. Two photons are shown alongside a single stand-alone electron-hole pair. At the start, incidental light with the PEC/PV configuration enters the broad bandgap p-$GaInP_2$ layer that undertakes absorption of the more vigorous photons. This then stimulates the electron hole, generating photovoltage Vph1, and less vigorous photons enter through the $GaInP_2$ with the GaA bottom p/n junction absorbing them for generation of photovoltage Vph2 (Fig. 22.4). A grouping of electrons and holes recombines at the tunnel's junction. When the produced photovoltage is in excess of Vph = Vph1 + Vph2 required for photo-electrolysis with this configuration of cells, water will be driven toward the semiconductor electrode while water at the counter-electrode will only need two photons to create an electron within the external circuit and just four electrons will be needed for the generation of a single H_2 molecule, which represents the creation of clean energy.

Results/Discussion

Electrostatic Force Analysis

The efforts to create mathematically a proliferation of electric forces around our plastic tank to attract water downwards require the resolving of the issue of dynamic photon proliferation by integrating the manufactured HSEF electrical field. Thus the local U(1) gauge invariant permitted adding a mass term related to gauge particles from $\varnothing' \rightarrow e^{i\alpha(x)}\varnothing$. A covariant derivative employing a special rule for transformation offers a detailed explanation for the scalar field as shown in [24] (Fig. 22.6):

$$\partial_\mu \rightarrow D_\mu = \partial_\mu = ieA_\mu \quad \text{[covariant derivatives]}$$

$$A'_\mu = A_\mu + \frac{1}{e}\partial_\mu \alpha \quad [A_\mu \text{ derivatives}] \tag{22.13}$$

in which local U(1) gauge invariant HSEF matched with the complex scalar field as shown below:

$$\mathfrak{h} = (D^\mu)^\dagger (D_\mu \emptyset) - \frac{1}{4} F_{\mu v} F^{\mu v} - V(\emptyset) \tag{22.14}$$

Significantly, the function $\frac{1}{4} F_{\propto v} F^{\propto v}$ is representative of the kinetic term for the gauge field, i.e., heating photon, with $V(\varnothing)$ representing the extra term within the HSEF. Thus $V(\varnothing^*\varnothing) = \mu^2(\varnothing^*\varnothing) + \lambda(\varnothing^*\varnothing)^2$.

So the HSEF ($\mathfrak{h}$), being subjected to agitation in the quantum field, begins with massive scalar particles ϕ_1 and ϕ_2 combined with a mass μ. Subsequently $\mu^2 < 0$ implies measureless quantum and all of them satisfy $\phi_1^2 + \phi_2^2 = -\mu^2 / \lambda = v^2$ and the $\mathfrak{H}$ via the covariant derivatives that also employ shifted fields η and ξ defined by the quantum field as $\phi_0 = \frac{1}{\sqrt{2}}\left[(\upsilon+\eta)+i\xi\right]$.

Kinetic term is shown as $\mathfrak{H}_{kin}(\eta,\xi) = (D^{\mu}\phi)^{\dagger}(D^{\mu}\phi)$

$$= (\partial^{\mu} + ieA^{\mu})\phi^{*}(\partial_{\mu} - ieA_{\mu})\phi \qquad (22.15)$$

Potential term encompasses $V(\eta,\xi) = \lambda\, \upsilon^2\eta^2$ and is applicable as far as second order in a number of fields. Thus the full HSEF may be shown as

$$\mathfrak{H}_{kin}(\eta,\xi) = \tfrac{1}{2}(\partial_{\mu}\eta)^2 - \lambda v^2\eta^2 + \tfrac{1}{2}(\partial_{\mu}\xi)^2 - \tfrac{1}{4}F_{\mu\nu}F^{\mu\nu} + \tfrac{1}{2}e^2v^2A_{\mu}^2 - evA_{\mu}(\partial^{\mu}\xi) + \text{int. terms} \qquad (22.16)$$

Here, massive is shown as η and massless as ξ (as per previous) with a mass term for quantum A_{μ} that is constant in number as far as $\partial_{\mu}\alpha$; as Eq. (22.14) shows A_{μ} and ϕ will undergo variations at the same time and so they can be redefined, incorporating the heating photon particle spectrum falling in the quantum field; thus

$$\mathfrak{H}_{scalar} = (D^{\mu}\phi)^{\dagger}(D^{\mu}\phi) - V(\phi^{\dagger}\phi)$$

$$= (\partial^{\mu} + ieA^{\mu})\frac{1}{\sqrt{2}}(v+h)(\partial_{\mu} - ieA_{\mu})\frac{1}{\sqrt{2}}(v+h) - V(\phi^{\dagger}\phi)$$

$$= \tfrac{1}{2}(\partial_{\mu}h)^2 + \tfrac{1}{2}e^2A_{\mu}^2(v+h)^2 - \lambda v^2h^2 - \lambda vh^3 - \tfrac{1}{4}\lambda h^4 + \tfrac{1}{4}\lambda h^4 \qquad (22.17)$$

Thus, the term for expansion in the $\mathfrak{H}$ connected with the scalar field implies that HSEF electric field is prepared to propagate static electricity for creation of a quantum field able to pull water toward itself.

Confirmation that the water has been pulled down demands calculations for the isotropic spread of movement on the differential cone accounting for the angle θ and staying in a range from θ and $\theta + d\theta$ constitutes $\frac{1}{2}\sin\theta\, d\theta$ and the differential static electric force density is found through energy $\in$ and angle θ; thus

$$dn = \frac{1}{2}n(\in)\sin\theta\, d\in d\theta. \qquad (22.18)$$

So calculation of the function the high static electricity force in order to encompass the directional form of $c\ (1 - \cos\theta)$, with the absorption level of water vapor it has been integrated by the following equation

$$\frac{d\tau_{\text{abs}}}{dx} = \int\int \frac{1}{2}\sigma n(\in)(1-\cos\theta)\sin\theta\, d\in d\theta. \qquad (22.19)$$

For modification of the functions to achieve an integration instead of s or θ by (22.3) and (22.5), the following calculation is used to determine the accuracy:

$$\frac{d\tau_{abs}}{dx} = \pi r_0^2 \left(\frac{m^2c^4}{E}\right)^2 \int_{\frac{m^2c^4}{E}}^{\infty} \in^{-2} n(\in)\bar{\varphi}\left[s_0(\in)\right]de, \tag{22.20}$$

in which

$$\bar{\varphi}\left[s_0(\in)\right] = \int_1^{s_0(\in)} s\bar{\sigma}(s)ds, \quad \bar{\sigma}(s) = \frac{2\sigma(s)}{\pi r_0^2} \tag{22.21}$$

The outcome appeared as a dimension variable $\bar{\varphi}$ in combination with dimensionless cross section $\bar{\sigma}$. To calculate the variable $\bar{\varphi}\,[s_0]$ a detailed graphical frame for $1 < s_0 < 10$ was employed. Reliability of the functional asymptotic calculation for $\bar{\varphi}$ was essential so $s_0 - 1 \ll 1$ and $s_0 \gg 1$ this is expressed as

$$\bar{\varphi}[s_0] = \frac{1+\beta_0^2}{1-\beta_0^2}\ln\omega_0 - \beta_0^2\ln\omega_0 - \ln^2\omega_0 - \frac{4\beta_0}{1-\beta_0^2} + 2\beta_0 + 4\ln\omega_0\ln(\omega_0+1) - L(\omega_0),$$

in which

$$\beta_0^2 = \frac{1-1}{s_0}, \quad \omega_0 = \frac{(1+\beta_0)}{(1-\beta_0)} \tag{22.22}$$

$$L(\omega_0) = \int_1^{\omega_0} \omega^{-1}\ln(\omega+1)d\omega.$$

The final integral can appropriately be written as

$$(\omega+1) = \omega\left(\frac{1+1}{\omega}\right), \quad L(\omega_0) = \frac{1}{2}\ln^2\omega_0 + L'(\omega_0),$$

with

$$\begin{aligned} L'(\omega_0) &= \int_1^{\omega_0} \omega^{-1}\ln\left(1+\frac{1}{\omega}\right)d\omega, \\ &= \frac{\pi^2}{12} - \sum_{n=1}^{\infty}(-1)n^{-1}n^{-2}\omega_0^{-n}. \end{aligned} \tag{22.23}$$

The ultimate expression of the water pulled down through the use of the static electricity force may easily be shown through implementation of the calculations of

$\bar{\varphi}$ [s_0] for confirmation of the predicted s_0 value in harvesting water vapor. For this reason this research employs the corrective functional asymptotic formulas as follows:

$$\bar{\varphi}[s_0] = 2s_0(\ln 4s_0 - 2) + \ln 4s_0(\ln 4s_0 - 2) - \frac{(\pi^2 - 9)}{3} + s_0^{-1}\left(\ln 4s_0 + \frac{9}{8}\right) + \cdots + (s_0 \gg 1), \tag{22.24}$$

$$\bar{\varphi}[s_0] = \left(\frac{2}{3}\right)(S_0 - 1)^{\frac{3}{2}} + \left(\frac{5}{3}\right)(S_0 - 1)^{\frac{5}{2}} - \left(\frac{1507}{420}\right)(S_0 - 1)^{\frac{7}{2}} + \cdots + (s_0 - 1 \ll 1). \tag{22.25}$$

So the illustration of the function $\frac{\bar{\varphi}[s_0]}{(s_0 - 1)}$ is $1 < s_0 < 10$. With the larger s_0, there is a natural logarithmic, i.e., s_0 for confirmation of the attraction of 100% water vapor directed to thc plastic tank by HSEF.

For everyday living, an individual needs around 100 gallons of water per diem. This means that an average family of four would need 146,000 gallons of water per annum (100_{gallons}/day/person × 4_{persons} × 365_{days}). An average oak tree emits around 40,000 gallons of water (151,000 L) per annum, and so the requirements of such a family could be met by just four trees if the HSEF accomplished 100% water harvesting. With six average trees, the family could use the water from four of them for their domestic needs and the other two could generate green energy to power their home via electrolysis.

Hydrogen Energy Production Through Electrolysis

The proposed model produces H_2 using water-splitting reactions which is simple as H_2 devolution reactions generally show the smallest overvoltage losses. This reduces the requirement to employ catalysts and so allows for the ideal counter electrode to be used for more complex O_2 evolution reaction; it also demands cathodic protection of the semiconductor surface when eliminated. In Fig. 22.4 we can see the photocard-voltage curves that apply to *p*-GaInP$_2$(Pt)/TJ/GaAs and *p*-GaInP$_2$(Pt) electrodes viewed in a double-electrode configuration. In darkness, there is a predicted open-circuit voltage of −0.75 V for the *p*-GaInP$_2$(Pt)/TJ/GaAs and *p*-GaInP$_2$(Pt) electrode and ~−0.64 for the *p*-GaInP$_2$(Pt) electrode. Significantly, the dark reduction current rests within the microamp range for both electrodes. When illuminated, the *p*-GaInP$_2$(Pt) begins producing hydrogen at a 500 mV negative (0 V bias). This suggests that if the semiconductor is going to split the water it will require extra voltage from an external source. However, when the *p*-GaInP$_2$(Pt)/TJ/GaAs electrode is illuminated it shows an open-circuit voltage of ~0.55 V, implying that additional voltage has been generated by the GaAs cell. H_2 devolution

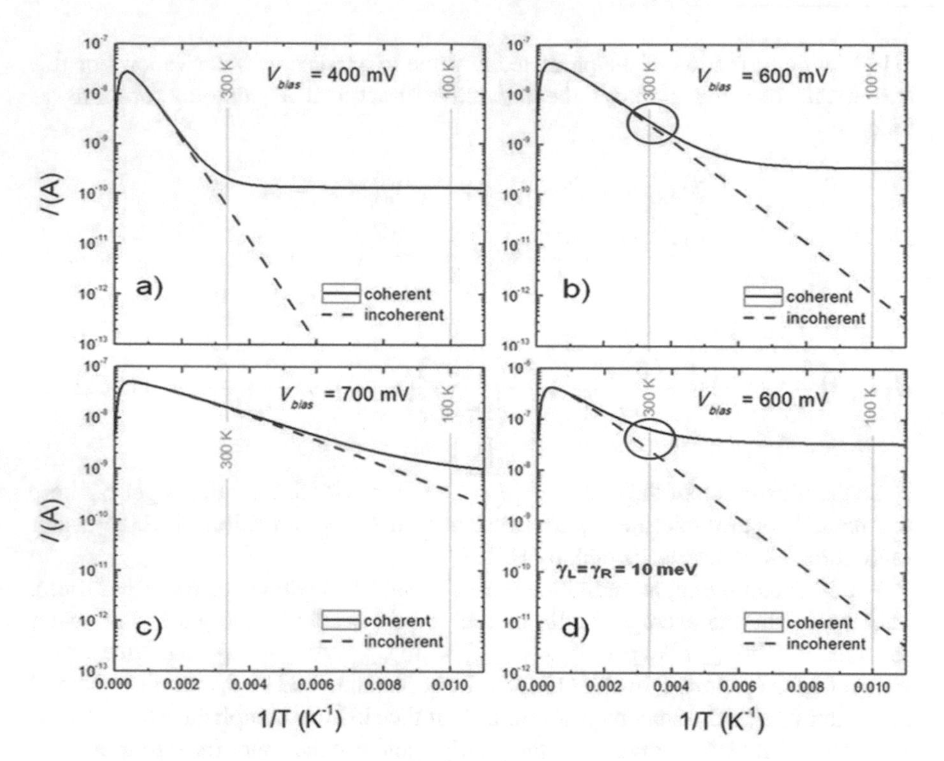

Fig. 22.5 Electrical current through a one-level molecular junction as shown in Eq. (22.18) (broadened molecular level) and Eq. (22.19) (zero-width molecular level). The calculations employ parameters including $\varepsilon = 0.4\,\text{eV}$, $\eta = 0.5$, $V_G = 0$, and $\gamma_L = \gamma_R = 1$ meV for (**a–c**) and $\gamma_L = \gamma_R = 10$ meV for (**d**). Such parameters are standard values in the molecular junction in this research

commences directly when at 400 mV positive of a short circuit. Photocurrent density may be met with limiting values of 120 mA/cm$_2$ and at ~0.15 V. This will remain constant as bias rises. As gas bubbles can grow to a size that makes them capable of becoming miniature lenses that could pit the semiconductor electrode, 0.0 1 M of the surfactant Triton X-100 is selected for the solution to allow smaller bubbles to form that would more rapidly vacate the sample surface. The photocurrent with lower saturation in the *p*-GaInP$_2$(Pt)/TJ/GaAs electrode compared to the *p*-GaInP$_2$ electrode is taken into consideration in demonstrating that the *p*/*n* GaA bottom cell creates a junction that limits the current (Fig. 22.5).

Collecting and analyzing of photo-electrolysis output will be undertaken using mass spectrometry. How efficiently H_2 is generated will be calculated using the formula efficiency = (power out)/(power in). In this equation, input power comprises an incident light intensity of around 1190 mW/cm^2 (Fig. 22.6a). Regarding output power, assuming that the photocurrent electrolysis is 100% efficient, a current for generating H_2 of 120 mA/cm^2 × 1.23 V represents the optimum fuel cell limits for 25 °C that will allow for the most efficient generation of energy from H_2. 25 °C represents the lowest value for heating with hydrogen (Fig. 22.6b).

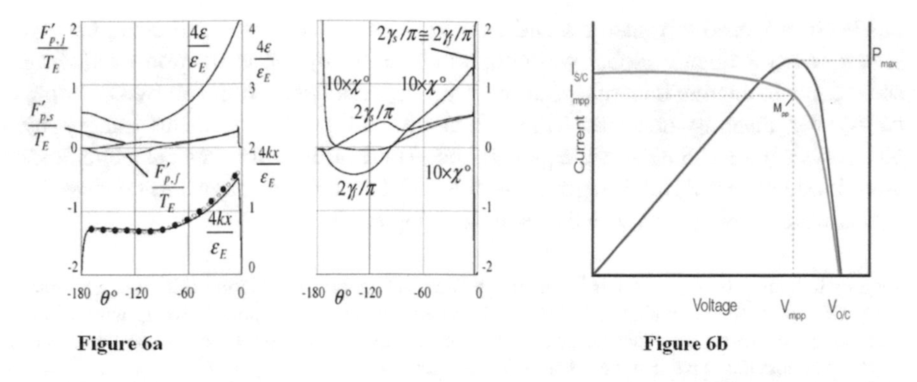

Fig. 22.6 (**a**) Current–voltage attributes of curve 1 (p-GaInP$_2$(Pt)/TJ/GaAs) and curve 2 (p-GaInP$_2$) electrodes in 3 M H_2SO_4 illuminated under white light. (**b**) Photocurrent time profile recorded at short circuit for PEC/PV tandem cell in 3 M H_2SO_4 using 0.01 M Triton X-100 illuminated under tungsten-halogen white light. An efficiency of 5120 mA/cm^2 31.23 V 3100/1190 mW/cm^2 512.4% at current density was recorded

The calculations produce an estimate that a gallon of water can produce 0.42 kg of hydrogen. With the PEM electrolyzers suggested, the significant figure is 44.5 kWh/kg H_2, producing a total yield of 16.7 kWh. Significantly, a kilowatt hour comprises 3600 kJ with the enthalpy for production of hydrogen energy deriving from liquid water at 25 °C amounting to −296 kg/mol, with a gallon of water constituting 3785 mL. Accepting that water should be 18 mL/mol, representing around 210 mol of water, the calculation of 286 × 210.3/3600 would produce an energy of 116.7 kWh. An average four-person family needs 30 kWh of energy per diem, and so 2 gallons of water would suffice to fulfill their daily energy requirements if undertaken every day, and so the quantity of water vapor produced by a small oak tree would suffice for these purposes when using electrolysis.

Conclusion

Energy, water, and environmental vulnerability are the top three problems for global environmental equilibrium where trees play a significant role in creating these problems by the process of transpiration. The hydraulic conductivity of the soil and the magnitude of the pressure gradient through the soil force the excessive flow of water moving into the plant from the roots to the leaves. Interestingly only 0.5% of this water is used for the plants' metabolism and thus the remaining water is evaporated by transpiration. As H2O molecule transpires from the surface of the plant's leaf, it pulls on the adjacent water molecule, confirming a continuous excessive flow of water through the roots. Ultimately this process causes the loss of tremendous amount of groundwater since it has been finally released as water vapor into the air which causes global warming. To mitigate these problems, transpiration mechanism has been proposed to transform and convert it into clean water and clean energy, subsequently reducing global warming significantly by the utilization of electrostatic force to capture this transpiration water vapor and treat it in situ by UV application for meeting domestic water demand and producing clean energy by electrolysis to meet the daily energy demand. Application of this natural technology would indeed be a novel, integrated, and innovative field in science to console the global water, global energy, and global warming crisis.

Acknowledgements Green Globe Technology offered support in preparing this research under the grant RD-02019-06 aiming to build a better environment. The authors came up with all the findings, assumptions, conclusions, and prediction in this chapter and there was no conflict of interest in choosing to publish the research in a journal.

References

1. Andreas Reinhard. "Strongly correlated photons on a chip", Nature Photonics, 12/18/2011.
2. Douglas, J. S., H. Habibian, C.-L. Hung, A. V. Gorshkov, H. J. Kimble, and D. E. Chang. "Quantum many-body models with cold atoms coupled to photonic crystals", Nature Photonics, 2015.
3. Zhu, Yu, Xiaoyong Hu, Hong Yang, and Qihuang Gong. "On-chip plasmon-induced transparency based on plasmonic coupled nanocavities", Scientific Reports, 2014.
4. Reed M. Maxwell, Laura E. Condon. Connections between groundwater flow and transpiration partitioning. Science 22 Jul 2016: Vol. 353, Issue 6297, pp. 377-380.
5. Leijing Yang, Sheng Wang, Qingsheng Zeng, Zhiyong Zhang, Tian Pei, Yan Li & Lian-Mao Peng (2011). Efficient photovoltage multiplication in carbon nanotubes – Nature Photonics pp. 672 – 676.
6. Tu, M. W. Y. & Zhang, W. M. Non-Markovian decoherence theory for a double-dot charge qubit. Phys. Rev. B 78, 235311 (2008).
7. Xiao, Y. F. et al. Asymmetric Fano resonance analysis in indirectly coupled microresonators. Phys. Rev. A 82, 065804 (2010).
8. Kamal, E., A. Aitouche, R. Ghorbani, and M. Bayart. "Fault Tolerant Control of Wind Energy System subject to actuator faults and time varying parameters", 2012 20th Mediterranean Conference on Control & Automation (MED), 2012.
9. G. Baur, K. Hencken, D. Trautmann. Revisiting unitarity corrections for electromagnetic processes in collisions of relativistic nuclei. Phys. Rep. 453, 1 (2007)
10. G. Baur, K. Hencken, D. Trautmann, S. Sadovsky, Y. Kharlov. Dense laser-driven electron sheets as relativistic mirrors for coherent production of brilliant X-ray and γ-ray beams. Phys. Rep. 364, 359 (2002)
11. J. Eichler, Th. Stöhlker. Radiative electron capture in relativistic ion-atom collisions and the photoelectric effect in hydrogen-like high-Z systems, Phys. Rep. 439, 1 (2007).
12. Kai Hencken. "Transverse Momentum Distribution of Vector Mesons Produced in Ultraperipheral Relativistic Heavy Ion Collisions", Physical Review Letters, 01/2006.
13. L. Langer, S. V. Poltavtsev I. A. Yugova, M. Salewski, D. R. Yakovlev, G. Karczewski, T. Wojtowicz, I. A. Akimov & M. Bayer. Access to long-term optical memories using photon echoes retrieved from semiconductor spins. Nature Photonics 8, 851–857 (2014).
14. Masujima. "Calculus of Variations: Applications", Applied Mathematical Methods in Theoretical Physics, 08/19/2009.
15. Najjari, A. B. Voitkiv, A. Artemyev, A. Surzhykov. Simultaneous electron capture and bound-free pair production in relativistic collisions of heavy nuclei with atoms, Phys. Rev. A 80, 012701 (2009).

16. Tame, M. S., K. R. McEnery, Ş. K. Özdemir, J. Lee, S. A. Maier, and M. S. Kim. "Quantum plasmonics", Nature Physics, 2013.
17. Md. Faruque Hossain. Theory of global cooling. 2016. Energy, Sustainability and Society. 6-24.
18. Zhang, W. M., Lo, P. Y., Xiong, H. N., Tu, M. W. Y. & Nori, F. General Non-Markovian Dynamics of Open Quantum Systems. Phys. Rev. Lett. 109, 170402 (2012).
19. T. Pregnolato, E. H. Lee, J. D. Song, S. Stobbe, and P. Lodahl. "Single-photon non-linear optics with a quantum dot in a waveguide", Nature Communications, 2015.
20. Yan, Wei-Bin, and Heng Fan. "Single-photon quantum router with multiple output ports", Scientific Reports, 2014.
21. O. Khaselev. "A Monolithic Photovoltaic-Photoelectrochemical Device for Hydrogen Production via Water Splitting", Science, 04/17/1998.
22. U. Becker, N. Grun, W. Scheid. K-shell ionisation in relativistic heavy-ion collisions, J. Phys. B: At. Mol. Phys. 20, 2075 (1987).
23. Scott Jasechko, Zachary D. Sharp, Peter J. Fawcett. Terrestrial water fluxes dominated by transpiration. 2013. Nature 496, 347–350.
24. Tobias D. Wheeler & Abraham D. Stroock. The transpiration of water at negative pressures in a synthetic tree. 2008. Nature 455, 208–212.

Index

A
Albanian photon charge, 99
Accumulation, 400
Acid precipitation, 106
Activated carbon, 241, 242
ADSORBSIA™, 245
Adsorption
  ADSORBSIA™, 245
  ArsenXnp, 245
  CNTs, 239, 241, 242
  conventional, 239
  fixed-bed reactors, 245
  metal-based nano adsorbents, 242–244
  nano-adsorbents, 239, 245
  polymeric nano-adsorbents, 244
  remove organic/inorganic contaminants, 239
Advanced design technology, 8
Advanced oxidation processes (AOPs), 262, 267, 303
Aerodynamic power, 181, 182, 191, 192, 218
Aerodynamic subsystem, 187
Aerodynamic system, 181
Aerodynamics force, 367, 368, 374
Afforestation, 405
Air gap flux, 208, 219, 374
Air gap flux leakage, 191
Air gas flux, 217
Air quality
  CO_2 concentration, 381
  EOF pattern, 382
Airstream energy generation module, 215
Airstream turbine, 215
Albanian electric fields, 35
Aminofullerenes, 251
Anaerobic bioreactor, 137
Anthropocentric functions, 338
Antimicrobial mechanisms, 253, 254
Antimicrobial nanomaterials, 246, 253–255
Aquaporins, 248
Aromatic compounds, 265
ArsenXnp, 245
Artificial neural network (ANN), 69
Ascertaining sediment toxicity, 400
Astrodynamical calculations, 27
Asymptotic formulas, 98
Asynchronous model, 207
Atmospheric CO_2, 130
Atmospheric pressures, 226
Automated energy torque, 374
Azimuth angle, 25

B
Barotropic (depth-averaged) water modeling, 226
Battery modeling, 195–196
  flying car design, 209, 219, 220
Benthic organisms, 399
Bernoulli-Euler beam equation, 158, 366
Betz's coefficient, 161, 372
Bioenergy, 130, 133, 134, 137, 140
Biogas, 130, 132, 137
Biological membranes, 248, 249
Biomass, 400
Bioreactor
  biogas, 137
  biowaste chamber, 131
  biowaste transformation rate, 138
  biowaste, electricity energy, 130

M. F. Hossain, *Global Sustainability in Energy, Building, Infrastructure, Transportation, and Water Technology*,
https://doi.org/10.1007/978-3-030-62376-0

Bioreactor (*cont.*)
LRFD, 130
sludge, 131
solid biowaste, 130
water dynamic force, 130
Bioswales, 152
Boltzmann's constant, 20, 47, 83, 109, 110, 120
Bose-Einstein (*B-E*) discrete photon structure, 82, 101
Bose-Einstein discrete photonic distribution, 82
Bose-Einstein photon dynamics, 418
Boussinesq approximation, 226
Bremsstrahlung, 98
Building sector
atmospheric CO_2 level, 130
environmental vulnerability, 129
fossil energy, 129
fossil fuel, 129
Building skin solar panel, 109, 112, 123

C

Capturing water vapor, 452
Carbon-based nano-adsorbents
heavy metal removal, 242
organic removal, 239, 241
regeneration and reuse cycles, 242
Carbon nanotubes (CNT), 131
Cartesian coordinate system, 15, 111, 112
Catalyst fixation, 270
Celestial body, 111
Cell's short-circuit current motion, 109
Ceramic membranes, 258
Chemical vapor deposition (CVD), 248
Chick–Watson (C–W) model, 293, 294, 296
Chlorination, 262
Chlorofluorocarbons (CFCs), 81, 418
Clean energy, 126, 337, 440, 443–446, 452
Clean energy production, 348, 349
Climate change, 149, 238, 382
CNT membranes, 248
CO_2 absorption, 385–386
CO_2 emission
and atmospheric impact, 389, 392
atmospheric, 382
balance, 383
burning, 382
fossil fuel, 383, 384, 387, 388, 390
growth rate of atmospheric CO_2 concentration, 386
land-use change, 384, 387, 388, 390
ocean CO_2 sink, 384, 385, 387, 389, 391
sequestration and absorption, 382
terrestrial vegetation and earth, 385–387, 389, 391
Cohesion-tension mechanism, 439
Colorimetric assays, 257
Commercial nanofiltration membranes, 248
Commercial utilization, 178
Community, *see* Sustainable cities
Computational fluid dynamic (CFD) models, 202
Consuming rate, 234
Contaminant bacterial ignition, 399
Contaminants, 285, 286
Control design, 178
Conventional adsorbents, 239
Conventional energy, 14, 106
Conventional indicator systems, 256
Conventional infrastructure system, 149, 152
Conventional roof, 153
Conventional water treatment, 396, 401
Cooling and heating conversion mechanism, 418
Cooling mechanism
application, 81
cooling-state photon, 82
cool-state region, 89
DOS, 89–92
energies and discharges CO_2, 82
PBG, 82, 92
PDOS, 89, 92
photodynamic modules, 82
photon element, 84
photon model, 83
photon structure, 91
photonic motion, 93
single-diode cooling state, 83, 84
solar energy, 82
Cooling mechanism, earth surface
EDOS, 426
1D and 2D earth surface, 424
photonic proliferations, 424
2D and 1D earth surface plane, 425
Cooling photon mode, 420
Cooling-state photon, 82, 418, 419, 421
Cost estimation, 167
Cost reduction, 172
Cost saving, 172
CoTe QDs, 258
Counteractive function, 96
COVID-19, 3–5, 8
Current–voltage relationship, 34
Curtain wall skin, 112
Custodial analysis, 226
Cyclobutane pyrimidine dimers (CPD), 6

D
Debye mass, 99
Defuzzification, 215
Degenerated scalar circle, 113
Degussa P-25 TiO_2, 286, 294
Demand-free condition, 293
Dendrimers, 244
Dendrimer-ultrafiltration system, 244
Densities of states (DOS), 89–92, 116, 117, 122, 424, 426
Derating factor, 359
Diode-induced constant, 109
Direct-driven dynamo, 374
Disinfection (DIS) system, 131
Disinfection and microbial control
  antimicrobial mechanisms, 253, 254
  antimicrobial nanomaterials, 253–255
  CNTs, 254
  coating techniques, 254
  DBPs, 252
  MWNTs, 254
  nano-Ag, 254
  nanomaterials, 253, 254
  nanotechnology-enabled disinfection, 254
Disinfection by-products (DBPs), 252
Disinfection process, 131
Dissipative integral–differential equation, 91
Dissolved organic carbon (DOC), 285
Dissolved oxygen (DO), 265, 283, 284
Domestic biowaste
  anaerobic co-digestion, 137
  bioenergy, 130
  building wastewater, 130
  human stool, 130
  methanogenesis process, 137
DOS dimensional modes, 29
Doubly fed induction generator (DFIG), 161, 178, 181–184, 186–189, 207, 210, 212–216
Dyes, 274
Dynabead®, 256
Dynamic electricity energy, 72
Dynamic global vegetation modeling (DGVM), 384
Dynamic photon capture, 118

E
ECs, 396, 397
Efficiency ratio, 123
Electric field, 93
Electrical rotor angular speed, 191
Electrical subsystem, 181–186, 189–192
  flying car design, 216–218
Electrical system, 181
Electrical voltage formations, 183
Electricity, 357–359
Electricity (EF) vector, 27
Electricity energy, 23, 35, 69, 135, 142
Electricity generation, 121, 136
Electricity transformation, 120
Electrochemical carbon nanotube filters, 132
Electrolysis, 443, 449–451
Electromagnetic field (EMF) vector, 117
Electromagnetic force, 158, 161
Electromagnetic force induction, 371
Electromagnetic induction, 164
Electromagnetic radiation (EM), 4–6
Electromagnetic torque, 183, 213
Electromagnetic wave, 349
Electronic torque, 190, 217, 370
Electron-state hydrogen energy, 444
Electron-state photon, 119
Electrospinning, 246
Electrospun nanofibers, 246, 258
Electrostatic force
  analysis, 123–125
  formation, 117
  generation, 113–117
  H_2O, 106
  positive charge, 106
Electrostatic force analysis, 446–449
Electrostatic force-generating plastic tank, 106
EM field, 113
Emerging contaminant, 396, 398, 399
Energy consumption, 14, 105, 158
Energy conversion, 91, 178, 181, 182, 188, 189, 191, 215
Energy cost, 359
Energy crisis, 55
Energy generation rate, 72
Environment, 339, 340
Environmental impacts, 106
Environmental pollution, 395, 401
Environmental sustainability, 152, 154, 336, 339, 340
Environmental vulnerability, 82, 129, 331, 383
Environmental water use, 233
Environmental waters, 396
Evaporation, 227
Evapotranspiration, 227
Expected density of state (EDOS), 426

F
Fiber-optic light source, 445
Field intensity, 119

Fifth Assessment Report of the Intergovernmental Panel on Climate Change (IPCC AR5), 382
Fixed-bed reactors, 245
Flux orientation system, 212
Flying car design
 aerodynamic characteristics and features, 210
 battery modeling, 209, 219, 220
 car's speed, 211
 CFD, 202
 conceptual model, 203
 DFIG and MPPT, 210
 electrical subsystem, 216–218
 generator modeling, 207, 208, 217–219
 9618 NACA series airfoils, 204
 performance in air, 202
 principles, 210
 stream velocity, 210
 3D CFD, 202
 3D *k-omega* turbulence model, 203, 204
 voltage dips, 210
 WECS, 207, 215, 216
 wind energy modeling, 204–206, 212–215
 wind speed signals, 210
 wing, 202
 wing positions and attack angles, 209
Flying transportation technology, *see* Flying car design
Fock cooling, 93, 428
Fock states, 119
Fock-state cooling photons, 428
Fock-state photon number, 119
Forward osmosis (FO), 249
Fossil fuel, 105, 106, 129, 202, 383, 384, 387, 388, 390
Fourth-order algorithms, 229
Freeze-drying of food, 226
Functional integral equation, 117
Functional materials, 246
Functional nanomaterials, 246, 258
Fuzzifications, 188
Fuzzy inference system (FIS), 188
Fuzzy logic controller (FLC), 187–189
Fuzzy lucidity checker (FLC), 215, 216

G

Gearbox, 213
Generator modeling
 flying car design, 207, 208, 217–219
Geothermal energy
 air gap flux link, 66
 algorithm, 68
 climate changing, 60
 direct-driven synchronous force, 64
 DOD and temperature, 73
 earth and biosphere, 60
 electricity energy, 65, 66
 electricity energy generation, 73
 electricity power output, 67
 energy calculation, 68
 energy charge rate, 66
 energy cofactor, 66
 energy flow calculation, 66
 energy generation, 60, 69
 feedwater pump piping, 68
 full-scale conversion, 64
 generator system, 63
 geothermal heat, 60
 greenhouse gas emission, 59
 ground-source heat pumps, 62
 in situ technology, 60
 kinetic and potential energy flow, 60
 mass fraction, 61, 62
 mechanism technology, 61
 motor rotation rate, 66
 MPP, 71
 net energy generation, 61
 one-mass model calculation, 63
 P-s diagram, 67
 shaft frequency, 64
 turbine rotor moment, 63
 volt function, 72
Glazing wall surface, 4, 6
Global carbon budget, 382
Global CO_2 concentration, 382
Global energy, 55, 82
Global environmental crisis, 59
Global environmental equilibrium
 clean energy, 440, 443–446
 electrostatic force analysis, 446–449
 hydrogen energy production, 449–451
 on-site water treatment, 442, 443
 photosynthesis, 439
 plants, 439
 potable water, 440
 static electric forces, 440–443
 transpiration, 439, 440
Global fossil fuel energy consumption, 14
Global heating and cooling mechanism, 417
Global mean temperature (GMT), 14
Global solar energy, 14, 22, 23
Global warming, 158, 325, 326
Green alleys, 150, 152–154
Green infrastructure system
 climate change, 149, 152
 conventional infrastructure system, 149, 150

conventional roof, urban area, 153
environmental sustainability, 152
green alleys, 153, 154
green roof, 150, 151, 153
smart growth practice, 151, 152
stormwater management, 149–152
stormwater runoff, 150
sustainable urban design, 151
urban forestry management, 151, 153
Green roof construction, 150, 153, 154
Greenhouse gas (GHG) emission, 106, 149, 178, 359
Greenshield's model, 162, 163
Grid-connected hybrid, 54
Groundwater analysis, 228
Groundwater loss
global warming, 325, 326
global water demand, 326
transpiration mechanism (*see* Transpiration mechanism)
Groundwater recharge, 227
Guidance force, 371
Guidance kinetic force, 367
Guideway model, 158, 366, 367

H

Heating mechanism
active variables, 95
application, 81
asymptotic formulas, 98
counteractive function, 96
Debye mass, 99
deformation mechanism, 95
electrically formed photon, 95
functional divergent, 94
heat-photon counts, 97
Higgs boson electrodynamics, 93
Higgs boson electro-quantum field, 86
internuclear distance, 100
Lagrangian function, 86–88, 94
photon deliberation, 99
photon density, 95
photon emission, 101
photonic particle, 99
positive and negative variables, 98
power-law spectrum, 97
quantum field, 94
quantum mode, 86
radial field, 87
scalar field, 86, 87
Heating mechanism, earth surface
cooling-state photon, 422
cool-state photons, 428
covariant derivative, 423
heating energy-level photon activation, 425
heating photons, 424
heating strategy, 423
Higgs boson electromagnetic field, 422
kinetic mechanism, 429
photon receptor, 429
scalar field, 429
variable scalars, 423
Heating photoemission rates, 435
Heating photon, 446
Heating photon energy emission, 434
Heating photon production, 431
Heat-photon counts, 97
Heavy metals, 244, 302, 303
Heterogeneous TiO_2 photocatalysis
aromatic compounds, 265
conduction-band hole, 264
DO, 265
electron scavengers, 264
$HO_2^{\bullet}$ radical, 265
liquid-phase organic compounds, 265
microorganisms, 266
organic compounds, 265
oxidative-reductive reactions, 263
photodegradation, 265, 266
photoinactivation, 266
photoinduced formation mechanism, electron/hole pair, 263, 264
photonic excitation, 263
reaction steps, 265, 266
reductive and oxidative reactions, 263
surface-trapped valence-band electron, 264
surrogate organic compounds, 265
water molecules, 265
Higgs boson ($H \rightarrow \gamma\gamma^-$) electro-quantum, 82
Higgs boson BR ($H \rightarrow \gamma\gamma^-$) quantum mechanics, 82, 86, 93, 95, 100, 101
Higgs boson electro-quantum field, 86
Higgs boson quantum dynamics, 428
Higgs bosons ($H \rightarrow \gamma\gamma^-$), 418
High-energy photons, 430
High-speed flying train
aerodynamic force, 368
air gap flux, 374
airstream-driven energy generation, 373
and flight height, 371
automated energy torque, 374
Betz's coefficient, 372
free body diagram, 366
generator modeling, 370
guideway model, 366, 367
guideway system, 368
innovative transportation technology, 366

High-speed flying train (*cont.*)
kinetic energy, 373
k-omega turbulence model, 369
levitation and lateral force, 371, 372
levitation force, 371
power control mechanism, 373
RANS, 366
SCIG, 374
sky smoothly, 372
squirrel cage induction generator, 370
TSR, 373
velocity, 374
wind velocity, 372
Hom model, 294–296
Hom-power model, 295
Hospital-dispensed chemicals, 398
Hossain cooling photon (HcP^-), 83, 89, 101
Hossain static electric force (HSEF), 113, 116, 123–125, 326, 440, 442, 446, 447, 449
Hybrid photocatalytic-membrane reactor system, 276
Hydrogen chalcogenide molecule, 225
Hydrogen energy production, 449–451
Hydroperoxyl radical ($HO_2^{\bullet}$), 265
Hydrophilic coating, 249
Hydrostatic momentum, 226

I

Immobilizer fibers, 271
Incremental conductance (IC), 70
Induction generators, 204, 207, 218
Industrial emissions, 153
Industrial water use, 232
Infectious microorganism, 3
Infrastructure, 338, 341, 342
Infrastructure development
and rehabilitation processes, 202
Inorganic contaminants, 395
Inorganic ions
Cl^-, 301
Cu^{2+}, 300
Fe^{2+}, 300
Mn^{2+}, 301
modification, TiO_2 surface, 302
nitrogen-containing molecules, 300
NO_3^- ion, 301
pH, 300
photocatalyst, 299
photocatalytic water treatment, 300, 302
photodegradation rates, 301
photo-disinfection reactions, 300
photomineralization reaction rates, 300
photonic efficiency, 301
PO_4^{3-}, 301
pollutants containing sulfur atoms, 300
salts, 301
surrogate model organic compound, 300
TiO_2 photoactivity, 301
TiO_2 photocatalyst, 302
TiO_2 photomineralization, 300
TiO_2 water treatment process, 299
water, 299
water pretreatment, 302
In site biowaste treatment technology
anaerobic bacteria, 132
bioenergy, 134
biogas, 132, 142
bioreactor, 131, 137, 138
cell mode, 141
circuit panel, 132–134, 140
CNT, 131
DIS system, 131
electricity energy, 134–137, 140
electricity energy production, 141, 142
electricity generation model, 136, 137
electrochemical carbon nanotube filters, 132
electrolysis, 131
landscaping/sludge, 132
LRFD bioreactor, 130
methane generation efficiency, 138
methanogenesis, 133
microorganisms, 131
open-circuit voltage, 135, 141
oxidation process, 131
photo-physics radiation application, 131
sludge, 132
two-chamber bioreactor, 131
UV application, 131
water dynamic force, 130
In situ water treatment, 125, 329
Integrated building design technology
cooling mechanism (*see* Cooling mechanism)
heating mechanism (*see* Heating mechanism)
Intergovernmental Panel on Climate Change (IPCC), 153
International Geosphere and Biosphere Program (IGBP), 403
International Human Dimension Program (IHDP), 403
Internuclear distance, 100
Invisible infrastructure transportation
Betz's coefficient, 161
construction cost estimate comparison, 167–171

cost, 171–173
electromagnetic induction, 164
energy consumption, 158
global warming, 157, 158
Greenshield's model, 162, 163
guidance factors, 160
guideway modeling, 158
LRFD, 164
maglev (*see* Maglev infrastructure)
magnetic bearing system, 164, 165
magnetic forces, 159
optimal rotor speed, 162
polarization of coil, 164, 165
Standard Simulink/Sim Power Systems, 162
turbine speed and mechanical powers, 162, 163
urban and suburban area, 157
velocity, 161
wind speed, 162
wind turbine, 161, 167
Iron oxide nanoparticles, 243
Irrigation, 233, 234

K
Kepler's second law, 27
Killing pathogens, 4
Kinetic energy, 371
Kinetic energy conversion mechanism, 178
Kinetics and modeling
heavy/noble metals, 302, 303
inorganic ions, 299–302
photo-disinfection kinetics (*see* Photo-disinfection kinetics)
photomineralization, 290, 291, 293
photoreactor systems, 290
turbidity, 299
water quality, 299
k-omega turbulence model, 369

L
Lagrangian function, 86–88, 94
Land
activities, 404
agroecosystem, 409
biodiversity, 405
biophysical condition, 404
climate changes, 411
clustered sampling, 408
cover and use, 404
deforestation, 411
degradation, 405
diversification, 411
earth surface, 411
elasticity, 407
erodibility, 404
essential earth system properties, 404
farming systems, 406
function, 404
human activities, 411
human measurements, 406
identification, 405
intensification analysis, 406
intuitions, 404
loss of variety, 404
management, 405, 409
mapping and spatial process, 405
nonagricultural sources, 411
nonspatial and spatial methods, 409
political action, 406
population pressure, 404
primary data variability, 407
road construction effect, 404
root-cause analysis, 409
root-cause pattern, 411
scales, 405
social/ecological system, 411
socio-environment interaction dynamics, 406
soil erosion, 409, 410
spatial analysis, 405
spatial distribution, 406, 408
spatial sampling methods, 407
surface, 405
systematic/random-purpose samples, 408
use, 409
water resource, 403
Land-use change, 384, 387, 388, 390
Langmuir–Hinshelwood (L–H) mechanism, 283
Levitation force, 371
Life cycle assessment (LCA)
environmental impact, 304
heterogeneous photocatalysis, 303, 304
impact categories, photocatalytic water treatment process, 304
large-scale water application, 304
photocatalytic water/wastewater treatment processes, 303
photo-Fenton plants, 303
pilot plants, 304
ROS-based oxidation processes, 303
solar irradiation, 305
Light emission, 98

Light intensity, 287, 288
Light wavelength, 286, 287
Load-resistant factor design (LRFD), 130, 164, 167, 170, 171
Load-side converter, 183
Localized surface plasmon resonance (LSPR), 256

M

Maglev infrastructure
  cost, 171
  4-DOF rigid maglev vehicle, 161
  greener and cleaner, 158
  guideway coupling systems, 158
  guideway *vs.* vehicle force, 159
  heavy-duty waterproofing membrane, 164
  MATLAB software, 158
  propulsion, 164
  superconducting force, 160
  vehicle's force and directional diagram, 164, 166
  wind profile, 167
  wind turbine generation system, 166
Magnetic bearing system, 164, 165
Magnetic forces, 158, 159
Magnetic nanoparticles, 243, 249, 256
Magnetic suspension systems, 158
Magnetism, 243
Magnetizing inductance, 191
Markovian unique equation, 116
Mass balances, 226
Mass spectrometry, 450
MATLAB Simulink, 158, 178
MATLAB software, 61, 202, 226, 382, 383, 418
Maximum power peak (MPP), 52, 68, 358
Maximum power point track (MPPT), 19, 49, 53, 55, 107, 108, 180, 186–188, 193, 206, 210, 212–214
Maxwell's rules, 116
Maxwell's theory, 425
Mechanical power control, 372
Mechanistic models, 296–298
Melting point, 226
Membrane distillation (MD), 277, 278
Membranes and membrane processes
  biological membranes, 248, 249
  energy consumption, 245
  FO, 249
  functional nanomaterials, 245
  nanocomposite, 246, 247
  nanofiber, 246
  TFN, 247, 248
  unconventional water sources, 245
Mesoporous clays, 270
Metal-based nano-adsorbents
  arsenic removal, 244
  heavy metals, 244
  iron oxide nanoparticles, 243
  magnetic nanoparticles, 243
  metal oxide nanocrystals, 244
  metal oxide nanomaterials, 244
  nanoscale effect, 243
  particle size, nano-magnetite, 243
  regeneration and reuse cycles, 244
  sorption, 242
Metal nanomaterials, 256
Metal oxide nanocrystals, 244
Metal oxide nanomaterials, 252
Metal oxide nanoparticles, 246
Methanogenesis, 130, 133, 137
Microamp range, 449
Microbial contaminants, 397
Microfiltration (MF), 270
Microorganism, 3, 297, 397, 398
Motor winding resistance, 191
Multifunctional devices, 258

N

Nano-adsorbents, 239, 245
Nano-Ag, 246, 253, 254
Nano-Au, 258
Nanocomposite, 249
Nanocomposite membranes, 246, 247
Nanofiber membranes, 246
Nanofiltration (NF), 277
Nanomaterials, 238, 239, 252, 254, 256, 259, 260
Nano-photocatalyst optimization, 250, 251
Nanophotocatalysts, 247
Nanoscale effect, 243
Nanotechnology
  membrane, 246
  water/wastewater treatment (*see* Water/ wastewater treatment)
Nano-TiO_2, 247
Nano-zeolites, 247
Natural clays, 270
Natural water resources, 228, 234
Negative and positive indexes, 432
Nephelometric turbidity units (NTU), 299
Net electricity energy generation
  Boltzmann's constant, 33
  calculation, 22
  electricity energy, 33, 36
  formation, 23
  global earth surface, 22

I–V earth surface, 34
photon-induced current, 34
solar energy, 21
Net energy formation, 110
Net photo-physical current generation, 19
Net realistic energy, 189
Net solar energy
constant, 25, 26
on earth surface, 24
PB frequency, 28
solar radiation, 25
Net solar energy, earth
ambient temperature, 19
equatorial angles, 15
latitude, 16
photon flux, 16
spherical law of cosines, 24
Net water resources, 228
Niobium-titanium alloy, 166
Noble metals, 302, 303
Nonequilibrium condition, 119
Nonequilibrium photon scattering theorem, 118
Nonmetal doping, 275
Nonorthogonal chain rules, 227
Non-series (Ns) cells, 34, 84, 121
Non-series (Ns) earth surface, 421
Non-Weisskopf-Wigner mode, 89, 117, 425

O
Ocean CO_2 sink, 384, 385, 387, 389, 391
One-level molecular junction, 450
On-site water treatment, 442, 443
Open-circuit point, 357
Open-circuit voltage, 21, 54, 110
Open-circuit voltage-based MPP techniq*ue*, 72
Operational parameters, photocatalytic reactors
DO, 283, 284
pH, 281
catalyst particles *vs.* microorganisms, 282
particle size, 282
photocatalytic reaction, 281
photo-disinfection, microorganisms, 281
PZC, TiO, 281
TiO_2 catalyst, 281, 282
wastewater treatment, 282
temperature, 282, 283
TiO_2 loadings, 280
Optimal rotor speed, 162
Organic contaminants, 395
Organic removal, 239, 241
Organophilic membrane, 277
Oxidation process, 131
Oxidative stress, 266, 287
Oxidized CNTs, 242
Ozone layer, 81

P
Paris protocol, 149, 153
Particle agglomeration, 263
Particulate matter, 153
Particulate-phase concentrations, 399
Particulate-phase determination, 399
Pathogen detection
CNTs, 256, 257
colorimetric assays, 257
conventional indicator systems, 256
diagnosis-based water disinfection approach, 256
Dynabead®, 256
LSPR, 256
magnetic nanoparticles, 256
metal nanomaterials, 256
nanomaterials, 256
public health, 256
QDs, 256
Raman scattering, 257
recognition agents, 256
semiconducting CNTs, 257
SERS, 257
silica nanoparticles, 257
SWNT, 257
PEC Schottky-style junction, 444
Permanent magnet synchronous generator (PMSG), 189
Permanent synchronous generator (PMSG), 216
Pervaporation, 277
Peukert's coefficient, 209
Peukert's law, 162, 195, 219
p-$GaInP_2$(Pt)/TJ/GaAs electrode, 449
p-$GaInP_2$/GaAs, 445
pH, 281, 282
Photoactivity kinetics, 275
Photocard-voltage curves, 449
Photocatalysis
challenges, 252
environmental friendly, 252
metal oxide nanomaterials, 252
microbial pathogens removal, 249
nanomaterials, 252
nano-photocatalyst optimization, 250, 251
photoreactor, 252

Photocatalysis (*cont.*)
Purifics Photo-Cat™ system, 252
recalcitrant organic compounds, 249
SODIS system, 252
sustainable water treatment technology, 252
TiO_2 photocatalyst optimization, 250
Photocatalyst modification and doping, 272, 275
Photocatalyst particles, 275
Photocatalytic membrane (PMs), 272
Photocatalytic membrane reactors (PMRs), 276–278
See also Photocatalytic reactors
Photocatalytic nanoparticle-incorporated membranes, 247
Photocatalytic oxidation, 249
Photocatalytic reaction kinetics, 249
Photocatalytic reactors
configurations, 275
direct photon utilization, 275
hybrid photocatalysis-MD process, 278
hybrid photocatalytic-membrane reactor system, 276
MD, 278
MF membrane, 277
operational parameters (*see* Operational parameters, photocatalytic reactors)
organophilic membrane, 277
parabolic light deflectors, 275
pervaporation, 277
photocatalyst particles, 275, 277
photocatalysts, 277
Photo-Cat™ system, 278, 279
photooxidation efficiency, 277
PMRs, 276
retentate TiO_2 stream, 279
slurry-type, 275
submerged membrane, 276
types, 275
water treatment, 275
water-phase degradation, 277
Photocatalytic water treatment technology
adsorption/coagulation, 262
AOPs, 262
challenges, 269, 270
chlorination, 262
clean water shortage, 261
commercialization, 305
contaminants, 285, 286
conventional water treatment methods, 262
environmental friendly, 261
features, 262
kinetics and modeling (*see* Kinetics and modeling)
LCA, 303–305
light intensity, 287, 288
light wavelength, 286, 287
mesoporous clays, 270
nanofibers, 271
nanorods, 271, 272
nanowires, 271
nonbiodegradable water-soluble pollutants, 305
particle agglomeration, 263
photocatalyst modification and doping, 272, 275
photocatalytic reactors (*see* Photocatalytic reactors)
photo-disinfection, 261, 263
photomineralization, 261, 263
photoreactor operation parameters, 261
pilot plant investigations, 306
PMs, 272
rainwater storage, 261
recalcitrant organics, 305
response surface analysis, 289
ROS, 262
semiconductor catalysts, 262
solar energy, 306
stand-alone treatment system, 305
sustainable treatment technology, 261
technical constraints, 305
TiO_2, 262
TiO_2 crystal deposite, clay materials, 271
TiO_2 nanowire membrane, 273
TiO_2 photocatalysis (*see* TiO_2 photocatalysis)
UV light/solar, 305
water industries, 261
Photo-Cat™ system, 278, 279
Photocurrent density, 450
Photocurrent electrolysis, 450
Photocurrents, 110
Photo-disinfection kinetics
catalyst–bacterium interaction, 298
C–W model, 293, 294, 296
Degussa P-25 TiO_2, 294
demand-free condition, 293
E. coli cells *vs.* catalyst agglomerates, 298
Hom model, 294–296
inactivation regions, 294
L–H model, 298
mechanistic models, 296, 297
microbial cell, 296
microorganisms, 296, 297
oxidation products, 298

parameters, 298
pseudo-adsorption constant, 298
rational model, 295
ROS, 298
Selleck model, 295, 296
TiO_2 catalyst, 294
water treatment process, 293
Photodynamic modules, 82
Photoelectric sunshine recorder, 17
Photo-electrochemical (PEC)/PV device, 443, 445, 446
Photo-electrolysis, 445, 450
Photo-fenton reaction
ferrous/ferric ions, 267
H_2O_2, 268
light irradiation, 267
light source, 267
mechanism, 267
pH, 268
photoactivity, 268
photoassisted reaction, 267
TiO_2 photocatalysis, 267, 268
water pollutants, 267
Photoinactivation, 266
Photomineralization, 261, 280, 283–286
Photomineralization kinetics
first-order kinetics, 290
K-parameter, 291
K-value, 291
L–H model, 290
lump-sum L–H saturation kinetics profile, 292
$OH^{\bullet}$ radicals, 290
organic compounds, 290
organic concentrations, 290
organic dye molecules degradation, 292
photocatalytic reaction rate, 291, 292
pseudo-first-order model, 292
TOC degradation, 293
Photomineralization rate, 289
Photon absorbance, 88
Photon band edges (PBEs), 418
Photon bandgap (PBG) waveguides, 418
Photon cell
cooling temperature, 86
Photon cell's unit area, 83
Photon deliberation, 99
Photon density, 95, 430
Photon dissipation proliferation, 93
Photon distribution, 119
Photon electromagnetic wave, 107
Photon element, 84
Photon emission, 99, 101
Photon energy, 107, 117, 120
Photon energy emission, 30
Photon energy frequency module, 30
Photon energy generation, 31, 32
Photon flux, 113
Photon frequency module, 118
Photon generation, 83, 116, 118, 119
Photon irradiance emission, 99
Photon nanostructure
nano-point-break modules, 82
Photon proliferation dynamics, 118
Photon radiation, 107
Photon structure, 91
Photon's transformation mechanism, 88
Photonic band (PB), 89, 91–93
Photonic band structure, 91
Photonic bandgap (PBG), 82, 91, 92
Photonic bandgap energy, 20
Photonic electromagnetic field generation, 4
Photonic electron-state energy, 114
Photonic motion, 93
Photonic particle, 99
Photonic probable density, 419
Photonic structures
current–voltage (*I–V*) features, 83
Photonic wavelengths, 4, 6
Photon-induced current, 122
Photo-physical current generation, 108
Photo-physics radiation, 4, 5, 330, 443
Photoreaction, 107
Photoreactor system, 252, 272
Photosynthesis, 439
Photovoltage, 446
Photovoltaic (PV) panel, 350
Boltzmann's constant, 109
Cartesian coordinate system, 111, 112
cell current, 109
DOS, 116, 117
electrostatic force (*see* Electrostatic force)
EMF vector, 117
energy deliberation, 108
exterior curtain wall, 107
functional circuit diagram, 121
mechanism, 107
nanocrystalline materials and films, 111
one-diode circuit diagram, 108
open-circuit voltage, 110
peak energy generation, 110
photon flux, 113
photonic electron-state energy, 114
quantum technology, 113
solar energy, 111
solar panel, 112
temperature condition, 108
tornado-resistant, 111

Photovoltaic (PV) panel (*cont.*)
vicinity, 107
volt relationship, 109
wind load resistance, 111
Point-of-zero charge (PZC), 281
Polar molecule, 225
Polymeric nano-adsorbents, 244
Potable water, 125, 126, 440
agricultural use, 231, 232
characteristic roots of algorithms, 230
civilization and generation, 226
domestic use, 231
earth surface, 227
environmental use, 233
geometrical interpretation, 228
graphical distribution, 230, 231
groundwater analysis, 228
industrial use, 232
irrigation, 233, 234
MATLAB software, 226
recreational use, 233
surface water analysis, 227, 228
surface water elevation, 229
topography, 229
water stability limit, 229
Power backup model, 209
Power coefficient, 181, 191, 206, 372
Power control mechanism, 373
Power generation, 193
Powering the transportation vehicles, 178, 179, 182, 184, 186, 193, 196
Power-law spectrum, 97
Projected DOS (PDOS), 89, 92
Propulsion coils, 164
Propulsion force, 371
Proton-proton (PP) chain reaction, 14
Purifics Photo-Cat™ system, 252
PV panel
MPPT, 108
Pyranometer, 17
Pyrheliometer, 17

Q
Quantum dots (QDs), 256
Quantum electrodynamics (QED), 82, 418
Quantum field, 94, 447

R
Radar interference, 177
Radial field, 87
Radial static electricity force, 115
Rainwater harvesting design system, 150
Rapid communication, 366
Rational model, 295
Raw water pretreatment, 259
Reactive oxygen species (ROS), 250, 262
Realistic static electricity fields, 115
Recognition agents, 256
Recreational water use, 233
Regeneration, 242, 244
Renewable power generation, 232
Reproductive disorder, 81
Response surface analysis, 289
Reverse osmosis (RO), 246
Reynolds-averaged Navier–Stokes (RANS), 204, 366
Rotor angular velocity, 181
Rotor current, 212
Rotor leakage inductance, 191
Rural areas, 365, 366, 374, 375
Rural development
building, 337, 338, 341
communications technology, 335
energy, 337, 340, 341
environment, 336, 339, 340
environmental perplexity, 336
global environmental crisis, 335
housing, 337, 338, 341
infrastructure, 338, 341, 342
massive urbanization, 336
transportation, 338, 341, 342
urban congestion, 335, 336
water, 338, 339, 342

S
Scalar field, 86, 87, 123, 124, 446, 447
Sediments, 399
Self-generated energy source, 71
Semiconductor catalysts, 262
Sensing and monitoring
challenges, 259
cost-effectiveness, 259, 260
multifunctional devices, 258
nanotechnologies, 259, 260
pathogen detection, 256, 257
potential hazards, 260
research needs, 259
retention and reuse, nanomaterials, 258, 259
trace contaminant detection, 257, 258
Separation process, 258, 269
Severe dreadful respiratory syndrome (SARS-CoV-2), 4
Silica chemistry, 257
Silica nanoparticles, 257

Silver ions, 253
SimPower Systems, 191
Simulink power systems, 217
Simulink-MATLAB, 208, 371
Single-compound acute toxicity testing, 396
Single-diode
  cooling state, 83, 84
  I–V clarification, 85
  semiconductor, 82, 101
Single-diode mode, 109
Single-diode solar cell, 120
Skin cancer, 81
Sky smoothly, 369, 372
Sludge, 132
Slurry-type photocatalytic reactor, 275
Smart building technology
  conventional energy, 106
  dynamic photon capture, 118
  electricity transformation, 120–122
  energy consumption, 105
  energy requirement, 105
  environmental impacts, 106
  GHG, 106
  in situ water treatment, 125
  Markovian unique equation, 116
  photon dynamic rate, 118
  photon energy, 107
  photon frequency module, 118
  photon generation, 116
  photon proliferation dynamics, 118
  photon radiation, 107
  photoreaction, 107
  PV panel (*see* Photovoltaic (PV) panel)
  solar radiation, 106
  urban cloud, 106
  water supply, 106
  Weisskopf-Wigner approximation mechanism, 116
Smart transportation technology, *see* Flying car design
SODIS framework, 125
Solar disinfection (SODIS), 252, 287, 442
Solar electricity, 113
Solar energy, 6, 8, 14, 16, 19, 82, 86, 106, 111, 306, 349, 359
  fossil fuel, 14
  hydrogen atoms, 14
  proliferation, 107
  radioactive CO_2, 14
Solar energy emission, 30
Solar energy formation, 113
Solar energy generation, 28
Solar energy utilization, 14
Solar irradiance, 30, 85, 107, 122
  speed and working temperature, 83
Solar irradiation, 272, 287
Solar panel, 112
Solar radiation, 16, 17, 106, 121, 348
Solid biowaste, 130
Sorption, 242
Squirrel cage induction generator, 208, 370
Squirrel-confined initiation generator (SCIG), 214
Stand-alone treatment system, 305
Standard Simulink/Sim Power Systems, 162
Static electric force, 326–330, 440–443
Static electrical force, 440
Static electricity, 326
Static electricity force mechanism, 115
Static electricity forces, 441, 443, 447, 448
Static equilibrium, 368
Stator active and reactive powers, 185
Stator and rotor flux, 183
Stator leakage inductance, 191
Stator power, 212
Stator voltage vectors, 184
Stator winding resistance, 191
Stator's constant flux, 184, 185
Stefan–Boltzmann laws, 349
Stormwater management, 149–152, 154
Stormwater runoff, 150
Stratospheric ozone depletion, 106
Sun-oriented vitality, 42
Superconducting guideway beam, 159
Supercritical steam, 226
Supersynchronous operation, 167, 212
Surface-enhanced Raman spectroscopy (SERS), 257
Surface oxidation, 242
Surface water analysis, 227, 228
Surface water elevation, 229
Survival period of life on earth, 383
Suspension control laws, 158
Sustainable building technology, 106
Sustainable cities
  Boltzmann's constant, 353
  buildings and houses, 349
  cartesian coordinate, 350
  current–voltage calculation, 357
  diode current, 353
  electricity, 357–359
  electromagnetic wave, 349
  electron charge, 353
  energy consumption, 348
  energy cost, 359
  energy density, 355
  environmental and climate change disaster, 347

Sustainable cities (*cont.*)
environmental perplexity, 347
fossil energy consumption, 348
functional temperature, 356
graphene silicon surface, 349
innovative technology, 349, 360
inverse saturation current, 353
mathematical model, 351
non-series (Ns) cells, 357
open-circuit voltage, 354
parameters, 350
photocurrent, 353
photon flux, 355
photonic thermal activation mechanism, 352
photonic thermal energy conversion modes, 356
photovoltaic array, 350
photovoltaic modules, 350
power conditioning efficiency, 352
radiation energy density, 351
relativistic temperature conditions, 357
semiconductor, 353
series-connected cells, 357
short-circuit current, 353
single-diode mode, 355
solar cell, 353
solar energy, 349
solar irradiance, 356, 357
solar panel, 349
solar radiation, 348, 352
solar thermal energy, 351
standard single diode, 353
thermal energy density, 355
three parallel circuits, 350
world energy consumption, 348
Sustainable environment, 343
Sustainable infrastructure, 154
SVPWM, 207
Synchronous generator, 218
Synchronous models, 207

T

Taguchi-DOE approach, 289
Temperature, 282, 283
Thermal photon production, 433
Thermal relativistic condition, 118
Thin-film composite (TFC) membranes, 247
Thin-film nanocomposite (TFN) membranes, 247–249
Time-stepping algorithms, 230
TiO_2 photocatalysis
advancements, 269
heterogeneous (*see* Heterogeneous TiO_2 photocatalysis)
homogeneousphoto-Fenton reaction, 267
TiO_2 photo-disinfection, 283
TiO_2 slurry reactor, 284
Tip speed ratio (TSR), 179
Tip speed ratio equation, 205
Titanium dioxide (TiO_2), 250, 251, 262, 280
Topography, 229
Tornado-resistant PV panel, 111
Trace contaminant detection, 257, 258
Transformation mechanism, 420
Transmitted solar irradiance
emission and absorption, 83
Transpiration, 439, 440, 451, 452
Transpiration mechanism
filtration, 330
groundwater loss, 325
in situ water treatment, 329
static electric force, 326–330
total water demand, 326
UV application, 330
water vapor, 326, 331
Transportation, 338, 341, 342
Transportation sector
turbine modeling, 178–181
Tunnel junction (TJ), 445
Turbidity, 299
Turbine drive force, 180
Turbine modeling
transportation sector, 178–181
Turbulence function, 181

U

Ultraviolet germicidal irradiation (UVGI)
DNA structure, 7
DNA/RNA, fatal pathogens, 6
EM, 4–6
exterior glazing wall, 5
glazing wall surface, 6
pathogens, 4, 5, 7
photonic electromagnetic field generation, 4
photonic irradiance structure, 4
photo-physics radiation application, 5
sunlight, 4–6
Unconventional water sources, 238, 245
United Nations Conferences on Environment and Development (UNCED), 403

Urban areas, 365, 366, 374, 375
Urban cloud, 106
Urban congestion, 335, 336, 374
Urban forestry management, 151, 153
Urban mobility, 366
Urban particulate pollution, 153
Urban vehicle emissions, 153
US National Primary Drinking Water Standard code, 125
UV irradiation, 286
UV radiation, 287, 329
UV technology, 125, 126, 442–444, 452

V
Vehicle stability, 158
Volatile organic compounds (VOCs), 153
Volatile power, 213

W
Wastewater, 234
Wastewater treatment, 400
Wastewater treatment plants (WwTPs), 396
Wastewater treatment process (WwTP), 397
Water, 238, 325, 337–339, 342
  attributes, 226
  benthic organisms, 399
  characteristic, 226
  chemicals, 398
  ECs, 396, 397
  emerging contaminant, 398
  emerging contaminants, 399
  hospital-dispensed chemicals, 398
  microorganism, 397, 398
  organic and inorganic chemicals, 398
  particulate-phase concentrations, 399
  particulate-phase determination, 399
  physicochemical properties, 396
  rainfall, 398
  sampling, 397
  single-compound acute toxicity testing, 396
  toxicological effect, 400
  WwTP, 396, 397
Water contaminants, 285, 398
Water molecule, 225
Water quality, 285, 299
Water reserve, 229
Water stability, 229, 230
Water supply, 238
Water surface *S*-coordinate, 227
Water topography, 227
Water vapor, 226, 439, 440, 442, 447, 449, 451, 452
Water/wastewater treatment
  adsorption (*see* Adsorption)
  challenges, 260
  disinfection and microbial control, 252–255
  membranes and membrane processes (*see* Membranes and membrane processes)
  nanomaterials, 238, 239, 260
  nanotechnology, 238, 240–241, 260
  photocatalysis (*see* Photocatalysis)
  sensing and monitoring (*see* Sensing and monitoring)
  water quality, 238
Weisskopf-Wigner approximation mechanism, 25, 116
Weisskopf-Wigner assumption, 89
Weisskopf-Wigner theory, 424
Wind energy, 54, 158, 162, 173, 181–182, 188–189
  ageostrophic wind, 42
  air pressure, 42
  application, 193
  beach front zones, 42
  capture and utilization, 178
  Coriolis power, 42
  diode model, 47
  dynamic mode, 54, 190
  energy demand and environmental crisis, 42
  energy system modeling, 48
  energy torque, 44
  induction and synchronous, 189
  kinetic energy, 43
  modeling and simulation, 46
  MPPT, 53
  net wind energy, 45
  optimum rotor angular speed, 50
  output, 48
  power coefficient, 46
  renewable energy, 42
  shaft angular speed, 51
  solid breezes, 41
  spatial scale, 41
  technology development, 177–178
  TSR control, 52
  turbine, 49, 52
  velocity, 43
Wind energy conversion system (WECS)
  flying car design, 207, 215, 216
Wind energy electric system (WEES), 193

Wind energy generation, 43
Wind energy modeling
  flying car design, 204–206, 212–215
Wind load resistance, 111
Wind power, 177, 178, 180, 186, 193, 196
  in car, 193–194
  technologies, 178
  transformation, 194
Wind profile, 167
Wind regulation, 187
Wind speed, 162, 181, 193, 205, 212
  and power, 206
  signal and voltage dips, 186
Wind turbine, 44, 49, 161, 167, 186–187
  construction, 177
  control subsystem, 183
  defuzzification, 188
  design, 193
  energy production, 204
  functional areas, 179
  generation, 186
  power mechanism, 178
  speed, 178
  TSR, 179
Wind turbine generation system, 166
Wind velocity, 179, 180, 193, 372
  nonlinear characteristics, 189
Wing designing phase, 202
WT schemes, 217

Z

Zeolite TFN technology, 247
Zero-emission vehicles
  electrical subsystem, 182–186
  turbine modeling, 178–182
  wind energy (*see* Wind energy)
  wind turbine (*see* Wind turbine)

Printed in the United States
by Baker & Taylor Publisher Services